16

CRM
SERIES

The Seventh European Conference on Combinatorics, Graph Theory and Applications

EuroComb 2013

edited by
Jaroslav Nešetřil and Marco Pellegrini

EDIZIONI
DELLA
NORMALE

ISBN 978-88-7642-474-8
e-ISBN 978-88-7642-475-5

Contents

Foreword

We are presenting in this volume the collection of extended abstracts that were selected for presentation at the European Conference on Combinatorics, Graph Theory and Applications - Eurocomb 2013, held in Pisa on September 9-13, 2013. Eurocomb conferences are organized biannually, the series was started in Barcelona 2001, and continued with Prague 2003, Berlin 2005, Seville 2007, Bordeaux 2009 and Budapest 2011. Since Prague Eurocomb 2003 the European Prize in Combinatorics is awarded at the meeting. The prize is established to recognize excellent contributions in Combinatorics, Discrete Mathematics and their Applications by young European researchers (eligibility of EU) not older than 35. It is supported by DIMATIA and by private sources.

The role of combinatorics is growing in the contemporary research in mathematics, computer science, engineering, also in physics, life and social sciences. This is reflected in the growing interest in Eurocomb conferences, which became an established forum of the frontier research in these fields. This year a total of 181 contributions were submitted and the Program Committee selected 95 extended abstracts and 9 posters to be presented at the conference. These figures, similar to those of the previous Eurocomb conference, indicate that Eurocomb meeting and its format got firmly established in the spectrum of high level European conferences. We thank all members of the Program Committee for their work. We also thank to all members of the Organizing Committee of Eurocomb 2013 for all their work in preparing the meeting in lovely Pisa.

The conference is organized by the Institute for Informatics and Telematics of the National Research Council of Italy (CNR), with the generous support of the Scuola Normale Superiore - Centro di Ricerca Matematica Ennio De Giorgi, the Dipartimento di Informatica - Università di Pisa, and the Center for Discrete Mathematics, Theoretical Computer Science and Applications (DIMATIA) of Charles University.

Jaroslav Nešetřil and Marco Pellegrini
(PC co-chairs)

PC members

Béla Bollobás
Stefan Felsner
Jacob Fox
Anna Galluccio
Jarek Grytczuk
Ervin Györi
Tommy Jensen
Ken-ichi Kawarabayashi
Peter Keevash
Daniel Král'
Daniela Kühn
Alberto Marchetti-Spaccamela
Mickaël Montassier
Patrice Ossona de Mendez
André Raspaud
Alex Scott
Oriol Serra
Martin Škoviera
József Solymosi
Benjamin Sudakov
Jan Arne Telle

Organization Committee

Bruno Codenotti
Paola Favati
Filippo Geraci
Roberto Grossi
Adriana Lazzaroni
Marco Pellegrini (General Chair)
Giovanni Resta

Organizing Secretariat

Patrizia Andronico
Raffaella Casarosa
Petra Milstainova

Invited Speakers

Endre Szemerédi
Packing Trees into a Graph
Alfréd Rényi Institute of Mathematics, Hungarian Academy of Sciences and Rutgers University
szemered@cs.rutgers.edu

Andrew Thomason
Containers for independent sets
University of Cambridge,
A.G.Thomason@dpmms.cam.ac.uk

Michel Goemans
Polynomiality for Packing and Scheduling in the High Multiplicity Setting
Massachusetts Institute of Technology,
goemans@math.mit.edu

Dániel Marx
Decomposition Theorems for Graphs Excluding Structure
Institute for Computer Science and Control, Hungarian Academy of Sciences (MTA SZTAKI), Budapest, Hungary
dmarx@cs.bme.hu

Alexandr Kostochka
Lower Bounds on the Size of k-Critical n-Vertex Graphs and their Applications
University of Illinois at Urbana-Champaign, Urbana, USA
and
Sobolev Institute of Mathematics, Novosibirsk, Russia
kostochk@math.uiuc.edu

Asaf Shapira
Ramsey Theory, Integer Partitions and a New Proof of the Erdös-Szekeres Theorem
Tel-Aviv University
asafico@post.tau.ac.il

Luca Trevisan
Spectral graph theory
Stanford University
trevisan@stanford.edu

Michele Conforti
Integer Programming, Lattices and Polyhedra
Dipartimento di Matematica, Università di Padova
conforti@math.unipd.it

Mathias Schacht
Extremal Combinatorics in Random Structures
University of Hamburg
schacht@math.uni-hamburg.de

Erdős problems

A problem of Erdős and Sós on 3-graphs

Roman Glebov[1], Daniel Král'[2] and Jan Volec[3]

Abstract. We show that for every $\varepsilon > 0$ there exist $\delta > 0$ and $n_0 \in \mathbb{N}$ such that every 3-uniform hypergraph on $n \geq n_0$ vertices with the property that every k-vertex subset, where $k \geq \delta n$, induces at least $\left(\frac{1}{4}+\varepsilon\right)\binom{k}{3}$ edges, contains K_4^- as a subgraph, where K_4^- is the 3-uniform hypergraph on 4 vertices with 3 edges. This question was originally raised by Erdős and Sós. The constant 1/4 is the best possible.

1 Introduction

One of the most influential results in the extremal graph theory is the celebrated theorem of Turán [14] that determines the largest possible number of edges in an n-vertex graph without a complete subgraph of a given size. Erdős, Simonovits, and Stone [5,7] generalized Turán's theorem by showing that the *extremal number* of an arbitrary graph F, defined as

$$\mathrm{ex}(n,F) := \max\{|E(G)| : G \text{ is a graph on } n \text{ vertices with no copy of } F\},$$

is asymptotically determined by its chromatic number. Specifically, for every graph F with at least one edge,

$$\mathrm{ex}(n,F) = \left(\frac{\chi(F)-2}{\chi(F)-1} + o(1)\right)\binom{n}{2}.$$

[1] Mathematics Institute and DIMAP, University of Warwick, Coventry CV4 7AL, UK. Email: R.Glebov@warwick.ac.uk

[2] Mathematics Institute, DIMAP and Department of Computer Science, University of Warwick, Coventry CV4 7AL, UK. Email: D.Kral@warwick.ac.uk

[3] Mathematics Institute and DIMAP, University of Warwick, Coventry CV4 7AL, UK. Email: honza@ucw.cz

The work leading to this invention has received funding from the European Research Council under the European Union's Seventh Framework Programme (FP7/2007-2013)/ERC grant agreement no. 259385.
The first author was also supported by DFG within the research training group "Methods for Discrete Structures".
The third author was also supported by the student grant GAUK 601812.

Note that this problem is dual to determining the minimum number of edges $m(F)$ that guarantees an n-vertex graph G with at least $m(n, F)$ edges to contain a copy of F, since $m(n, F) = ex(n, F) + 1$.

In Ramsey-Turán type problems, which were introduced by Sós in [13], we are again interested in the largest possible number of edges of an n-vertex graph without a copy of F, but under the additional restriction that the graph has to be somewhat "far" from the Turán graph. Typically, such a restriction is expressed by requiring that the graph has only sublinear independence number. For more details on Ramsey-Turán theory, see the survey by Simonovits and Sós [12].

In our problem, we are dealing with an even stronger restriction; we require every linear-size subset of vertices not only to induce at least one edge, but actually to induce a positive proportion of all possible edges. We define the δ-*linear density* of a graph G to be the smallest density induced on an δ-fraction of vertices, *i.e.*,

$$d(G, \delta) := \min \left\{ \frac{|E(G[A])|}{\binom{|A|}{2}} \; : \; A \subseteq V(G), |A| \geq \delta |V(G)| \right\}.$$

Note that for any graph G the function $d(G, \delta)$ is a non-decreasing function of δ taking values in $[0, 1]$.

However, requiring a positive δ-linear density immediately forces large graphs to behave similarly to random graphs in the sense that they contain every given graph as a subgraph, which follows, *e.g.*, from [11, Theorem 1].

Observation 1.1. For every $\varepsilon > 0$ and a fixed graph F, there exist $\delta > 0$ and $n_0 \in \mathbb{N}$ such that every graph G on at least n_0 vertices with $d(G, \delta) \geq \varepsilon$ contains F as an subgraph.

The notion of the δ-linear density has a natural generalization to a k-uniform hypergraph H, where we define

$$d(H, \delta) := \min \left\{ \frac{|E(H_n[A])|}{\binom{|A|}{k}} \; : \; A \subseteq V(H), |A| \geq \delta |V(H)| \right\}.$$

We now focus on 3-uniform hypergraphs. Let K_4 be the complete hypergraph on 4 vertices, and K_4^- the hypergraph on 4 vertices with 3 edges. Erdős and Sós [6, Problem 5] asked whether, analogously to Observation 1.1 for graphs, a sufficiently large hypergraph with positive δ-linear density contains a copy of K_4, or at least a copy of K_4^-. However, Füredi observed that the following construction of Erdős and Hajnal [4] shows that the situation in 3-uniform hypergraphs is completely different (hence giving a negative answer to the question above).

Example 1.2. Consider a random tournament T_n on n vertices. Let H_n be the 3-uniform hypergraph on the same vertex set consisting of exactly those triples that span an oriented cycle in T_n.

One can check that in every hypergraph obtained in this way, any four vertices span at most two edges, *i.e.*, H_n does not contain K_4^- for every n. On the other hand, for every $\delta > 0$, the δ-linear density of H_n tends to $1/4$ when n goes to infinity. Additional information about this problem and its history can be found in [12, Section 5]. It is also worth noting that in [11], Rödl presented a sequence of K_4-free (but not K_4^--free) 3-uniform hypergraphs with δ-linear density tending to $1/2$.

Our main result is the following theorem stating that the bound $1/4$ on δ-linear density for K_4^--free sequences is in fact the best possible, which answers the explicit question of Erdős [3] positively.

Theorem 1.3. *For every $\varepsilon > 0$ there exist $\delta > 0$ and $n_0 \in \mathbb{N}$ such that every 3-uniform hypergraph H on at least n_0 vertices with $d(H,\delta) \geq 1/4 + \varepsilon$ contains K_4^- as a subgraph.*

2 Sketch of the proof

Suppose that Theorem 1.3 were false, then the following would be true.

There exists $\varepsilon > 0$ and a sequence $(H_n)_{n\in\mathbb{N}}$ of K_4^--free hypergraphs with order tending to infinity such that $d(H_n, 1/n) \geq \frac{1}{4} + \varepsilon$. $(\star)$

We say that a sequence of 3-uniform hypergraphs $(H_n)_{n\in\mathbb{N}}$ is *convergent* if for every fixed hypergraph F the probabilities $p(F, H_n)$ that random $|F|$ vertices of $V(H_n)$ induce a copy of F converge. We refer to $p(F, H_n)$ as to the *induced density* of F in H_n. By a standard compactness argument, every sequence of hypergraphs have a convergent subsequence. Therefore, we may assume that the sequence given in $(\star)$ is convergent.

Our aim is to show that the edge density of the sequence given in $(\star)$ must tend to zero, which would contradict having positive δ-linear density.

The main tool used for the proof of Theorem 1.3 is the framework of flag algebras. The framework was introduced by Razborov [10], who was inspired by the theory of dense graph limits from Borgs *et al.* [1,2] and Lovász and Szegedy [9]. However, unlike just applying the standard flag algebra approach, in order to prove Theorem 1.3 we need to find a way of expressing the δ-linear density condition in this framework. Instead of using the assumption on δ-linear density itself, we establish a certain type of inequalities that are valid for any convergent sequence of K_4^--free hypergraphs (H_n) such that $d(H_n, f(n)) \geq 1/4$, where $f(n) \to 0$. Let us be more specific about that. Fix a K_4^--free hypergraph F, and consider an arbitrary K_4^--free superhypergraph F' of F on $|F|+1$ vertices. For any such a choice of a pair (F, F'), we obtain one inequality as follows. Let S be a fixed copy of F in H_n, and let $U(S)$ be the set of vertices v of $V(H_n) \setminus V(S)$ such that $S \cup \{v\}$ induces F' in H_n. Let T be now three random vertices from H_n not contained in S. The following two outcomes are possible:

- If the size of $U(S)$ is small, *i.e.*, $o(|V(H_n)|)$, then both the probability that T is a subset of $U(S)$ and the probability that T is a subset of $U(S)$ and spans an edge in H_n, are $o(1)$.
- If the size of $U(S)$ is large, *i.e.*, $\Omega(|V(H_n)|)$, then, by the assumption on the δ-linear density, the probability that $T \subseteq U(S)$ and span an edge is at least quarter of the probability that $T \subseteq U(S)$.

Therefore, we get that the following holds for every choice of S

$$4 \cdot P[T \subseteq U(S) \wedge T \in E(H_n)] - P[T \subseteq U(S)] \geq -o(1). \qquad (2.1)$$

Note that the little o goes to zero with n tending to infinity. The sought inequality corresponding to the pair (F, F') is then obtained by taking the expectation of (2.1), where the randomness comes from picking a copy S of F in H_n. This expectation can be expressed using the language of flag algebras.

It turned out, that for the proof of Theorem 1.3, it is enough to consider the inequalities for the pairs (F_2, F_3^0) and (F_2, F_3^1), where F_2 is the only 2-vertex 3-uniform hypergraph, F_3^0 is the empty 3-vertex hypergraph, and F_3^1 is the 3-vertex hypergraph with one edge.

Recall that our aim is to show that edge density of the sequence given in ($\star$) tends to zero. Unfortunately, we do not know how to apply flag algebras directly to derive this claim. Instead, we use them only to show that induced density of F_5 tends to zero, where F_5 is the 5-vertex hypergraph in Figure 2.1.

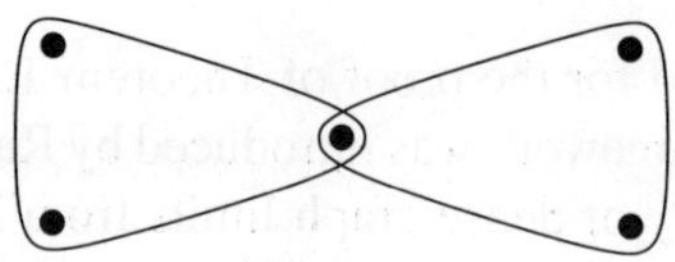

Figure 2.1. The hypergraph F_5.

To cope with this, we *sparsify* each hypergraph H_n given in ($\star$), *i.e.*, we remove each edge of H_n independently with a sufficiently small but fixed positive probability, say $\varepsilon/2$. Therefore, we obtain a new sequence of hypergraphs (I_n) such that $d\,(I_n, 1/n) \geq 1/4$ almost surely. Hence, the induced density of F_5 in the sequence (I_n) tends to zero. However, since the hypergraphs I_n were obtained from H_n by a random sparsification, the non-induced density of F_5 in H_n has to tend to zero. But that implies, that almost every vertex in H_n has degree at most n. Therefore, the number of edges in H_n is $O(n^2)$, which finishes the proof of Theorem 1.3.

References

[1] C. BORGS, J. T. CHAYES, L. LOVÁSZ, V. T. SÓS and K. VESZTERGOMBI, *sequences of dense graphs I: Subgraph frequencies, metric properties and testing*, Advances in Mathematics **219** (2008), 1801–1851.

[2] C. BORGS, J. T. CHAYES, L. LOVÁSZ, V. T. SÓS and K. VESZTERGOMBI, *Convergent sequences of dense graphs II: Multiway cuts and statistical physics*, Annals of Mathematics **176** (2012), 151–219.

[3] P. ERDŐS, *Problems and results on graphs and hypergraphs: Similarities and differences*, In: "Mathematics of Ramsey Theory", J. J. Nešetřil and V. Rödl (eds.), Springer-Verlag, 1990, 223–233.

[4] P. ERDŐS and A. HAJNAL, *On Ramsey like theorems. Problems and results*, In: "Combinatorics: being the proceedings of the Conference on Combinatorial Mathematics held at the Mathematical Institute", Oxford, 1972, Southend-on-Sea: Institute of Mathematics and Its Applications, 1972, 123–140.

[5] P. ERDŐS and M. SIMONOVITS, *An extremal graph problem*, Acta Mathematica Academiae Scientiarum Hungaricae **22** (1971), 275–282.

[6] P. ERDŐS and V. T. SÓS, *On Ramsey-Turán type theorems for hypergraphs*, Combinatorica **2** (1982), 289–295.

[7] P. ERDŐS and A. H. STONE, *On the structure of linear graphs*, Bulletin of the American Mathematical Society **52** (1946), 1089–1091.

[8] P. FRANKL and V. RÖDL, *Some Ramsey-Turán type results for hypergraphs*, Combinatorica **8** (1988), 323–332.

[9] L. LOVÁSZ and B. SZEGEDY, *Limits of dense graph sequences*, Journal of Combinatorial Theory, Series B **96** (2006), 933–957.

[10] A. RAZBOROV, *Flag algebras*, Journal of Symbolic Logic, **72** (2007), 1239–1282.

[11] V. RÖDL, *On universality of graphs with uniformly distributed edges*, Discrete Mathematics **59** (1986), 125–134.

[12] M. SIMONOVITS and V. T. SÓS, *Ramsey-Turán theory*, Discrete Mathematics **229** (2001), 293–340.

[13] V. T. SÓS, *On extremal problems in graph theory*, In: "Proceedings of the Calgary International Conference on Combinatorial Structures and their Application", 1969, 407–410.

[14] P. TURÁN, *Eine Extremalaufgabe aus der Graphentheorie* (in Hungarian), Matematikai és Fizikai Lapok **48** (1941), 436–452, see also: *On the theory of graphs*, Colloquium Mathematicum **3** (1954), 19–30.

An analogue of the Erdős-Ko-Rado theorem for multisets

Zoltán Füredi[1], Dániel Gerbner[1] and Máté Vizer[1]

Abstract. We verify a conjecture of Meagher and Purdy [4] by proving that if $1 \le t \le k$, $2k - t \le n$ and $\mathcal{F}$ is a family of *t-intersecting k-multisets* of $[n]$, then

$$|\mathcal{F}| \le AK(n + k - 1, k, t),$$

where $AK(n, k, t) := max_i |\mathcal{A}_{n,k,t,i}|$ with $\mathcal{A}_{n,k,t,i} := \{A : A \subseteq [n], |A| = k, |A \cap [t + 2i]| \ge t + i\}$.

1 Introduction

1.1 Definitions, notation

Let n and l be positive integers and let

$$M(n, l) = \{(i, j) : 1 \le i \le n, 1 \le j \le l\}$$

be an $n \times l$ rectangle. We call $A \subseteq M(n, l)$ a *k-multiset* if the cardinality of A is k and $(i, j) \in A$ implies $(i, j') \in A$ for all $j' \le j$. We think of multisets as sets with multiplicities, but it helps finding short and precise notation if we identify them with these special subsets of the rectangle. We denote the multiplicity of i in F by $m(i, F)$, *i.e.* $m(i, F) := \max\{s : (i, s) \in F\}$ ($m(i, F) \le l$ by definition).

Let $\mathcal{F}$ be a family of k-*multisets* of $M(n, l)$. We call $\mathcal{F}$ *t-intersecting* if $t \le |F_1 \cap F_2|$ for all $F_1, F_2 \in \mathcal{F}$. Let

$$\mathcal{M}(n, l, k, t) = \{\mathcal{F} : \mathcal{F} \textit{ is t-intersecting set of k-multisets of } M(n, l)\},$$

i.e. the class of t-intersecting families of k-multisets.

Let $\mathcal{F} \in \mathcal{M}(n, l, k, t)$. We call $T \subseteq M(n, l)$ a *t-kernel* for $\mathcal{F}$ if $|F_1 \cap F_2 \cap T| \ge t$ for all $F_1, F_2 \in \mathcal{F}$.

[1] Alfréd Rényi Institute of Mathematics, P.O.B. 127, Budapest H-1364, Hungary. Email: z-furedi@illinois.edu, gerbner.daniel@renyi.mta.hu, vizer.mate@renyi.mta.hu

1.2 History

Let us briefly summarize some results using our notation.

Theorem 1.1 (Erdős, Ko, Rado, [3]). *If $n \geq 2k$ and $\mathcal{F} \in \mathcal{M}(n, 1, k, 1)$ then*

$$|\mathcal{F}| \leq \binom{n-1}{k-1}.$$

If $n > 2k$, then equality holds if and only if all members of $\mathcal{F}$ contain a fixed element of $[n]$.

They also proved that if n is large enough, every member of the largest t-intersecting family of sets contains a fixed t-element set, but did not give the optimal threshold. Frankl [5] showed for $t \geq 15$ and Wilson [6] for every t that the optimal threshold is $n = (k-t+1)(t+1)$. Finally, Ahlswede and Khachatrian [1] determined the maximum families for all values of n.

Theorem 1.2 (Ahlswede, Khachatrian [1]). *Let $t \leq k \leq n$ and $\mathcal{A}_{n,k,t,i} = \{A : A \subseteq [n], |A| = k, |A \cap [t+2i]| \geq t+i\}$. If $\mathcal{F} \in \mathcal{M}(n, 1, k, t)$, then*

$$|\mathcal{F}| \leq max_i |\mathcal{A}_{n,k,t,i}| = AK(n, k, t).$$

Theorem 1.3 (Meagher, Purdy [4]). *If $n \geq k+1$ and $\mathcal{F} \in \mathcal{M}(n, k, k, 1)$, then*

$$|\mathcal{F}| \leq \binom{n+k-2}{k-1}.$$

If $n > k+1$, then equality holds if and only if all members of $\mathcal{F}$ contain a fixed element of $M(n, k)$.

1.3 Conjectures

Brockman and Kay stated the following conjecture [2]:

Conjecture 1.4 ([2], Conjecture 5.2). There is $n_0(k, t)$ such that if $n \geq n_0(k, t)$ and $\mathcal{F} \in \mathcal{M}(n, k, k, t)$, then

$$|\mathcal{F}| \leq \binom{n+k-t-1}{k-t}.$$

Furthermore, equality is achieved if and only if each member of $\mathcal{F}$ contains a fixed t-multiset of $M(n, k)$.

Meagher and Purdy also gave a possible candidate for the threshold $n_0(k, t)$.

Conjecture 1.5 ([4], Conjecture 4.1). Let k, n and t be positive integers with $t \leq k$, $t(k-t)+2 \leq n$ and $\mathcal{F} \in \mathcal{M}(n,k,k,t)$, then

$$|\mathcal{F}| \leq \binom{n+k-t-1}{k-t}.$$

Moreover, if $n > t(k-t)+2$, then equality holds if and only if all members of $\mathcal{F}$ contain a fixed t-multiset of $M(n,k)$.

Note that if $n < t(k-t)+2$, then the family consisting of all multisets which contain a fixed t-multiset of $M(n,k)$ still has cardinality $\binom{n+k-t-1}{k-t}$, but cannot be the largest. Indeed, if we fix a $t+2$-multiset T and consider the family of the multisets F with$|F \cap T| \geq t+1$, we get a larger family.

1.4 Results

The main idea of our proof is the following: instead of the well-known *left-compression* operation, which is a usual method in the theory of intersecting families, we define (in two different ways) an operation on $\mathcal{M}(n,l,k,t)$ which can be called a kind of *down-compression.*

Theorem 1.6. *Let* $1 \leq t \leq k$, $2k-t \leq n$ *and* l *be arbitrary. There exists*

$$f : \mathcal{M}(n,l,k,t) \to \mathcal{M}(n,l,k,t)$$

satisfying the following properties:

(i) $|\mathcal{F}| = |f(\mathcal{F})|$ *for all* $\mathcal{F} \in \mathcal{M}(n,l,k,t)$;

(ii) $M(n,1)$ *is a* t*-kernel for* $f(\mathcal{F})$.

Using Theorem 1.6 we prove the following theorem which not only verifies Conjecture 1.5, but also gives the maximum cardinality of t-intersecting families of multisets in the case $2k-t \leq n \leq t(k-t)+2$.

Theorem 1.7. *Let* $1 \leq t \leq k$ *and* $2k-t \leq n$. *If* $\mathcal{F} \in \mathcal{M}(n,k,k,t)$ *then*

$$|\mathcal{F}| \leq AK(n+k-1,k,t).$$

2 Concluding remarks

Note that for a family $\mathcal{A}_{n+k-1,k,t,i}$ we can define a t-intersecting family of k-multisets in $M(n,k)$, hence the bound given in Theorem 1.7 is sharp. However, we do not know any nontrivial bounds in case $n < 2k-t$.

Another interesting problem is the case $l < k$. Theorem 1.6 gives us a small t-kernel, but the proof of Theorem 1.7 does not work in this case.

References

[1] R. AHLSWEDE and L. H. KHACHATRIAN, *The complete intersection theorem for systems of finite sets*, European J. Combin. **18** (2) (1997), 125–136.

[2] G. BROCKMAN and B. KAY, *Elementary Techniques for Erdős-Ko-Rado-like Theorems*, http://arxiv.org/pdf/0808.0774v2.pdf.

[3] P. ERDŐS, C. KO and R. RADO, *Intersection theorems for systems of finite sets*, Quart. J. Math. Oxford Ser. **12** (2) (1961), 313–320.

[4] K. MEAGHER and A. PURDY, *An Erdős-Ko-Rado theorem for multisets*, Electron. J. Combin. **18** (1) (2011), Paper 220.

[5] P. FRANKL, *The shifting technique in extremal set theory*, In: "Surveys in Combinatorics", Lond. Math. Soc. Lect. Note Ser., 123 (1987), 81–110.

[6] R. M. WILSON, *The exact bound on the Erdős-Ko-Rado Theorem*, Combinatorica **4** (1984), 247–257.

Polynomial gap extensions of the Erdős–Pósa theorem

Jean-Florent Raymond[1] and Dimitrios M. Thilikos[2]

Abstract. Given a graph H, we denote by $\mathcal{M}(H)$ all graphs that can be contracted to H. The following extension of the Erdős–Pósa Theorem holds: for every h-vertex planar graph H, there exists a function $f_H : \mathbb{N} \to \mathbb{N}$ such that every graph G, either contains k disjoint copies of graphs in $\mathcal{M}(H)$, or contains a set of $f_H(k)$ vertices meeting every subgraph of G that belongs in $\mathcal{M}(H)$. In this paper we prove that f_H can be polynomially (upper) bounded for every graph H of pathwidth at most 2 and, in particular, that $f_H(k) = 2^{O(h^2)} \cdot k^2 \cdot \log k$. As a main ingredient of the proof of our result, we show that for every graph H on h vertices and pathwidth at most 2, either G contains k disjoint copies of H as a minor or the treewidth of G is upper-bounded by $2^{O(h^2)} \cdot k^2 \cdot \log k$. We finally prove that the exponential dependence on h in these bounds can be avoided if $H = K_{2,r}$. In particular, we show that $f_{K_{2,r}} = O(r^2 \cdot k^2)$.

1 Introduction

In 1965, Paul Erdős and Lajos Pósa proved that every graph that does not contain k disjoint cycles, contains a set of $O(k \log k)$ vertices meeting all its cycles [6]. Moreover, they gave a construction asserting that this bound is tight. This classic result can be seen as a "loose" min-max relation between covering and packing of combinatorial objects. Various extensions of this result, referring to different notions of packing and covering, attracted the attention of many researchers in modern Graph Theory (see, *e.g.* [1, 11]).

Given a graph H, we denote by $\mathcal{M}(H)$ the set of all graphs that can be contracted to H (*i.e.* if $H' \in \mathcal{M}(H)$, then H can be obtained from H' after contracting edges). We call the members of $\mathcal{M}(H)$ *models* of

[1] LIRMM, Montpellier, France. Email: jean-florent.raymond@ens-lyon.org

[2] Department of Mathematics, National and Kapodistrian University of Athens and CNRS (LIRMM). Email: sedthilk@thilikos.info

Co-financed by the E.U. (European Social Fund - ESF) and Greek national funds through the Operational Program "Education and Lifelong Learning" of the National Strategic Reference Framework (NSRF) - Research Funding Program: "Thales. Investing in knowledge society through the European Social Fund".

H. Then the notions of covering and packing can be extended as follows: we denote by $\mathbf{cover}_H(G)$ the minimum number of vertices that meet every model of H in G and by $\mathbf{pack}_H(G)$ the maximum number of mutually disjoint models of H in G. We say that a graph H *has the Erdős–Pósa Property* if there exists a function $f_H : \mathbb{N} \to \mathbb{N}$ such that for every graph G,

$$\text{if } k = \mathbf{pack}_H(G), \text{ then } k \leqslant \mathbf{cover}_H(G) \leqslant f_H(k) \tag{1.1}$$

We will refer to f_H as the *gap* of the Erdős–Pósa Property. Clearly, if $H = K_3$, then (1.1) holds for $f_{K_3} = O(k \log k)$ and the general question is to find, for each instantiation of H, the best possible estimation of the gap f_H, if it exists.

It turns out that H has the Erdős–Pósa Property if and only if H is a planar graph. This beautiful result appeared as a byproduct of the Graph Minors series of Robertson and Seymour. In particular, it is a consequence of the grid-exclusion theorem, proved in [14] (see also [3]).

Proposition 1.1. *There is a function $g : \mathbb{N} \to \mathbb{N}$ such that if a graph excludes an r-vertex planar graph R as a minor, then its treewidth is bounded by $g(r)$.*

In [14] Robertson, Seymour, and Thomas conjectured that g is a low degree polynomial function. Currently, the best known bound for g is $g(k) = 2^{O(k \log k)}$ and follows from [4] and [13] (see also [12, 14] for previous proofs and improvements). As the function g is strongly used in the construction of the function f_H in (1.1), the best, so far, estimation for f_H is far from being exponential in general. This initiated a quest for detecting instantiations of H where a polynomial gap f_H can be proved.

The first result in the direction of proving polynomial gaps for the Erdős–Pósa Property appeared in [9] where H is the graph θ_c consisting of two vertices connected by c multiple edges (also called *c-pumpkin graph*). In particular, in [9] it was proved that $f_{\theta_c}(k) = O(c^2k^2)$. More recently Fiorini, Joret, and Sau optimally improved this bound by proving that $f_{\theta_c}(k) \leqslant c_t \cdot k \cdot \log k$ for some computable constant c_t depending on c [8]. In [15] Fiorini, Joret, and Wood proved that if T is a tree, then $f_T(k) \leqslant c_T \cdot k$ where c_T is some computable constant depending on T. Finally, very recently, Fiorini [7] proved that $f_{K_4} = O(k \log k)$.

Our main result is a polynomial bound on f_H for a broad family of planar graphs, namely those of pathwidth at most 2. We prove the following:

Theorem 1.2. *If H is an h-vertex graph of pathwidth at most 2 and $h > 5$, then* (1.1) *holds for $f_H(k) = 2^{O(h^2)} \cdot k^2 \cdot \log k$.*

Note that the contribution of h in f_H is exponential. However, such a dependence can be waived when we restrict H to be $K_{2,r}$. Our second result is the following:

Theorem 1.3. *If $H = K_{2,r}$, then* (1.1) *holds for* $f_H(k) = O(r^2 \cdot k^2)$.

Both results above are based on a proof of Proposition 1.1, with polynomial g, for the cases where R consists of k disjoint copies of H and H is either a graph of pathwidth at most 2 or $H = K_{2,3}$ (Theorems 2.1 and 2.2 respectively). For this, we follow an approach that makes strong use of the k-mesh structure introduced by Diestel *et al.* [4] in their proof of Proposition 1.1. Our proof indicates that, when excluding copies of some graph of pathwidth at most 2, the entangled machinery of [4] can be partially modified so that polynomial bounds on treewidth are possible. Finally, these bounds are then "translated" to polynomial bounds for the Erdős–Pósa gap using a technique developed in [10] (see also [9]).

Definitions and preliminaries. All graphs in this paper are simple, finite and undirected and logarithms are binary. We use standard notation in Graph Theory. We define $k \cdot H$ as the graph obtained if we take k disjoint copies of H.

Treewidth. A *tree decomposition* of a graph G is a tree T whose vertices are some subsets of $V(G)$ such that: (i) $\bigcup_{X \in V(T)} X = V(G)$, (ii) for every edge e of G there is a vertex of T containing both end of e, and (iii) for all $v \in V(G)$, the subgraph of T induced by $\{X \in V(T),\ v \in X\}$ is connected. The *width* of a tree decomposition T is defined as equal to $\max_{X \in V(T)} |X| - 1$. The *treewidth* of G, written $\mathbf{tw}(G)$, is the minimum width of any of its tree decompositions. The *pathwidth* of G, written $\mathbf{pw}(G)$, is defined as the treewidth if we consider paths instead of trees.

2 Excluding packings of planar graphs

Theorems 1.2 and 1.3 follow combining the two following results with the machinery introduced in [10] (see also [9]). They have independent interest as they detect cases of Theorem 1.1 where g depends polynomially on k. In this extended abstract we only sketch the proof of Theorem 2.1.

Theorem 2.1. *Let H be a graph of pathwidth at most 2 on $r > 5$ vertices. If G does not contain k disjoint copies of H as minor then* $\mathbf{tw}(G) \leqslant 2^{O(r^2)} \cdot k^2 \cdot \log 2k$.

Theorem 2.2. *For every positive integer r, if G does not contain k disjoint copies of $K_{2,r}$ as minor then* $\mathbf{tw}(G) = O(r^2k^2)$.

Proof of theorem 2.1. (sketch) We prove the contrapositive. Let k be a integer, H a graph on $r > 5$ vertices and of pathwidth at most 2 and G a graph. It can easily be proved that $H \leqslant_{\mathrm{m}} \Xi_r$ where Ξ_r is the graph obtained by a $(r \times 2)$ if we subdivide once the "horizontal" edges. If we show that G contains k disjoint copies of Ξ_r as minors then we are done. Let $g\colon \mathbb{N}\to\mathbb{N}$ such that $g(k,r)=k^2 \log 2k \left(180\cdot 2^{r(r-2)} - 24 \cdot 2^{\frac{1}{2}r(r-2)}\right)+ 6 \cdot 2^{\frac{1}{2}r(r-2)} - 1$. We prove that for all graph G, $\mathbf{tw}(G) \geqslant g(k,r)$ implies that $G \geqslant_{\mathrm{m}} k \cdot \Xi_r$. Let k and $r > 5$ be two positive integers and assume that $\mathbf{tw}(G) \geqslant g(k,r)$. We examine below the case where $\delta_c(G) < c \cdot 3rk\sqrt{\log 3rk}$ (the case $\delta_c(G) \geqslant c \cdot 3rk\sqrt{\log 3rk}$ is ommited). Observe that $c \cdot 3rk\sqrt{\log 3rk} < c \cdot 3r\sqrt{\log 6r} \cdot k\sqrt{\log 2k}$. Let $k_0 = k\sqrt{\log 2k}$ and $r_0 = 3 \cdot 2^{\frac{r(r-2)}{2}}$, and remark that $k_0 \geqslant k$ and, $r_0 \geqslant c \cdot 3r\sqrt{\log 6r}$ (remember that $c \leqslant 648$ and $r > 5$). With these notations, we have $\delta_c(G) < 2k_0r_0$. We will show that $G \geqslant_{\mathrm{m}} k_0 \cdot \Xi_r$ from which yields that $G \geqslant_{\mathrm{m}} k \cdot \Xi_r$. By assumption, $\mathbf{tw}(G) \geqslant g(k,r)$. Using the results of [4], we can prove that G contains $2k_0$ subsets $X_1, \dots, X_{2k_0}$ of $V(G)$ and a set $\mathcal{P}$ of $k_0r_0 = 3k_0 \cdot 2^{\frac{r(r-2)}{2}}$ disjoint paths of length at least 2 in G such that (i) $\forall i \in [\![1, 2k_0]\!]$, X_i is of size $r_0 = 3 \cdot 2^{\frac{r(r-2)}{2}}$ and is connected in G by a tree T_i using the elements of some set $A \subseteq V(G)$, (ii) any path in $\mathcal{P}$ has one of its ends in some X_i with $i \in [\![1, k_0]\!]$, the other end in X_{2i} and its internal vertices are in none of the X_l, for all $l \in [\![1, 2k_0]\!]$, nor in A, and (iii) $\forall i, j \in [\![1, 2k_0]\!]$, $i \neq j \Rightarrow T_i \cap T_j = \emptyset$.

We assume that for all $i \in [\![1, 2k_0]\!]$, $X_i = \{v \in V(T_i),\ \deg_T(v) \leqslant 2\}$. It is easy to come down to this case by considering the minor of G obtained after deleting in T_i the leaves that are not in X_i and contracting one edge meeting a vertex of degree 2 which is not in X while such a vertex exists. As T_i is a ternary tree, one can easily prove that for all $i \in [\![1, 2k_0]\!]$, T_i contains a path containing $2 \log \frac{2}{3} |X_i| = (r-1)^2 + 1$ vertices of X_i. Let us call P_i such a path whose two ends are in X_i. Let us consider now the paths $\{P_i\}_{i \in [\![1, 2k_0]\!]}$ and the paths that link the elements of different P_i's. For each path $i \in [\![1, 2k_0]\!]$, we choose in P_i one end vertex (remember that both are in X_i) that we name $p_{i,0}$. We follow P_i from this vertex and we denote the other vertices of $P_i \cap X_i$ by $p_{i,1}, p_{i,1}, \dots, p_{i,(r-1)^2}$ in this order. The *corresponding vertex* of some vertex $p_{i,j}$ of $P_i \cap X_i$ (for $i \in [\![1, k_0]\!]$) is defined as the vertex of $P_{2i} \cap X_{2i}$ to which $p_{i,j}$ is linked to by a path of $\mathcal{P}$. As said before, the sets $\{P_i \cap X_i\}_{i \in [\![1, 2k_0]\!]}$ are of size $(r-1)^2 + 1$. According to [5], one can find for all $i \in [\![1, k_0]\!]$ a subsequence of length r in $p_{i,0}, p_{i,1}, \dots, p_{i,(r-1)^2}$, such that the corresponding vertices in X_{2i} of this sequence are either in the same order (with respect to the subscripts of the names of the vertices), or in reverse order. For all $i \in [\![1, k_0]\!]$, this subsequence, its corresponding vertices and the vertices of the paths that link them together forms a Ξ_r model. We have thus k_0 models of Ξ_r in G, that gives us k disjoint models of Ξ_r in G (since $k \leqslant k_0$). We showed that for all k and $r > 5$ positive integers, if a graph G has $\mathbf{tw}(G) \geqslant g(k, r)$, then $G \geqslant_{\mathrm{m}} k \cdot \Xi_r$. Consequently, if G has treewidth at least $g(k, r)$, then G contains k disjoint copies of H and we are done. □

Postscript. Very recently, the general open problem of estimating $f_H(k)$ when H is a general planar graph has been tackled in [2]. Moreover, very recently, using the results of [13] we were able to improve both Theorems 2.1 and 1.3 by proving polynomial (on both k and $|V(H)|$) bounds for more general instantiations of H.

References

[1] E. BIRMELÉ, J. A. BONDY and B. A. REED, *The Erdős–Pósa property for long circuits*, Combinatorica **27** (2007), 135–145.

[2] C. CHEKURI and J. CHUZHOY, *Large-treewidth graph decompositions and applications*, In: “45st Annual ACM Symposium on Theory of Computing”, (STOC 2013), 2013.

[3] R. DIESTEL, “Graph Theory”, volume 173 of *Graduate Texts in Mathematics*, Springer-Verlag, Heidelberg, fourth edition, 2010.

[4] R. DIESTEL, T. R. JENSEN, K. YU. GORBUNOV and C. THOMASSEN, *Highly connected sets and the excluded grid theorem*, J. Combin. Theory Ser. B **75** (1) (1999), 61–73.

[5] P. ERDŐS and G. SZEKERES, *A combinatorial problem in geometry*, In: “Classic Papers in Combinatorics”, Ira Gessel and Gian-Carlo Rota (eds.), Modern Birkhäuser Classics, Birkhäuser Boston, 1987, 49–56.

[6] P. ERDŐS and L. PÓSA, *On independent circuits contained in a graph*, Canad. J. Math. **17** (1965), 347–352.

[7] S. FIORINI, T. HUYHN and G. JORET, personal communication, 2013.

[8] S. FIORINI, G. JORET and I. SAU, *Optimal Erdős–Pósa property for pumpkins*, Manuscript, 2013.

[9] F. V. FOMIN, D. LOKSHTANOV, N. MISRA, G. PHILIP and S. SAURABH, *Quadratic upper bounds on the Erdős–Pósa property for a generalization of packing and covering cycles*, Journal of Graph Theory, to appear in 2013.

[10] F. V. FOMIN, S. SAURABH and D. M. THILIKOS, *Strengthening Erdős–Pósa property for minor-closed graph classes*, Journal of Graph Theory **66** (3) (2011), 235–240.

[11] J. GEELEN and K. KABELL, *The Erdős–Pósa property for matroid circuits*, J. Comb. Theory Ser. B **99** (2) (2009), 407–419.

[12] KEN-ICHI KAWARABAYASHI and Y. KOBAYASHI, *Linear min-max relation between the treewidth of H-minor-free graphs and its largest grid*, In: "29th Int. Symposium on Theoretical Aspects of Computer Science (STACS 2012)", Vol. 14 of *LIPIcs*, Dagstuhl, Germany, 2012, 278–289.

[13] A. LEAF and P. SEYMOUR, *Treewidth and planar minors*, Manuscript, 2012.

[14] N. ROBERTSON and P. D. SEYMOUR, *Graph minors. V. Excluding a planar graph*, J. Combin. Theory Series B **41** (2) (1986), 92–114.

[15] D. R. WOOD SAMUEL FIORINI and G. JORET, *Excluded forest minors and the Erdős-Pósa property*, Technical report, Cornell University, 2012.

The Erdős-Pósa property for long circuits

Dirk Meierling[1], Dieter Rautenbach[1] and Thomas Sasse[1]

Abstract. For an integer ℓ at least 3, we prove that if G is a graph containing no two vertex-disjoint circuits of length at least ℓ, then there is a set X of at most $\frac{5}{3}\ell + \frac{29}{2}$ vertices that intersects all circuits of length at least ℓ. Our result improves the bound $2\ell + 3$ due to Birmelé, Bondy, and Reed (The Erdős-Pósa property for long circuits, Combinatorica 27 (2007), 135-145) who conjecture that ℓ vertices always suffice.

1 Introduction

A family $\mathcal{F}$ of graphs is said to have the *Erdős-Pósa property* if there is a function $f_{\mathcal{F}} : \mathbb{N} \to \mathbb{N}$ such that for every graph G and every $k \in \mathbb{N}$, either G contains k vertex-disjoint subgraphs that belong to $\mathcal{F}$ or there is a set X of at most $f_{\mathcal{F}}(k)$ vertices of G such that $G - X$ has no subgraph that belongs to $\mathcal{F}$. The origin of this notion is [3] where Erdős and Pósa prove that the family of all circuits has this property.

Let ℓ be an integer at least 3. Let $\mathcal{F}_\ell$ denote the family of circuits of length at least ℓ. In [2] Birmelé, Bondy, and Reed show that $\mathcal{F}_\ell$ has the Erdős-Pósa property with

$$f_{\mathcal{F}_\ell}(k) \leq 13\ell(k-1)(k-2) + (2\ell+3)(k-1), \tag{1.1}$$

which improves an earlier doubly exponential bound on $f_{\mathcal{F}_\ell}(k)$ obtained by Thomassen [5]. The main contribution of Birmelé, Bondy, and Reed [2] is to prove (1.1) for $k = 2$, that is, to show

$$f_{\mathcal{F}_\ell}(2) \leq 2\ell + 3. \tag{1.2}$$

For $k \geq 3$, an inductive argument allows to deduce (1.1) from (1.2).

Birmelé, Bondy, and Reed [2] conjecture that

$$f_{\mathcal{F}_\ell}(2) \leq \ell, \tag{1.3}$$

[1] Institut für Optimierung und Operations Research, Universität Ulm, Ulm, Germany.
Email: dirk.meierling@uni-ulm.de, dieter.rautenbach@uni-ulm.de, thomas.sasse@uni-ulm.de

that is, for every graph G containing no two vertex-disjoint circuits of length at least ℓ, there is a set X of at most ℓ vertices such that $G - X$ has no circuit of length at least ℓ. In view of the complete graph of order $2\ell - 1$, (1.3) would be best possible. For $\ell = 3$, (1.3) was shown by Lovász [4] and for $\ell \in \{4, 5\}$, (1.3) was shown by Birmelé [1].

Our contribution in the present paper is the following result.

Theorem 1.1. *Let ℓ be an integer at least* 3. *Let G be a graph containing no two vertex-disjoint circuits of length at least ℓ.*

There is a set X of at most $\frac{5}{3}\ell + \frac{29}{2}$ vertices that intersects all circuits of length at least ℓ, that is, $f_{\mathcal{F}_\ell}(2) \leq \frac{5}{3}\ell + \frac{29}{2}$.

While Theorem 1.1 is a nice improvement of (1.2), for $k \geq 3$, the above-mentioned inductive argument still leads to an estimate of the form $f_{\mathcal{F}_\ell}(k) = O(\ell k^2)$.

The rest of this paper is devoted to the proof of Theorem 1.1.

2 Proof of Theorem 1.1

With respect to notation and terminology we follow [2] and recall some specific notions. All graphs are finite, simple, and undirected. We abbreviate *vertex-disjoint* as *disjoint*. If A and B are sets of vertices of a graph G, then an *(A, B)-path* is a path P in G between a vertex in A and a vertex in B such that no internal vertex of P belongs to $A \cup B$. If P is a path and x and y are vertices of P, then $P[x, y]$ denotes the subpath of P between x and y. Similarly, if C is a circuit endowed with an orientation and x and y are vertices of C, then $C[x, y]$ denotes the segment of C from x to y following the orientation of C. In all figures of circuits the orientations will be counterclockwise.

We fix an integer ℓ at least 3 and call a circuit of length at least ℓ *long*. If C is a circuit and P and P' are disjoint $(V(C), V(C))$-paths such that P is between u and v and P' is between u' and v', then

- P and P' are called *parallel (with respect to C)* if u, u', v', v appear in the given cyclic order on C and
- P and P' are called *crossing (with respect to C)* if u, u', v, v' appear in the given cyclic order on C.

See Figure 2.1.

In the proof of Theorem 1.1 below we consider three cases according to the length L of a shortest long circuit. If L is less than $3\ell/2$, the result is trivial. For L between $3\ell/2$ and 2ℓ the following lemma implies the desired bound. Finally, for L larger than 2ℓ, Lemma 2.2 implies the desired bound.

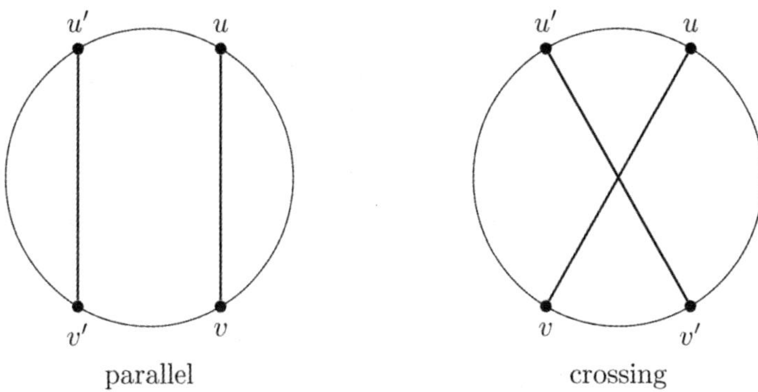

Figure 2.1. Parallel and crossing pairs of paths.

Lemma 2.1. *Let G be a graph containing no two disjoint long circuits. If the shortest long circuit of G has length L with* $L \geq 3\left(\left\lceil \frac{1}{2}\ell \right\rceil - 2\right)$*, then there is a set X of at most* $\frac{1}{3}L + \ell + \frac{14}{3}$ *vertices that intersects all long circuits.*

Proof. Let C be a shortest long circuit of G. We endow C with an orientation. We decompose C into 6 cyclically consecutive and internally disjoint segments $C_1, \ldots, C_6$ such that C_1, C_3, and C_5 have length $\left\lceil \frac{1}{2}\ell \right\rceil - 2$ and C_2, C_4, and C_6 have lengths between $\left\lfloor \frac{1}{3}L - \left(\left\lceil \frac{1}{2}\ell \right\rceil - 2\right) \right\rfloor$ and $\left\lceil \frac{1}{3}L - \left(\left\lceil \frac{1}{2}\ell \right\rceil - 2\right) \right\rceil$, that is, the six segments cover all of C and C_i and C_{i+1} overlap in exactly one vertex for every $i \in [6]$ where we identify indices modulo 6.

Let $X_1 = V(C_1) \cup V(C_3) \cup V(C_5)$. See the left part of Figure 2.2.

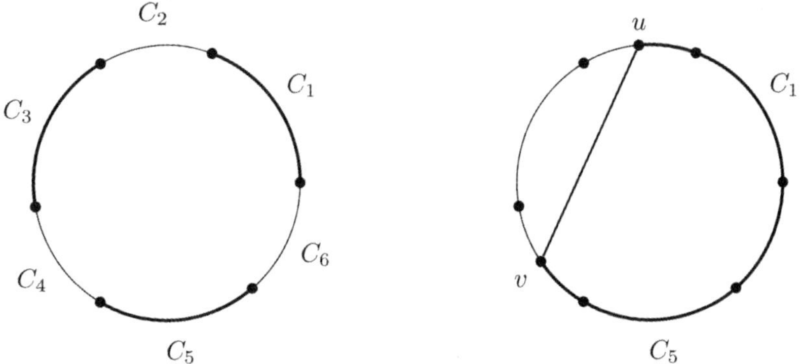

Figure 2.2. On the left the six segments of C and the set X in bold. On the right a long circuit formed by a $(V(C_2), V(C_4))$-path between u and v in $G - (X_1 \cup V(C_6))$.

Let $i \in [6]$ be even. Let $\mathcal{P}_i$ denote the set of $(V(C_i), V(C_{i+2}))$-paths in $G - (X_1 \cup V(C_{i+4}))$. The choice of C implies that every path P in $\mathcal{P}_i$

has length at least $\frac{1}{2}\ell$; otherwise P together with a segment of C avoiding $V(C_{i+1})$ forms a long circuit that is shorter than C. See the right part of Figure 2.2. This implies that for every path P in $\mathcal{P}_i$, P together with a segment of C containing $V(C_{i+1})$ forms a long circuit.

Let $\mathcal{P} = \mathcal{P}_2 \cup \mathcal{P}_4 \cup \mathcal{P}_6$. Since G has no two disjoint long circuits, it follows that $\mathcal{P}$ contains no two disjoint parallel paths and no four disjoint crossing paths. See Figure 2.3.

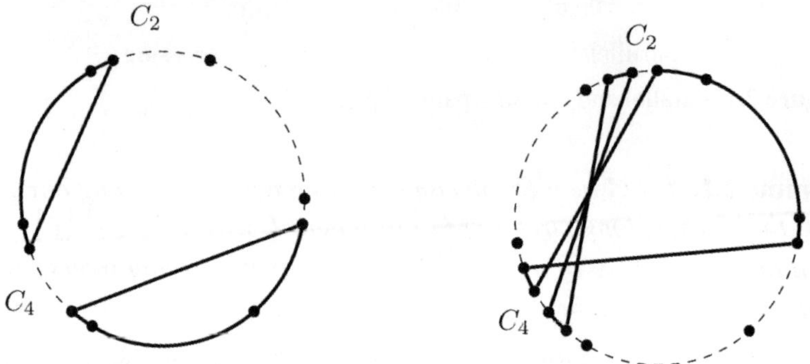

Figure 2.3. Two disjoint long circuits formed by two disjoint parallel paths in $\mathcal{P}$ or by four disjoint crossing paths in $\mathcal{P}$.

Let X_2 be a smallest set of vertices separating $V(C_2)$ and $V(C_4) \cup V(C_6)$ in $G - X_1$. Let X_3 be a smallest set of vertices separating $V(C_4)$ and $V(C_6)$ in $G - (X_1 \cup X_2)$. By the above observations and Menger's theorem, $|X_2| \leq 3$ and $|X_3| \leq 3$.

There is some even $j \in [6]$ such that in $G - (X_1 \cup X_2 \cup X_3)$, all long circuits intersect C only in $V(C_j)$; otherwise there is a $(V(C_i), V(C_{i+2}))$-path in $G - (X_1 \cup X_2 \cup X_3)$ for some even $i \in [6]$. This implies that $X_1 \cup X_2 \cup X_3 \cup V(C_j)$ intersects all long circuits of G. Since

$$\begin{aligned}|X_1 \cup X_2 \cup X_3 \cup V(C_j)| &\leq 3\left(\left\lceil \frac{1}{2}\ell \right\rceil - 2\right) + 3 + 3 \\ &\quad + \left\lceil \frac{1}{3}L - \left(\left\lceil \frac{1}{2}\ell \right\rceil - 2\right)\right\rceil + 1 \\ &\leq \frac{1}{3}L + \ell + \frac{14}{3}\end{aligned}$$

we obtain the desired result. □

Lemma 2.2. *Let G be a graph containing no two disjoint long circuits.*

If the shortest long circuit of G has length at least $2\ell - 3$, then there is a set X of at most $\frac{3}{2}\ell + \frac{29}{2}$ vertices that intersects all long circuits.

Sketch of the proof. We only sketch the proof. Specifically, we do not give the proofs of the claims. Let C be shortest long circuit of G. Let L denote the length of C. We endow C with an orientation.

As in [2], a path between two vertices x and y of C that is internally disjoint from C is called *long*, if the segments $C[x, y]$ and $C[y, x]$ both have length at least $\frac{1}{2}\ell$.

Claim A. *Every long path has length at least* $\ell - 1$.

Choose a long circuit D of G distinct from C and a segment $C[x, y]$ of C such that $C[x, y]$ contains $V(C) \cap V(D)$ and has minimum possible length. Note that $x, y \in V(C) \cap V(D)$.

We consider the two cases $x \neq y$ and $x = y$. Here we only sketch the more general case $x \neq y$. Let X_1 denote the set of $\lceil \frac{1}{2}\ell \rceil - 1$ vertices immediately preceeding x and let X_2 denote the set of $\lceil \frac{1}{2}\ell \rceil - 1$ vertices immediately following y. Let $A = V(C) \setminus (X_1 \cup X_2 \cup V(C[x, y]))$ and $B = V(C[x, y])$. In $G - (X_1 \cup X_2)$, there are no two disjoint parallel (A, B)-paths and no four disjoint crossing (A, B)-paths; otherwise there would be two disjoint long circuits. Hence, by Menger's theorem, there is a set X_3 of at most 3 vertices separating A and B in $G-(X_1 \cup X_2)$. The circuit D uniquely decomposes into a set $\mathcal{P}$ of at least two (B, B)-paths of length at least 1.

Claim B. *If $\mathcal{P}$ contains a path P between x and y, then in $G - (X_1 \cup X_2 \cup X_3)$, there are at most $\lceil \frac{1}{2}\ell \rceil + 1$ disjoint $(A, V(P))$-paths.*

Claim C. *If $\mathcal{P}$ contains two paths, say P and P', between x and y, then in $G-(X_1 \cup X_2 \cup X_3)$, there are at most $\lceil \frac{1}{2}\ell \rceil + 1$ disjoint $(A, V(D))$-paths.*

Claim D. *If P is a path in $\mathcal{P}$ that is not a path between x and y, then in $G - (X_1 \cup X_2 \cup X_3)$, there are at most* 3 *disjoint $(A, V(P))$-paths.*

Claim E. *If $P_1, \ldots, P_4$ are four distinct paths in $\mathcal{P}$ that are no paths between x and y, then in $G - (X_1 \cup X_2 \cup X_3)$, there are no four disjoint paths $Q_1, \ldots, Q_4$ such that Q_i is a $(A, V(P_i))$-path for $i \in [4]$.*

Let V_1 denote the set of vertices r of D such that $\mathcal{P}$ contains a path between x and y that contains r and let V_2 denote the set of vertices s of D such that $\mathcal{P}$ contains a path not between x and y that contains s. Clearly, $V_1 \cup V_2 = V(D)$. By Claims B and C and Menger's theorem, there is a set X_4 of at most $\lceil \frac{1}{2}\ell \rceil + 1$ vertices separating A and V_1 in $G-(X_1 \cup X_2 \cup X_3)$. By Claims D and E and Menger's theorem, there is a set X_5 of at most 9 vertices separating A and V_2 in $G-(X_1 \cup X_2 \cup X_3)$. Let $X = \{x, y\} \cup X_1 \cup X_2 \cup X_3 \cup X_4 \cup X_5$.

If $G - X$ contains a long circuit, say D', then D' intersects A. Since D and D' intersect, there is an $(A, V(D))$-path P in $G - X$. In view of

X_3, P cannot end in B; in view of X_4, P cannot end in V_1; and, in view of X_5, P cannot end in V_2, which is a contradiction. Hence X intersects all long circuits. Since

$$|X| \leq 2 + |X_1| + |X_2| + |X_3| + |X_4| + |X_5| \leq \frac{3}{2}\ell + \frac{29}{2},$$

this completes the proof in the case $x \neq y$. □

Proof of Theorem 1.1. Let C be shortest long circuit of G. Let L denote the length of C.

If L is at most $\frac{5}{3}\ell + \frac{29}{2}$, then let $X = V(C)$. If L is larger than $\frac{5}{3}\ell + \frac{29}{2}$ but less than $2\ell - 4$, then Lemma 2.1 implies the existence of a set X with the desired properties. If L is at least $2\ell - 3$, then Lemma 2.2 implies the existence of a set X with the desired properties. □

Our main interest was to improve the factor of ℓ in the bound in Theorem 1.1 and not the additive constant, which can easily be improved slightly.

The main open problem remains the conjectured inequality (1.3). We believe that further ideas are needed for its proof. Furthermore, it is unclear whether the quadratic dependence on k in (1.1) is best possible. For $\ell = 3$, that is, the classical case considered by Erdős and Pósa [3], it is known that $f_{\mathcal{F}_3}(k) = O(k \log k)$.

References

[1] E. Birmelé, Thèse de doctorat, Université de Lyon 1, 2003.

[2] E. Birmelé, J. A. Bondy and B. A. Reed, *The Erdős-Pósa property for long circuits*, Combinatorica **27** (2007), 135–145.

[3] P. Erdős and L. Pósa, *On independent circuits contained in a graph*, Canad. J. Math. **17** (1965), 347–352.

[4] L. Lovász, *On graphs not containing independent circuits (Hungarian)*, Mat. Lapok **16** (1965), 289–299.

[5] C. Thomassen, *On the presence of disjoint subgraphs of a specified type*, J. Graph Theory **12** (1988), 101–111.

Hypergraphs

A hypergraph Turán theorem via Lagrangians of intersecting families

Dan Hefetz[1] and Peter Keevash[2]

Abstract. Let $\mathcal{K}^3_{3,3}$ be the 3-graph with 15 vertices $\{x_i, y_i : 1 \leq i \leq 3\}$ and $\{z_{ij} : 1 \leq i, j \leq 3\}$, and 11 edges $\{x_1, x_2, x_3\}$, $\{y_1, y_2, y_3\}$ and $\{\{x_i, y_j, z_{ij}\} : 1 \leq i, j \leq 3\}$. We show that for large n, the unique largest $\mathcal{K}^3_{3,3}$-free 3-graph on n vertices is a balanced blow-up of the complete 3-graph on 5 vertices. Our proof uses the stability method and a result on Lagrangians of intersecting families that has independent interest.

1 Introduction

The *Turán number* $\text{ex}(n, F)$ is the maximum number of edges in an F-free r-graph on n vertices. It is a long-standing open problem in Extremal Combinatorics to develop some understanding of these numbers for general r-graphs F. For ordinary graphs ($r = 2$) the picture is fairly complete, but for $r \geq 3$ there are very few known results. Turán [5] posed the natural question of determining $\text{ex}(n, F)$ when $F = K^r_t$ is a complete r-graph on t vertices. To date, no case with $t > r > 2$ of this question has been solved, even asymptotically. For a summary of progress on hypergraph Turán problems we refer the reader to the survey [2].

In this paper, we determine the Turán number of the 3-graph $\mathcal{K}^3_{3,3}$ with vertices $\{x_i, y_i : 1 \leq i \leq 3\}$ and $\{z_{ij} : 1 \leq i, j \leq 3\}$, and edges $\{x_1, x_2, x_3\}$, $\{y_1, y_2, y_3\}$ and $\{\{x_i, y_j, z_{ij}\} : 1 \leq i, j \leq 3\}$. For an integer $n \geq 5$, let $T^3_5(n)$ denote the balanced blow-up of K^3_5 on n vertices, that is, we partition the vertices into 5 parts of sizes $\lfloor n/5 \rfloor$ or $\lceil n/5 \rceil$, and take

[1] School of Mathematics, University of Birmingham, Edgbaston, Birmingham, B15 2TT, UK. Email: d.hefetz@bham.ac.uk.

[2] School of Mathematical Sciences, Queen Mary University of London, Mile End Road, London E1 4NS, England. Email: p.keevash@qmul.ac.uk.

Research supported in part by ERC grant 239696 and EPSRC grant EP/G056730/1.

as edges all triples in which the vertices belong to 3 distinct parts. Write $t_5^3(n) := e(T_5^3(n))$. Our main result is as follows.

Theorem 1.1. *$ex(n, \mathcal{K}_{3,3}^3) = t_5^3(n)$ for sufficiently large n. Moreover, if n is sufficiently large and G is a $\mathcal{K}_{3,3}^3$-free 3-graph with n vertices and $t_5^3(n)$ edges, then $G \cong T_5^3(n)$.*

We prove Theorem 1.1 by the stability method and Lagrangians. Given an r-graph G on $[n] = \{1, \ldots, n\}$, we define a polynomial in the variables $x = (x_1, \ldots, x_n)$ by

$$p_G(x) := \sum_{e \in E(G)} \prod_{i \in e} x_i.$$

The *Lagrangian* of G is

$$\lambda(G) = \max\{p_G(x) : x_i \geq 0 \text{ for } 1 \leq i \leq n \text{ and } \sum_{i=1}^{n} x_i = 1\}.$$

A key tool in the proof will be the following result that determines the maximum possible Lagrangian among all intersecting 3-graphs: it is uniquely achieved by K_5^3, which has $\lambda(K_5^3) = \binom{5}{3}(1/5)^3 = 2/25$.

Theorem 1.2. *Let G be an intersecting 3-graph. If $G \neq K_5^3$, then $\lambda(G) \leq \lambda(K_5^3) - 10^{-3}$.*

2 Lagrangians of intersecting 3-graphs

In this section we provide a rough sketch of the proof of Theorem 1.2. Our proof consists of the following two steps:

(1) Prove that it suffices to bound from above the Lagrangian of a small number of specific 3-graphs.
(2) Bound from above the Lagrangian of the 3-graphs determined in step (1).

Step (1) consists of the following sub-steps:

(a) Prove that it suffices to consider intersecting 3-graphs which satisfy certain desirable properties (e.g. the set of hyperedges is maximal with respect to inclusion and covers pairs of vertices whose weight in the assignment under consideration is strictly positive).
(b) Apply a *down shifting operation* (or just shifting for brevity) to the set of hyperedges of the 3-graphs considered and prove that there is just a small number of resulting hypergraphs and that these have some desirable properties (all of these are depicted in Figure 2.1; note that

every such hypergraph has at most 7 vertices). Given an intersecting 3-graph $\mathcal{F}$ with vertex set $[n]$, a *shift* of $\mathcal{F}$ is any hypergraph obtained by applying the following rule: as long as there exist $i \in A \in \mathcal{F}$ such that $\mathcal{F}' = (\mathcal{F} \setminus \{A\}) \cup \{A \setminus \{i\}\}$ is intersecting, replace $\mathcal{F}$ by $\mathcal{F}'$ and repeat.

(c) For every 3-graph $\mathcal{F}$ as in (a) let $S(\mathcal{F})$ be the hypergraph obtained from it by shifting in (b). We define a 3-graph $G = Gen(n, S(\mathcal{F}))$ such that $|V(G)| = n$ and $G \supseteq \mathcal{F}$. The family $\mathcal{L}$ of all such G is the family of 3-graphs we will consider in step (2).

Is step (2) we go through all 3-graphs in $\mathcal{L}$ and prove a suitable upper bound on the Lagrangian of each of its elements. The details vary from 3-graph to 3-graph but the general idea is to use the method of lagrange multipliers while making use of a combinatorial weight shifting argument.

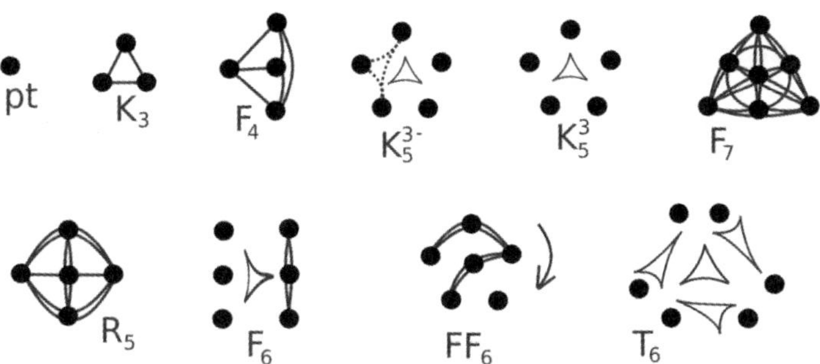

Figure 2.1. The family of hypergraphs resulting from down shifting.

3 An application to a hypergraph Turán problem

In this section we apply Theorem 1.2 to prove our main theorem on the Turán number of $\mathcal{K}^3_{3,3}$, namely that for large n, the unique extremal example is $T^3_5(n)$, i.e. the balanced blow-up of K^3_5. First we note some simple facts about $T^3_5(n)$. It is $\mathcal{K}^3_{3,3}$-free, as for any attempted embedding of $\mathcal{K}^3_{3,3}$ in $T^3_5(n)$, there must be some $1 \leq i, j \leq 3$ such that x_i and y_j lie in the same part, but then $x_i y_j z_{ij}$ cannot be an edge. The number of edges satisfies

$$t^3_5(n) = \sum_{0 \leq i < j < k \leq 4} \left\lfloor \frac{n+i}{5} \right\rfloor \cdot \left\lfloor \frac{n+j}{5} \right\rfloor \cdot \left\lfloor \frac{n+k}{5} \right\rfloor = \frac{2}{25} n^3 + O(n^2).$$

Also, the minimum degree satisfies

$$\delta_5^3(n) = t_5^3(n) - t_5^3(n-1) = \sum_{0 \le i < j \le 3} \left\lfloor \frac{n+i}{5} \right\rfloor \cdot \left\lfloor \frac{n+j}{5} \right\rfloor = \frac{6}{25} n^2 + O(n).$$

We start by showing that the asymptotic result follows quickly from Theorem 1.2. First we need some definitions. Suppose F and G are r-graphs. The *Turán density* of F is $\pi(F) = \lim_{n\to\infty} \binom{n}{r}^{-1} \mathrm{ex}(n, F)$. Given r-graphs F and G we say $f : V(F) \to V(G)$ is a homomorphism if it preserves edges, i.e. $f(e) \in E(G)$ for all $e \in E(F)$. We say that G is F-*hom-free* if there is no homomorphism from F to G. The *blow-up density* is $b(G) = r!\lambda(G)$. We say G is *dense* if every proper subgraph G' satisfies $b(G') < b(G)$. We also need the following two standard facts (see e.g. [2, Section 3]):

(i) $\pi(F)$ is the supremum of $b(G)$ over F-hom-free dense G,
(ii) dense r-graphs cover pairs.

Theorem 3.1. *ex*$(n, \mathcal{K}_{3,3}^3) = \frac{2}{25} n^3 + o(n^3)$.

Proof. An equivalent formulation is that $\pi(\mathcal{K}_{3,3}^3) = \frac{12}{25}$. The lower bound is given by the construction $T_5^3(n)$. For the upper bound, by fact (i) above, it suffices to show that $b(G) \le 12/25$ for any $\mathcal{K}_{3,3}^3$-hom-free dense G. Suppose for a contradiction that G is $\mathcal{K}_{3,3}^3$-hom-free, dense, and has $\lambda(G) > 2/25$. By Theorem 1.2, G is not intersecting, so we can choose disjoint edges $\{x_1, x_2, x_3\}$ and $\{y_1, y_2, y_3\}$. Then by fact (ii), G covers pairs, so for every $1 \le i, j \le 3$ there exists an edge $\{x_i, y_j, z_{ij}\}$. However, this defines a homomorphism from $\mathcal{K}_{3,3}^3$ to G, which contradicts G being $\mathcal{K}_{3,3}^3$-hom-free. □

3.1 Stability

In order to prove Theorem 1.1, we will first prove the following stability result.

Theorem 3.2. *For any $\varepsilon > 0$ there exist $\delta > 0$ and an integer n_0 such that if $\mathcal{F}$ is a $\mathcal{K}_{3,3}^3$-free 3-graph with $n \ge n_0$ vertices and at least $(\frac{2}{25} - \delta)n^3$ edges, then there exists a partition $V(\mathcal{F}) = A_1 \cup \ldots \cup A_5$ of the vertex set of $\mathcal{F}$ such that $\sum_{1 \le i < j \le 5} e(A_i \cup A_j) < \varepsilon n^3$.*

Part of our proof follows the main ideas of [3]. The proof consists of two main stages. In the first stage we gradually change $\mathcal{F}$ (as well as some other related structure) by iterating a process which is called *Symmetrization.* This process consists of two parts: *Cleaning* and *Merging.*

We refer to the basic object with which we operate as a *pointed partitioned* 3-*graph*; by this we mean a triple $(\mathcal{G}, \mathcal{P}, U)$ where $\mathcal{G} = (V, E)$ is a 3-graph, $\mathcal{P} = \{P_u : u \in V\}$ is a partition of V such that $u \in P_u$ for every $u \in V$, and $U \subseteq V$ is a transversal of $\mathcal{P}$. The ultimate goal of this process is to end up with a pointed partitioned 3-graph $(\mathcal{F}_{sym}, \mathcal{P}, U)$ such that $\mathcal{F}_{sym}[U] \cong K_5^3$ and $(1/5 - \beta)n \leq |A| \leq (1/5 + \beta)n$ for every part $A \in \mathcal{P}$, where $\beta > 0$ is a sufficiently small real number. Note that Theorem 1.2 plays a crucial role in achieving this goal.

The second stage is to show that $\mathcal{F}[V(\mathcal{F}_{sym})]$ is a subgraph of a blow-up of K_5^3. To do so, we will reverse the Merging steps performed during Symmetrization (this process was called Splitting in [3]). Since the first stage ensures that $|V(\mathcal{F})| - |V(\mathcal{F}_{sym})|$ is very small, this will complete the proof of Theorem 3.2.

Using Theorem 3.2 we can now prove Theorem 1.1. Let $\mathcal{F} = (V, E)$ be a maximum size $\mathcal{K}_{3,3}^3$-free 3-graph on n vertices, where n is sufficiently large. Clearly $|E| \geq t_5^3(n)$. The main steps of the proof are as follows:

(1) Show that it suffices to prove Theorem 1.1 under the additional assumption that the minimum degree of $\mathcal{F}$ is at least $\delta_5^3(n)$.

(2) Let $V = A_1 \cup \ldots \cup A_5$ be a partition of the vertex set of $\mathcal{F}$ which minimizes

$$\Sigma := \sum_{1 \leq i < j \leq 5} e(A_i \cup A_j) - 2\sum_{i=1}^{5} e(A_i).$$

Using Theorem 3.2 we can assume that $\Sigma < \varepsilon n^3$.

(3) Observe that $||A_i| - n/5| \leq \delta n$ for every $1 \leq i \leq 5$ and some small real number $\delta > 0$ as otherwise we would have the contradiction $|E| \leq 2n^3/25 - \delta^2 n^3/40 + \varepsilon n^3 < t_5^3(n)$.

(4) Using (1), the fact that $\mathcal{F}$ is $\mathcal{K}_{3,3}^3$-free and the minimality of the partition in (2), prove that for every $u \in V$ there are at most αn^2 edges $e \in E$ such that $u \in e$ and $|e \cap A_i| > 1$ holds for some $1 \leq i \leq 5$, where $\alpha > 0$ is a small real number.

(5) Using (1), the fact that $\mathcal{F}$ is $\mathcal{K}_{3,3}^3$-free and the minimality of the partition in (2), prove that $|e \cap A_i| \leq 1$ holds for every $1 \leq i \leq 5$ and every $e \in E$.

4 Concluding remarks and open problems

The natural open problem is to extend our results from 3-graphs to general r-graphs. We would like to determine the Turán number of the r-graph $\mathcal{K}_{r,r}^r$ with vertex set

$$V(\mathcal{K}_{r,r}^r) = \{x_i, y_i : 1 \leq i \leq r\} \cup \{z_{ijk} : 1 \leq i, j \leq r, 1 \leq k \leq r-2\}$$

and edge set

$$E(\mathcal{K}^r_{r,r}) = \{\{x_1,\ldots,x_r\},\{y_1,\ldots,y_r\}\} \cup \{\{x_i,y_j,z_{ij1},\ldots,z_{ij(r-2)}\} : 1 \le i,j \le r\}.$$

The main difficulty seems to be in obtaining the analogue of Theorem 1.2, i.e. determining the maximum Lagrangian of an intersecting r-graph. At first, one might think that K^r_{2r-1} should be optimal, since this is the case when $r = 3$. However, this has Lagrangian $\binom{2r-1}{r}\left(\frac{1}{2r-1}\right)^r$, whereas stars (in which edges consist of all r-tuples containing some fixed vertex) give Lagrangians that approach $\frac{1}{r!}\left(1-\frac{1}{r}\right)^{r-1}$, which is better for $r \ge 4$. We conjecture that stars are optimal for $r \ge 4$ and that their blow-ups are extremal. Namely, we conjecture that the following hypergraph Turán result holds. Let $S^r(n)$ be the r-graph on n vertices with parts A and B, where the edges consist of all r-tuples with 1 vertex in A and $r-1$ vertices in B, and the sizes of A and B are chosen to maximise the number of edges (so $|A| \sim n/r$). Write $s^r(n) = e(S^r(n))$.

Conjecture 4.1. $\mathrm{ex}(n, \mathcal{K}^r_{r,r}) = s^r(n)$ for $r \ge 4$ and sufficiently large $n > n_0(r)$. Moreover, if n is sufficiently large and G is a $\mathcal{K}^r_{r,r}$-free r-graph with n vertices and $s^r(n)$ edges, then $G \cong S^r(n)$.

More generally, our work suggests a direction of investigation in Extremal Combinatorics, namely to determine the maximum Lagrangian for any specified property of r-graphs. For this paper, the property was that of being intersecting. This direction was already started by Frankl and Füredi [1], who considered the question of maximising the Lagrangian of an r-graph with a specified number of edges. They conjectured that initial segments of the colexicographic order are extremal. Many cases of this have been proved by Talbot [4], but the full conjecture remains open.

References

[1] P. FRANKL and Z. FÜREDI, *Extremal problems whose solutions are the blow-ups of the small Witt-designs*, J. Combin. Theory Ser. A **52** (1989), 129–147.

[2] P. KEEVASH, "Hypergraph Turán Problems", Surveys in Combinatorics, 2011.

[3] O. PIKHURKO, *An Exact Turán Result for the Generalized Triangle*, Combinatorica **28** (2008), 187–208.

[4] J. TALBOT, *Lagrangians of hypergraphs*, Combin. Probab. Comput. **11** (2002), 199–216.

[5] P. TURÁN, *Research problem*, Közl MTA Mat. Kutató Int. **6** (1961), 417–423.

Tight minimum degree conditions forcing perfect matchings in uniform hypergraphs

Andrew Treglown[1] and Yi Zhao[2]

Abstract. Given positive integers k and ℓ where $k/2 \leq \ell \leq k-1$, we give a minimum ℓ-degree condition that ensures a perfect matching in a k-uniform hypergraph. This condition is best possible and improves on work of Pikhurko [12] who gave an asymptotically exact result, and extends work of Rödl, Ruciński and Szemerédi [15] who determined the threshold for $\ell = k-1$. Our approach makes use of the absorbing method.

1 Introduction

A central question in graph theory is to establish conditions that ensure a (hyper)graph H contains some spanning (hyper)graph F. Of course, it is desirable to fully characterize those (hyper)graphs H that contain a spanning copy of a given (hyper)graph F. Tutte's theorem [18] characterizes those graphs with a perfect matching. (A *perfect matching* in a (hyper)graph H is a collection of vertex-disjoint edges of H which cover the vertex set $V(H)$ of H.) However, for some (hyper)graphs F it is unlikely that such a characterization exists. Indeed, for many (hyper)graphs F the decision problem of whether a (hyper)graph H contains F is NP-complete. For example, in contrast to the graph case, the decision problem whether a k-uniform hypergraph contains a perfect matching is NP-complete for $k \geq 3$ (see [4, 6]). Thus, it is desirable to find sufficient conditions that ensure a perfect matching in a k-uniform hypergraph.

Given a k-uniform hypergraph H with an ℓ-element vertex set S (where $0 \leq \ell \leq k-1$) we define $d_H(S)$ to be the number of edges containing S. The *minimum ℓ-degree* $\delta_\ell(H)$ of H is the minimum of $d_H(S)$ over all

[1] School of Mathematical Sciences, Queen Mary, University of London, Mile End Road, London, E1 4NS, UK. Email: treglown@maths.qmul.ac.uk

[2] Department of Mathematics and Statistics, Georgia State University, Atlanta, Georgia 30303, USA. Email: yzhao6@gsu.edu

This research was undertaken whilst the first author was a researcher at Charles University, Prague. The work leading to this invention has received funding from the European Research Council under the European Union's Seventh Framework Programme (FP7/2007-2013)/ERC grant agreement no. 259385.
The second author was partially supported by NSA grants H98230-10-1-0165 and H98230-12-1-0283.

ℓ-element sets S of vertices in H. Clearly $\delta_0(H)$ is the number of edges in H. We also refer to $\delta_1(H)$ as the *minimum vertex degree* of H and $\delta_{k-1}(H)$ the *minimum codegree* of H.

Over the last few years there has been a strong focus in establishing minimum ℓ-degree thresholds that force a perfect matching in a k-uniform hypergraph. See [13] for a survey on matchings (and Hamilton cycles) in hypergraphs. In particular, Rödl, Ruciński and Szemerédi [15] determined the minimum codegree threshold that ensures a perfect matching in a k-uniform hypergraph on n vertices for all $k \geq 3$. The threshold is $n/2 - k + C$, where $C \in \{3/2, 2, 5/2, 3\}$ depends on the values of n and k. This improved bounds given in [9, 14].

Less is known about minimum vertex degree thresholds that force a perfect matching. One of the earliest results on perfect matchings was given by Daykin and Häggkvist [3], who showed that a k-uniform hypergraph H on n vertices contains a perfect matching provided that $\delta_1(H) \geq (1 - 1/k)\binom{n-1}{k-1}$. Hàn, Person and Schacht [5] determined, asymptotically, the minimum vertex degree that forces a perfect matching in a 3-uniform hypergraph. Kühn, Osthus and Treglown [10] and independently Khan [7] made this result exact. Khan [8] has also determined the exact minimum vertex degree threshold for 4-uniform hypergraphs. For $k \geq 5$, the precise minimum vertex degree threshold that ensures a perfect matching in a k-uniform hypergraph is not known.

The situation for ℓ-degrees where $1 < \ell < k - 1$ is also still open. Hàn, Person and Schacht [5] provided conditions on $\delta_\ell(H)$ that ensure a perfect matching in the case when $1 \leq \ell < k/2$. These bounds were subsequently lowered by Markström and Ruciński [11]. Alon et al. [1] gave a connection between the minimum ℓ-degree that forces a perfect matching in a k-uniform hypergraph and the minimum ℓ-degree that forces a *perfect fractional matching*. As a consequence of this result they determined, asymptotically, the minimum ℓ-degree which forces a perfect matching in a k-uniform hypergraph for the following values of (k, ℓ): $(4, 1), (5, 1), (5, 2), (6, 2)$, and $(7, 3)$.

Pikhurko [12] showed that if $\ell \geq k/2$ and H is a k-uniform hypergraph whose order n is divisible by k then H has a perfect matching provided that $\delta_\ell(H) \geq (1/2 + o(1))\binom{n}{k-\ell}$. This result is best possible up to the $o(1)$-term (see the constructions in $\mathcal{H}_{\text{ext}}(n, k)$ below).

In [16, 17] we make Pikhurko's result exact. In order to state this result, we need some more definitions. Fix a set V of n vertices. Given a partition of V into non-empty sets A, B, let $E^k_{\text{odd}}(A, B)$ ($E^k_{\text{even}}(A, B)$) denote the family of all k-element subsets of V that intersect A in an odd (even) number of vertices. (Notice that the ordering of the vertex classes A, B is important.) Define $\mathcal{B}_{n,k}(A, B)$ to be the k-uniform hypergraph with vertex set $V = A \cup B$ and edge set $E^k_{\text{odd}}(A, B)$. Note that the complement $\overline{\mathcal{B}}_{n,k}(A, B)$ of $\mathcal{B}_{n,k}(A, B)$ has edge set $E^k_{\text{even}}(A, B)$.

Suppose $n, k \in \mathbb{N}$ such that k divides n. Define $\mathcal{H}_{\text{ext}}(n, k)$ to be the collection of the following hypergraphs: $\mathcal{H}_{\text{ext}}(n, k)$ contains all hypergraphs $\overline{\mathcal{B}}_{n,k}(A, B)$ where $|A|$ is odd. Further, if n/k is odd then $\mathcal{H}_{\text{ext}}(n, k)$ also contains all hypergraphs $\mathcal{B}_{n,k}(A, B)$ where $|A|$ is even; if n/k is even then $\mathcal{H}_{\text{ext}}(n, k)$ also contains all hypergraphs $\mathcal{B}_{n,k}(A, B)$ where $|A|$ is odd.

It is easy to see that no hypergraph in $\mathcal{H}_{\text{ext}}(n, k)$ contains a perfect matching. Indeed, first assume that $|A|$ is even and n/k is odd. Since every edge of $\mathcal{B}_{n,k}(A, B)$ intersects A in an odd number of vertices, one cannot cover A with an odd number of disjoint odd sets. Similarly $\mathcal{B}_{n,k}(A, B)$ does not contain a perfect matching if $|A|$ is odd and n/k is even. Finally, if $|A|$ is odd then since every edge of $\overline{\mathcal{B}}_{n,k}(A, B)$ intersects A in an even number of vertices, $\overline{\mathcal{B}}_{n,k}(A, B)$ does not contain a perfect matching.

Given $\ell \in \mathbb{N}$ such that $k/2 \le \ell \le k-1$ define $\delta(n, k, \ell)$ to be the maximum of the minimum ℓ-degrees among all the hypergraphs in $\mathcal{H}_{\text{ext}}(n, k)$. For example, it is not hard to see that

$$\delta(n,k,k-1) = \begin{cases} n/2 - k + 2 & \text{if } k/2 \text{ is even and } n/k \text{ is odd} \\ n/2 - k + 3/2 & \text{if } k \text{ is odd and } (n-1)/2 \text{ is odd} \\ n/2 - k + 1/2 & \text{if } k \text{ is odd and } (n-1)/2 \text{ is even} \\ n/2 - k + 1 & \text{otherwise.} \end{cases} \tag{1.1}$$

In [16, 17] we prove the following exact version of Pikhurko's result.

Theorem 1.1. *Let $k, \ell \in \mathbb{N}$ such that $k \ge 3$ and $k/2 \le \ell \le k-1$. Then there exists an $n_0 \in \mathbb{N}$ such that the following holds. Suppose H is a k-uniform hypergraph on $n \ge n_0$ vertices where k divides n. If*

$$\delta_\ell(H) > \delta(n, k, \ell)$$

then H contains a perfect matching.

In [16] we prove Theorem 1.1 for k divisible by 4 and then in [17] we extend this result to all values of k. Independent to our work, Czygrinow and Kamat [2] have proven Theorem 1.1 in the case when $k = 4$ and $\ell = 2$.

As explained before, the minimum ℓ-degree condition in Theorem 1.1 is best possible. Theorem 1.1 and (1.1) together give the aforementioned result of Rödl, Ruciński and Szemerédi [15].

In general, the precise value of $\delta(n, k, \ell)$ is unknown because it is not known what value of $|A|$ maximizes the minimum ℓ-degree of $\mathcal{B}_{n,k}(A, B)$ (or $\overline{\mathcal{B}}_{n,k}(A, B)$). (See [16] for a discussion on this.) However, in [16] we gave a tight upper bound on $\delta(n, 4, 2)$.

2 Overview of the proof of Theorem 1.1

The proof of Theorem 1.1 follows the so-called *stability approach.* We first prove that

(i) H has a perfect matching or;
(ii) H is 'close' to one of the hypergraphs $\mathcal{B}_{n,k}(A, B)$ or $\overline{\mathcal{B}}_{n,k}(A, B)$ in $\mathcal{H}_{\text{ext}}(n, k)$.

The extremal situation (ii) is then dealt with separately. For example, suppose H is 'close' to an element $\mathcal{B}_{n,k}(A, B)$ from $\mathcal{H}_{\text{ext}}(n, k)$. (So we can view A, B as a partition of $V(H)$.) The minimum ℓ-degree condition on H ensures that H contains an edge e that intersects A in an even number of vertices. Recall that no such edge exists in $\mathcal{B}_{n,k}(A, B)$; this is the 'reason' why $\mathcal{B}_{n,k}(A, B)$ does not have a perfect matching. Thus, e acts as a 'parity breaking' edge and can be used to form part of a perfect matching in H.

Almost perfect matchings

To show that (i) or (ii) holds, we apply the following result of Markström and Ruciński [11] to ensure an 'almost' perfect matching in H.

Theorem 2.1 (Lemma 2 in [11]). *For each integer $k \geq 3$, every $1 \leq \ell \leq k-2$ and every $\gamma > 0$ there exists an $n_0 \in \mathbb{N}$ such that the following holds. Suppose that H is a k-uniform hypergraph on $n \geq n_0$ vertices such that*

$$\delta_\ell(H) \geq \left(\frac{k-\ell}{k} - \frac{1}{k^{(k-\ell)}} + \gamma\right)\binom{n-\ell}{k-\ell}.$$

Then H contains a matching covering all but at most $\sqrt{n}$ vertices.

(In [11], Markström and Ruciński only stated Theorem 2.1 for $1 \leq \ell < k/2$. In fact, their proof works for all values of ℓ such that $1 \leq \ell \leq k-2$.) In the case when $\ell = k - 1$, we need a result of Rödl, Ruciński and Szemerédi [15, Fact 2.1]: Suppose H is a k-uniform hypergraph on n vertices. If $\delta_{k-1}(H) \geq n/k$, then H contains a matching covering all but at most k^2 vertices in H. Note that this minimum codegree condition is substantially smaller than the corresponding condition in Theorem 1.1. Further, if $k/2 \leq \ell < k - 1$ then the minimum ℓ-degree condition in Theorem 2.1 is also substantially smaller than the minimum ℓ-degree in Theorem 1.1.

Absorbing sets

Given a k-uniform hypergraph H, a set $S \subseteq V(H)$ is called an *absorbing set for $Q \subseteq V(H)$*, if both $H[S]$ and $H[S \cup Q]$ contain perfect matchings.

If the hypergraph H in Theorem 1.1 contains a 'small' set S which is an absorbing set for *any* set $Q \subseteq V(H)$ where $|Q| \leq \sqrt{n}$ is divisible by k, then it is easy to find a perfect matching in H. Indeed, in this case the minimum ℓ-degree of $H - S$ satisfies the hypothesis of Theorem 2.1 (or the hypothesis of Fact 2.1 in [15] if $\ell = k - 1$). Thus, $H - S$ contains a matching M covering all but a set Q of at most $\sqrt{n}$ vertices. Then since $H[S \cup Q]$ contains a perfect matching M', $M \cup M'$ is a perfect matching in H.

We give two conditions that ensure such an absorbing set S exists in H (and thus guarantee a perfect matching in H). Roughly speaking, the first condition asserts that $V(H)$ contains 'many' ℓ-tuples whose ℓ-degree is 'significantly' larger than $\delta(n, k, \ell)$. The second condition concerns a certain 'common neighbourhood' property. (Fixing $r := \lceil k/2 \rceil$, this condition roughly asserts that for any r-tuple $P \in \binom{V(H)}{r}$, more than half of the r-tuples P' in $\binom{V(H)}{r}$ are such that P and P' have a common neighbourhood which is not 'too small'.) We will refer to these properties as (α) and (β) respectively.

The auxiliary graph G

We then show that if H does not satisfy (α) and (β), then (ii) must be satisfied. That is, H is 'close' to one of the hypergraphs $\mathcal{B}_{n,k}(A, B)$ or $\overline{\mathcal{B}}_{n,k}(A, B)$ in $\mathcal{H}_{\text{ext}}(n, k)$. For this, we consider an auxiliary bipartite graph G defined as follows: Set $r := \lceil k/2 \rceil$, $r' := \lfloor k/2 \rfloor$, $X^r := \binom{V(H)}{r}$ and $Y^{r'} := \binom{V(H)}{r'}$. Further, let $N := \binom{n}{r}$ and $N' := \binom{n}{r'}$. G has vertex classes X^r and $Y^{r'}$. Two vertices $x_1 \ldots x_r \in X^r$ and $y_1 \ldots y_{r'} \in Y^{r'}$ are adjacent in G if and only if $x_1 \ldots x_r y_1 \ldots y_{r'} \in E(H)$.

We show that, since H fails to satisfy (α) and (β), G is 'close' to the disjoint union of two copies of $K_{N/2,N'/2}$. Once we have this information, we give direct arguments on G to show that this implies that (ii) is indeed satisfied.

References

[1] N. ALON, P. FRANKL, H. HUANG, V. RÖDL, A. RUCIŃSKI and B. SUDAKOV, *Large matchings in uniform hypergraphs and the conjectures of Erdős and Samuels*, J. Combin. Theory Ser. A **119** (2012), 1200–1215.

[2] A. CZYGRINOW and V. KAMAT, *Tight co-degree condition for perfect matchings in 4-graphs*, Electron. J. Combin. **19** (2012), P20.

[3] D. E. DAYKIN and R. HÄGGKVIST, *Degrees giving independent edges in a hypergraph*, Bull. Austral. Math. Soc. **23** (1981), 103–109.

[4] M. R. GAREY and D. S. JOHNSON, "Computers and Intractability", Freeman, 1979.

[5] H. HÀN, Y. PERSON and M. SCHACHT, *On perfect matchings in uniform hypergraphs with large minimum vertex degree*, SIAM J. Discrete Math. **23** (2009), 732–748.

[6] R. M. KARP, *Reducibility among combinatorial problems*, In: "Complexity of Computer Computations" (Proc. Sympos., IBM Thomas J. Watson Res. Center, Yorktown Heights, N.Y., 1972), Plenum, New York, 1972, 85–103.

[7] I. KHAN, *Perfect Matching in* 3 *uniform hypergraphs with large vertex degree*, arXiv:1101.5830.

[8] I. KHAN, *Perfect Matchings in 4-uniform hypergraphs*, arXiv:1101.5675.

[9] D. KÜHN and D. OSTHUS, *Matchings in hypergraphs of large minimum degree*, J. Graph Theory **51** (2006), 269–280.

[10] D. KÜHN, D. OSTHUS and A. TREGLOWN, *Matchings in 3-uniform hypergraphs*, J. Combin. Theory Ser. B **103** (2013), 291–305.

[11] K. MARKSTRÖM and A. RUCIŃSKI, *Perfect matchings (and Hamilton cycles) in hypergraphs with large degrees*, European J. Combin. **32** (2011), 677–687.

[12] O. PIKHURKO, *Perfect matchings and K_4^3-tilings in hypergraphs of large codegree*, Graphs Combin. **24** (2008), 391–404.

[13] V. RÖDL and A. RUCIŃSKI, *Dirac-type questions for hypergraphs – a survey (or more problems for Endre to solve)*, An Irregular Mind (Szemerédi is 70), Bolyai Soc. Math. Studies **21** (2010), 1–30.

[14] V. RÖDL, A. RUCIŃSKI and E. SZEMERÉDI, *Perfect matchings in uniform hypergraphs with large minimum degree*, European J. Combin. **27** (2006), 1333–1349.

[15] V. RÖDL, A. RUCIŃSKI and E. SZEMERÉDI, *Perfect matchings in large uniform hypergraphs with large minimum collective degree*, J. Combin. Theory Ser. A **116** (2009), 613–636.

[16] A. TREGLOWN and Y. ZHAO, *Exact minimum degree thresholds for perfect matchings in uniform hypergraphs*, J. Combin. Theory Ser. A **119** (2012), 1500–1522.

[17] A. TREGLOWN and Y. ZHAO, *Exact minimum degree thresholds for perfect matchings in uniform hypergraphs II*, J. Combin. Theory Ser. A **120** (2013), 1463–1482.

[18] W. T. TUTTE, *The factorisation of linear graphs*, J. London Math. Soc. **22** (1947), 107–111.

Fractional and integer matchings in uniform hypergraphs

Daniela Kühn[1], Deryk Osthus[2] and Timothy Townsend[3]

Abstract. A conjecture of Erdős from 1965 suggests the minimum number of edges in a k-uniform hypergraph on n vertices which forces a matching of size t, where $t \leq n/k$. Our main result verifies this conjecture asymptotically, for all $t < 0.48n/k$. This gives an approximate answer to a question of Huang, Loh and Sudakov, who proved the conjecture for $t \leq n/3k^2$. As a consequence of our result, we extend bounds of Bollobás, Daykin and Erdős by asymptotically determining the minimum vertex degree which forces a matching of size $t < 0.48n/(k-1)$ in a k-uniform hypergraph on n vertices. We also obtain further results on d-degrees which force large matchings. In addition we improve bounds of Markström and Ruciński on the minimum d-degree which forces a perfect matching in a k-uniform hypergraph on n vertices. Our approach is to inductively prove fractional versions of the above results and then translate these into integer versions.

Large matchings in hypergraphs with many edges

A *k-uniform hypergraph* is a pair $G = (V, E)$ where V is a finite set of vertices and the edge set E consists of unordered k-tuples of elements of V. A *matching* (or *integer matching*) M in G is a set of disjoint edges of G. The *size* of M is the number of edges in M. M is *perfect* if it has size $|V|/k$.

A classical theorem of Erdős and Gallai [6] determines the number of edges in a graph which forces a matching of a given size. In 1965, Erdős [5] made a conjecture which would generalize this to k-uniform hypergraphs.

Conjecture 1. Let $n, k \geq 2$ and $1 \leq s \leq n/k$ be integers. The minimum number of edges in a k-uniform hypergraph on n vertices which forces a matching of size s is

$$\max\left\{\binom{ks-1}{k}, \binom{n}{k} - \binom{n-s+1}{k}\right\} + 1.$$

It is easy to see that the conjecture would be best possible: the first expression in the lower bound is obtained by considering the k-uniform

[1] School of Mathematics, University of Birmingham, Birmingham B15 2TT, United Kingdom. Email: d.kuhn@bham.ac.uk

[2] School of Mathematics, University of Birmingham, Birmingham B15 2TT, United Kingdom. Email: d.osthus@bham.ac.uk

[3] School of Mathematics, University of Birmingham, Birmingham B15 2TT, United Kingdom. Email: txt238@bham.ac.uk

clique $K_{ks-1}^{(k)}$ (complemented by $n - ks + 1$ isolated vertices); the second expression in the lower bound is obtained as follows. Let $H(s)$ be a k-uniform hypergraph on n vertices with edge set consisting of all k-element subsets of $V(H(s))$ intersecting a given subset of $V(H(s))$ of size $s-1$, that is $H(s) = K_n^{(k)} - K_{n-s+1}^{(k)}$.

The case $s = 2$ of Conjecture 1 corresponds to the Erdős-Ko-Rado Theorem on intersecting families [7]. The conjecture also has applications to the Manickam-Mikós-Singhi conjecture in number theory (for details see *e.g.* [2]). Despite its seeming simplicity Conjecture 1 is still wide open in general. For the cases $k \leq 4$, it was verified asymptotically by Alon, Frankl, Huang, Rödl, Ruciński and Sudakov [1]. For $k = 3$, it was recently proved by Frankl [8]. Bollobás, Daykin and Erdős [3] proved Conjecture 1 for general k whenever $s < n/(2k^3)$, which extended earlier results of Erdős [5]. Huang, Loh and Sudakov [10] proved it for $s < n/(3k^2)$. The main result in this paper verifies Conjecture 1 asymptotically for matchings of any size up to almost half the size of a perfect matching. This gives an asymptotic answer to a question in [10].

Theorem 2. [15] *Let n, $k \geq 2$ and $0 \leq a < 0.48/k$ be such that n, k, $an \in \mathbb{N}$. The minimum number of edges in a k-uniform hypergraph on n vertices which forces a matching of size an is*

$$\left(1 - (1-a)^k + o(1)\right)\binom{n}{k}.$$

Large matchings in hypergraphs with large degrees

It is also natural to consider degree conditions that force matchings in uniform hypergraphs. Given a k-uniform hypergraph $G = (V, E)$ and $S \in \binom{V}{d}$, where $0 \leq d \leq k-1$, let $deg_G(S) = |\{e \in E : S \subseteq e\}|$ be the *degree* of S in G. Let $\delta_d(G) = \min_{S \in \binom{V}{d}}\{deg_G(S)\}$ be the *minimum d-degree* of G. When $d = 1$, we refer to $\delta_1(G)$ as the *minimum vertex degree* of G. Note that $\delta_0(G) = |E|$.

For integers n, k, d, s satisfying $0 \leq d \leq k-1$ and $0 \leq s \leq n/k$, we let $m_d^s(k, n)$ denote the minimum integer m such that every k-uniform hypergraph G on n vertices with $\delta_d(G) \geq m$ has a matching of size s. So the results discussed in the previous section correspond to the case $d = 0$. The following degree condition for forcing perfect matchings has been conjectured in [9, 14] and also received much attention recently.

Conjecture 3. Let n and $1 \leq d \leq k-1$ be such that $n, d, k, n/k \in \mathbb{N}$. Then

$$m_d^{n/k}(k, n) = \left(\max\left\{\frac{1}{2}, 1 - \left(\frac{k-1}{k}\right)^{k-d}\right\} + o(1)\right)\binom{n-d}{k-d}.$$

The lower bound here is given by the hypergraph $H(n/k)$ defined after Conjecture 1 and the following parity-based construction from [13]. For any integers n, k, let H' be a k-uniform hypergraph on n vertices with vertex partition $A \cup B = V(H')$, such that $||A| - |B|| \leq 2$ and $|A|$ and n/k have different parity. Let H' have edge set consisting of all k-element subsets of $V(H')$ that intersect A in an odd number of vertices. Observe that H' has no perfect matching, and that for every $1 \leq d \leq k-1$ we have that $\delta_d(H') = (1/2 + o(1))\binom{n-d}{k-d}$.

For $d = k-1$, $m_{k-1}^{n/k}(k, n)$ was determined exactly by Rödl, Ruciński and Szemerédi [19]. This was generalized by Treglown and Zhao [20], who determined the extremal families for all $d \geq k/2$. The extremal constructions are similar to the parity based one of H' above. For $d < k/2$ less is known. In [1] Conjecture 3 was proved for $k-4 \leq d \leq k-1$, by reducing it to a probabilistic conjecture of Samuels. In particular, this implies Conjecture 3 for $k \leq 5$. Khan [11], and independently Kühn, Osthus and Treglown [16], determined $m_1^{n/k}(k, n)$ exactly for $k = 3$. Khan [12] also determined $m_1^{n/k}(k, n)$ exactly for $k = 4$. As a consequence of these results, $m_1^s(k, n)$ is determined exactly whenever $s \leq n/k$ and $k \leq 4$ (for details see the concluding remarks in [16]). More generally, we propose the following version of Conjecture 3 for non-perfect matchings.

Conjecture 4. For all $\varepsilon > 0$ and all integers n, d, k, s with $1 \leq d \leq k-1$ and $0 \leq s \leq (1-\varepsilon)n/k$ we have

$$m_d^s(k, n) = \left(1 - \left(1 - \frac{s}{n}\right)^{k-d} + o(1)\right)\binom{n-d}{k-d}.$$

In fact it may be that the bound holds for all $s \leq n - C$, for some C depending only on d and k. The lower bound here is given by $H(s)$. The case $d = k-1$ of Conjecture 4 follows easily from the determination of $m_{k-1}^s(k, n)$ for s close to n/k in [19]. Bollobás, Daykin and Erdős [3] determined $m_1^s(k, n)$ for small s, *i.e.* whenever $s < n/2k^3$. As a consequence of our main result, for $1 \leq d \leq k-2$ we are able to determine $m_d^s(k, n)$ asymptotically for non-perfect matchings of any size less than $0.48n/(k-d)$. Note that this proves Conjecture 4 in the case $0.53k \leq d \leq k-2$, say.

Theorem 5. [15] *Let* $\varepsilon > 0$ *and let* n, k, d *be integers with* $1 \leq d \leq k-2$, *and let* $0 \leq a < \min\{0.48/(k-d), (1-\varepsilon)/k\}$ *be such that* $an \in \mathbb{N}$. *Then*

$$m_d^{an}(k, n) = \left(1 - (1-a)^{k-d} + o(1)\right)\binom{n-d}{k-d}.$$

We now focus again on the case $s = n/k$, *i.e.* perfect matchings. It was shown by Hàn, Person and Schacht [9] that for $k \geq 3$, $1 \leq d < k/2$ we have $m_d^{n/k}(k,n) \leq ((k-d)/k + o(1))\binom{n-d}{k-d}$. (The case $d = 1$ of this is already due to Daykin and Häggkvist [4].) These bounds were slightly improved by Markström and Ruciński [17], using similar techniques, to

$$m_d^{n/k}(k,n) \leq \left(\frac{k-d}{k} - \frac{1}{k^{k-d}} + o(1)\right)\binom{n-d}{k-d}.$$

Using similar methods to those developed to prove Theorem 5, we are also able to slightly improve on this bound.

Theorem 6 ([15]). *Let n and $1 \leq d < k/2$ be such that $n, k, d, n/k \in \mathbb{N}$. Then*

$$m_d^{n/k}(k,n) \leq \left(\frac{k-d}{k} - \frac{k-d-1}{k^{k-d}} + o(1)\right)\binom{n-d}{k-d}.$$

Large fractional matchings

Our approach to proving these results uses the concepts of fractional matchings and fractional vertex covers. A *fractional matching* in a k-uniform hypergraph $G = (V, E)$ is a function $w : E \to [0,1]$ of weights of edges, such that for each $v \in V$ we have $\sum_{e \in E: v \in e} w(e) \leq 1$. The *size* of w is $\sum_{e \in E} w(e)$. w is *perfect* if it has size $|V|/k$. A *fractional vertex cover* in G is a function $w : V \to [0,1]$ of weights of vertices, such that for each $e \in E$ we have $\sum_{v \in e} w(v) \geq 1$. The *size* of w is $\sum_{v \in V} w(v)$.

A key idea (already used *e.g.* in [1, 18]) is that we can switch between considering the largest fractional matching and the smallest fractional vertex cover of a hypergraph. The determination of these quantities are dual linear programming problems, and hence by the Duality Theorem they have the same size.

For $s \in \mathbb{R}$ we let $f_d^s(k,n)$ denote the minimum integer m such that every k-uniform hypergraph G on n vertices with $\delta_d(G) \geq m$ has a fractional matching of size s. It was shown in [18] that $f_{k-1}^{n/k}(k,n) = \lceil n/k \rceil$. Similarly to [1], we now formulate the fractional version of Conjecture 1.

Conjecture 7. For all integers n,k,s with $k \geq 2$ and $1 \leq s \leq n/k$ we have

$$f_0^s(k,n) = \max\left\{\binom{ks-1}{k}, \binom{n}{k} - \binom{n-s+1}{k}\right\} + 1.$$

As discussed in [1], this conjecture has applications to a problem on information storage and retrieval. To prove Theorems 2 and 5, we first prove Conjecture 7 asymptotically for fractional matchings of any size up to $0.48n/k$.

Theorem 8 ([15]). *Let $n, k \geq 2$ be integers and let $0 \leq a \leq 0.48/k$. Then*

$$f_0^{an}(k,n) = \left(1 - (1-a)^k + o(1)\right)\binom{n}{k}.$$

We use Theorem 8, along with methods similar to those developed in [1], to convert our edge-density conditions for the existence of fractional matchings into corresponding minimum degree conditions. For $1 \leq d \leq k-2$ the following theorem asymptotically determines $f_d^s(k,n)$ for fractional matchings of any size up to $0.48n/(k-d)$. Note that this determines $f_d^s(k,n)$ asymptotically for all $s \in (0, n/k)$ whenever $d \geq 0.52k$.

Theorem 9 ([15]). *Let n, $k \geq 3$, and $1 \leq d \leq k-2$ be integers and let $0 \leq a \leq \min\{0.48/(k-d), 1/k\}$. Then*

$$f_d^{an}(k,n) = \left(1 - (1-a)^{k-d} + o(1)\right)\binom{n-d}{k-d}.$$

We then use Theorem 8 and a variant of Theorem 9, along with the Weak Hypergraph Regularity Lemma, to prove Theorems 2 and 5 respectively, by converting our fractional matchings into integer ones. We prove Theorem 6 in a similar fashion, via the following two theorems.

Theorem 10 ([15]). *Let n, $k \geq 2$, $d \geq 1$ be integers. Then*

$$f_0^{n/(k+d)}(k,n) \leq \left(\frac{k}{k+d} - \frac{k-1}{(k+d)^k} + o(1)\right)\binom{n}{k}.$$

Theorem 11 ([15]). *Let n, $k \geq 3$, $1 \leq d \leq k-2$ be integers. Then*

$$f_d^{n/k}(k,n) \leq \left(\frac{k-d}{k} - \frac{k-d-1}{k^{k-d}} + o(1)\right)\binom{n-d}{k-d}.$$

References

[1] N. ALON, P. FRANKL, H. HUANG, V. RÖDL, A. RUCIŃSKI and B. SUDAKOV, *Large matchings in uniform hypergraphs and the conjectures of Erdős and Samuels*, J. Combin. Theory A **119** (2012), 1200–1215.

[2] N. ALON, H. HUANG and B. SUDAKOV, *Nonnegative k-sums, fractional covers, and probability of small deviations*, J. Combin. Theory B **102** (2012), 784–796.

[3] B. BOLLOBÁS, D. E. DAYKIN and P. ERDŐS, *Sets of independent edges of a hypergraph*, Quart. J. Math. Oxford **27** (1976), 25–32.

[4] D.E. DAYKIN and R.HÄGGKVIST, *Degre giving independent edges in a hypergraph*, Bull. Austral. Math. Soc. **23** (1981), 103–109.

[5] P. ERDŐS, *A problem on independent r-tuples*, Ann. Univ. Sci. Budapest. Eőtvős Sect. Math **8** (1965), 93–95.

[6] P. ERDŐS and T. GALLAI, *On maximal paths and circuits of graphs*, Acta Math. Acad. Sci. Hung. **10** (1959), 337–356.

[7] P. ERDŐS, C. KO and R. RADO, *Intersection theorems for systems of finite sets*, Quart. J. Math. Oxford **12** (1961), 313–320.

[8] P. FRANKL, *On the maximum mumber of edges in a hypergraph with given matching number*, arXiv:1205.6847v1 (2012).

[9] H. HÀN, Y. PERSON and M. SCHACHT, *On perfect matchings in uniform hypergraphs with large minimum vertex degree*, SIAM J. Discrete Math. **23** (2009), 732–748.

[10] H. HUANG, P. LOH and B. SUDAKOV, *The size of a hypergraph and its matching number*, Combinatorics, Probability and Computing **21** (2012), 442–450.

[11] I. KHAN, *Perfect matchings in 3-uniform hypergraphs with large vertex degree*, arXiv:1101.5830v3 (2011).

[12] I. KHAN, *Perfect matchings in 4-uniform hypergraphs*, arXiv:1101.5675v2 (2011).

[13] D. KÜHN and D. OSTHUS, *Matchings in hypergraphs of large minimum degree*, J. Graph Theory **51** (2006), 269–280.

[14] D. KÜHN and D. OSTHUS, *Embedding large subgraphs into dense graphs*, In: S. Huczynka, J. Mitchell, C. Roney-Dougal (eds.), "Surveys in Combinatorics", London Math. Soc. Lecture Note Ser. **365**, Cambridge University Press, 2009, 137–167.

[15] D. KÜHN, D. OSTHUS and T. TOWNSEND, *Fractional and integer matchings in uniform hypergraphs*, submitted.

[16] D. KÜHN, D. OSTHUS and A. TREGLOWN, *Matchings in 3-uniform hypergraphs*, J. Combin. Theory B **103** (2013), 291–305.

[17] K. MARKSTRÖM and A. RUCIŃSKI, *Perfect matchings and Hamiltonian cycles in hypergraphs with large degrees*, European J. Combin. **32** (2011), 677–687.

[18] V. RÖDL, A. RUCIŃSKI and E. SZEMERÉDI, *Perfect matchings in uniform hypergraphs with large minimum degree*, European J. Combin. **27** (2006), 1333–1349.

[19] V. RÖDL, A. RUCIŃSKI and E. SZEMERÉDI, *Perfect matchings in large uniform hypergraphs with large minimum collective degree*, J. Combin. Theory A **116** (2009), 616–636.

[20] A. TREGLOWN and Y. ZHAO, *Exact minimum degree thresholds for perfect matchings in uniform hypergraphs II*, J. Combin. Theory A **120** (2013), 1463–1482

Cubic graphs

On cubic bridgeless graphs whose edge-set cannot be covered by four perfect matchings

Louis Esperet[1] and Giuseppe Mazzuoccolo[2]

Abstract. The problem of establishing the number of perfect matchings necessary to cover the edge-set of a cubic bridgeless graph is strongly related to a famous conjecture of Berge and Fulkerson. In this paper we show that deciding whether this number is at most 4 for a given cubic bridgeless graph is NP-complete. Our proof makes heavy use of small cuts, so an interesting problem is to construct large *snarks* (cyclically 4-edge-connected cubic graphs of girth at least five and chromatic index four) whose edge-set cannot be covered by 4 perfect matchings. A well-known example is the Petersen graph and the unique other known examples were recently found by Hägglund using a computer program. In this paper we construct an infinite family $\mathcal{F}$ of snarks whose edge-set cannot be covered by 4 perfect matchings. It turns out that the family $\mathcal{F}$ also has interesting properties with respect to the shortest cycle cover problem. The Petersen graph and one of the graphs constructed by Hägglund are the only known snarks with m edges and no cycle cover of length $\frac{4}{3}m$ (indeed their shortest cycle covers have length $\frac{4}{3}m + 1$). We show that all the members of $\mathcal{F}$ satisfy the former property, and we construct a snark with no cycle cover of length less than $\frac{4}{3}m + 2$.

1 Introduction

Throughout this paper, a graph G always means a simple connected finite graph (without loops and parallel edges). A *perfect matching* of G is a 1-regular spanning subgraph of G. In this context, a *cover*, or a k-*cover* of G is a set of k perfect matchings of G such that each edge of G belongs to at least one of the perfect matchings. Following the terminology introduced in [2], the *excessive index* of G, denoted by $\chi'_e(G)$, is the least integer k such that G has a k-cover.

A famous conjecture of Berge and Fulkerson [6] states that the edge-set of every cubic bridgeless graph can be covered by 6 perfect matchings,

[1] Laboratoire G-SCOP (Grenoble-INP, CNRS), Grenoble, France.
Email: louis.esperet@g-scop.inpg.fr. Partially supported by the French *Agence Nationale de la Recherche* under reference ANR-10-JCJC-0204-01.

[2] Laboratoire G-SCOP (Grenoble-INP, CNRS), Grenoble, France.
Email: mazzuoccolo@unimore.it. Research supported by a fellowship from the European Project "INdAM fellowships in mathematics and/or applications for experienced researchers cofunded by Marie Curie actions".

such that each edge is covered precisely twice. The second author recently proved that this conjecture is equivalent to another conjecture of Berge stating that every cubic bridgeless graph has excessive index at most five [10].

Note that a cubic bridgeless graph has excessive index 3 if and only if it is 3-edge-colorable, and deciding the latter is NP-complete. Hägglund [7, Problem 3] asked if it is possible to give a characterization of all cubic graphs with excessive index 5. We show that a characterization of all cubic graphs with excessive index at least 5 is unlikely unless P=NP.

Theorem 1.1. *Determining whether a cubic bridgeless graph G satisfies* $\chi'_e(G) \leq 4$ *(resp.* $\chi'_e(G) = 4$*) is an NP-complete problem.*

The gadgets used in the proof of NP-completeness have many 2-edge-cuts, so our first result does not say much about 3-edge-connected cubic graphs. A *snark* is a non 3-edge-colorable cubic graph with girth (length of a shortest cycle) at least five that is cyclically 4-edge-connected. A question raised by Fouquet and Vanherpe [5] is whether the Petersen graph is the unique snark with excessive index at least five. This question was answered by the negative by Hägglund using a computer program [7]. He proved that the smallest snark distinct from the Petersen graph having excessive index at least five is a graph $\mathring{H}$ on 34 vertices. We show that the graph $\mathring{H}$ found by Hägglund is a special member of an infinite family $\mathcal{F}$ of snarks with excessive index precisely five.

We also show that our family $\mathcal{F}$ has interesting properties with respect to shortest cycle covers. A *cycle cover* of a graph G is a covering of the edge-set of G by cycles (connected subgraphs with all degrees even), such that each edge is in at most one cycle. The *length* of a cycle cover is the sum of the number of edges in all cycles of the cover. The *Shortest Cycle Cover Conjecture* of Alon and Tarsi [1] states that every bridgeless graph G has a cycle cover of length at most $\frac{7}{5}|E(G)|$. Note that it was recently proved by Steffen [11] and Hou, Lai, and Zhang [8] that it is enough to prove the Cycle Double Cover conjecture (a famous conjecture due to Seymour and Szekeres, and implied by the Shortest Cycle Cover Conjecture) for snarks with excessive index at least five. For a cubic bridgeless graph G there is a trivial lower bound of $\frac{4}{3}|E(G)|$ on the length of a cycle cover. On the other hand, the best known upper bound, $\frac{34}{21}|E(G)|$, was obtained by Kaiser, Král', Lidický, Nejedlý, and Šámal [9] in 2010.

We show that no graph G in our infinite family $\mathcal{F}$ has a cycle cover of length $\frac{4}{3}|E(G)|$. We also find the first known snark with no cycle cover of length less than $\frac{4}{3}|E(G)| + 2$ (it has 106 vertices).

The reader is referred to the full version of the paper [4] for more results and proofs.

2 An infinite family of snarks with excessive index 5

In this section we show how to construct a snark G with excessive index at least 5 from three snarks G_0, G_1, G_2 each having excessive index at least 5. Taking $G_0 = G_1 = G_2$ to be the Petersen graph, we obtain the graph $\mathring{H}$ found by Hägglund using a computer program [7], and for which no combinatorial proof showing that its excessive index is 5 was known. Our proof holds for any graph obtained using this construction, thus we exhibit an infinite family $\mathcal{F}$ of snarks with excessive index 5. This answers a question of Fouquet and Vanherpe [5] in a very strong sense.

The construction For $i = 0, 1, 2$, consider a snark G_i with an edge $x_i y_i$. Let x_i^0 and x_i^1 (resp. y_i^0 and y_i^1) be the neighbors of x_i (resp. y_i) in G_i different from y_i (resp. x_i). For $i = 0, 1, 2$, let H_i be the graph obtained from G_i by removing vertices x_i and y_i. We construct a new graph G from the disjoint union of H_0, H_1, H_2 and a new vertex u as follows. For $i = 0, 1, 2$, we introduce a set $A_i = \{a_i, b_i, c_i\}$ of vertices such that a_i is adjacent to x_{i+1}^0 and y_{i-1}^0, b_i is adjacent to x_{i+1}^1 and y_{i-1}^1, and c_i is adjacent to a_i, b_i and u (here and in the following all indices i are taken modulo 3). This construction is depicted in Figure 2.1.

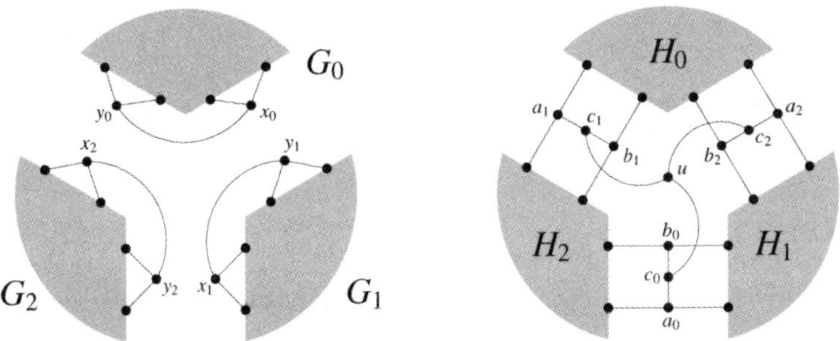

Figure 2.1. The construction of G (right) from G_0, G_1, and G_2.

Theorem 2.1. *Let G_0, G_1, G_2 be snarks such that $\chi'_e(G_i) \geq 5$ for $i = 0, 1, 2$. Then the graph G obtained from G_0, G_1, G_2 by the construction above is a snark with $\chi'_e(G) \geq 5$.*

3 Shortest cycle cover

The length of a shortest cycle cover of a bridgeless graph G is denoted by $\mathrm{scc}(G)$. Note that for any cubic graph G, $\mathrm{scc}(G) \geq \frac{4}{3}|E(G)|$. We show that for any graph G in the family $\mathcal{F}$, $\mathrm{scc}(G) > \frac{4}{3}|E(G)|$.

Theorem 3.1. *Let G_0, G_1, G_2 be snarks such that $\mathrm{scc}(G_i) > \frac{4}{3}|E(G_i)|$ for $i = 0, 1, 2$. Then the graph G obtained from G_0, G_1, G_2 by the construction of Section 2 is a snark with $\mathrm{scc}(G) > \frac{4}{3}|E(G)|$.*

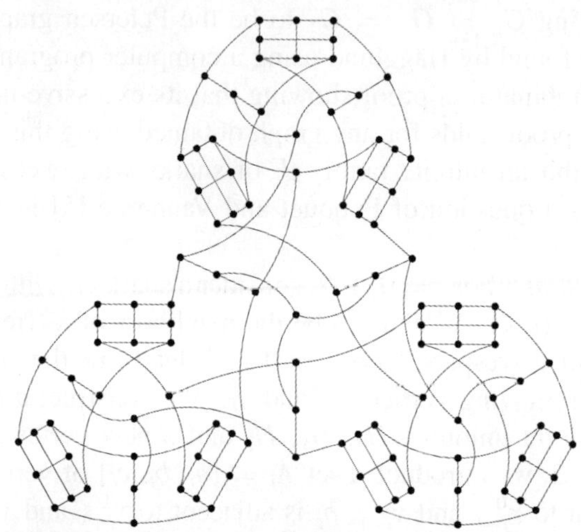

Figure 3.1. A snark G with $\mathrm{scc}(G) \geq \frac{4}{3}|E(G)| + 2$.

Furthermore, we exhibit the first known snark G satisfying $\mathrm{scc}(G) > \frac{4}{3}|E(G)| + 1$.

Theorem 3.2. *Any cycle cover of the graph G depicted in Figure* 3.1 *has length at least $\frac{4}{3}|E(G)| + 2$.*

4 Open problems

Hägglund proposed the following two problems (Problems 3 and 4 in [7]):

- Is it possible to give a simple characterization of cubic graphs G with $\chi'_e(G) = 5$?
- Are there any cyclically 5-edge-connected snarks G with excessive index at least five distinct from the Petersen graph?

If Berge-Fulkerson conjecture is true, then the first problem has a negative answer by Theorem 1.1 (unless P=NP). However, even assuming the correctness of Berge-Fulkerson conjecture, the second problem is still open, since each element of the infinite family $\mathcal{F}$ contains cyclic 4-edge-cuts. Furthermore, the gadgets we use in the proof of Theorem 1.1

have many 2-edge-cuts. Hence we leave open the problem of establishing whether it is possible to give a simple characterization of 3-edge-connected or cyclically 4-edge-connected cubic graphs with excessive index (at least) 5.

Theorem 3.2 proves the existence of a snark $G \in \mathcal{F}$ with no cycle cover of length less than $\frac{4}{3}|E(G)| + 2$. We believe that there exist snarks in $\mathcal{F}$ for which the constant 2 can be replaced by an arbitrarily large number. On the other hand, Brinkmann, Goedgebeur, Hägglund, and Markström [3] conjectured that every snark G has a cycle cover of size at most $(\frac{4}{3} + o(1))|E(G)|$.

References

[1] N. ALON and M. TARSI, *Covering multigraphs by simple circuits*, SIAM J. Algebraic Discrete Methods **6** (1985), 345–350.

[2] A. BONISOLI and D. CARIOLARO, *Excessive Factorizations of Regular Graphs*, In: Graph theory in Paris, A. Bondy *et al.* (eds.), Birkhäuser, Basel (2007), 73–84.

[3] G. BRINKMANN, J. GOEDGEBEUR, J. HÄGGLUND and K. MARKSTRÖM, *Generation and properties of snarks*, J. Combin. Theory Ser. B, to appear.

[4] L. ESPERET and G. MAZZUOCCOLO, *On cubic bridgeless graphs whose edge-set cannot be covered by four perfect matchings*, Manuscript, 2013. arXiv:1301.6926

[5] J. L. FOUQUET and J. M. VANHERPE, *On the perfect matching index of bridgeless cubic graphs*, Manuscript, 2009. arXiv:0904.1296.

[6] D. R. FULKERSON, *Blocking and anti-blocking pairs of polyhedra*, Math. Programming **1** (1971), 168–194.

[7] J. HÄGGLUND, *On snarks that are far from being 3-edge colorable*, Manuscript, 2012. arXiv:1203.2015

[8] X. HOU, H. J. LAI and C. Q. ZHANG, *On matching coverings and cycle coverings*, Manuscript, 2012.

[9] T. KAISER, D. KRÁL', B. LIDICKÝ, P. NEJEDLÝ and R. ŠÁMAL, *Short cycle covers of cubic graphs and graphs with minimum degree three*, SIAM J. Discrete Math. **24** (2010), 330–355.

[10] G. MAZZUOCCOLO, *The equivalence of two conjectures of Berge and Fulkerson*, J. Graph Theory **68** (2011) 125–128.

[11] E. STEFFEN, *1-factor and cycle covers of cubic graphs*, Manuscript, 2012. arXiv:1209.4510

Relating ordinary and total domination in cubic graphs of large girth

Simone Dantas[1], Felix Joos[2], Christian Löwenstein[2], Dieter Rautenbach[2] and Deiwison S. Machado[1]

Abstract. For a cubic graph G of order n, girth at least g, and domination number $\left(\frac{1}{4}+\epsilon\right)n$ for some $\epsilon \geq 0$, we show that the total domination number of G is at most $\frac{13}{32}n + O\left(\frac{n}{g}\right) + O(\epsilon n)$, which implies $\frac{\gamma_t(G)}{\gamma(G)} \leq 1.92472 + O\left(\frac{n}{g}\right)$.

For a finite, simple, and undirected graph G, a set D of vertices of G is a *dominating set of* G if every vertex in $V(G) \setminus D$ has a neighbor in D. Similarly, a set T of vertices of G is a *total dominating set of* G if every vertex in $V(G)$ has a neighbor in T. Note that a graph has a total dominating set exactly if it has no isolated vertex. The minimum cardinalities of a dominating and a total dominating set of G are known as the *domination number* $\gamma(G)$ *of* G and the *total domination number* $\gamma_t(G)$ *of* G, respectively. These two parameters are among the most fundamental and well studied parameters in graph theory [3, 4, 6]. In view of their computational hardness especially upper bounds were investigated in great detail.

The two parameters are related by some very simple inequalities. Let G be a graph without isolated vertices. Since every total dominating set of G is also a dominating set of G, we immediately obtain

$$\gamma_t(G) \geq \gamma(G). \tag{1}$$

Similarly, if D is a dominating set of G, then adding, for every isolated vertex u of the subgraph $G[D]$ of G induced by D, a neighbor of u in G

[1] Instituto de Matemática e Estatística, Universidade Federal Fluminense, Niterói, Brazil. Email: sdantas@im.uff.br, dws.sousa@gmail.com

[2] Institute of Optimization and Operations Research, Ulm University, Ulm, Germany. Email: felix.joos@uni-ulm.de, christian.loewenstein@uni-ulm.de, dieter.rautenbach@uni-ulm.de

Simone Dantas acknowledges support by FAPERJ and CNPq. Christian Löwenstein and Dieter Rautenbach acknowledge support by the CAPES/ DAAD Probral project "Cycles, Convexity, and Searching in Graphs" (PPP Project ID 56102978).

to the set D, results in a total dominating set of G, which implies

$$\gamma_t(G) \leq 2\gamma(G). \tag{2}$$

The complete bipartite graph $K_{n/2,n/2}$ and the complete graph K_n show that (1) and (2) are sharp, respectively. In [2, 5] the trees that satisfy (1) or (2) with equality are characterized constructively.

While numerous very deep results concerning bounds on the domination number and the total domination number under various conditions have been obtained, the relation of these two parameters is not really well understood. The characterization of the extremal graphs for (1) and (2) and/or improvements of (1) and (2) even under strong additional assumptions appear to be very difficult. If the graph G arises, for instance, by subdividing every edge of the complete graph K_n with $n \geq 3$ twice, then $\gamma(G) = n$ and $\gamma_t(G) = 2n - 1$, that is, forbidding cycles of length up to 8 does not allow to improve (2) by much. Similarly, if the graph G has vertex set $\bigcup_{i\in[k]}(A_i \cup B_i \cup C_i)$, where

- the sets A_i, B_i, and C_i for all $i \in [k]$ are disjoint,
- $|A_i| = a$, $|B_i| = a + 1$, and $|C_i| = ka$ for every $i \in [k]$,
- the closed neighborhood $N_G[u]$ of a vertex u in A_j for $j \in [k]$ is $B_j \cup \bigcup_{i\in[k]} A_i$,
- the closed neighborhood $N_G[v]$ of a vertex v in B_j for $j \in [k]$ is $A_j \cup \{v\} \cup C_j$, and
- the closed neighborhood $N_G[w]$ of a vertex w in C_j for $j \in [k]$ is $B_j \cup C_j$,

then G is regular of degree $(k + 1)a$, has connectivity a, diameter 5, $\gamma(G) = k + 1$, and $\gamma_t(G) = 2k$, that is, a large minimum degree, a large degree of regularity, a large connectivity, a small diameter, and a large value of the domination number do not force any serious improvement of (2).

In the present extended abstract we consider the relation of the domination number and the total domination number for cubic graphs of large girth.

Let G be a cubic graph of order n and girth at least g, that is, G has no cycles of length less than g. Clearly, $\gamma(G) \geq \frac{1}{4}n$ and $\gamma_t(G) \geq \frac{1}{3}n$. The best known upper bound on the domination number of G, improving earlier results from [10, 11], is due to Král *et al.* [9], who show

$$\gamma(G) \leq 0.299871n + O\left(\frac{n}{g}\right). \tag{3}$$

Combining the above bounds, we immediately obtain the following improvement of (1).

Corollary 1. *If G is a cubic graph of order n and girth at least g, then*

$$\frac{\gamma_t(G)}{\gamma(G)} \geq 1.111589 - O\left(\frac{1}{g}\right).$$

For a graph G of order n, minimum degree at least 2, and girth at least g, Henning and Yeo [7,8] show $\gamma_t(G) \leq \frac{1}{2}n + O\left(\frac{n}{g}\right)$. Applying a trick from [10], this result leads to the following corollary.

Corollary 2. *If G is a cubic graph of order n and girth at least g, then*

$$\gamma_t(G) \leq \frac{121}{248}n + O\left(\frac{n}{g}\right) \leq 0.488n + O\left(\frac{n}{g}\right). \tag{4}$$

Proof. Let G be as in the statement. In view of the desired bound, we may assume that g is sufficiently large. Since the 5th power of the line graph of G is neither an odd cycle nor complete, has order $\frac{3}{2}n$, and maximum degree 124, the theorem of Brooks implies that there is a set M of at least $\frac{3}{248}n$ edges of G at pairwise distance at least 5 in G. Let T_0 denote the set of $2|M|$ vertices incident with the edges in M and let $G_1 = G \setminus N_G[T_0]$. By construction, the graph G_1 has order $n - 6|M|$, minimum degree at least 2, and girth at least g. By the above result of Henning and Yeo, the graph G_1 has a total dominating set T_1 of order at most $\frac{1}{2}(n - 6|M|) + O\left(\frac{n}{g}\right)$. Since $T_0 \cup T_1$ is a total dominating set of G, we obtain

$$\begin{aligned}
\gamma_t(G) &\leq 2|M| + \frac{1}{2}(n - 6|M|) + O\left(\frac{n}{g}\right) \\
&\leq \frac{1}{2}n - |M| + O\left(\frac{n}{g}\right) \\
&\leq \frac{1}{2}n - \frac{3}{248}n + O\left(\frac{n}{g}\right) \\
&= \frac{121}{248}n + O\left(\frac{n}{g}\right),
\end{aligned}$$

which completes the proof. □

Combining the above bounds, we immediately obtain the following improvement of (2).

Corollary 3. *If G is a cubic graph of order n and girth at least g, then*

$$\frac{\gamma_t(G)}{\gamma(G)} \leq \frac{121}{62} + O\left(\frac{1}{g}\right) \leq 1.952 + O\left(\frac{1}{g}\right).$$

Note that Corollary 1 can only be close to the truth if the total domination number is close to $\frac{1}{3}n$. Similarly, Corollary 3 can only be close to the truth if the domination number is close to $\frac{1}{4}n$. Our main result shows that, for a cubic graph G of order n and girth at least g, for which the domination number is close to $\frac{1}{4}n$, the total domination number is smaller than guaranteed by (4). Specifically, we prove the following result.

Theorem 4. *If G is a cubic graph of order n, girth at least g, and domination number* $\left(\frac{1}{4}+\epsilon\right)n$ *for some* $\epsilon \geq 0$, *then*

$$\gamma_t(G) \leq \frac{13}{32}n + \frac{15}{4(g-1)}n + \frac{187}{8}\epsilon n \leq 0.40625n + O\left(\frac{n}{g}\right) + O(\epsilon n).$$

Because of the space restriction, we omit the proof of Theorem 4 in the present extended abstract.

This result allows to improve Corollary 3 as follows.

Corollary 5. *If G is a cubic graph of order n and girth at least g, then*

$$\frac{\gamma_t(G)}{\gamma(G)} \leq \frac{701437}{364436} + O\left(\frac{n}{g}\right) \leq 1.92472 + O\left(\frac{n}{g}\right).$$

Proof. Let G be as in the statement and let $\gamma(G) = \left(\frac{1}{4}+\epsilon\right)n$ for some $\epsilon \geq 0$. By Corollary 2 and Theorem 4, we obtain

$$\frac{\gamma_t(G)}{\gamma(G)} \leq \frac{\min\left\{\frac{13}{32} + \frac{187}{8}\epsilon,\ \frac{121}{248}n\right\}}{\frac{1}{4}+\epsilon} + O\left(\frac{n}{g}\right).$$

Since $\left(\frac{13}{32} + \frac{187}{8}\epsilon\right) / \left(\frac{1}{4}+\epsilon\right)$ is increasing as a function of $\epsilon \geq 0$ and $\frac{13}{32} + \frac{187}{8}\epsilon = \frac{121}{248}n$ for $\epsilon = \frac{81}{23188}$, the desired result follows. □

While the constants in our results improve previous estimates, they are clearly far from the truth and should be improved. Suitably modifying the proof strategy of Theorem 4, it is possible to show an upper bound on the domination number of a cubic graph G of order n and girth at least g, for which the total domination number is close to $\frac{1}{3}n$. Unfortunately, this bound is weaker than the result of Král *et al.* [9].

References

[1] B. BOLLOBÁS and E. J. COCKAYNE, *Graph-theoretic parameters concerning domination, independence, and irredundance*, J. Graph Theory **3** (1979), 241–249.

[2] M. DORFLING, W. GODDARD, M. A. HENNING and C. M. MYNHARDT, *Construction of trees and graphs with equal domination parameters*, Discrete Math. **306** (2006), 2647–2654.

[3] T. W. HAYNES, S. T. HEDETNIEMI and P. J. SLATER, "Fundamentals of Domination in Graphs", Marcel Dekker, Inc. New York, 1998.

[4] T. W. HAYNES, S. T. HEDETNIEMI and P.J. SLATER, "Domination in Graphs: Advanced Topics", Marcel Dekker, Inc. New York, 1998.

[5] M. A. HENNING, *Trees with large total domination number*, Util. Math. **60** (2001), 99–106.

[6] M. A. HENNING, *A survey of selected recent results on total domination in graphs*, Discrete Math. **309** (2009), 32–63.

[7] M. A. HENNING and A. YEO, *Total domination in graphs with given girth*, Graphs Combin. **24** (2008), 333–348.

[8] M. A. HENNING and A. YEO, *Girth and total domination in graphs*, Graphs Combin. **28** (2012), 199–214.

[9] D. KRÁL, P. ŠKODA and J. VOLEC, *Domination number of cubic graphs with large girth*, J. Graph Theory **69** (2012), 131–142.

[10] C. LÖWENSTEIN and D. RAUTENBACH, *Domination in graphs with minimum degree at least two and large girth*, Graphs Combin. **24** (2008), 37–46.

[11] D. RAUTENBACH and B. REED, Domination in cubic graphs of large girth, *LNCS* **4535** (2008), 186–190.

Snarks with large oddness and small number of vertices

Robert Lukot'ka[1], Edita Máčajová[2], Ján Mazák[1] and Martin Škoviera[2]

Abstract. We estimate the minimum number of vertices of a cubic graph with given oddness and cyclic connectivity. We show that a 2-connected cubic graph G with oddness $\omega(G)$ different from the Petersen graph has order at least $5.41\,\omega(G)$, and for any integer k with $2 \leq k \leq 6$ we construct an infinite family of cubic graphs with cyclic connectivity k and small oddness ratio $|V(G)|/\omega(G)$. For cyclic connectivity 2, 4, 5, and 6 we improve the upper bounds on the oddness ratio of snarks to 7.5, 13, 25, and 99 from the known values 9, 15, 76, and 118, respectively. We also construct a cyclically 4-connected snark of girth 5 with oddness 4 and order 44, improving the best previous value of 46.

1 Introduction

Snarks are connected bridgeless cubic graphs with chromatic index 4, sometimes required to satisfy additional conditions, such as cyclic 4-edge-connectivity and girth at least five, to avoid triviality. There are several important conjectures in graph theory where snarks are the principal obstacle: if true for snarks, they would hold for all graphs. Some of the conjectures have been verified for snarks that are close to 3-edge-colourable graphs. For example, the 5-flow conjecture is known to hold for snarks with oddness at most 2, and the cycle double cover conjecture has been verified for snarks with oddness at most 4 (see [4,5]). However, snarks with large oddness remain potential counterexamples to these and other conjectures, and therefore merit further investigation.

[1] Trnavská univerzita. Priemyselná 4, 918 43 Trnava.
Email: robert.lukotka@truni.sk, jan.mazak@truni.sk

[2] Univerzita Komenského, Mlynská dolina, 842 48 Bratislava.
Email: macajova@dcs.fmph.uniba.sk, skoviera@dcs.fmph.uniba.sk

The authors acknowledge partial support from the APVV grant ESF-EC-0009-10 within the EUROCORES Programme EUROGIGA (project GReGAS) of the European Science Foundation, from APVV-0223-10, and from VEGA 1/1005/12.

The *oddness* of a cubic graph G, denoted by $\omega(G)$, is the minimum number of odd circuits in a 2-factor of G. A cubic graph is 3-edge-colourable if and only if its oddness is 0, so oddness provides a natural measure of uncolourability of a cubic graph. Another common measure of uncolourability is the *resistance* of G, $\rho(G)$, the minimum number of edges whose removal from G yields a 3-edge-colourable graph. Since $\rho(G) \leq \omega(G)$, resistance provides a practical lower bound for oddness [8].

Our aim to provide bounds on the the ratio $|V(G)|/\omega(G)$ for a snark G within the class of cyclically k-connected snarks. So far, only trivial lower bounds for oddness ratio have been known; as regards the upper bounds, there are various constructions, probably not optimal. Since the oddness ratio of the Petersen graph is 5, it is meaningless to attempt improving this absolute lower bound. In Section 22 we therefore take an asymptotic approach similar to that found in [3] and [8]. We summarise the known results and our improvements in Table 11; we only consider cyclic connectivity $k \leq 6$ since no cyclically 7-connected snarks are known. In fact, they are believed not to exist.

connectivity k	LB	current UB	previous UB
2	5.41	7.5	9 (Steffen [8])
3	5.52	9	9 (Steffen [8])
4	5.52	13	15 (Hägglund [3])
5	5.83	25	76 (Steffen [8])
6	7	99	118 (Kochol [6])

Table 1. Upper (UB) and lower (LB) bounds on oddnes ratio $|V|/\omega$.

Besides general bounds, we are also interested in identifying the smallest snarks with oddness 4, addressing a long-standing open problem restated as Problem 4 in [1]. Our best results in this direction are shown in Figure 4.1 and described in Section 44. For more details and full proofs see [7].

2 Oddness and resistance ratios

The *oddness ratio* of a snark G is the quantity $|V(G)|/\omega(G)$, and its *resistance ratio* is $|V(G)|/\rho(G)$. We also examine asymptotic quantities

$$A_\omega = \liminf_{|V(G)|\to\infty} \frac{|V(G)|}{\omega(G)} \quad \text{and} \quad A_\rho = \liminf_{|V(G)|\to\infty} \frac{|V(G)|}{\rho(G)}.$$

Since the oddness ratio of a graph is at least as large as its resistance ratio, we have $A_\omega \leq A_\rho$. The oddness and resistance ratios heavily depend on the cyclic connectivity of a graph in question. This suggests to study analogous values A^k_ω and A^k_ρ obtained under the assumption that the class of snarks is restricted to those with cyclic connectivity at least k. Note that $A^2_\omega = A_\omega$, $A^2_\rho = A_\rho$, and $A^k_\omega \leq A^k_\rho$ for every $k \geq 2$. Similar ideas were pursued by Steffen [8] who proved that $8 \leq A_\rho \leq 9$ and therefore $A_\omega \leq A_\rho \leq 9$. Since snarks constructed in [8] are cyclically 3-connected, we also have $A_\omega = A^2_\omega \leq A^3_\omega \leq A^3_\rho \leq 9$.

3 Lower bounds on oddness ratio

A snark with girth at least 5 has oddness ratio at least 5. This bound is best possible because of the Petersen graph, but it can be improved for any other graph. Our approach is based on the following key observation.

Proposition 3.1. *Let $\mathcal{C}$ be a set of 5-circuits of a bridgeless cubic graph G. Then G has a 2-factor that contains at most $1/6$ of 5-circuits from $\mathcal{C}$.*

Proof. Let $\mathcal{P}$ be the perfect matching polytope of G. For a vector $\mathbf{x} \in \mathbb{R}^{|E(G)|}$ let $\mathbf{x}(e)$ denote the entry corresponding to an edge $e \in E(G)$, and let $\delta(U)$ be the set of all edges with precisely one end in the subgraph U of G. Since G is cubic and bridgeless, we have $\mathcal{P} \neq \emptyset$. Note that the vector $\mathbf{t} = (1/3, 1/3, \ldots, 1/3)$ always belongs to $\mathcal{P}$. Consider the function

$$f(\mathbf{x}) = \sum_{C \in \mathcal{C}} \sum_{e \in \delta(C)} \mathbf{x}(e)$$

defined for each $\mathbf{x} \in \mathcal{P}$. Since f is linear, there is a vertex of $\mathcal{P}$ where f reaches its minimum. Since $\mathbf{t} \in \mathcal{P}$, we have $f(\mathbf{x}_0) \leq 5/3 \cdot |\mathcal{C}|$.

Let M be the perfect matching corresponding to $\mathbf{x}_0$ and let F be the 2-factor complementary to M. Assume that F contains k circuits from $\mathcal{C}$. If a 5-circuit $C \in \mathcal{C}$ belongs to F, it adds 5 to the sum in $f(\mathbf{x}_0)$. If C does not belong to F, it adds at least 1. Altogether $f(\mathbf{x}_0) \geq 5k + (|\mathcal{C}| - k) = |\mathcal{C}| + 4k$. Since $f(\mathbf{x}_0) \leq 5/3 \cdot |\mathcal{C}|$, we obtain $k \leq |\mathcal{C}|/6$. □

Proposition 3.1 has an interesting corollary: If G is a snark different from the Petersen graph, then for every vertex v of G there exists a 2-factor F of G such that every 5-circuit of F misses v. This gives a much shorter alternative way of proving the result of DeVos [2] that the Petersen graph is the only cubic graph having each 2-factor composed only of 5-cycles.

The following lemma provides the main tool for bounding oddness from above.

Lemma 3.2. *Let G be a snark of order n with girth at least 4. If G has q circuits of length 5, then $\omega(G) \leq 3n + q/21$.*

4 Constructions

Cyclic connectivity 2. Figure 4.1 (top) displays the smallest snarks of oddness 4. Their order is 28 and cyclic connectivity is 2 and 3, respectively. The proof of minimality is computer assisted. We also create an infinite family of snarks with oddness ratio approaching 7.5 from below and conjecture that every snark with oddness ω has at least $7.5\,\omega - 5$ vertices.

Cyclic connectivity 4. Let P_4^v and P_4^e be the Petersen graph with either two adjacent vertices removed or two non-adjacent edges disconnected, respectively, and the dangling edges retained. There are two pairs of dangling edges in both. By the Parity Lemma, in P_4^v the dangling edges of each pair must have the same colour for every 3-edge-colouring, while in P_4^e they must have different colours. If we join a pair of dangling edges from P_4^v to a pair of dangling edges from P_4^e, we get an uncolourable 4-pole N_1 with 18 vertices. The 4-pole N_2 with 26 vertices arises from P_4^e and two distinct copies of P_4^v by joining each pair of dangling edges of P_4^e to a pair of edges edges in a different copy of P_4^v. The 4-pole N_2 is uncolourable even after the removal of a vertex w. This is clearly true if w belongs to a copy of P_4^v. If it was false for some w from a copy of P_4^e, then P_4^e would have a 3-edge-colouring where the edges in both pairs of dangling edges have the same colour. This would yield a 3-edge-colouring of the Petersen graph minus a vertex, but no such colouring exists.

To construct a cyclically 4-connected snark with arbitrarily large oddness we take a number of copies of N_1 and a number of N_2, arrange them into a circuit, and join one pair of dangling edges from each copy to a pair of dangling edges of its predecessor and another pair of dangling edges to a pair of dangling edges of its successor. The way in which copies of N_1 and N_2 are arranged is not unique, therefore we may get several non-isomorphic graphs even if we only use copies of one of N_1 and N_2.

In this construction, each copy of N_1 adds 1 and each copy of N_2 adds 2 to the resistance of the resulting graph. Thus if we take r copies of N_2, we get a cyclically 4-connected snark of order $26r$ with resistance $2r$ and oddness at least $2r$. If we take r copies of N_2 and one copy of N_1 we get a cyclically 4-connected snark of order $26r + 18$ with resistance $2r + 1$ and oddness at least $2r + 2$. This shows, in particular, that $A_\omega^4 \leq A_\rho^4 \leq 13$.

For $r = 1$ we obtain a cyclically 4-connected snark of order 44 with resistance 3 and oddness 4 shown in Figure 4.1, currently the smallest known non-trivial snark of oddness 4, improving the previous value of 46 [3].

Cyclic connectivity 6. Let P_3 be the Petersen graph with one vertex removed and the dangling edges retained. Take r copies $Q_1, Q_2, \ldots, Q_r$ of P_3. Arrange them into a circuit and, for each $i \in \{1, 2, \ldots, r\}$, join one dangling edge of Q_i to a dangling edge of Q_{i-1} and do the same for Q_{i+1} (indices reduced modulo r). There are $r/2$ pairs of oppositely positioned copies of P_3; join the remaining dangling edges for each such pair. The result of is a cubic graph L_r of order $9r$ and resistance r.

To obtain a snark with cyclic connectivity 6 we apply superposition to L_r (see [6] for details). Nontrivial supervertices will be copies of the 7-pole X having a vertex incident with three edges belonging to different connectors, plus two more edges joining the first two connectors of size 3. Nontrivial superedges will be copies of a 6-pole Y with 18 vertices created from the flower snark J_5 by removing two nonadjacent vertices, one from the single 5-circuit of J_5. We choose a circuit C in L_r which passes through five vertices of each copy of P_3 and finish the superposition by replacing each vertex on C with a copy of X and each edge on C with a copy of Y; we use trivial supervertices and superedges everywhere outside C. The resulting graph has order $99r$. Its resistance is at least r due to the following proposition.

Proposition 4.1. *Let $\tilde{G}$ be a snark resulting from a proper superposition of a snark G. Then $\rho(\tilde{G}) \geq \rho(G)$.*

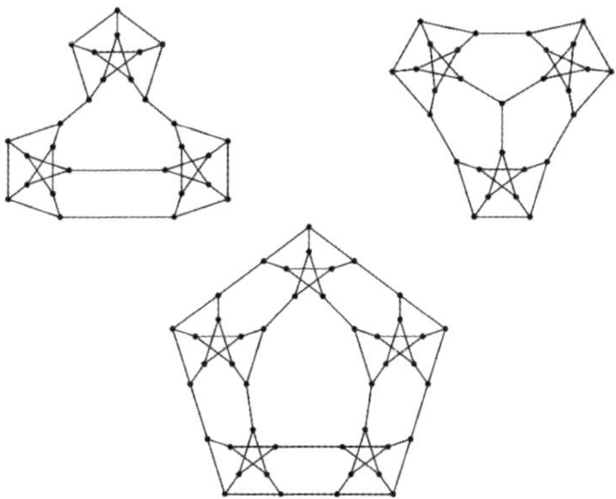

Figure 4.1. Smallest snarks of oddness 4 with cyclic connectivity 2, 3, 4.

References

[1] G. BRINKMANN, J. GOEDGEBEUR, J. HÄGGLUND and K. MARKSTRÖM, *Generation and properties of snarks*, accessed 7th June 2012, http://abel.math.umu.se/ klasm/Uppsatser/snarks genprop.pdf,

[2] M. DeVos, *Petersen graph conjecture*, http://garden.irmacs.sfu.ca/?q=op/petersen_graph_conjecture.

[3] J. Hägglund, *On snarks that are far from being 3-edge-colorable*, http://arxiv.org/pdf/1203.2015.pdf.

[4] R. Häggkvist and S. McGuinness, *Double covers of cubic graphs of oddness 4*, J. Combin. Theory Ser. B **93** (2005), 251–277.

[5] A. Huck and M. Kochol, *Five cycle double covers of some cubic graphs*, J. Combin. Theory Ser. B **64** (1995), 119–125.

[6] M. Kochol, *Superposition and constructions of graphs without nowhere-zero k-flows*, European J. Combin. **23** (2002), 281–306.

[7] R. Lukot'ka, E. Máčajová, J. Mazák and M. Škoviera, *Small snarks with large oddness*, http://arxiv.org/pdf/1212.3641v1.pdf.

[8] E. Steffen, *Measurements of edge-uncolorability*, Discrete Math. **280** (2004), 191–214.

Non-trivial snarks with given circular chromatic index

Robert Lukot'ka[1] and Ján Mazák[1]

Abstract. We introduce a new framework designed for constructing graphs with given circular chromatic index. This framework allows construction of graphs with arbitrary maximum degree and with additional properties, *e.g.* high connectivity or large girth. We utilize this framework to construct a cyclically 4-edge-connected cubic graph with girth 5 and circular chromatic index r for any rational $r \in (3, 3 + 1/4.5)$.

1 Introduction

Edge-colourings of graphs emerged more than a century ago among the first topics in graph theory. They are especially important for 3-regular graphs because of their deep connections to various other branches of graph theory. According to the Vizing theorem, every cubic graph has a 4-edge-colouring; most of them have a 3-edge-colouring. Bridgeless cubic graphs with chromatic index 4 are called *snarks*.

It transpired that certain snarks are in many settings trivial when compared to others. The most interesting snarks are those that are cyclically 4-edge-connected and have girth at least 5. Such snarks are called *non-trivial*. A discussion of several aspects of triviality of snarks can be found in [6].

For a real number $r \geq 2$, a circular r*-edge-colouring* of a given graph G is a mapping $c : E(G) \to [0, r)$ such that $1 \leq |c(e) - c(f)| \leq r - 1$

[1] Faculty of Education, Trnava University, Priemyselná 4, 918 43 Trnava.
Email: robert.lukotka@truni.sk, jan.mazak @truni.sk

The authors acknowledge partial support from the research grant APVV-0223-10 and from the APVV grant ESF-EC-0009-10 within the EUROCORES Programme EUROGIGA (project GReGAS) of the European Science Foundation. The second author also acknowledges partial support from the Trnava University grant 8/TU/13.

for any two adjacent edges e and f of G. The *circular chromatic index* of G is the infimum of all r such that G has a circular r-edge-colouring. In fact, this infimum is always attained and the value of circular chromatic index is always rational for a finite graph G. A proof of this and many other properties of circular colourings can be found in the surveys [7, 8]. It is known that if a graph G has a circular r-edge-colouring and r is an integer, then G also has an r-edge-colouring with colours being integers from the set $\{0, 1, \ldots, r-1\}$. Thus we may omit the word "circular" when referring to an r-edge-colouring.

Afshani *et al.* [1] proved that the circular chromatic index of a 2-edge-connected cubic graph lies in the interval $[3, 11/3]$. The only known bridgeless cubic graph with circular chromatic index greater than $7/2$ is the Petersen graph with index $11/3$. It was conjectured that there are no other such graphs; this folklore conjecture was proved for bridgeless cubic graphs with girth at least six [3]. For any rational $r \in (3, 10/3)$, there are infinitely many cubic graphs with circular chromatic index r; the only known construction giving an interval of realized values of χ'_c is described in [4]. All new snarks constructed in [4] contain 2-edge-cuts and are thus trivial; the purpose of the present paper is to improve the construction method and then use it to produce non-trivial snarks with given circular chromatic index.

In the original construction, one has to construct the desired graph with circular chromatic index r together with a feasible r-edge-colouring. Our first improvement separates the construction of the graph from the construction of the colouring; in fact, the colouring is obtained from the modified construction without any additional effort. The second improvement consists in simplifying the construction to its bare essentials. The concept of balanced schemes hides many of the technical details and helps both authors and prospective readers. Another promising aspect of balanced schemes is that they do not depend on degrees of vertices of the graph being constructed. We describe the framework in Section 2.

Although we are dealing only with cubic graphs throughout the paper, our framework can be extended to regular graphs with larger degree in a straightforward way. In fact, only the building blocks would be different—the rest of the machinery works with no change. Thus only the current lack of suitable blocks separates us from a construction of regular graphs with given circular chromatic index. We have also managed to construct snarks of large girth with given circular chromatic index r from some interval $(3, 3+\alpha)$, however, the relation between girth and α is yet to be determined.

2 Balanced schemes

Our new construction is based on the concept of monochromatic networks devised in [4]. We refer the reader to this article for all the details on monochromatic networks not covered here.

A *p-line monochromatic network* N is a subcubic graph with $2p$ vertices of degree one called *terminals*. The terminals of the network N are paired into p disjoint pairs in such a way that in every proper 3-edge-colouring of N, the edges incident to paired terminals have the same colour; we will denote this pairing by $P(N)$. The *terminal* edge incident to a terminal v of a network N is denoted by $e(v)$.

Two monochromatic networks can be naturally *joined* by identifying terminal edges. A *monochromatic line* of a network N is a set of edges having the same colour in every 3-edge-colouring of N and containing the two edges adjacent to the vertices in a pair of terminals from $P(N)$. By joining monochromatic networks we also merge monochromatic lines; in fact, we are only interested in *primary* monochromatic lines which are created from a starting terminal edge (which is a monochromatic line in itself) by repeatedly extending it by joining other monochromatic networks. Primary monochromatic lines are a crucial concept of our construction.

For $\varepsilon \in [0, 1]$, the *ε-changeability* of a network N is the maximal possible value of

$$\sum_{(x,y)\in P(N)} |c(e(x)) - c(e(y))|_{3+\varepsilon}$$

over all $(3 + \varepsilon)$-edge-colourings c of N, where $|q|_{3+\varepsilon} = \min\{|q|, |3 + \varepsilon - q|\}$. It is clear that if we join two monochromatic networks M and N, the ε-changeability of the resulting monochromatic network does not exceed the sum of ε-changeabilities of M and N.

Monochromatic networks with one or two lines can be easily constructed from any cubic graph by cutting an edge of any cubic graph or by removing two adjacent vertices and turning the dangling edges into terminal edges. This construction yields a 2-line monochromatic network B from the Petersen graph (Figure 2.1).

Our aim is to simplify the view of monochromatic networks composed of many smaller networks. We will represent such monochromatic networks by schemes. An *r-balanced scheme* is a pair $\mathcal{S} = (G, \varphi)$ such that

- G is a multigraph that may contain parallel edges and semiedges; we denote the end of an edge e incident with a vertex v by e^v;
- φ is a labeling by rational numbers that assigns a value to both ends of every edge and one value to each semiedge;

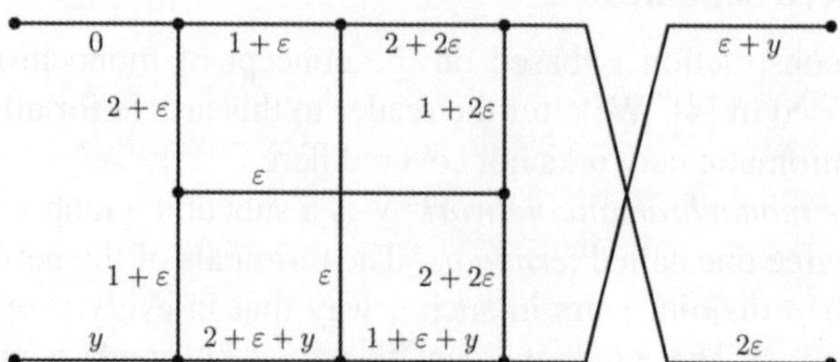

Figure 2.1. The monochromatic network B with a $(3+\varepsilon)$-edge-colouring.

- the sum of values at every vertex v of G (one value for every incident edge or semiedge) is r.

The set of vertices of the underlying multigraph G of a scheme S is denoted by $V(\mathcal{S})$. The vertices of $\mathcal{S}$ represent primary monochromatic lines of the network corresponding to the scheme $\mathcal{S}$ and the edges (semiedges) of $\mathcal{S}$ represent basic 2-line (1-line) monochromatic networks used in the construction.

We say that an edge $e = uv$ of $\mathcal{S}$ is *representable* if there exists a 2-line monochromatic network N_2 such that $P(N_2) = \{(u_1, u_2), (v_1, v_2)\}$ and

- the $(1/r)$-changeability of N_2 is $\varphi(e^u) + \varphi(e^v)$;
- there exists a circular $(3 + 1/r)$-edge-colouring c of N_2 such that $|c(e(u_1)) - c(e(u_2))|_{3+\varepsilon} = \varphi(e^u)$, $|c(e(v_1)) - c(e(v_2))|_{3+\varepsilon} = \varphi(e^v)$, and the colours of the edges incident to u_1, u_2, v_1, v_2 are all rational.

A network N_2 satisfying the above definition is called a *representative* of the edge e. The representability of a semiedge of $\mathcal{S}$ can be defined in a similar way; we use a 1-line monochromatic network as the representative. An r-balanced scheme is *representable* if all its edges and semiedges are representable.

We will describe how to construct a subcubic graph G with circular chromatic index $3 + 1/r$ from a representable r-balanced scheme $\mathcal{S}$. Let p be the number of vertices of $\mathcal{S}$. We start with a p-line monochromatic network N_0 isomorphic to a matching containing p edges. Let ℓ be an arbitrary bijection between $V(\mathcal{S})$ and the monochromatic lines of N_0. For each edge of N_0, we fix one of its endvertices; these fixed vertices will remain untouched until the very last step of the construction.

For each edge uv (and also for each semiedge) of $\mathcal{S}$ we do the following: According to our definitions, the edge uv has a representative N. We join N to the monochromatic lines $\ell(u)$ and $\ell(v)$. The resulting network has $(1/r)$-changeability at least p. Hence if we identify the paired vertices of the resulting network, we get a graph with circular chromatic index at least $3 + 1/r$.

To produce a graph with circular chromatic index exactly $3 + 1/r$, we do a "copying trick". Let d be the least common denominator of $1/r$ and all the denominators in the fractions of the labeling φ. we start with d primary monochromatic lines for each vertex of $\mathcal{S}$. Instead of joining N to the lines $\ell(u)$ and $\ell(v)$, we join d copies of N to all the lines $\ell(u)$ and $\ell(v)$ (note that both $\ell(u)$ and $\ell(v)$ contain d monochromatic lines) so that each line will be used exactly once. Apparently, the resulting network has $(1/r)$-changeability at least dp, thus if we identify endvertices paired according to the primary monochromatic lines, we get a graph with circular chromatic index at least $3 + 1/r$.

We will produce a $(3 + 1/r)$-colouring of our network alongside our construction. The idea is to guarantee that both before and after joining d copies of each network N for each vertex $v \in \mathcal{S}$ all the rational numbers $i/d \in [0, 3 + 1/r)$ are used as colours on the edges incident with non-fixed ends of the primary lines of N. The second property in the definition of a representable network guarantees that the total change of colour on each primary monochromatic line will be exactly 1. Therefore at the last step we may identity endvertices of primary monochromatic lines. The resulting graph will have the circular chromatic index equal to $3 + 1/r$.

We are now ready to sketch the proof of main result.

Theorem 2.1. *For each rational $t \in (3, 3+1/4.5)$, there is a non-trivial snark with circular chromatic index t.*

Proof. For each rational $r \in (4.5, \infty)$, we construct a representable r-balanced scheme that uses only the network B as a representative. After carrying out the construction described above we get a subcubic graph G with desired circular chromatic index. If G is not connected, we can take an arbitrary component of G. To make this graph cubic we just take two copies of it and connect the corresponding vertices of degree 2. If we use the same edge-colouring on both of the copies, then the circular chromatic index does not increase (note that the difference of colours incident to vertices of degree 2 is exactly 1 and therefore we can colour the third edge). By performing the construction a bit more carefully we can guarantee the resulting graph to be cyclically 4-edge connected. □

References

[1] P. Afshani, M. Ghandehari, M. Ghandehari, H. Hatami, R. Tusserkani and X. Zhu, *Circular chromatic index of graphs of maximum degree* 3, J. Graph Theory **49** (2005), 325–335.

[2] M. GHEBLEH, *Circular chromatic index of generalized Blanusa snarks*, Electron. J. Combin. **15** (2008).
[3] D. KRÁL', E. MÁČAJOVÁ, J. MAZÁK and J.-S. SERENI, *Circular edge-colourings of cubic graphs with girth six*, J. Combin. Theory Ser. B **100** (2010), 351–358.
[4] R. LUKOT'KA and J. MAZÁK, *Cubic graphs with given circular chromatic index*, SIAM J. Discrete Math. **24** (3) (2011), 1091–1103.
[5] J. MAZÁK, *Circular chromatic index of type 1 Blanuša snarks*, J. Graph Theory **59** (2008), 89–96.
[6] R. NEDELA and M. ŠKOVIERA, *Decompositions and reductions of snarks*, J. Graph Theory **22** (1996), 253–279.
[7] X. ZHU, *Circular chromatic number: a survey*, Discrete Math. **229** (2001) 371–410.
[8] X. ZHU, *Recent developments in circular colourings of graphs*, In: "Topics in Discrete Mathematics", Springer, 2006, 497–550.

Graphs

The graph formulation of the union-closed sets conjecture

Henning Bruhn[1], Pierre Charbit[2] and Jan Arne Telle[3]

Abstract. In 1979 Frankl conjectured that in a finite non-trivial union-closed collection of sets there has to be an element that belongs to at least half the sets. We show that this is equivalent to the conjecture that in a finite non-trivial graph there are two adjacent vertices each belonging to at most half of the maximal stable sets. In this graph formulation other special cases become natural. The conjecture is trivially true for non-bipartite graphs and we show that it holds also for the class of chordal bipartite graphs and the class of bipartitioned circular interval graphs.

A set $\mathcal{X}$ of sets is *union-closed* if $X, Y \in \mathcal{X}$ implies $X \cup Y \in \mathcal{X}$. The following conjecture was formulated by Peter Frankl in 1979 [4].

Union-closed sets conjecture. *Let $\mathcal{X}$ be a finite union-closed set of sets with $\mathcal{X} \neq \{\emptyset\}$. Then there is a $x \in \bigcup_{X \in \mathcal{X}} X$ that lies in at least half of the members of $\mathcal{X}$.*

In spite of a great number of papers, see e.g. the good bibliography of Marković [7] for papers up to 2007, this conjecture is still wide open, allthough several special cases are known to hold. Various equivalent formulations are known, in particular by Poonen [9] who among other things translates the conjecture into the language of lattice theory.

In this paper we give a formulation of the conjecture in the language of graph theory. A set of vertices in a graph is *stable* if no two vertices of the set are adjacent. A stable set is *maximal* if it is maximal under inclusion, that is, every vertex outside has a neighbour in the stable set.

Conjecture 1. Let G be a finite graph with at least one edge. Then there will be two adjacent vertices each belonging to at most half of the maximal stable sets.

Note that Conjecture 1 is true for non-bipartite graphs. Indeed, if vertices u and v are adjacent there is no stable set containing them both and

[1] Université Pierre et Marie Curie (Paris 6), Paris, France. Email: bruhn@math.jussieu.fr

[2] LIAFA, Université Paris 7, France. Email: pierre.charbit@liafa.univ-paris-diderot.fr

[3] Department of Informatics, University of Bergen, Norway. Email: telle@ii.uib.no

so one of them must belong to at most half of the maximal stable sets. An odd cycle will therefore imply the existence of two adjacent vertices each belonging to at most half of the maximal stable sets. The conjecture is for this reason open only for bipartite graphs. Moreover, in a connected bipartite graph, for any two vertices u and v in different bipartition classes we have a path from u to v containing an odd number of edges, so that if u and v each belongs to at most half the maximal stable sets there will be two adjacent vertices each belonging to at most half the maximal stable sets. Conjecture 1 is therefore equivalent to the following.

Conjecture 2. *Let G be a finite bipartite graph with at least one edge. Then each of the two bipartition classes contains a vertex belonging to at most half of the maximal stable sets.*

In this paper we show that Conjectures 1 and 2 are equivalent to the union-closed sets conjecture. The merit of this graph formulation is that other special cases become natural, in particular subclasses of bipartite graphs. We show that the conjecture holds for the class of chordal bipartite graphs and the class of bipartitioned circular interval graphs. Moreover, the reformulation allows to test Frankl's conjecture in a probabilistic sense: Bruhn and Schaudt [1] show that almost every random bipartite graph satisfies Conjecture 2 up to any given $\varepsilon > 0$, that is, almost every such graph contains in each bipartition class a vertex for which the number of maximal stable sets containing it is at most $\frac{1}{2} + \varepsilon$ times the total number of maximal stable sets.

Stable sets are also called independent sets, with the maximal stable sets being exactly the independent dominating sets. The set of all maximal stable sets of a bipartite graph was studied by Prisner [10] who gave upper bounds on the size of this set, also when excluding certain subgraphs. More recently, Duffus, Frankl and Rödl [3] and Ilinca and Kahn [6] investigate the number of maximal stable sets in certain regular and biregular bipartite graphs. In work related to the graph parameter boolean-width, Rabinovich, Vatshelle and Telle [11] study balanced bipartitions of a graph that bound the number of maximal stable sets.

For a subset S of vertices of a graph we denote by $N(S)$ the set of vertices adjacent to a vertex in S. We need two easy lemmas. The proof of the first is trivial.

Lemma 3. *Let G be a bipartite graph with bipartition U, W, and let S be a maximal stable set. Then* $S = (U \cap S) \cup (W \setminus N(U \cap S))$.

Lemma 4. *Let G be a bipartite graph with bipartition U, W, and let S and T be maximal stable sets. Then* $(U \cap S \cap T) \cup (W \setminus N(S \cap T))$ *is a maximal stable set.*

Proof. Clearly, $R = (U \cap S \cap T) \cup (W \setminus N(S \cap T))$ is stable. Trivially, any vertex in $W \setminus R$ has a neighbour in R. A vertex u in $U \setminus R$ does not lie in S or not in T (perhaps, it is not contained in either), let us say that $u \notin T$. As T is maximal, u has a neighbour $w \in W \cap T$. This neighbour w cannot be adjacent to any vertex in $U \cap S \cap T$ as T is stable. So, w belongs to R as well, which shows that R is a maximal stable set. □

For a fixed graph G let us denote by $\mathcal{A}$ the set of all maximal stable sets, and for any vertex v let us write $\mathcal{A}_v$ for the sets of $\mathcal{A}$ that contain v and $\mathcal{A}_{\overline{v}}$ for the sets of $\mathcal{A}$ that do not contain v. Let us call a vertex v *rare* if $|\mathcal{A}_v| \leq \frac{1}{2}|\mathcal{A}|$.

Theorem 5. *Conjecture 2 is equivalent to the union-closed sets conjecture.*

Proof. Let us consider first a union-closed set $\mathcal{X} \neq \{\emptyset\}$, which, without restricting generality, we may assume to include $\emptyset$ as a member. We put $U = \bigcup_{X \in \mathcal{X}} X$ and we define a bipartite graph G with vertex set $U \cup \mathcal{X}$, where we make $X \in \mathcal{X}$ adjacent with all $u \in X$.

Now we claim that $\tau : S \mapsto U \setminus S$ is a bijection between $\mathcal{A}$ and $\mathcal{X}$. First note that indeed $\tau(S) \in \mathcal{X}$ for every maximal stable set: Set $A = U \cap S$ and $\mathcal{B} = \mathcal{X} \cap S$. If $U \subseteq S$ then $U \setminus S = \emptyset \in \mathcal{X}$, by assumption. So, assume $U \not\subseteq S$, which implies $\mathcal{B} \neq \emptyset$. As S is a maximal stable set, it follows that $U \setminus S = U \setminus A = N(\mathcal{B})$. On the other hand, $N(\mathcal{B})$ is just the union of the $X \in S \cap \mathcal{X} = \mathcal{B}$, which is by the union-closed property equal to a set X' in $\mathcal{X}$. To see that τ is injective note that, by Lemma 3, S is determined by $U \cap S$, which in turn determines $U \setminus S$. For surjectivity, consider $X \in \mathcal{X}$. We set $A = U \setminus N(X)$ and observe that $S = A \cup (\mathcal{X} \setminus N(A))$ is a stable set. Moreover, as $X \in \mathcal{X} \setminus N(A)$ every vertex in $U \setminus A$ is a neighbour of $X \in S$, which means that S is maximal.

Now, assuming that Conjecture 2 is true, there is a rare $u \in U$, that is, it holds that $|\mathcal{A}_u| \leq \frac{1}{2}|\mathcal{A}|$. Clearly $\mathcal{A}$ is the disjoint union of $\mathcal{A}_u$ and of $\mathcal{A}_{\overline{u}}$, so that

$$|\tau(\mathcal{A}_{\overline{u}})| = |\mathcal{A}_{\overline{u}}| \geq \frac{1}{2}|\mathcal{A}| = \frac{1}{2}|\mathcal{X}|.$$

As $u \in \tau(S) \in \mathcal{X}$ for every $S \in \mathcal{A}_{\overline{u}}$, the union-closed sets conjecture follows.

For the other direction, consider a bipartite graph with bipartition U, W and at least one edge. Define $\mathcal{X} := \{U \setminus S : S \in \mathcal{A}\}$, and note that $\mathcal{X} \neq \{\emptyset\}$ as G has at least two distinct maximal stable sets. By Lemma 3, there is a bijection between $\mathcal{X}$ and $\mathcal{A}$. Moreover, it is a direct consequence of Lemma 4 that $\mathcal{X}$ is union-closed. From this, it is straightforward that Conjecture 2 follows from the union-closed sets conjecture. □

Let us say that a bipartite graph *satisfies the union-closed sets conjecture* if each of its bipartition classes contains a rare vertex. For a set X of vertices we define $\mathcal{A}_X$ to be the set of maximal stable sets containing all of X. As before, we abbreviate $\mathcal{A}_{\{x\}}$ to $\mathcal{A}_x$.

Lemma 6. *Let x be a vertex of a bipartite graph G. Then there is an injection $\mathcal{A}_{N(x)} \to \mathcal{A}_x$.*

Proof. We define

$$i : \mathcal{A}_{N(x)} \to \mathcal{A}_x,\ S \mapsto S \setminus L_1 \cup \{x\} \cup (L_2 \setminus N(S \cap L_3)),$$

where L_i denotes the set of vertices at distance i to x. That $i(S)$ is stable and maximal is a direct consequence of the definition. Moreover, $i(S) = i(T)$ for $S, T \in \mathcal{A}_{N(x)}$ implies that S and T are identical outside $L_1 \cup L_2$. Moreover, S and T are also identical on $L_1 \cup L_2$: First, $L_1 = N(x)$ shows that L_1 lies in both S and T. Second, since every vertex in L_2 is a neighbour of one in $L_1 \subseteq S \cap T$, no vertex of L_2 can lie in either of S or T. Thus, $S = T$, and we see that i is an injection. □

We denote by $N^2(x) = N(N(x))$ the second neighbourhood of x.

Lemma 7. *Let x, y be two adjacent vertices in a bipartite graph G with $N^2(x) \subseteq N(y)$. Then y is rare.*

Proof. From $N^2(x) \subseteq N(y)$ it follows that every maximal stable set containing y must contain all of $N(x)$. Thus, $\mathcal{A}_y = \mathcal{A}_{N(x)}$, which means by Lemma 6 that $|\mathcal{A}_y| \leq |\mathcal{A}_x|$ and as $|\mathcal{A}_y| + |\mathcal{A}_x| \leq |\mathcal{A}|$ the lemma is proved. □

We now apply the lemma to the class of *chordal bipartite* graphs. This is the class of bipartite graphs in which every cycle with length at least six has a chord. A vertex v in a bipartite graph is *weakly simplicial* if the neighbourhoods of its neighbours form a chain under inclusion. Hammer, Maffray and Preissmann [5], and also Pelsmajer, Tokaz and West [8] prove the following:

Theorem 8. *A bipartite graph with at least one edge is chordal bipartite if and only if every induced subgraph has a weakly simplicial vertex. Moreover, such a vertex can be found in each of the two bipartition classes.*

Theorem 9. *Any chordal bipartite graph with at least one edge satisfies the union-closed sets conjecture.*

Proof. For a given bipartition class, let x be a weakly simplicial vertex in it. Among the neighbours of x denote by y the one whose neighbourhood includes the neighbourhoods of all other neighbours of x. Then y is rare, by Lemma 7. □

For two vertices u, v let us denote by $\mathcal{A}_{uv}$ the set of $S \in \mathcal{A}$ containing both of u and v, by $\mathcal{A}_{u\overline{v}}$ the set of $S \in \mathcal{A}$ containing u but not v, and by $\mathcal{A}_{\overline{uv}}$ the set of $S \in \mathcal{A}$ containing neither of u and v.

Lemma 10. *Let G be a bipartite graph. Let y and z be two neighbours of a vertex x so that $N^2(x) \subseteq N(y) \cup N(z)$. Then one of y and z is rare.*

Proof. We may assume that $|\mathcal{A}_{y\overline{z}}| \leq |\mathcal{A}_{\overline{y}z}|$. Now, from $N^2(x) \subseteq N(y) \cup N(z)$ we deduce that $\mathcal{A}_{yz} = \mathcal{A}_{N(x)}$. Thus, by Lemma 6, we obtain $|\mathcal{A}_{yz}| \leq |\mathcal{A}_x|$. Since $\mathcal{A}_x \subseteq \mathcal{A}_{\overline{yz}}$ it follows that $|\mathcal{A}_y| = |\mathcal{A}_{y\overline{z}}| + |\mathcal{A}_{yz}| \leq |\mathcal{A}_{\overline{y}z}| + |\mathcal{A}_{\overline{yz}}| = |\mathcal{A}_{\overline{y}}|$. As $|\mathcal{A}| = |\mathcal{A}_y| + |\mathcal{A}_{\overline{y}}|$, we see that y is rare. □

We give an application of Lemma 10. The class of circular interval graphs plays a fundamental role in the structure theorem of claw-free graphs of Chudnovsky and Seymour [2] and are defined as follows: Let a finite subset of a circle be the vertex set, and for a given set of subintervals of the circle consider two vertices to be adjacent if there is an interval containing them both. We may obtain a rich class of bipartite graphs from circular interval graphs: For any circular interval graph, partition its vertex set and delete every edge with both its endvertices in the same class. We call any graph arising in this manner a *bipartitioned circular interval graph.* The proof that these graphs satisfy the union-closed sets conjecture has been left out of this extended abstract.

Theorem 11. *Every bipartitioned circular interval graph with at least one edge satisfies the union-closed sets conjecture.*

References

[1] H. BRUHN and O. SCHAUDT, *The union-closed sets conjecture almost holds for almost all random bipartite graphs*, CoRR, arXiv:1302.7141 [math.CO], 2013.

[2] M. CHUDNOVSKY and P. D. SEYMOUR, *Claw-free graphs. III. Circular interval graphs*, J. Combin. Theory (Series B) **98** (4) (2008), 812–834.

[3] D. DUFFUS, P. FRANKL and V. RÖDL, *Maximal independent sets in bipartite graphs obtained from boolean lattices*, Eur. J. Comb. **32** (1) (2011), 1–9.

[4] P. FRANKL, *Handbook of combinatorics (vol. 2)*, MIT Press, Cambridge, MA, USA, 1995, 1293–1329.

[5] P. L. HAMMER, F. MAFFRAY and M. PREISSMANN, "A Characterization of Chordal Bipartite Graphs", Rutcor research report, Rutgers University, New Brunswick, NJ, 1989.

[6] L.ILINCA and J. KAHN, *Counting maximal antichains and independent sets*, CoRR **abs/1202.4427** (2012).

[7] P. MARKOVIĆ, *An attempt at Frankl's conjecture.*, Publications de l'Institut Mathématique. Nouvelle Série **81** (95) (2007), 29–43.

[8] M. J. PELSMAJER, J. TOKAZ and D. B. WEST, *New proofs for strongly chordal graphs and chordal bipartite graphs*, preprint 2004.

[9] B. POONEN, *Union-closed families*, J. Combin. Theory (Series A) **59** (1992), 253–268.

[10] E. PRISNER, *Bicliques in graphs I: Bounds on their number*, Combinatorica **20** (1) (2000), 109–117.

[11] Y. RABINOVICH, J. A. TELLE and M. VATSHELLE, *Upper bounds on the boolean width of graphs with an application to exact algorithms*, submitted, 2012.

The union-closed sets conjecture almost holds for almost all random bipartite graphs

Henning Bruhn[1] and Oliver Schaudt[1]

Abstract. Frankl's union-closed sets conjecture states that in every finite union-closed set of sets, there is an element that is contained in at least half of the member-sets (provided there are at least two members). The conjecture has an equivalent formulation in terms of graphs: In every bipartite graph with least one edge, both colour classes contain a vertex belonging to at most half of the maximal stable sets.

We prove that, for every fixed edge-probability, almost every random bipartite graph almost satisfies Frankl's conjecture.

1 Introduction

A full paper version of this extended abstract is available at the arXiv [4]. One of the most basic conjectures in extremal set theory is Frankl's conjecture on union-closed set systems. A set $\mathcal{X}$ of sets is *union-closed* if $X \cup Y \in \mathcal{X}$ for all $X, Y \in \mathcal{X}$.

Union-closed sets conjecture. *Let $\mathcal{X} \neq \{\emptyset\}$ be a finite union-closed set of sets. Then there is a $x \in \bigcup_{X \in \mathcal{X}} X$ that lies in at least half of the members of $\mathcal{X}$.*

While Frankl [6] dates the conjecture to 1979, it apparently did not appear in print before 1985, when it was mentioned as an open problem in Rival [8]. Despite being widely known, there is only little substantial progress on the conjecture. For a survey on the conjecture, see [3].

Recently, Bruhn, Charbit, Schaudt and Telle [2] gave an equivalent formulation in terms of graphs. For this, let us say that a vertex set S in a graph is *stable* if no two of its vertices are adjacent, and that it is *maximally stable* if, in addition, every vertex outside S has a neighbour in S.

Conjecture 1.1 (Bruhn, Charbit, Schaudt and Telle [2]). *Let G be a bipartite graph with at least one edge. Then each of the two biparti-*

[1] Equipe Combinatoire et Optimisation, Université Pierre et Marie Curie (Paris 6), 4 place Jussieu, 75252 Paris. Email: bruhn@math.jussieu.fr, schaudt@math.jussieu.fr

tion classes contains a vertex belonging to at most half of the maximal stable sets.

So far, the graph formulation is only verified for chordal-bipartite graphs, bipartite subcubic graphs, bipartite series-parallel graphs and bipartitioned circular interval graphs [2].

We prove a slight weakening of Conjecture 1.1 for random bipartite graphs. For $\delta > 0$, we say that a bipartite graph *satisfies the union-closed sets conjecture up to* δ if each of its two bipartition classes has a vertex for which the number of maximal stable sets containing it is at most $\frac{1}{2} + \delta$ times the total number of maximal stable sets. A *random bipartite graph* is a graph on bipartition classes of cardinalities m and n, where any two vertices from different classes are independently joined by an edge with probability p. We say that *almost every* random bipartite graph has property P if for every $\varepsilon > 0$ there is an N such that, whenever $m + n \geq N$, the probability that a random bipartite graph on $m + n$ vertices has P is at least $1 - \varepsilon$.

Our main result is the following.

Theorem 1.2. *Let* $p \in (0,1)$ *be a fixed edge-probability. For every* $\delta > 0$*, almost every random bipartite graph satisfies the union-closed sets conjecture up to* δ.

2 Discussion of averaging

Many of the partial results on Frankl's conjecture are based on a technique called *averaging*. It consists in taking the average of the number of member sets containing a given element, where the average ranges over the set $U = \bigcup_{X \in \mathcal{X}} X$ of all elements. If that average is at least $\frac{1}{2}|U|$ then clearly $\mathcal{X}$ satisfies the conjecture. Averaging was used successfully by Balla, Bollobàs and Eccles [1] when $n \geq \lceil \frac{1}{3} 2^{m+1} \rceil$. Reimer [7] showed that the average is always at least $\log_2(|U|)$.

Averaging does not always work. It is easy to construct union-closed set systems in which the average is too low. Czédli, Maróti and Schmidt [5] even found such set systems of size $|\mathcal{X}| = \lfloor 2^{|U|+1}/3 \rfloor$. Nevertheless, we see that, in the graph formulation, averaging almost always allows us to conclude that the union-closed sets conjecture is satisfied (up to any $\delta > 0$).

To describe the averaging technique for bipartite graphs, let us write $\mathcal{A}(G)$ for the set of maximal stable sets of a bipartite graph G. Conjecture 1.1 is satisfied if G contains a *rare* vertex in both bipartition classes, that is, a vertex that lies in at most half of the maximal stable sets.

We consider a bipartite graph G to have a fixed bipartition, which we denote by $(L(G), R(G))$. When discussing the bipartition classes, we often refer to $L(G)$ as the *left side* and to $R(G)$ as the *right side* of the graph. Let us consider a fixed edge probability $p \in (0, 1)$ and put $q = 1 - p$. We denote by $\mathcal{B}(m, n; p)$ the probability space whose elements are the random bipartite graphs G with $|L(G)| = m$ and $|R(G)| = n$. We note first that exchanging the sides turns a random bipartite graph $G \in \mathcal{B}(m, n; p)$ into a member of $\mathcal{B}(n, m; p)$, which means that it suffices to show the existence of a rare vertex in $L(G)$. All the discussion that follows focuses on the left side $L(G)$.

That a vertex v is rare means that $|\mathcal{A}_v(G)|$, the number of maximal stable sets containing v, is at most $\frac{1}{2}|\mathcal{A}(G)|$. Thus, if for the average

$$\sum_{v \in L(G)} \frac{|\mathcal{A}_v(G)|}{|\mathcal{A}(G)|} \leq \frac{1}{2}|L(G)|$$

then $L(G)$ contains a rare vertex. Double-counting shows that the above average is equal to

$$\text{left-avg}(G) := \sum_{A \in \mathcal{A}(G)} \frac{|A \cap L(G)|}{|\mathcal{A}(G)|},$$

and thus our aim is to show that when $m + n$ is very large, it follows with high probability that left-avg$(G) \leq \frac{m}{2}$ for any $G \in \mathcal{B}(m, n; p)$.

Unfortunately, we do not reach this aim. While we show for large parts of the parameter space (m, n) that the average is, with high probability, small enough, we also see that when n is roughly $q^{-\frac{m}{2}}$ the average becomes very close to $\frac{m}{2}$, so close that our tools are not sharp enough to separate the average from slightly above $\frac{m}{2}$. Therefore, we provide for a bit more space by settling on bounding the average away from $(\frac{1}{2} + \delta)m$ for any positive δ, which then only allows us to deduce the existence of a vertex $v \in L(G)$ that is *almost rare*, in the sense that v lies in at most $(\frac{1}{2} + \delta)|\mathcal{A}(G)|$ maximal stable sets.

Much of the previous discussion is subsumed in the following lemma.

Lemma 2.1. *Let G be a bipartite graph, and let $\delta \geq 0$. If*

$$\text{left-avg}(G) \leq \left(\tfrac{1}{2} + \delta\right) |L(G)|$$

then there exists a vertex in $L(G)$ that lies in at most $\left(\frac{1}{2} + \delta\right) |\mathcal{A}(G)|$ maximal stable sets.

The following result is the heart of our main result, Theorem 1.2:

Theorem 2.2. *For all $\delta > 0$ and all $\varepsilon > 0$ there is an integer N so that for $G \in \mathcal{B}(m, n; p)$*

$$\Pr\left[\text{left-avg}(G) \leq \left(\tfrac{1}{2} + \delta\right) m\right] \geq 1 - \varepsilon$$

for all m, n with $m + n \geq N$ and $n \geq \max\{20, (\lceil 3 \log_{1/q}(2) \rceil + 2)^2\} + 1$.

In order to show how Theorem 1.2 follows from Theorem 2.2, we need to deal with the special case when one side is of constant size while the other becomes ever larger. Indeed, in this case averaging might fail—for a trivial reason. If we fix a constant right side $R(G)$, while $L(G)$ becomes ever larger, then $L(G)$ contains many isolated vertices. Since the isolated vertices lie in every maximal stable set they may push up left-avg(G) to above $\frac{m}{2}$.

However, isolated vertices are never a threat to Frankl's conjecture: A bipartite graph satisfies the union-closed sets conjecture if and only if it satisfies the conjecture with all isolated vertices deleted. More generally, it turns out that the special case of a constant right side is easily taken care of:

Lemma 2.3. *Let c be a positive integer, and let $\varepsilon > 0$. Then there is an N so that for $G \in \mathcal{B}(m, n; p)$*

$$\Pr\left[L(G) \text{ contains a rare vertex}\right] \geq 1 - \varepsilon,$$

for all m, n with $m \geq N$ and $n \leq c$.

Proof of Theorem 1.2. Let $\delta > 0$ be given. By symmetry, it is enough to show that the left side $L(G)$ of almost every random bipartite graph G in $\mathcal{B}(m, n; p)$ contains a vertex that lies in at most $(\frac{1}{2} + \delta)|\mathcal{A}|$ maximal stable sets. For this, consider a $\varepsilon > 0$, and let N be the maximum of the N given by Theorem 2.2 and Lemma 2.3 with $c = \max\{20, (\lceil 3 \log_{1/q}(2) \rceil + 2)^2\}$. Consider a pair m, n of positive integers with $m + n \geq N$. If $n \leq \max\{20, (\lceil 3 \log_{1/q}(2) \rceil + 2)^2\}$ then Lemma 2.3 yields a rare vertex in $L(G)$ with probability at least $1 - \varepsilon$. If, on the other hand, $n \geq \max\{20, (\lceil 3 \log_{1/q}(2) \rceil + 2)^2\} + 1$, Theorem 2.2 becomes applicable, which is to say that with probability at least $1 - \varepsilon$ we have left-avg$(G) \leq \left(\frac{1}{2} + \delta\right) m$. Now, Lemma 2.1 yields the desired vertex in $L(G)$. □

3 Sketch of the proof of Theorem 2.2

In order to prove Theorem 2.2, we distinguish several cases, depending on the relative sizes, m and n, of the two sides of the random bipartite graph G. In each of the cases we need a different method.

Our general strategy follows the observation that if there are many more maximal stable sets with small left side than with large left side, then the average over the left sides is small, too:

Lemma 3.1. *Let $\nu > 0$ and $\delta \geq 0$, and let G be a bipartite graph with $|L(G)| = m$. Let $\mathcal{L}$ be the maximal stable sets A of G with $|A \cap L(G)| \geq (\frac{1}{2} + \delta)m$, and let $\mathcal{S}$ be those maximal stable sets B with $|B \cap L(G)| \leq (1 - \nu)\frac{m}{2}$. If $|\mathcal{S}| \geq \frac{1}{\nu}|\mathcal{L}|$ then*

$$\text{left-avg}(G) \leq \left(\tfrac{1}{2} + \delta\right) m.$$

We bound the number of maximal stable sets with large left side, usually counted by a random variable $\mathcal{L}_G$, and at the same time we show that there are many maximal stable sets with a small left side; those we count with $\mathcal{S}_G$.

Up to $n < q^{-\frac{m}{2}}$ we are able to use the same bound for the number $\mathcal{L}_G$ of maximal stable sets whose left sides are of size at least $\frac{m}{2}$: We prove that with high probability $\mathcal{L}_G$ is bounded by a polynomial in n. For right sides that are much larger than the left side, i.e. $m \gg n$, we even extend such a bound to maximal stable sets with left side $\geq \frac{m}{3}$.

For the maximal stable sets with small left side, counted by $\mathcal{S}_G$, we need to distinguish several cases. When the left side of the graph is much larger than the right side, namely $m \geq q^{-\sqrt[5]{n}}$, we find with high probability a large induced matching in G. This in turn implies that the total number of maximal stable sets is high, and thus clearly also the number of those with small left side.

When the sides of the graph do not differ too much in size, $m \leq q^{-\sqrt[5]{n}}$ and $n \leq q^{-\sqrt[5]{m}}$, the variance of the number of maximal stable sets with small left side is moderate enough to apply Chebychev's inequality. Since the expectation of $\mathcal{S}_G$ is high, we again can use Lemma 3.1 to deduce Theorem 2.2.

However, when the left side of the graph becomes much larger than the right side, we cannot control the variance of $\mathcal{S}_G$ anymore. Instead, for $q^{-\sqrt[5]{m}} \leq n \leq q^{-\frac{m}{16}}$, we cut the right side into many pieces each of large size and apply Hoeffding's inequality to each of the pieces together with the left side. The inequality ensures that we find on at least one of the pieces a large number of maximal stable sets of small left side. Surpassing $n \geq q^{-\frac{m}{16}}$, we have to refine our estimations but we can still use this strategy up to slightly below $n = q^{-\frac{m}{2}}$.

In the interval $q^{-\frac{m}{2}} \leq n \leq q^{-m^3}$, we encounter a serious obstacle. There, we have to cope with an average that is very close to $\frac{m}{2}$. It is precisely for this reason that, overall, we only prove that left-avg$(G) \leq$

$\left(\frac{1}{2}+\delta\right)m$ instead of left-avg$(G) \leq \frac{m}{2}$. To keep below the slightly higher average, we only need to bound the number of maximal stable sets with left side $> (\frac{1}{2}+\delta)m$. This number we almost trivally bound by $2^{\lambda m}$, with some $\lambda < 1$. On the other hand, we see that the number $\mathcal{S}_G$ of maximal stable sets of small left side is $2^{\lambda' m}$ with a λ' as close to 1 as we want.

In the remaining case, we are dealing with an enormous right side: $n \geq q^{-m^3}$. Then, it is easy to see that with high probability there is an induced matching that covers all of the left side, which implies that every subset of $L(G)$ is the left side of a maximal stable set. This immediately gives us left-avg$(G) = \frac{m}{2}$.

References

[1] I. BALLA, B. BOLLOBÀS and T. ECCLES, *Union-closed families of sets*, J. Combin. Theory (Series A) **120** (2013), 531–544.

[2] H. BRUHN, P. CHARBIT, O. SCHAUDT and J. A. TELLE, *The graph formulation of the union-closed sets conjecture*, preprint, 2012.

[3] H. BRUHN and O. SCHAUDT, *The journey of the union-closed sets conjecture*, preprint.

[4] H. BRUHN and O. SCHAUDT, *The union-closed sets conjecture almost holds for almost all random bipartite graphs*, arXiv:1302.7141 [math.CO], 2013.

[5] G. CZÉDLI, M. MARÓTI and E. T. SCHMIDT, *On the scope of averaging for Frankl's conjecture*, Order **26** (2009), 31–48.

[6] P. FRANKL, "Handbook of Combinatorics", Vol. 2, R. L. Graham, M. Grötschel, and L. Lovász (eds.), MIT Press, Cambridge, MA, USA, 1995, 1293–1329.

[7] D. REIMER, *An average set size theorem*, Comb., Probab. Comput. (2003), 89–93.

[8] I. RIVAL (ed.), "Graphs and Order", NATO ASI Series, Vol. 147, Springer Netherlands, 1985.

The robust component structure of dense regular graphs

Daniela Kühn[1], Allan Lo[1], Deryk Osthus[1] and Katherine Staden[1]

Abstract. We study the large-scale structure of dense regular graphs. This involves the notion of robust expansion, a recent concept which has already been used successfully to settle several longstanding problems. Roughly speaking, a graph is robustly expanding if it still expands after the deletion of a small fraction of its vertices and edges. Our main result states that every dense regular graph can be partitioned into 'robust components', each of which is a robust expander or a bipartite robust expander. We apply our result to obtain the following.

(i) We prove that whenever $\varepsilon > 0$, every sufficiently large 3-connected D-regular graph on n vertices with $D \geq (1/4 + \varepsilon)n$ is Hamiltonian. This asymptotically confirms the only remaining case of a conjecture raised independently by Bollobás and Häggkvist in the 1970s.
(ii) We prove an asymptotically best possible result on the circumference of dense regular graphs of given connectivity. The 2-connected case of this was conjectured by Bondy and already proved by Wei.

1 Introduction

Our main result states that any dense regular graph G is the vertex-disjoint union of boundedly many 'robust components'. Each such component has a strong expansion property that is highly 'resilient' and almost all edges of G lie inside these robust components. In other words, the result implies that the large scale structure of dense regular graphs is remarkably simple. This can be applied *e.g.* to Hamiltonicity problems in dense regular graphs. Note that the structural information obtained in this way is quite different from that given by Szemerédi's regularity lemma.

The crucial notion in our partition is that of robust expansion. This is a structural property which has close connections to Hamiltonicity. Given a graph G on n vertices, $S \subseteq V(G)$ and $0 < \nu \leq \tau < 1$, we define the *ν-robust neighbourhood* $RN_{\nu,G}(S)$ of S to be the set of all those vertices of G with at least νn neighbours in S. We say G is a *robust* (ν, τ)-*expander* if, for every $S \subseteq V(G)$ with $\tau n \leq |S| \leq (1 - \tau)n$, we have that $|RN_{\nu,G}(S)| \geq |S| + \nu n$.

[1] School of Mathematics, University of Birmingham, Birmingham B15 2TT, United Kingdom. Email: d.kuhn@bham.ac.uk, s.a.lo@bham.ac.uk, d.osthus@bham.ac.uk, kls103@bham.ac.uk

There is an analogous notion of robust outexpansion for digraphs. Kühn, Osthus and Treglown showed in [13] that every sufficiently large dense robust outexpander contains a directed Hamilton cycle. The notion has been crucial in several other recent papers.

Let G be a bipartite graph with vertex classes A and B. Then clearly G is not a robust expander. However, we can obtain a bipartite analogue of robust expansion by only considering sets $S \subseteq A$ with $\tau|A| \leq |S| \leq (1-\tau)|A|$. This notion extends in a natural way to graphs which are 'close to bipartite'.

Our main result implies that for fixed $r \in \mathbb{N}$, $\varepsilon > 0$ and n sufficiently large, any D-regular graph on n vertices with $D \geq (\frac{1}{r+1} + \varepsilon)n$ has a partition into at most r (bipartite) robust expander components, so that the number of edges between these is very small. We now give a more formal statement of this.

Definition 1.1. *(Robust partitions)* Let $n, D, k, \ell \in \mathbb{N}$ and let $0 < \rho \leq \nu \leq \tau < 1$. Suppose that G is a D-regular graph on n vertices. We say that $\mathcal{V}$ is a *robust partition of G with parameters* ρ, ν, τ, k, ℓ if the following hold:

- $\mathcal{V} = \{V_1, \dots, V_k, W_1, \dots, W_\ell\}$ is a partition of $V(G)$;
- for all $X \in \mathcal{V}$ we have that $|X| \geq D - \rho n$ and $e(X, V(G) \setminus X) \leq \rho n^2$;
- for all $1 \leq j \leq \ell$, W_j has bipartition A_j, B_j such that $G[W_j]$ can be made into a balanced bipartite graph with respect to this bipartition by removing at most ρn vertices and at most ρn^2 edges;
- for all $1 \leq i \leq k$, $G[V_i]$ is a robust (ν, τ)-expander and for all $1 \leq j \leq \ell$, $G[W_j]$ is a bipartite robust (ν, τ)-expander with vertex classes A_j, B_j;
- for all $X, X' \in \mathcal{V}$ and all $x \in X$ we have $d_X(x) \geq d_{X'}(x)$. In particular, $d_X(x) \geq D/(k+\ell)$;
- for all $1 \leq j \leq \ell$ and for all $u \in A_j$ we have $d_{B_j}(u) \geq d_{A_j}(u)$, and analogously for $v \in B_j$;
- for all $X \in \mathcal{V}$ and all but at most ρn vertices $x \in X$ we have $d_X(x) \geq D - \rho n$;
- $k + 2\ell \leq \lfloor (1+\rho^{1/3})n/D \rfloor$.

Note that the last property implies there are only a small number of possible choices for k and ℓ when D is large. Our main result is the following.

Theorem 1.2 ([12]). *For all $\alpha, \tau > 0$ and every non-decreasing function $f : (0,1) \to (0,1)$, there exists $n_0 \in \mathbb{N}$ such that the following holds. For all D-regular graphs G on $n \geq n_0$ vertices where $D \geq \alpha n$, there exist $k, \ell \in \mathbb{N}$ and ρ, ν with $1/n_0 \leq \rho \leq \nu \leq \tau$; $\rho \leq f(\nu)$ and $1/n_0 \leq f(\rho)$ such that G has a robust partition $\mathcal{V}$ with parameters ρ, ν, τ, k, ℓ.*

In the special case of dense vertex-transitive graphs (which are always regular) a related partition result was obtained by Christofides, Hladký and Máthé [6], who resolved the dense case of a question of Lovász [16] on Hamilton paths (and cycles) in vertex-transitive graphs. It would be interesting to obtain such robust partition results for further classes of graphs. In particular, it might be possible to generalise Theorem 1.2 to sparser graphs.

2 Two applications to longest cycles in regular graphs

Consider the classical result of Dirac that every graph on $n \geq 3$ vertices with minimum degree at least $n/2$ contains a Hamilton cycle. Suppose we wish to strengthen this by reducing the degree threshold at the expense of introducing some other condition(s). The two extremal examples for Dirac's theorem (*i.e.* the disjoint union of two cliques and the almost balanced complete bipartite graph) make it natural to consider regular graphs with some connectivity property, see *e.g.* the recent survey of Li [14] and handbook article of Bondy [3].

In particular, Szekeres (see [9]) asked for which D every 2-connected D-regular graph G on n vertices is Hamiltonian. Jackson [9] showed that $D \geq n/3$ suffices. This improved earlier results of Nash-Williams [17], Erdős and Hobbs [7] and Bollobás and Hobbs [2]. Hilbig [8] improved the degree condition to $n/3 - 1$, unless G is the Petersen graph or another exceptional graph. Bollobás [1] as well as Häggkvist (see [9]) independently made the natural and far more general conjecture that any t-connected regular graph on n vertices with degree at least $n/(t+1)$ is Hamiltonian. However, the following counterexample (see Figure 2.1), due to Jung [11] and independently Jackson, Li and Zhu [10], disproves this conjecture for $t > 3$.

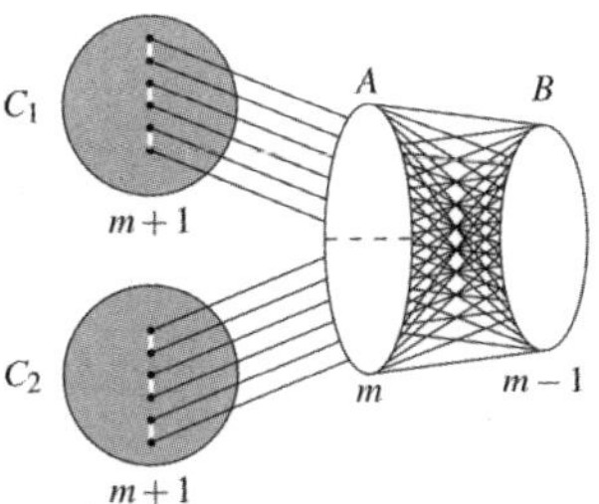

Figure 2.1. An extremal example for Conjecture 2.1.

For m divisible by four, construct G as follows. Let C_1, C_2 be two disjoint copies of K_{m+1} and let A, B be two disjoint independent sets of orders $m, m-1$ respectively. Add every edge between A and B. Add a set of $m/2$ independent edges from each of C_1 and C_2 to A so that together these edges form a matching of size m. Delete $m/4$ independent edges in each of C_1, C_2 so that G is m-regular. Then G has $4m + 1$ vertices and is $m/2$-connected. However G is not Hamiltonian since $G \setminus A$ has $|A| + 1$ components (in other words, G is not 1-tough). Note that G has a robust partition $C_1, C_2, A \cup B$ (so $(k, \ell) = (2, 1)$ in the definition of robust partitions in Section 1).

Jackson, Li and Zhu believe that the conjecture of Bollobás and Häggkvist is true in the remaining open case when $t = 3$.

Conjecture 2.1 ([10]). Let G be a 3-connected D-regular graph on $n \geq 13$ vertices such that $D \geq n/4$. Then G contains a Hamilton cycle.

The above example shows that the degree condition in Conjecture 2.1 cannot be reduced. One can also show that none of the other conditions can be relaxed.

There have been several partial results towards this conjecture. For instance, Li and Zhu [15] proved it in the case when $D \geq 7n/22$and Broersma, van den Heuvel, Jackson and Veldman [5] proved it for $D \geq 2(n + 7)/7$. In [10] it is proved that, if G satisfies the conditions of the conjecture, any longest cycle in G is dominating provided that n is not too small. (Here, a subgraph H of a graph G is *dominating* if $G \setminus V(H)$ is an independent set.) By considering robust partitions, we are able to prove an approximate version of the conjecture. We hope that our methods can be used to obtain the exact bound $D \geq n/4$ for large n. This is work in progress.

Theorem 2.2 ([12]). *For all $\varepsilon > 0$, there exists $n_0 \in \mathbb{N}$ such that every 3-connected D-regular graph on $n \geq n_0$ vertices with $D \geq (1/4 + \varepsilon)n$ is Hamiltonian.*

More generally, one can consider the circumference of dense regular graphs of given connectivity. Bondy [4] conjectured that, for $r \geq 3$, every sufficiently large 2-connected D-regular graph G on n vertices with $D \geq n/r$ has circumference $c(G) \geq 2n/(r-1)$. (Here the circumference $c(G)$ of G is the length of the longest cycle in G.) This was confirmed by Wei [18], who proved the conjecture for all n and in fact showed that $c(G) \geq 2n/(r - 1) + 2(r - 3)/(r - 1)$, which is best possible. We are able to extend this (asymptotically) to t-connected dense regular graphs.

Theorem 2.3. *Let $t, r \in \mathbb{N}$. For all $\varepsilon > 0$ there exists $n_0 \in \mathbb{N}$ such that the following holds. Whenever G is a t-connected D-regular graph on $n \geq n_0$ vertices where $D \geq (1/r + \varepsilon)n$, the circumference of G is at least $\min\{t/(r - 1), 1 - \varepsilon\}n$.*

This is asymptotically best possible. Moreover, as discussed above, the extremal example in Figure 2.1 shows that in general $\min\{t/(r - 1), 1 -$

$\varepsilon\}n$ cannot be replaced by $\min\{t/(r-1), 1\}n$. Theorem 2.3 shows that the conjecture of Bollobás and Häggkvist is in fact close to being true after all – any t-connected regular graph with degree slightly higher than $n/(t+1)$ contains an almost spanning cycle.

We are also confident that our robust partition result (Theorem 1.2) will have applications to other problems.

3 Sketch of the proof of Theorems 1.2 and 2.2

3.1 Sketch proof of Theorem 1.2

The basic proof strategy is to successively refine an appropriate partition of G. So let G be a D-regular graph on n vertices, where D is linear in n. Suppose that G is not a (bipartite) robust expander. Then $V(G)$ contains a set S such that N is not much larger than S, where $N := RN_{\nu,G}(S)$ for appropriate ν. Consider a minimal S with this property. Since G is regular, N cannot be significantly smaller than S. One can use this to show that there are very few edges between $S \cup N$ and $X := V(G) \setminus (S \cup N)$. Moreover, one can show that S and N are either almost identical or almost disjoint. In the former case, $G[S \cup N]$ is a robust expander and in the latter $G[S \cup N]$ is close to a bipartite robust expander. So in both cases, $S \cup N$ is a (bipartite) robust expander component. Similarly, if X is non-empty, it is either a (bipartite) robust expander component or we can partition it further along the above lines. In this way, we eventually arrive at the desired partition.

3.2 Sketch proof of Theorem 2.2

Let $\varepsilon > 0$ and let G be a 3-connected D-regular graph on n vertices, where $D \geq (1/4+\varepsilon)n$. Theorem 1.2 gives us a robust partition $\mathcal{V}$ of G containing exactly k robust expander components and ℓ bipartite robust expander components where $k + 2\ell \leq 3$, so there are only five possible choices of (k, ℓ). Assume for simplicity that $\mathcal{V}$ consists of three robust expander components G_1, G_2, G_3. So $(k, \ell) = (3, 0)$. (The cases when $\ell > 0$ are harder.) The result of [13] mentioned in Section 1 implies that G_i contains a Hamilton cycle for $i = 1, 2, 3$. In fact, it can be used to show that G_i is Hamilton p-linked for each bounded p. (Here a graph G is *Hamilton p-linked* if, whenever $x_1, y_1, \ldots, x_p, y_p$ are distinct vertices, there exist vertex-disjoint paths $P_1, \ldots, P_p$ such that P_j connects x_j to y_j, and such that together these paths cover all vertices of G.) This means that the problem of finding a Hamilton cycle in G can be reduced to finding only a suitable set of *external edges*, where an edge is external if it has endpoints in different G_i. We use the assumption of 3-connectivity to find these external edges.

References

[1] B. Bollobás, "Extremal Graph Theory", 167, Academic Press, 1978.

[2] B. Bollobás and A. M. Hobbs, *Hamiltonian cycles in regular graphs*, Advances in Graph Theory **3** (1978), 43–48.

[3] J. A. Bondy, *Basic graph theory: paths and circuits*, Handbook of Combinatorics, Vol. 1 (1995), 3–110.

[4] J. A. Bondy, *Hamilton cycles in graphs and digraphs*, Congress Numerantium **21** (1978), 3–28.

[5] H. J. Broersma, J. van den Heuvel, B. Jackson and H. J. Veldman, *Hamiltonicity of regular 2-connected graphs*, J. Graph Theory **22** (1996), 105–124.

[6] D. Christofides, J. Hladký and A. Máthé, *Hamilton cycles in dense vertex-transitive graphs*, preprint.

[7] P. Erdős and A. M. Hobbs, *A class of Hamiltonian regular graphs*, J. Combin. Theory B **37** (1978), 129–135.

[8] F. Hilbig, "Kantenstrukturen in nichthamiltonischen Graphen", Ph.D. Thesis, Technical University Berlin 1986.

[9] B. Jackson, *Hamilton cycles in regular 2-connected graphs*, J. Combin. Theory B **29** (1980), 27–46.

[10] B. Jackson, H. Li and Y. Zhu, *Dominating cycles in regular 3-connected graphs*, Discrete Mathematics **102** (1991), 163–176.

[11] H. A. Jung, *Longest circuits in 3-connected graphs*, Finite and infinite sets, Vol I, II, Colloq. Math. Soc. János Bolyai **37** (1984), 403–438.

[12] D. Kühn, A. Lo, D. Osthus and K. Staden, *The robust component structure of dense regular graphs and applications*, preprint.

[13] D. Kühn, D. Osthus and A. Treglown, *Hamiltonian degree sequences in digraphs*, J. Combin. Theory B **100** (2010), 367–380.

[14] H. Li, *Generalizations of Dirac's theorem in Hamiltonian graph theory – a survey*, Discrete Mathematics (2012), in press.

[15] H. Li and Y. Zhu, *Hamilton cycles in regular 3-connected graphs*, Discrete Mathematics **110** (1992), 229–249.

[16] L. Lovász, *Problem 11*, In: "Combinatorial Structures and their Applications", Gordon and Breach Science Publishers, 1970.

[17] C. St. J. A. Nash-Williams, *Hamiltonian arcs and circuits*, Recent trends in Graph Theory **186** (1971), 197–210.

[18] B. Wei, *On the circumferences of regular 2-connected graphs*, J. Combin. Theory B **75** (1999), 88–99.

The (Δ, D) and (Δ, N) problems in double-step digraphs with unilateral diameter

Cristina Dalfó[1] and Miquel Àngel Fiol[2]

1 Preliminaries

We study the (Δ, D) and (Δ, N) problems for double-step digraphs considering the unilateral distance, which is the minimum between the distance in the digraph and the distance in its converse digraph, obtained by changing the directions of all the arcs.

The first problem consists of maximizing the number of vertices N of a digraph, given the maximum degree Δ and the unilateral diameter D^*, whereas the second one consists of minimizing the unilateral diameter given the maximum degree and the number of vertices. We solve the first problem for every value of the unilateral diameter and the second one for some infinitely many values of the number of vertices.

Miller and Sirán [4] wrote a comprehensive survey about (Δ, D) and (Δ, N) problems. In particular, for the double-step graphs considering the standard diameter, the first problem was solved by Fiol, Yebra, Alegre and Valero [3], whereas Bermond, Iliades and Peyrat [2], and also Beivide, Herrada, Balcázar and Arruabarrena [1] solved the (Δ, N) problem. In the case of the double-step digraphs, also with the standard diameter, Morillo, Fiol and Fàbrega [5] solved the (Δ, D) problem and provided some infinite families of digraphs which solve the (Δ, N) problem for their corresponding numbers of vertices.

[1] Departament de Matemàtica Aplicada IV. Universitat Politècnica de Catalunya, BarcelonaTech. Email: cdalfo@ma4.upc.edu

[2] Departament de Matemàtica Aplicada IV. Universitat Politècnica de Catalunya, BarcelonaTech. Email: fiol@ma4.upc.edu

Research supported by the Ministerio de Educación y Ciencia, Spain, and the European Regional Development Fund under project MTM2011-28800-C02-01 and by the Catalan Research Council under project 2009SGR1387.

1.1 Double-step digraphs

A *double-step digraph* $G(N; a, b)$ has set of vertices $\mathbb{Z}_N = \mathbb{Z}/N\mathbb{Z}$ and arcs from every vertex i to vertices $i + a \mod N$ and $i + b \mod N$, for $0 \leq i \leq N - 1$, where a, b are some integers called *steps* such that $1 \leq a < b \leq N - 1$. Because of the automorphisms $i \mapsto i + \alpha$ for $1 \leq \alpha \leq N - 1$, the double-step digraphs are vertex-transitive. Moreover, they are strongly connected if and only if $\gcd(N, a, b) = 1$. It is known that the maximum order N of a double-step digraph with diameter k is upper bounded by the Moore-like bound $N \leq M_{DSD}(2, k) = \binom{k+2}{2}$, where the equality would hold if all the numbers $ma + nb$ were different modulo N, with $m, n \geq 0$ and $m + n \leq k$. In fact, this bound cannot be attained for $k > 1$.

Every double-step digraph has an L-shaped form associated, which tessellates the plane. If one of the steps, say a, equals 1, we can choose an L-shaped tile with dimensions $\ell = b$, h being the quotient obtained dividing N by ℓ, $w = \ell - s$ with s being the remainder of such a division, and $y = 1$. Then, $N = \ell h + s$ with $0 \leq s < \ell$ (see Figure 1.1).

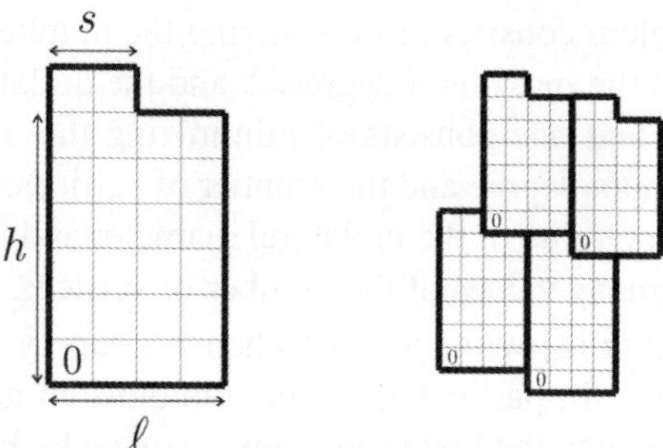

Figure 1.1. An L-shaped form and its tessellation.

1.2 Unilateral distance

Given a digraph $G = (V, A)$, the *unilateral distance* between two vertices $u, v \in V$ is defined as

$$\text{dist}^*_G(u,v) = \min\{\text{dist}_G(u,v), \text{dist}_G(v,u)\} = \min\{\text{dist}_G(u,v), \text{dist}_{\overline{G}}(u,v)\},$$

where dist_G is the standard distance in digraph G and $\text{dist}_{\overline{G}}$ is the distance in its *converse* digraph $\overline{G}$, that is, the digraph obtained by changing the directions of all the arcs of G. From this concept, we can define the *unilateral eccentricity* ecc* from vertex u, the *unilateral radius* r^* of G, and the *unilateral diameter* D^* of G as follows:

$$\text{ecc}^*(u) = \max_{v \in V}\{\text{dist}^*_G(u,v)\}, \quad r^* = \min_{u \in V}\{\text{ecc}^*(u)\}, \quad \text{and } D^* = \max_{u \in V}\{\text{ecc}^*(u)\}.$$

As an example, if we have $G = C_N$, the directed cycle on N vertices, then $D^* = \lfloor N/2 \rfloor$.

Note that, obviously, these parameters have as lower bounds the ones corresponding to the underlying graph, obtained from digraph G by changing the arcs for edges without direction.

2 The unilateral diameter of double-step digraphs with step *a*=1

In this section we study the unilateral diameter of the double-step digraphs with $a = 1$ having 'small' b. Although we have not been able to prove that the optimal results can be obtained always by taking such values of the steps, computational experiments seem to support this claim. In fact, as we see in the next section, this approach allows us to solve the (Δ, D^*) problem for every value of D^*, and also to solve the $(\Delta, N)^*$ problem for infinitely many values of N.

As we have already seen, a double-step digraph $G(N; 1, b)$ with $N = \ell h + s$ and $0 \leq s < \ell$, can be described by an L-shaped form with dimensions $\ell = b$, $h = \lfloor N/\ell \rfloor$, $y = 1$, and $w = \ell - s$. See again Figure 1.1. In this context we have the following result for the unilateral diameter D^*.

Proposition 2.1. *For $N = \ell h + s$, where $1 < \ell \leq \lceil \sqrt{N} \rceil$ and $0 \leq s \leq \ell - 1$, a double-step digraph $G(N; a, b)$ with $a = 1$ and $b = \ell$ has unilateral diameter*

$$D^* = \begin{cases} \left\lfloor \dfrac{\ell + h + s - 1}{2} \right\rfloor & \text{if } 0 \leq s \leq \ell - 2, \\ \left\lfloor \dfrac{\ell + h - 1}{2} \right\rfloor & \text{if } s = \ell - 1. \end{cases} \tag{2.1}$$

If there is not restriction for the value of ℓ, then the values in Eq. (2.1) are an upper bound for the unilateral diameter. For instance, is for $N = 430$, the unilateral diameter is 22 and the upper bound given by Equation (2.1) is 23.

3 The (Δ, D^*) and $(\Delta, N)^*$ problems for double-step digraphs with unilateral diameter

3.1 The (Δ, D^*) problem

In our context, the (Δ, D^*) problem consists of finding the double-step digraph $G(N; a, b)$ with maximum number of vertices given a unilateral diameter D^* and the maximum degree $\Delta = 2$, that is, to find the steps that maximize the number of vertices for such a unilateral diameter. To get a Moore-like bound (see Miller and Sirán [4]), notice that at distance $k = 1, 2, \ldots, D^*$ from vertex 0 there are at most $2(k+1)$ vertices ($k+1$ of them going forward and the other $k+1$ going backwards). Then, this gives

$$N \leq M(2, D^*) = 2(1+2+\cdots+D^*+1)-1 = (D^*)^2+3D^*+1. \quad (3.1)$$

Moreover, if the maximum is attained, we get an 'optimal' X-shaped tile which tessellates the plane, and this allows us to solve the (Δ, D^*) problem, as shown in the following result.

Proposition 3.1. *For each integer value $k \geq 0$, the double-step digraph $G(N; 1, b)$, with $N = M(2, k) = k^2+3k+1$ and $b = k+1$ has unilateral diameter $D^* = k$.*

3.2 The $(\Delta, N)^*$ problem

In our context, the $(\Delta, N)^*$ problem consists of finding the minimum unilateral diameter D^* in double-step digraphs given a number of vertices N and their maximum degree $\Delta = 2$, that is, to find the steps that minimize the unilateral diameter for such a number of vertices. We begin with the following general upper bound for the unilateral diameter.

Proposition 3.2. *Given any number of vertices $N \geq 5$, there exists a double-step digraph with unilateral diameter D^* satisfying*

$$D^* \leq \left\lceil \sqrt{2(N+2)} \right\rceil - 2.$$

To solve the $(\Delta, N)^*$ problem for double-step digraphs with minimum unilateral diameter we consider the case $s = \ell - 1$ of Proposition 2.1. Moreover, to keep track of the excluded vertices from the maximum $M(2, k)$, we define r as the subindex of the triangular number $T_r = 1+2+\cdots+r = \binom{r+1}{2}$.

Proposition 3.3.

(a) *If $0 \leq r < \frac{1}{2}(\sqrt{8k+9}-1)$, the double-step digraph $G(N; a, b)$, with number of vertices $N = k^2+3k+1-r(r+1)$ and steps $a = 1$ and $b = \ell = k-r+1$, has minimum unilateral diameter $D^* = k$.*

(b) *If $0 \leq r < \sqrt{k+1}$, the double-step digraph $G(N; a, b)$, with number of vertices $N = k^2+2k-r^2$ and steps $a = 1$ and $b = \ell = k-r+1$, has minimum unilateral diameter $D^* = k$.*

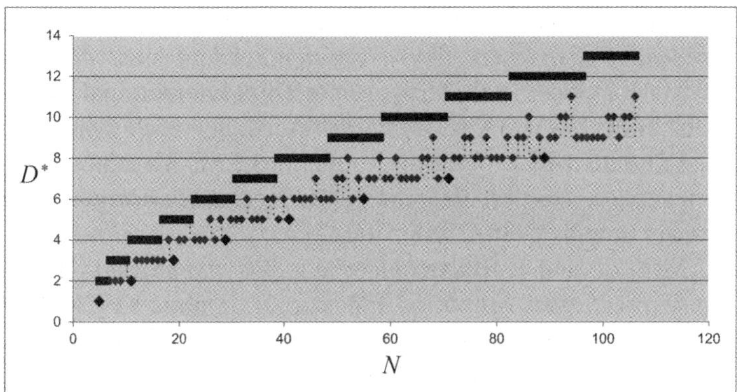

Figure 3.1. The minimum unilateral diameter D^* with respect to the number of vertices N, for $5 \leq N \leq 106$. (The largest points correspond to the (Δ, D^*) problem, and the thick lines to the upper bound given in Proposition 3.2.)

Table 1. Some results of the (Δ, D^*) and $(\Delta, N)^*$ problems solved with Proposition 3.3.

Problem	$\ell + h$	r	ℓ	h	$N = \ell h + \ell - 1$	D^*
(Δ, D^*)	even	0	$k+1$	$k+2$	k^2+3k+1	k
$(\Delta, N)^*$	even	1	k	$k+3$	k^2+3k-1	k
$(\Delta, N)^*$	even	2	$k-1$	$k+4$	k^2+3k-5	k
$(\Delta, N)^*$	even	3	$k-2$	$k+5$	$k^2+3k-11$	k
...	...	...	...	...	...	...
$(\Delta, N)^*$	odd	0	$k+1$	$k+1$	k^2+2k	k
$(\Delta, N)^*$	odd	1	k	$k+2$	k^2+2k-1	k
$(\Delta, N)^*$	odd	2	$k-1$	$k+3$	k^2+2k-4	k
$(\Delta, N)^*$	odd	3	$k-2$	$k+4$	k^2+2k-9	k
...	...	...	...	...	...	...

As shown in Figure 3.1, the unilateral diameter D^* does not increase monotonously with the number of vertices N.

Note that if we fix r for any k large enough, we get an infinite family of digraphs with minimum unilateral diameter for each number of vertices. See some examples of the cases of Proposition 3.3 in Table 1.

References

[1] R. BEIVIDE, E. HERRADA, J. L. BALCÁZAR and A. ARRUABARRENA, *Optimal distance networks of low degree for parallel computers*, IEEE T. Comput. **40** (1991), 1109–1124.

[2] J. C. Bermond, G. Iliades and C. Peyrat, *An optimization problem in distributed loop computer networks*, In: "Combinatorial Mathematics: Proceedings of the Third International Conference (New York, 1985)", Ann. New York Acad. Sci. **555** (1989), 45–55.

[3] M.A. Fiol, J.L.A. Yebra, I. Alegre, M. Valero, *A discrete optimization problem in local networks and data alignment*, IEEE Trans. Comput. **C-36** (1987), 702–713.

[4] M. Miller and J. Sirán, *Moore graphs and beyond: A survey of the degree/diameter problem*, Electron. J. Combin. **11** (2012), #R00.

[5] P. Morillo, M. A. Fiol and J. Fàbrega, *The diameter of directed graphs associated to plane tesselations*, Ars Combin. **20A** (1985), 17–27.

Critical groups of generalized de Bruijn and Kautz graphs and circulant matrices over finite fields

Swee Hong Chan[1], Henk D. L. Hollmann[1] and Dmitrii V. Pasechnik[1]

Abstract. We determine the critical groups of the generalized de Bruijn graphs $\mathrm{DB}(n, d)$ and generalized Kautz graphs $\mathrm{Kautz}(n, d)$, thus extending and completing earlier results for the classical de Bruijn and Kautz graphs. Moreover, for a prime p the critical groups of $\mathrm{DB}(n, p)$ are shown to be in close correspondence with groups of $n \times n$ circulant matrices over $\mathbb{F}_p$, which explains numerical data in [11] and suggests the possibility to construct normal bases in $\mathbb{F}_{p^n}$ from spanning trees in $\mathrm{DB}(n, p)$.

1 Introduction

The *critical group* of a directed graph G is an abelian group obtained from the Laplacian matrix Δ of G; it determines and is determined by the Smith Normal Form (SNF) of Δ. (For precise definitions of these and other terms, we refer to the next section.) The *sandpile group* $S(G, v)$ of G at a vertex v is an abelian group obtained from the reduced Laplacian Δ_v of G; its order is equal to the *complexity* $\kappa(G)$ of G, the number of directed trees rooted at v, a fact that is related to the Matrix Tree Theorem, see for example [8] and its references. If G is Eulerian, then $S(G, v)$ does not depend on v, and is then simply written as $S(G)$; in that case, it is equal to the critical group of G. The critical group has been studied in other contexts under several other names, such as group of components, Picard or Jacobian group, and Smith group. For more details and background, see, *e.g.*, [6].

Critical groups have been determined for a large number of graph families. For some examples, see the references in [1]. Here, we determine the critical group of the generalized de Bruijn graphs $\mathrm{DB}(n, d)$ and generalized Kautz graphs $\mathrm{Kautz}(n, d)$, thus extending and completing the results from [8] for the binary de Bruijn graphs $\mathrm{DB}(2^\ell, 2)$ and Kautz graphs

[1] School of Physical and Mathematical Sciences, Nanyang Technological University, 21 Nanyang Link, Singapore 637371.
Email: sweehong@ntu.edu.sg, henk.hollmann@ntu.edu.sg, dima@ntu.edu.sg

Kautz($(p-1)p^{\ell-1}, p$) (with p prime), and [3] for the classical de Bruijn graphs DB(d^ℓ, d) and Kautz graphs Kautz($(d-1)d^{\ell-1}, d$). Unlike the classical case, the generalized versions are not necessarily iterated line graphs, so to obtain their critical groups, different techniques have to be applied.

Our original motivation for studying these groups stems from their relations to some algebraic objects, such as the groups $C(n, p)$ of invertible $n \times n$-circulant matrices over $\mathbb{F}_p$ (mysterious numerical coincidences were noted in the OIES entry A027362 [11] by the third author, computed with the help of [12,15]), and *normal bases* (*cf. e.g.* [9]) of the finite fields $\mathbb{F}_{p^n}$. The latter were noted to be closely related to circulant matrices and to *necklaces* by Reutenauer [13, Sect. 7.6.2], see also [5], and the related numeric data collected in [2]. Here, we show that $C(n, p)/(\mathbb{Z}_{p-1} \times \mathbb{Z}_n)$ is isomorphic to the critical group of DB(n, p). Although we were not able to construct an explicit bijection between the former and the latter, we could speculate that potentially one might be able to design a new deterministic way to construct normal bases of $\mathbb{F}_{p^n}$.

2 Preliminaries

Let M be an $m \times n$ integer matrix of rank r. For a ring F, we write $R_F(M) = M^\top F^n$, the F-module generated by the rows of M. The *Smith group* [14] of M is defined as $\Gamma(M) = \mathbb{Z}^n/R_{\mathbb{Z}}(M)$. The submodule $\overline{\Gamma}(M) = \mathbb{Z}^n/R_{\mathbb{Q}}(M) \cap \mathbb{Z}^n$ of $\Gamma(M)$ is a finite abelian group called the *finite part* of $\Gamma(M)$. Indeed, if M has rank r, then $\Gamma(M) = \mathbb{Z}^{n-r} \oplus \overline{\Gamma}(M)$ with $\overline{\Gamma}(M) = \oplus_{i=1}^r \mathbb{Z}_{d_i}$, where $d_1, \ldots, d_r$ are the nonzero invariant factors of M, so that $d_i | d_{i+1}$ for $i = 1, \ldots, r-1$. For invariant factors and the Smith Normal Form, we refer to [10]. See [14] for further details and proofs.

Let $G = (V, E)$ be a directed graph on $n = |V|$ vertices. The indegree $d^-(v)$ and outdegree $d^+(v)$ is the number of edges ending or starting in $v \in V$, respectively. The *adjacency matrix* of G is the $n \times n$ matrix $A = (A_{v,w})$, with rows and columns indexed by V, where $A_{v,w}$ is the number of edges from v to w. The *Laplacian* of G is the matrix $\Delta = D - A$, where D is diagonal with $D_{v,v} = d_v^-$. The *critical group* $K(G)$ of G is the finite part of the Smith group of the Laplacian Δ of G. The *sandpile group* $S(G, v)$ of G at a $v \in V$ is the finite part of the Smith group of the $(n-1) \times (n-1)$ *reduced Laplacian* Δ_v, obtained from Δ by deleting the row and the column of Δ indexed by v. Note that by the Matrix Tree Theorem for directed graphs, the order of $S(G, v)$ equals the number of directed spanning trees rooted at v.

2.1 Generalized de Bruijn and Kautz graphs

Generalized de Bruijn graphs and generalized Kautz graphs [4] are known to have a relatively small diameter and attractive connectivity properties, and have been studied intensively due to their applications in interconnection networks. The generalized Kautz graphs were first investigated in [7], and are also known as *Imase-Itoh digraphs*. Both classes of graphs are Eulerian.

We will determine the critical group, or, equivalently, the sandpile group, of a generalized de Bruijn or Kautz graph on n vertices by embedding this group as a subgroup of index n in a group that we will refer to as the *sand dune* group of the corresponding digraph. Let us now turn to the details.

The *generalized de Bruijn graph* $\mathrm{DB}(n, d)$ has vertex set $\mathbb{Z}_n$, the set of integers modulo n, and (directed) edges $v \to dv + i$ for $i = 0, \dots, d-1$ and all $v \in \mathbb{Z}_n$. The *generalized Kautz graph* $\mathrm{Kautz}(n, d)$ has vertex set $\mathbb{Z}_n$ and directed edges $v \to -d(v+1) + i$ for $i = 0, \dots, d-1$ and all $v \in \mathbb{Z}_n$. Note that both $\mathrm{DB}(n, d)$ and $\mathrm{Kautz}(n, d)$ are Eulerian. In what follows, we will focus on the generalized de Bruijn graph; the generalized Kautz graph can be handled in a similar way, essentially by replacing d by $-d$ in certain places.

Let $\mathcal{Z}_n = \{a(x) \in \mathbb{Z}[x] \bmod x^n - 1 \mid a(1) = 0\}$. With each vertex $v \in \mathbb{Z}_n$, we associate the polynomial $f_v(x) = dx^v - x^{dv} \sum_{i=0}^{d-1} x^i \in \mathcal{Z}_n$. Since $f_v(x)$ is the associated polynomial of the vth row of the Laplacian $\Delta^{(n,d)}$ of the generalized de Bruijn graph $\mathrm{DB}(n, d)$, the Smith group $\Gamma(\Delta^{(n,d)})$ of the Laplacian of $\mathrm{DB}(n, d)$ is the quotient of $\mathbb{Z}[x] \bmod x^n - 1$ by the $\mathbb{Z}_n$-span $\langle f_v(x) \mid v \in \mathbb{Z}_n \rangle_{\mathbb{Z}_n}$ of the polynomials $f_v(x)$. Now note that $\mathbb{Z}[x] \bmod x^n - 1 \cong \mathbb{Z} \oplus \mathcal{Z}_n$, so since $\sum_{v \in \mathbb{Z}_n} f_v(x) = 0$, we have that

$$\begin{aligned} \Gamma(\Delta^{(n,d)}) &= (\mathbb{Z}[x] \bmod x^n - 1)/\langle f_v(x) \mid v \in \mathbb{Z}_n \rangle_{\mathbb{Z}_n} \\ &\cong \mathbb{Z} \oplus \mathcal{Z}_n/\langle f_v(x) \mid v \in \mathbb{Z}'_n \rangle_{\mathbb{Z}_n} \end{aligned} \tag{2.1}$$

where $\mathbb{Z}'_n = \mathbb{Z}_n \setminus \{0\}$. It is easily checked that the polynomials $f_v(x)$ with $v \in \mathbb{Z}'_n$ are independent over $\mathbb{Q}$, hence they constitute a basis for $\mathcal{Z}_n$ over $\mathbb{Q}$. As a consequence, each element in the quotient group

$$S(n, d) = S_{\mathrm{DB}}(n, d) = \mathcal{Z}_n/\langle f_v(x) \mid v \in \mathbb{Z}'_n \rangle_{\mathbb{Z}_n} \tag{2.2}$$

has finite order, and so $S(n, d)$ is the critical group, or, equivalently, the sandpile group, of the generalized de Bruijn graph $\mathrm{DB}(n, d)$. We define the *sand dune group* $\Sigma(n, d) = \Sigma_{\mathrm{DB}}(n, d)$ of $\mathrm{DB}(n, d)$ as $\Sigma(n, d) = \mathcal{Z}_n/\langle g_v(x) \mid v \in \mathbb{Z}'_n \rangle_{\mathbb{Z}_n}$, where $g_v(x) = (x-1)f_v(x) = dx^v(x-1) - x^{dv}(x^d - 1)$. Now let $e_v = x^v - 1$; we have that $e_0 = 0$, and $\mathcal{Z}_n =$

$\langle e_v \mid v \in \mathbb{Z}'_n \rangle_{\mathbb{Z}}$, the $\mathbb{Z}$-span of the polynomials e_v. Furthermore, let $\epsilon_v = de_v - e_{dv}$. The span in $\mathcal{Q}_n = \{a(x) \in \mathbb{Q}[x] \bmod x^n - 1 \mid a(1) = 0\}$ of the polynomials $g_v(x)$ with $v \in \mathbb{Z}_n$ is the set of polynomials of the form $dc(x) - c(x^d)$ with $c(1) = 0$; since $\epsilon_v = g_0(x) + \cdots g_{v-1}(x)$ for all $v \in \mathbb{Z}_n$, we conclude that

$$\Sigma(n, d) = \mathcal{Z}_n / \mathcal{E}_{n,d}, \tag{2.3}$$

where $\mathcal{Z}_n = \langle e_v \mid v \in \mathbb{Z}'_n \rangle_{\mathbb{Z}}$ and $\mathcal{E}_{n,d} = \langle \epsilon_v \mid v \in \mathbb{Z}'_n \rangle_{\mathbb{Z}}$ is the $\mathbb{Z}$-submodule of $\mathcal{Z}_n$ generated by the polynomials $\epsilon_v = de_v - e_{dv}$. The next result is crucial: it identifies the elements of the sand dune group $\Sigma(n, d)$ that are actually contained in the sandpile group $S(n, d)$. (Due to lack of space, we omit the not too difficult proofs in the remainder of this section.)

Theorem 2.1. *If $a \in \Sigma(n, d)$ with $a = \sum_v a_v e_v$, then $a \in S(n, d)$ if and only if $\sum_v v a_v \equiv 0 \bmod n$.*

Corollary 2.2. *We have $\Sigma(n, d)/S(n, d) = \mathbb{Z}_n$ and so $|\Sigma(n, d)| = n|S(n, d)|$.*

The above descriptions of the sandpile group $S(n, d)$ and sand dune group $\Sigma(n, d)$, and the embedding of $S(n, d)$ as a subgroup of $\Sigma(n, d)$ are very suitable for the determination of these groups. In the process, repeatedly information is required about the order of various group elements. The following two results provide that information.

Lemma 2.3. *Let $a = \sum_v a_v \epsilon_v \in \Sigma(n, d)$. Then the order of a in $\Sigma(n, d)$ is the smallest positive integer m for which $m a_v \in \mathbb{Z}$ for each v.*

We say that $v \in \mathbb{Z}_n$ has *d-type* (f, e) in $\mathbb{Z}_n$ if $v, dv, \ldots, d^{e+f-1}v$ are all distinct, with $d^{e+f}v = d^f v$. Now, by expressing e_v in terms of the ϵ_v, we can determine the order of e_v. The result is as follows.

Lemma 2.4. *Supposing v has d-type (f, e), then $e_v = \sum_{i=0}^{f-1} d^{-i-1} \epsilon_{d^i v} + \sum_{j=0}^{e-1} d^{j-f} (d^e - 1)^{-1} \epsilon_{d^{f+j} v}$ in $\mathcal{Z}_n$, and hence e_v has order $d^f (d^e - 1)$ in $\Sigma(n, d)$.*

2.2 Invertible circulant matrices

Let Q_n be the $n \times n$ permutation matrix over a field F corresponding to the cyclic permutation $(1, 2, \ldots, n)$. An $n \times n$ *circulant matrix* over F is a matrix that can be written as $a_1 Q_n + a_2 Q_n^2 + \ldots + a_n Q_n^n$ with $a_i \in F$ for $1 \leq i \leq n$. All the invertible circulant matrices form a commutative group (w.r.t. matrix multiplication), namely, the centralizer o… in $\mathrm{GL}_n(F)$. In the case $F = \mathbb{F}_p$ we consider here we denote this …tative group by $C(n, p)$. Note that $C(n, p)$ contains a subgroup …hic to $\mathbb{Z}_{p-1} \oplus \mathbb{Z}_n$, namely the direct product of the group of scalar … $F_p^* I := \{\lambda I \mid \lambda \in \mathbb{F}_p^*\}$ and the cyclic subgroup generated by … circulant matrix has all-ones vector $\mathbf{1} := (1, \ldots, 1)^\top$ as an …r. Thus $C'(n, p) := \{g \in C(n, p) \mid g\mathbf{1} = \mathbf{1}\}$ is a subgroup of C… and we have the following formula.

$$C(n, p) = C'(n, p) \times F_p^* I. \tag{2.4}$$

3 Main results

Let $n, d > 0$ be fixed integers. The description of the sandpile group $S(n, d)$ and the sand-dune group $\Sigma(n, d)$ of the generalized the Bruin graph $\mathrm{DB}(n, d)$ involves a sequence of numbers defined as follows. Put $n_0 = n$, and for $i = 1, 2, \ldots$, define $g_i = \gcd(n_i, d)$ and $n_{i+1} = n_i / g_i$. We have $n_0 > \cdots > n_k = n_{k+1}$, where k is the smallest integer for which $g_k = 1$. We will refer to the sequence $n_0 > \cdots > n_k = n_{k+1}$ as the *d-sequence* of n. In what follows, we will write $m = n_k$ and $g = g_0 \cdots g_{k-1}$. Note that $n = gm$ with $\gcd(m, d) = 1$.

Since $\gcd(m, d) = 1$, the map $x \to dx$ partitions $\mathbb{Z}_m$ into orbits of the form $O(v) = (v, dv, \ldots, d^{o(v)-1}v)$. We will refer to $o(v) = |O(v)|$ as the *order* of v.

For every prime $p|m$, we define $\pi_p(m)$ to be the largest power of p dividing m. Let V be a complete set of representatives of the orbits $O(v)$ different from $\{0\}$, where we ensure that for every divisor p of m, all integers of the form m/p^j are contained in V.

Theorem 3.1. *With the above definitions and notation, we have that*

$$\Sigma(n, d) = \left[\bigoplus_{i=0}^{k-1} \mathbb{Z}_{d^{i+1}}^{n_i - 2n_{i+1} + n_{i+2}}\right] \oplus \left[\bigoplus_{v \in V} \mathbb{Z}_{d^{o(v)} - 1}\right], \tag{3.1}$$

and

$$S(n, d) = \left[\bigoplus_{i=0}^{k-1} \mathbb{Z}_{d^{i+1}/g_i} \oplus \mathbb{Z}_{d^{i+1}}^{n_i - 2n_{i+1} + n_{i+2} - 1}\right] \oplus \left[\bigoplus_{v \in V} \mathbb{Z}_{(d^{o(v)} - 1)/c(v)}\right], \tag{3.2}$$

where $c(v) = 1$ except in the following cases. For any $p|m$,

$$c(m/\pi_p(m)) = \begin{cases} \pi_p(m), & \text{if } p \neq 2 \text{ or } d \equiv 1 \bmod 4 \text{ or } 4 \nmid m; \\ \pi_2(m)/2, & \text{if } p = 2 \text{ and } d \equiv 3 \bmod 4 \text{ and } 4|m, \end{cases}$$

and if $4|m$ and $d \equiv 3 \bmod 4$, then $c(m/2) = 2$.

For the generalized Kautz graph, a similar result holds. For $v \in$ [illegible] we let $O'(v)$ denote the orbit of v under the map $x \to -dv$, and w[illegible] $o'(v) = |O'(v)|$. Now take V' to be a complete set of represen[illegible] the orbits on $\mathbb{Z}'_m$. Finally, define $c'(v)$ similar to $c(v)$, except t[illegible] is replaced by $-d$ (so the special case now involves $d \equiv 1$ mo[illegible] we have the following.

Theorem 3.2. *The sandpile group* $S_{\text{Kautz}}(n, d)$ *of the generali*[illegible] *Kautz graph* Kautz(n, d) *is obtained from* $S(n, d)$ *by replacing* V *by* V', $o(v)$ *by* $o'(v)$*, and* $c(v)$ *by* $c'(v)$ *in (3.2).*

The above results can be proved in a number of steps. In what follows, we outline the method for the generalized de Bruijn graphs; for the generalized Kautz graphs, a similar approach can be used. Furthermore, we note that many of the steps below repeatedly use Theorem 2.1 and Lemma 2.4. First, we investigate the "multiplication-by-d" map $d : x \to dx$ on the sandpile and sand-dune group. Let $\Sigma_0(n, d)$ and $S_0(n, d)$ denote the kernel of the map d^k on $\Sigma(n, d)$ and $S(n, d)$, respectively. It is not difficult to see that $\Sigma(n, d) \cong \Sigma_0(n, d) \oplus \Sigma(m, d)$ and $S(n, d) \cong S_0(n, d) \oplus S(m, d)$. Then, we use the map d to determine $\Sigma_0(n, d)$ and $S_0(n, d)$. It is easy to see that for *any* n, we have $d\Sigma(n, d) \cong \Sigma(n/(n, d), d)$ and $dS(n, d) \cong S(n/(n, d), d)$. With much more effort, it can be show that the kernel of the map d on $\Sigma(n, d)$ and $S(n, d)$ is isomorphic to $\mathbb{Z}_d^{n-n/(n,d)}$ and $\mathbb{Z}_{d/(n,d)} \oplus \mathbb{Z}_d^{n-1-n/(n,d)}$, respectively. Then we use induction over the length $k + 1$ of the d-sequence of n to show that $\Sigma_0(n, d)$ and $S_0(n, d)$ have the form of the left part of the right hand side in (3.1) and (3.2), respectively. This part of the proof, although much more complicated, resembles the method used by [8] and [3].

Now it remains to handle the parts $\Sigma(m, d)$ and $S(m, d)$ with $\gcd(m, d) = 1$. For the "helper" group $\Sigma(m, d)$ that embeds $S(m, d)$, this is trivial: it is easily seen that $\Sigma(m, d) = \oplus_{v \in V} \langle e_v \rangle$, and the order of e_v is equal to the size $o(v)$ of its orbit $O(v)$ under the map d, so (3.1) follows immediately. The e_v are not contained in $S(m, d)$, but we can try to modify them slightly to obtain a similar decomposition for $S(m, d)$. The idea is to replace e_v by a modified version $\tilde{e}_v = e_v - \sum_{p|m} \lambda_p(v) e_{\pi_p(v)m/\pi_p(m)}$, where the numbers $\lambda_p(v)$ are chosen such that $\tilde{e}_v \in S(m, d)$, or by a suitable multiple of e_v, in some exceptional cases (these are cases where $c(v) > 1$). It turns out that this is indeed possible, and in this way the proof of Theorem 3.1 can be completed.

Finally, with the notation from Subsect. 2.2, we have the following isomorphisms, connecting critical groups and circulant matrices.

Theorem 3.3. *Let d be a prime. Then*

$$S(n,d) \cong C'(n,d)/\langle Q_n\rangle, \qquad \text{and} \qquad \Sigma(n,d) \cong C'(n,d).$$

The proof of Theorem 3.3 is by reducing to the case $\gcd(n, p) = 1$ by an explicit construction, and then by diagonalizing $C(n, p)$ over an appropriate extension of $\mathbb{F}_p$. Essentially, as soon as $\gcd(n, p) = 1$, one can read off a decomposition of $C(n, p)$ into cyclic factors from the irreducible factors of the polynomial $x^n - 1$ over $\mathbb{F}_p$.

References

[1] C. A. ALFARO and C. E. VALENCIA, *On the sandpile group of the cone of a graph*, Linear Algebra and its Applications **436** (5) (2012), 1154–1176.

[2] J. ARNDT, "Matters Computational: Ideas, Algorithms, Source Code", Springer, 2010, http://www.jjj.de/fxt/fxtbook.pdf.

[3] H. BIDKHORI and S. KISHORE, *A bijective proof of a theorem of Knuth*, Comb. Probab. Comput. **20** (1) (2010), 11–25.

[4] D.-Z. DU, F. CAO and D. F. HSU, *De Bruijn digraphs, Kautz digraphs, and their generalizations*, In: " Combinatorial Network Theory" D.-Z. Du and D. F. Hsu (eds.), Kluwer Academic, 1996, 65–105.

[5] S. DUZHIN and D. PASECHNIK, *Automorphisms of necklaces and sandpile groups*, 2013. http://arxiv.org/abs/1304.2563.

[6] A. E. HOLROYD, L. LEVINE, K. MÉSZÁROS, Y. PERES, J. PROPP and D. B. WILSON, *Chip-firing and rotor-routing on directed graphs*, In: "In and out of equilibrium. 2", volume 60 of Progr. Probab., Birkhäuser, Basel, 2008, 331–364.

[7] M. IMASE and M. ITOH, *A design for directed graphs with minimum diameter*, Computers, IEEE Transactions on **C-32** (8) (1983), 782 –784.

[8] L. LEVINE, *Sandpile groups and spanning trees of directed line graphs*, J. Combin. Theory Ser. A **118** (2) (2011), 350–364.

[9] R. LIDL and H. NIEDERREITER, "Finite Fields", volume 20 of Encyclopedia of Mathematics and its Applications, Cambridge University Press, Cambridge, second edition, 1997.

[10] M. NEWMAN, "Integral Matrices" Number v. 45 in Pure and Applied Mathematics - Academic Press. Acad. Press, 1972.

[11] The on-line encyclopedia of integer sequences, entry A027362, 2004-2011. see http://oeis.org/A027362.

[12] D. PERKINSON, "Sage Sandpiles", 2012. see http://people.reed.edu/ davidp/sand/sage/sage.html.

[13] C. REUTENAUER, “Free Lie Algebras”, Oxford University Press, 1993.

[14] J. J. RUSHANAN, “Topics in Integral Matrices and Abelian Group Codes”, ProQuest LLC, Ann Arbor, MI, 1986. Thesis (Ph.D.)–California Institute of Technology.

[15] W. STEIN *et al.*, “Sage Mathematics Software” (Version 5.7), The Sage Development Team, 2012. http://www.sagemath.org.

[16] D. WAGNER, “The critical Group of a Directed Graph”, 2000. Available at http://arxiv.org/abs/math/0010241.

Two notions of unit distance graphs

Noga Alon[1] and Andrey Kupavskii[2]

Abstract. A *complete (unit) distance graph* in $\mathbb{R}^d$ is a graph whose set of vertices is a finite subset of the d-dimensional Euclidean space, and two vertices are adjacent if and only if the Euclidean distance between them is exactly 1. A *(unit) distance graph* in $\mathbb{R}^d$ is any subgraph of such a graph. We study various properties of both types of distance graphs. We show that for any fixed d the number of complete distance graphs in $\mathbb{R}^d$ on n labelled vertices is $2^{(1+o(1))dn\log_2 n}$, while the number of distance graphs in $\mathbb{R}^d$ on n labelled vertices is $2^{(1-1/\lfloor d/2\rfloor+o(1))n^2/2}$. This is used to study a Ramsey type question involving these graphs. Finally, we discuss the following problem: what is the minimum number of edges a graph must have so that it is not realizable as a complete distance graph in $\mathbb{R}^d$?

1 Introduction

This paper is devoted to the notion of a (unit) distance graph. There are two well-known definitions:

Definition 1.1. A graph $G = (V, E)$ is a *(unit) distance graph in* $\mathbb{R}^d$ if $V \subset \mathbb{R}^d$ and $E \subseteq \{(x, y), x, y \in \mathbb{R}^d, |x - y| = 1\}$, where $|x - y|$ denotes the Euclidean distance between x and y.

Definition 1.2. A graph $G = (V, E)$ is a *complete (unit) distance graph in* $\mathbb{R}^d$ if $V \subset \mathbb{R}^d$ and $E = \{(x, y), x, y \in \mathbb{R}^d, |x - y| = 1\}$.

Distance graphs arise naturally in the study of two well-known problems of combinatorial geometry. First one, posed by Erdős [3], is the fol-

[1] Sackler School of Mathematics and Blavatnik School of Computer Science, Tel Aviv University, Tel Aviv 69978, Israel. Email: nogaa@tau.ac.il. Research supported in part by an ERC Advanced grant, by a USA-Israeli BSF grant, by the Hermann Minkowski Minerva Center for Geometry at Tel Aviv University and by the Israeli I-Core program.

[2] Department of Discrete Mathematics, Moscow Institute of Physics and Technology, Dolgoprudniy 141700, Russia. Email: kupavskii@yandex.ru. Research supported in part by the grant N MD-6277.2013.1 of President of RF and by the grant N 12-01-00683 of the Russian Foundation for Basic Research.

lowing: determine the maximum number $f_2(n)$ of unit distances among n points on the plane. Second is the following question, posed by Nelson (see [2]): what is the minimum number $\chi(\mathbb{R}^2)$ of colours needed to color the points of the plane so that no two points at unit distance apart receive the same colour?

Let $\mathcal{D}(d)$ denote the set of all labeled distance graphs in $\mathbb{R}^d$, and let $\mathcal{D}_n(d)$ denote the set of all those of order n. Similarly, denote by $\mathcal{CD}(d)$ the set of all labeled complete distance graphs in $\mathbb{R}^d$, and let $\mathcal{CD}_n(d)$ denote the set of those of order n.

We reformulate the stated above questions in terms of distance graphs:

$$f_2(n) = \max_{G \in \mathcal{D}_n(2)} |E(G)| = \max_{G \in \mathcal{CD}_n(2)} |E(G)|.$$

$$\chi(\mathbb{R}^d) = \max_{G \in \mathcal{D}(d)} \chi(G) = \max_{G \in \mathcal{CD}(d)} \chi(G),$$

The second series of equalities follows from the well-known Erdős– de Bruijn theorem, which states that the chromatic number of the space $\mathbb{R}^d$ is equal to the chromatic number of some finite distance graph in $\mathbb{R}^d$. It is easy to see that it does not matter which definition of distance graph we use in the study of these two problems. However, sets $\mathcal{D}(d)$ and $\mathcal{CD}(d)$ differ greatly, and we discuss it in the next section.

2 Results

The first theorem shows that in any dimension d the number of distance graphs is far bigger that the number of complete distance graphs.

Theorem 2.1.

1. *For any $d \in \mathbb{N}$, we have $|\mathcal{CD}_n(d)| = 2^{(1+o(1))dn\log_2 n}$.*
2. *For any $d \in \mathbb{N}$ we have $|\mathcal{D}_n(d)| = 2^{\left(1-\frac{1}{\lceil d/2\rceil}+o(1)\right)\frac{n^2}{2}}$.*
3. *If $d = d(n) = o(n)$, then we have $|\mathcal{CD}_n(d)| = 2^{o(n^2)}$.*
4. *If $d = d(n) \geqslant c\frac{n}{\log_2 n}$, where $c > 4$, then $|\mathcal{D}_n(d)| = (1 + o(1))2^{\frac{n(n-1)}{2}}$. In other words, almost every graph on n vertices can be realized as a distance graph in $\mathbb{R}^d$.*

Sketch of the proof. Let $P_1, \ldots, P_m$ be m real polynomials in l real variables. For a point $x \in \mathbb{R}^l$ the *zero pattern* of the P_j's at x is the tuple $(\varepsilon_1, \ldots, \varepsilon_m) \in \{0, 1\}^m$, where $\varepsilon_j = 0$, if $P_j(x) = 0$ and $\varepsilon_j = 1$ if $P_j(x) \neq 0$. Denote by $z(P_1, \ldots, P_m)$ the total number of zero patterns of polynomials $P_1, \ldots, P_m$. Upper bounds from point 1 and 3 of the theorem are proved using the following proposition from real algebraic geometry ([7], Theorem 1.3 and Corollary 1.5):

Proposition 2.2 ([7]). *Let $P_1, \dots, P_m$ be m real polynomials in l real variables, and suppose the degree of each P_j does not exceed k. Then* $z(P_1, \dots, P_m) \leqslant \binom{km}{\ell} \leqslant (ekm/l)^l$.

Denote by $(v_1^i, \dots, v_d^i)$ the coordinates of the vertex v_i in the distance graph. For each unordered pair $\{i, j\}$ of vertices of the graph define the following polynomial P_{ij}:

$$P_{ij} = -1 + \sum_{r=1}^{d} (v_r^i - v_r^j)^2.$$

It is easy to see that each labeled distance graph in $\mathbb{R}^d$ corresponds to a zero pattern of the polynomials $P_{12}, \dots, P_{n-1n}$.

Lower bound in part 1 of the theorem follows from the fact that certain bipartite graphs with maximum degree d in one part are realizable as complete distance graphs in $\mathbb{R}^d$.

Lower bound in part 1 of the theorem is a corollary of the fact that any $[d/2]$-partite graph is realizable as a distance graph in $\mathbb{R}^d$. Upper bound has several proofs, with the simplest one based on the regularity lemma. It rests on the fact that complete $([d/2] + 1)$-partite graph with three vertices in each part is not realizable as distance graph in $\mathbb{R}^d$.

The fourth part of the theorem is an easy consequence of the famous theorem by Bollobás [1] concerning the chromatic number of the random graph $G(n, 1/2)$. □

Next, we study the following two Ramsey-type quantities.

Definition 2.3. The *(complete) distance Ramsey number* $R_D(s, t, d)$ $\big(R_{CD}(s, t, d)\big)$ is the minimum natural m such that for any graph G on m vertices the following holds: either G contains an induced s-vertex subgraph isomorphic to a (complete) distance graph in $\mathbb{R}^d$ or its complement $\bar{G}$ contains an induced t-vertex subgraph isomorphic to a (complete) distance graph in $\mathbb{R}^d$.

The quantity $R_D(s, s, d)$ was introduced in [6], and studied in several follow-up papers, while the quantity $R_{CD}(s, s, d)$ was not studied so far. The following theorem was proved in [4]:

Theorem 2.4 ([4]).

1. *For every fixed* $d \geqslant 2$ *we have*

$$R_D(s, s, d) \geqslant 2^{\left(\frac{1}{2[d/2]} + o(1)\right)s}.$$

2. *For any $d = d(s)$, $2 \leqslant d \leqslant s/2$ we have*

$$R_D(s, s, d) \leqslant d \cdot R\left(\left\lceil \frac{s}{[d/2]} \right\rceil, \left\lceil \frac{s}{[d/2]} \right\rceil\right),$$

where $R(k, \ell)$ is the classical Ramsey number: the minimum number n so that any graph on n vertices contains either a clique of size k or an independent set of size ℓ.

By the last theorem the bounds for $R_D(s, s, d)$ are roughly the same as for the classical Ramsey number $R\left(\left\lceil \frac{s}{[d/2]} \right\rceil, \left\lceil \frac{s}{[d/2]} \right\rceil\right)$:

$$\frac{s}{2[d/2]}(1 + o(1)) \leqslant \log R_D(s, s, d) \leqslant \frac{2s}{[d/2]}(1 + o(1)),$$

where the $o(1)$-terms tend to zero as s tends to infinity.

Using Theorem 2.1 we can show that $R_{CD}(s, s, d)$ is far larger than $R_D(s, s, d)$.

Theorem 2.5.

1. *For any $d = d(s) = o(s)$ we have $R_{CD}(s, s, d) \geqslant 2^{(1+o(1))s/2}$.*
2. *For $d = d(s) \leqslant cs$, where $c < 1/2$ and $H(c) < 1/2$, there exists a constant $\alpha = \alpha(c) > 0$ such that $R_{CD}(s, s, d) \geqslant 2^{(1+o(1))\alpha s}$.*

This theorem is proved via standard probabilistic approach used to obtain lower bounds on the Ramsey numbers.

The last (possible) difference between $\mathcal{D}(d)$ and $\mathcal{CD}(d)$ we point out is the following. Fix an $l \in \mathbb{N}$. The following theorem was proved in [5].

Theorem 2.6 ([5]). *For any $g \in \mathbb{N}$ there exists a sequence of distance graphs in $\mathbb{R}^d$, $d = 1, 2, \ldots$, with girth greater than g such that the chromatic number of the graphs in the sequence grows exponentially with d.*

Unfortunately, we cannot prove a similar theorem for complete distance graphs. All we can prove is the following

Proposition 2.7. *For any $g \in \mathbb{N}$ there exists a sequence of complete distance graphs in $\mathbb{R}^d$, $d = 1, 2, \ldots$, with girth greater than g such that the chromatic number of the graphs in the sequence grows as $\Omega\left(\frac{d}{\log d}\right)$.*

Every bipartite graph is realizable as a distance graph in R^4. However, for any fixed d there exists a bipartite graph that is not realizable as a complete distance graph in $\mathbb{R}^d$. In general, it seems difficult even for a bipartite graph G to decide whether G is realizable as a complete distance graph in $\mathbb{R}^d$ or not. In particular, we introduce the quantity $g_2(d)$, which is equal to the minimum possible number of edges in a bipartite graph K that is not realizable as a complete distance graph in $\mathbb{R}^d$. We obtained the following theorem.

Theorem 2.8. *For any $d \geqslant 4$ we have $\binom{d+2}{2} \leqslant g_2(d) \leqslant \binom{d+3}{2} - 6$.*

The lower bound in this theorem states that if a bipartite graph is not realizable as a complete distance graph in $\mathbb{R}^d$, $d \geqslant 4$, then it must have at least as many edges as the complete graph K_{d+2} on $d+2$ vertices, which is an obvious example of a graph that is not realizable as a distance graph in $\mathbb{R}^d$. This is not the case for $d = 3$, since the graph $K_{3,3}$ is not realizable as a distance graph in $\mathbb{R}^3$ and it has $9 < \binom{5}{2}$ edges. It is interesting to determine, whether an arbitrary graph G that is not realizable as a complete distance graph in $\mathbb{R}^d$, $d \geqslant 4$, must have at least $\binom{d+2}{2}$ edges, and to study the similar question for distance graphs.

Sketch of the proof of theorem 2.8. Both upper and lower bounds are based on linear-algebraic considerations, mostly dealing with the notion of affine dependence.

To obtain the upper bound, we prove that the bipartite graph K'' with the parts $A = \{a_1, \ldots, a_d\}$, $B = \{b_1, \ldots, b_d\}$ and with the set of edges $E = \{(a_i, b_j) : i > j\} \cup \{(a_i, b_j) : i \leqslant 3\}$ is not realizable as a complete distance graph in $\mathbb{R}^d$.

To prove the lower bound we provide sufficient conditions for a bipartite graph to be realizable as a complete distance graph in $\mathbb{R}^d$. The construction of the realization is algorithmic. □

References

[1] B. BOLLOBÁS, *The chromatic number of random graphs*, Combinatorica **8** (1988), 49–55.

[2] P. BRASS, W. MOSER and J. PACH, "Research Problems in Discrete Geometry", Springer, Berlin, 2005.

[3] P. ERDŐS, *On a set of distances of n points*, Amer. Math. Monthly **53** (1946), 248–250.

[4] A. KUPAVSKII, A. RAIGORODSKII and M. TITOVA, *New bounds for distance Ramsey numbers*, submitted.

[5] A. KUPAVSKII, *Distance graphs with large chromatic number and arbitrary girth*, Moscow Journal of Combinatorics and Number Theory **2** (2012), 52–62.

[6] A. M. Raigorodskii, *On a Series of Ramsey-type Problems in Combinatorial Geometry*, Doklady Mathematics **75** (2007), N2, 221–223.
[7] L. Ronyai, L. Babai and M. K. Ganapathy, *On the number of zero-patterns of a sequence of polynomials*, J. Amer. Math. Soc. **14** (2001), 717–735.

An interlacing approach for bounding the sum of Laplacian eigenvalues of graphs

Aida Abiad[1], Miquel A. Fiol[2], Willem H. Haemers[2] and Guillem Perarnau[2]

Abstract. We apply interlacing techniques for obtaining lower and upper bounds for the sums of Laplacian eigenvalues of graphs. Mainly, we generalize two theorems of Grone and Grone & Merris, providing tight bounds and studying the cases of equality. As a consequence, some results on well-known parameters of a graph, such as the maximum and minimum cuts, the edge-connectivity, the (almost) dominating number, and the edge isoperimetric number, are derived.

1 Introduction

Throughout this paper, $G = (V, E)$ is a finite simple graph with $n = |V|$ vertices and $e = |E|$ edges. Recall that the Laplacian matrix of G is $\boldsymbol{L} = \boldsymbol{D} - \boldsymbol{A}$ where $\boldsymbol{D}$ is the diagonal matrix of the vertex degrees and $\boldsymbol{A}$ is the adjacency matrix of G. Let us also recall the following result about interlacing (see [2]).

Theorem 1.1. *Let $\boldsymbol{A}$ be a real symmetric $n \times n$ matrix with eigenvalues $\lambda_1 \geq \cdots \geq \lambda_n$. For some $m < n$, let $\boldsymbol{S}$ be a real $n \times m$ matrix with orthonormal columns, $\boldsymbol{S}^\top \boldsymbol{S} = \boldsymbol{I}$, and consider the matrix $\boldsymbol{B} = \boldsymbol{S}^\top \boldsymbol{A}\boldsymbol{S}$, with eigenvalues $\mu_1 \geq \cdots \geq \mu_m$. Then,*

(a) *The eigenvalues of $\boldsymbol{B}$ interlace those of $\boldsymbol{A}$, that is,*

$$\lambda_i \geq \mu_i \geq \lambda_{n-m+i}, \qquad i = 1, \ldots, m. \tag{1.1}$$

[1] Tilburg University, Department of Econometrics and O.R., Tilburg, The Netherlands. Email: A.AbiadMonge@uvt.nl

[2] Universitat Politècnica de Catalunya, BarcelonaTech, Department de Matemàtica Aplicada IV, Barcelona, Catalonia. Email: fiol@ma4.upc.edu, Haemers@uvt.nl, guillem.perarnau@ma4.upc.edu

(b) *If the interlacing is tight, that is, for some* $0 \leq k \leq m$, $\lambda_i = \mu_i$, $i = 1, \ldots, k$, *and* $\mu_i = \lambda_{n-m+i}$, $i = k+1, \ldots, m$, *then* $\boldsymbol{SB} = \boldsymbol{AS}$.

Two interesting particular cases are when $\boldsymbol{B}$ is a principal submatrix of $\boldsymbol{A}$; and when $\boldsymbol{B}$ is the so-called *quotient matrix* of $\boldsymbol{A}$ with respect to a given partition $\mathcal{P} = \{I_1, \ldots, I_m\}$ of $\{1, \ldots, n\}$.

The first case gives useful conditions for an induced subgraph G' of a graph G, because the adjacency matrix of G' is a principal submatrix of the adjacency matrix of G. However, the Laplacian matrix $\boldsymbol{L}'$ of G' is in general not a principal submatrix of the Laplacian matrix $\boldsymbol{L}$ of G. But $\boldsymbol{L}' + \boldsymbol{D}'$ is a principal submatrix of $\boldsymbol{L}$ for some nonnegative diagonal matrix $\boldsymbol{D}'$. Therefore the left hand inequalities in (1.1) still hold for the Laplacian eigenvalues, because adding the positive semi-definite matrix $\boldsymbol{D}'$ decreases no eigenvalue.

In the case that $\boldsymbol{B}$ is a quotient matrix of $\boldsymbol{A}$, the element b_{ij} of $\boldsymbol{B}$ is the average row sum of the block of $\boldsymbol{A}$ with rows and columns indexed by I_i and I_j, respectively. Actually, the quotient matrix $\boldsymbol{B}$ does not need to be symmetric or equal to $\boldsymbol{S}^\top \boldsymbol{AS}$, but in this case $\boldsymbol{B}$ is similar to (and therefore has the same spectrum as) $\boldsymbol{S}^\top \boldsymbol{AS}$.

If the interlacing is tight, then (*b*) of Theorem 1.1 reflects that $\mathcal{P}$ is a *regular* (or *equitable*) partition of $\boldsymbol{A}$, that is, each block of the partition has constant row and column sums. Moreover, if $\boldsymbol{A}$ is the adjacency matrix of a graph G, then the bipartite induced subgraphs $G[I_i, I_j]$ are biregular, and $G[I_i]$ is regular (then we say that $\mathcal{P}$ is a *regular* partition of G.) Alternatively, if the interlacing is tight for the quotient matrix of the Laplacian matrix of G, then the first condition also holds, but not necessarily the second (now we speak about an *almost regular* partition of G.)

Assuming that G has n vertices, with degrees $d_1 \geq d_2 \geq \cdots \geq d_n$, and Laplacian matrix $\boldsymbol{L}$ with eigenvalues $\lambda_1 \geq \lambda_2 \geq \cdots \geq \lambda_n (= 0)$, it is known that, for $1 \leq m \leq n$,

$$\sum_{i=1}^{m} \lambda_i \geq \sum_{i=1}^{m} d_i. \tag{1.2}$$

This is a consequence of Schur's theorem [6] stating that the spectrum of any symmetric, positive definite matrix majorizes its main diagonal. In particular, note that if $m = n$ we have equality in (1.2), because both terms correspond to the trace of $\boldsymbol{L}$. To prove (1.2) by using interlacing, let $\boldsymbol{B}$ be a principal $m \times m$ submatrix of $\boldsymbol{L}$ indexed by the subindexes corresponding to the m largest degrees, with eigenvalues $\mu_1 \geq \mu_2 \geq \cdots \geq \mu_m$. Then, $\operatorname{tr} \boldsymbol{B} = \sum_{i=1}^{m} d_i = \sum_{i=1}^{m} \mu_i$, and, by interlacing,

$\lambda_{n-m+i} \le \mu_i \le \lambda_i$ for $i = 1, \ldots, m$, whence (1.2) follows. Similarly, reasoning with the principal submatrix $\boldsymbol{B}$ (of $\boldsymbol{L}$) indexed by the m vertices with minimum degrees we get:

$$\sum_{i=1}^{m} \lambda_{n-m+i} \le \sum_{i=1}^{m} d_{n-m+i}. \tag{1.3}$$

The next result, which is an improvement of (1.2), is due to Grone [4], who proved that if G is connected and $m < n$ then,

$$\sum_{i=1}^{m} \lambda_i \ge \sum_{i=1}^{m} d_i + 1. \tag{1.4}$$

In this paper, we give a generalization of this result by considering the degrees of vertices in a given subset. Note that if we take $m = 1$ in (1.4), we get $\lambda_1 \ge d_1 + 1$. Guo [3] conjectured another generalization looking at individual eigenvalues, which was proved by Brouwer and Haemers [1]. They showed that if λ_i is the i-th largest Laplacian eigenvalue, and d_i is the i-th largest degree of a connected graph G on n vertices (in fact, it is is enough to assume that $G \ne K_m \cup (n-m)K_1$), then $\lambda_i \ge d_i - i + 2$, $1 \le i \le n-1$.

2 New results

In this section we present the main results of the paper. In particular, we use interlacing for generalizing two results of Grone [4] and Grone and Merris [5].

2.1 A generalization of Grone's result

We begin with a basic result from where most of our bounds derive. Given a graph G with a vertex subset $U \subset V$, let ∂U be the *vertex-boundary* of U, that is, the set of vertices in $\overline{U} = V \backslash U$ which have some adjacent vertex in U. Also, let $\partial(U, \overline{U})$ denote the *edge-boundary* of U, which is the set of edges which connect vertices in U with vertices in $\overline{U} = V \backslash U$, with cardinality $e(U, \overline{U}) = |\partial(U, \overline{U})|$. For every vertex $v \in V$, let $d_v = d(v)$ stands for its degree.

Proposition 2.1. *Let G be a graph on $n = |V|$ vertices, having Laplacian matrix $\boldsymbol{L} = (l_{uv})$ with eigenvalues $\lambda_1 \ge \lambda_2 \ge \cdots \ge \lambda_n (= 0)$. For any given vertex subset $U = \{u_1, \ldots, u_m\}$ with $0 < m < n$, we have*

$$\sum_{i=1}^{m} \lambda_{n-i} \le \sum_{u \in U} d_u + \frac{e(U, \overline{U})}{|\overline{U}|} \le \sum_{i=1}^{m} \lambda_i. \tag{2.1}$$

The equality on either side of (2.1) implies that the interlacing is tight, and therefore that the partition of G is almost equitable. In other words, in case of equality every vertex $x \in U$ is adjacent to either all or 0 vertices in $\overline{U}$, whereas each vertex $x \in \overline{U}$ has precisely $b = e(U, \overline{U})/|\overline{U}|$ neighbors in U. Using this, it is straightforward to construct nontrivial examples with equality. (If $n = m + 1$, equality holds trivially on both sides.)

In this context, observe that there is no graph (with $n > 2$) satisfying (1.4) for every $0 < m < n$. However the complete graph K_n provides an example for which the inequalities in Proposition 2.1 are equalities for all $0 < m < n$. Indeed, in this case we have $\lambda_i = n$ for any $1 \le i < n$ and for any set U of size m, $e(U, \overline{U})/|\overline{U}| = m$, thus giving for any $0 < m < n$,

$$\sum_{i=1}^{m} \lambda_i = mn = m(n-1) + m = \sum_{u \in U} d_u + \frac{e(U, \overline{U})}{|\overline{U}|},$$

If the vertex degrees of G are $d_1 \ge d_2 \ge \cdots \ge d_n$, we can choose conveniently the m vertices of U (that is, those with maximum or minimum degrees) to obtain the best inequalities in (2.1). Namely,

$$\sum_{i=1}^{m} \lambda_i \ge \sum_{i=1}^{m} d_i + \frac{e(U, \overline{U})}{|\overline{U}|}, \tag{2.2}$$

and

$$\sum_{i=1}^{m} \lambda_{n-i} \le \sum_{i=1}^{m} d_{n-i+1} + \frac{e(U, \overline{U})}{|\overline{U}|}. \tag{2.3}$$

Note that, since $e(U, \overline{U}) \ge 1$, (2.2) is a slight improvement of (1.2). Moreover, (2.3), together with (1.3) for $m + 1$, yields

$$\sum_{i=1}^{m} \lambda_{n-m+i} = \sum_{i=1}^{m} \lambda_{n-i} \le \sum_{i=1}^{m} d_{n-i+1} + \min\left\{ d_{n-m}, \frac{e(U, \overline{U})}{|\overline{U}|} \right\}. \tag{2.4}$$

If we have more information on the structure of the graph, we can improve the above results by either bounding $e(U, \overline{U})$ or 'optimizing' the ratio $b = e(U, \overline{U})/|\overline{U}|$. In fact, the right inequality in (2.1) (and, hence, (2.2)) can be improved when $\overline{U} \ne \partial U$. Simply first delete the vertices (and corresponding edges) of $\overline{U} \setminus \partial U$, and then apply the inequality. Then $d_1, \ldots, d_m$ remain the same and $\lambda_1, \ldots, \lambda_m$ do not increase. Thus we obtain:

Theorem 2.2. *Let G be a connected graph on $n = |V|$ vertices, with Laplacian eigenvalues $\lambda_1 \geq \lambda_2 \geq \cdots \geq \lambda_n (= 0)$. For any given vertex subset $U = \{u_1, \ldots, u_m\}$ with $0 < m < n$, we have*

$$\sum_{i=1}^{m} \lambda_i \geq \sum_{u \in U} d_u + \frac{e(U, \overline{U})}{|\partial U|}. \tag{2.5}$$

Notice that, as a corollary, we get Grone's result [4] since $e(U, \overline{U}) \geq |\partial U|$.

2.2 A generalization of a bound by Grone and Merris

In [5], Grone and Merris gave another lower bound for the sum of the Laplacian eigenvalues, in the case when there is an induced subgraph consisting of isolated vertices and edges. If the induced subgraph of a subset $U \subset V$ with $|U| = m$ consists of r pairwise disjoint edges and $m - 2r$ isolated vertices, then

$$\sum_{i=1}^{m} \lambda_i \geq \sum_{u \in U} d_u + m - r. \tag{2.6}$$

An improvement of this result was given by Brouwer and Haemers in [2]. Let G be a (not necessarily connected) graph with a vertex subset U, with $m = |U|$, and let h be the number of connected components of $G[U]$ that are not connected components of G. Then,

$$\sum_{i=1}^{m} \lambda_i \geq \sum_{u \in U} d_u + h. \tag{2.7}$$

Using interlacing, the bound (2.6) can also be generalized as follows:

Theorem 2.3. *Let G be a connected graph of order $n > 2$ with Laplacian eigenvalues $\lambda_1 \geq \lambda_2 \geq \cdots \geq \lambda_n$. Given a vertex subset $U \subset V$, with $m = |U| < n$, let $G[U] = (U, E[U])$ and $G[\overline{U}]$ be the corresponding induced subgraphs. Let θ_1 be the largest Laplacian eigenvalue of $G[\overline{U}]$. Then,*

$$\sum_{i=1}^{m+1} \lambda_i \geq \sum_{u \in U} d_u + m - |E[U]| + \theta_1. \tag{2.8}$$

The previous bounds on the sum of Laplacian eigenvalues are used to provide meaningful results for the size of the maximum cut of a graph, the edge-connectivity of the graph, the minimum size of a dominating set and the (edge-)isoperimetric number. Given a graph G on n vertices, the *isoperimetric number* is defined as $i(G) = \min_{U \subset V} \left\{ e(U, \overline{U})/|U| : 0 < |U| \leq n/2 \right\}$.

Proposition 2.4. *Let G be a graph on n vertices, with vertex degrees $d_1 \geq d_2 \geq \cdots \geq d_n$, and Laplacian eigenvalues $\lambda_1 \geq \lambda_2 \geq \cdots \geq \lambda_n (=0)$. Let m such that $\frac{n}{2} \leq m < n$. Then,*

$$i(G) \leq \sum_{i=1}^{m} (\lambda_i - d_i). \tag{2.9}$$

References

[1] A. E. BROUWER and W. H. HAEMERS, *A lower bound for the Laplacian eigenvalues of a graph—proof of a conjecture by Guo*, Linear Algebra Appl. **429** (2008), 2131–2135.

[2] A. E. BROUWER and W. H. HAEMERS, "Spectra of Graphs", Springer Verlag (Universitext), Heidelberg, 2012; available online at http://homepages.cwi.nl/ aeb/math/ipm/.

[3] J.-M. GUO, *On the third largest Laplacian eigenvalue of a graph*, Linear Multilinear Algebra **55** (2007), 93–102.

[4] R. GRONE, *Eigenvalues and the degree seqences of graphs*, Linear Multilinear Algebra **39** (1995), 133–136.

[5] R. GRONE and R. MERRIS, *The Laplacian spectrum of a graph.* II, SIAM J. Discrete Math. **7** (2) (1994), 221–229.

[6] I. SCHUR, *Über eine Klasse von Mittelbildungen mit Anwendungen die Determinanten*, Theorie Sitzungsber, Berlin. Math. Gessellschaft **22** (1923), 9–20.

On the structure of the group of balanced labelings on graphs

Yonah Cherniavsky[1], Avraham Goldstein[2] and Vadim E. Levit[1]

1 Abstract

Let $G = (V, E)$ be an undirected graph with possible multiple edges and loops (a multigraph). Let A be an Abelian group. In this work we study the following topics:

1) A function $f : E \to A$ is called balanced if the sum of its values along every closed truncated trail of G is zero. By a truncated trail we mean a trail without the last vertex. The set $H(E, A)$ of all the balanced functions $f : E \to A$ is a subgroup of the free Abelian group A^E of all functions from E to A. We give a full description of the structure of the group $H(E, A)$, and provide an $O(|E|)$-time algorithm to construct a set of the generators of its cyclic direct summands.

2) A function $g : V \to A$ is called balanceable if there exists some $f : E \to A$ such that the sum of all the values of g and f along every closed truncated trail of G is zero. The set $B(V, A)$ of all balanceable functions $g : V \to A$ is a subgroup of the free Abelian group A^V of all the functions from V to A. We give a full description of the structure of the group $B(V, A)$.

3) A function $h : V \cup E \to A$ taking values on vertices and edges is called balanced if the sum of its values along every closed truncated trail of G is zero. The set $W(V \cup E, A)$ of all balanced functions $h : V \cup E \to A$ is a subgroup of the free Abelian group $A^{V \cup E}$ of all functions from $V \cup E$ to A. The group $H(E, A)$ is naturally isomorphic to the subgroup of $W(V \cup E, A)$ consisting of all functions taking every vertex to 0. So we, abusing the notations, treat $H(E, A)$ as that subgroup of $W(V \cup E, A)$.

[1] Computer Science and Mathematics, Ariel University, Israel.
Email: yonahch@ariel.ac.il, levitv@ariel.ac.il

[2] CUNY, USA. Email: avraham.goldstein.nyc@gmail.com

The full-text paper can be found in [2].

There is a natural epimorphism from $W(V \cup E, A)$ onto $B(V, A)$, which "forgets" the values of h on the edges. The kernel of this epimorphism is precisely $H(E, A)$. Thus, $W(V \cup E, A)/H(E, A) \cong B(V, A)$. We use this fact to give a full description of the group $W(V \cup E, A)$.

2 Introduction

For directed graphs, the study of integer-valued functions that vanish on all cycles of a graph, and the notion of the cycle space are classical in Graph Theory. For undirected graphs these questions where first addressed in [1]. In that work such functions are called cycle-vanishing edge valuations. It turns out that in the undirected case the dimensions of the cycle space and, dually, of the space of cycle vanishing edge valuations are closely related 3-edge connectivity. In a very recent article [4] the balanced functions from the union of the set of vertices and the set of edges of a graph to a finite Abelian group A are considered. In that work, which is closely related to [1], the number of such functions is calculated.

In [6] they study the triples (Γ, g, G), where Γ is a graph, G is a group, and g is a function from the set of edges of Γ to G. The edges of Γ are thought of as having arbitrary, but fixed, orientation and the equality $g(-e) = (g(e))^{-1}$ holds for every edge e, where $-e$ is e with the inverted orientation. Such triples are called gain or voltage graphs. A gain graph is called balanced if the product of values of g along every cycle equals the identity element. For the voltage graphs, being balanced is equivalent to satisfying Kirchhoff's voltage law. In [6] several strong criteria for the gain graphs to be balanced are obtained. Voltage graphs are discussed in [3]. For the survey on signed graphs, gain graphs, and related topics see [8].

3 Results

A truncated trail (ttrail) p from a vertex x to a vertex y is an alternating sequence $v_1, e_1, v_2, e_2, ..., v_n, e_n$ of vertices and edges such that $v_1 = x$, each e_j, for $j = 1, ..., n-1$, connects v_j and v_{j+1}, e_n connects v_n and y, and $e_i \neq e_j$ for $i \neq j$. Let $p = v_1, e_1, v_2, e_2, ..., v_n, e_n$ be a ttrail from a vertex x to a vertex y and $p' = v'_1, e'_1, v'_2, e'_2, ..., v'_{n'}, e'_{n'}$ be a ttrail from y to a vertex w. Then, if p and p' have no common edges, the ttrail $p + p'$ from x to w is defined as $v_1, e_1, v_2, e_2, ..., v_n, e_n, v'_1, e'_1, v'_2, e'_2, ..., v'_{n'}, e'_{n'}$. Let $p = v_1, e_1, v_2, e_2, ..., v_n, e_n$ be a ttrail from a vertex x to a vertex y. Then the ttrail p^{-1} from y to x is defined as $v'_1, e'_1, v'_2, e'_2, ..., v'_n, e'_n$ where $v'_1 = y$, $v'_i = v_{n+2-i}$ for $i = 2, ..., n$, and $e'_j = e_{n+1-j}$ for $j = 1, ..., n$. A ttrail p from a vertex x to itself is called a closed ttrail. A closed ttrail is called a cycle if it contains every vertex only one time.

The subgroup of all elements of A of order 2 is denoted A_2. The image $2A$ of the doubling self-map $a \mapsto 2a$ of A is a subgroup of A.

Definition 3.1. Vertices v and w of G are called k-edge-connected if there exist k different ttrails between v and w such that no two of these ttrails have any common edges. By definition, we say that a vertex v of G is k-edge-connected to itself for all k.

The edge version of the famous Menger's Theorem, which asserts that two distinct vertices v and w of G are k-edge-connected if and only if no deletion of any $k-1$ edges from G disconnects v and w, plays crucial role in our work. Its corollary is that k-edge-connectivity is an equivalence relation on the vertices of G.

We denote the number of equivalence classes of k-edge-connected vertices of G by $con_k(G)$. Note that $con_1(G) = con(G)$ is just the number of the connected components of G. We denote the equivalence class of k-edge-connected vertices of G, which contains a vertex v, by $Con_k(v)$.

Definition 3.2. A ttrail between two k-edge-connected vertices is called a k-weakly closed ttrail of G.

Clearly, every closed ttrail of G is also a k-weakly closed ttrail of G for every k.

In this work we mostly concentrate on the case when $k = 3$, since it plays a crucial role in the structure of groups $H(E, A)$, $B(V, A)$ and $W(V \cup E, A)$. It is shown in [7] that there exists a simple linear-time algorithm for finding all 3-edge-connected components of an undirected graph. Further, we introduce short ttrails, and obtain several related structural results, like Lemma 3.6 and Theorem 3.7. Then, we obtain the dual results, like Lemma 3.13 and Theorem 3.12. Finally, we use these results to obtain the group structure of $H(E, A)$, $B(V, A)$, and $W(V \cup E, A)$.

Definition 3.3. The boundary linear map $\delta : \mathbb{F}_2^E \to \mathbb{F}_2^V$ is defined by taking each edge to the sum of its two adjacent vertices.

Definition 3.4. The Cycle Space of G is defined as $Ker(\delta)$, and its elements are called the homological cycles of G.

Definition 3.5. A 3-weakly closed ttrail between vertices v and w, where v can be equal to w, is called short if no one of its inner vertices is 3-edge-connected to v.

Lemma 3.6. *Every 3-weakly closed ttrail is a unique sum of short 3-weakly closed ttrails.*

Let x, y, u, w be (not necessarily distinct) vertices belonging to the same 3-edge-connected component of G.

Theorem 3.7. *If two short 3-weakly closed ttrails, one from x to y and the other from u to w, have a common inner vertex, then $x = u$, $y = w$, and their first and last edges coincide.*

Definition 3.8. Two short 3-weakly closed ttrails are called twins if their first and last edges coincide.

Twinship is an equivalence relation between short 3-weakly closed ttrails.

Definition 3.9. For two vertices v, w of G, a homological path p between v and w is a vector $p \in \mathbb{F}_2^E$ such that $\delta(p) = v + w$.

Definition 3.10. A homological path between two (not necessarily distinct) k-edge-connected vertices of G is called a k-weak homological cycle of G.

Definition 3.11. The subspace of $\mathbb{F}_2^E$, spanned by all the k-weak homological cycles of G, is called the k-Weak Cycle Space of G.

The Cycle Space is a subspace of the k-Weak Cycle Space for each k. Next, we obtain the following dual results on functions from G to A.

Theorem 3.12. *If a function $f : E \to A$ takes the Cycle Space of G to 0 then f is balanced. If a function $f : E \to A$ is balanced then it takes the 3-Weak Cycle Space of G into A_2 and the Cycle Space of G to 0.*

Lemma 3.13. *A function $g : V \to A$ such that $g(V) \in 2A$ or that for some 3-edge-connected component W of V and any $a \in A$, $g(w) = a$, if $w \in W$, and $g(w) = 0$ otherwise, is balanceable. In the other direction, if a function $g : V \to A$ is balanceable then for any 3-edge-connected vertices v and w we must have $g(v) - g(w) \in 2A$.*

Using the above findings, we construct certain appropriate bases for the Cycle Space of G, the 3-Weak Cycle Space of G, $\mathbb{F}_2^E$, $\mathbb{F}_2^{con_3(G)}$, and $\mathbb{F}_2^V$, which we need to prove our three concluding theorems.

Theorem 3.14. *The Abelian group $H(E, A)$ of all the balanced functions $f : E \to A$ is isomorphic to $A^{con_3(G)-con(G)} \times A_2^{|V|-con_3(G)}$.*

Theorem 3.15. *The Abelian group $B(V, A)$ of all the balanceable functions $g : V \to A$ is isomorphic to $A^{con_3(G)} \times (2A)^{|V|-con_3(G)}$.*

Theorem 3.16. *The Abelian group $W(V \cup E, A)$ of all the balanced functions $h : V \cup E \to A$ is isomorphic to $A^{|V|+con_3(G)-con(G)}$.*

Notice that Theorem 4 in [4] follows from our Theorem 3.16 for a finite A.

4 Conclusions

When every $a \in A$ is of order 2, for example, if $A = Z_2$, then $2A = 0$ and $A_2 = A$. In that case our results for $H(E, A)$ coincide with the classical results for balanced gain graphs.

When A is the additive group of the real numbers then $2A = A$ and $A_2 = 0$, and our findings on $H(E, A)$ coincide with the results of [1].

When A is a finite group, Theorem 4 in [4] follows from our Theorem 3.16.

We have studied the group structure of the group $W(V \cup E, A)$, its subgroup $H(E, A)$, and its factor-group $B(V, A)$. The dual problem is to understand the structure of the subgroup $B'(V, A)$ of all balanced functions on vertices and the factor-group $H'(E, A)$ of all balanceable functions on edges. The elements of $B'(V, A)$ are, by abuse of notation, such $h \in W(V \cup E, A)$ that $h(e) = 0$ for every edge $e \in E$. In the case when $A = Z_2$ this is identical to describing the consistent marked (vertex-signed) graphs. These graphs were treated and characterized in [5].

References

[1] R. BALAKRISHNAN and N. SUDHARSANAM, *Cycle vanishing edge valuations of a graph*, Indian J. Pure Appl. Math. **13** (3) (1982), 313–316.

[2] Y. CHERNIAVSKY, A. GOLDSTEIN and V. E. LEVIT, *On the structure of the group of balanced labelings on graphs*, preprint available at http://arxiv.org/abs/1301.4206.

[3] J. L. GROSS and T. W. TUCKER, "Topological Graph Theory", Dover Publications, 1987.

[4] M. JOGLEKAR, N. SHAH and A. A. DIWAN, *Balanced group-labeled graphs*, Discrete Mathematics **312** (2012), 1542–1549.

[5] F. S. ROBERTS and S. XU, *Characterizations of consistent marked graphs*, Discrete Appl. Math **127** (2003), 357–371.

[6] K. RYBNIKOV and T. ZASLAVSKY, *Criteria for Balance in Abelian Gain Graphs, with Applications to Piecewise-Linear Geometry*, Discrete & Computational Geometry, Volume 34, Issue 2 (August 2005), 251–268.

[7] Y. H. TSIN, *Yet another optimal algorithm for 3-edge-connectivity*, Journal of Discrete Algorithms 7.1 (2009), 130–146.

[8] T. ZASLAVSKY, "Bibliography of Signed and Gain Graphs" The Electronic Journal of Combinatorics. Dynamical Surveys. DS8 (September 26, 1999).

The price of connectivity for feedback vertex set

Rémy Belmonte[1], Pim van 't Hof[1], Marcin Kamiński[2] and Daniël Paulusma[3]

Abstract. Let fvs(G) and cfvs(G) denote the cardinalities of a minimum feedback vertex set and a minimum connected feedback vertex set of a graph G, respectively. In general graphs, the ratio cfvs$(G)/$fvs(G) can be arbitrarily large. We study the interdependence between fvs(G) and cfvs(G) in graph classes defined by excluding one induced subgraph H. We show that the ratio cfvs$(G)/$fvs(G) is bounded by a constant for every connected H-free graph G if and only if H is a linear forest. We also determine exactly those graphs H for which there exists a constant c_H such that cfvs$(G) \leq$ fvs$(G) + c_H$ for every connected H-free graph G, as well as exactly those graphs H for which we can take $c_H = 0$.

1 Introduction

Numerous important graph parameters are defined as the cardinality of a smallest subset of vertices satisfying a certain property. Well-known examples of such parameters include the cardinality of a minimum vertex cover, a minimum dominating set, or a minimum feedback vertex set in a graph. In many cases, requiring the subset of vertices to additionally induce a connected subgraph defines a natural variant of the original parameter. The cardinality of a minimum connected vertex cover or a minimum connected dominating set are just two examples of such parameters that have received considerable interest from both the algorithmic and structural graph theory communities. An interesting question is what effect the additional connectivity constraint has on the value of the graph parameter in question.

---

[1] Department of Informatics, University of Bergen, Norway. Email: remy.belmonte@ii.uib.no, pim.vanthof@ii.uib.no
Supported by the Research Council of Norway (197548/F20).

[2] Département d'Informatique, Université Libre de Bruxelles, Belgium – Institute of Computer Science, University of Warsaw, Poland. Email: mjk@mimuw.edu.pl
Supported by EPSRC (EP/G043434/1) and Royal Society (JP100692).

[3] School of Engineering and Computing Sciences, Durham University, UK.
Email: daniel.paulusma@durham.ac.uk

One notable graph parameter that has been studied in this context is the vertex cover number $\tau(G)$, defined as the cardinality of a minimum vertex cover of a graph G. The connected variant of this parameter is the connected vertex cover number, denoted by $\tau_c(G)$ and defined as the cardinality of a minimum connected vertex cover in G. The following observation on the interdependence between $\tau(G)$ and $\tau_c(G)$ for connected graphs G is due to Camby et al. [2].

Observation 1.1 ([2]). For every connected graph G, it holds that $\tau_c(G) \leq 2 \cdot \tau(G) - 1$.

Given a graph class $\mathcal{G}$, the worst-case ratio $\tau_c(G)/\tau(G)$ over all connected graphs G in $\mathcal{G}$ is defined to be the *price of connectivity* for vertex cover for the class $\mathcal{G}$. Observation 1.1 implies that for general graphs, the price of connectivity for vertex cover is upper bounded by 2, and the class of all paths shows that the bound of 2 is asymptotically sharp [2]. Cardinal and Levy [4], who coined the term "price of connectivity for vertex cover", showed a stronger upper bound of $2/(1+\epsilon)$ for graphs with average degree ϵn. Camby et al. [2] provided forbidden induced subgraph characterizations of graph classes for which the price of connectivity for vertex cover is upper bounded by t, for $t \in \{1, 4/3, 3/2\}$.

The above idea applies to other graph parameters as well. The following observation, due to Duchet and Meyniel [5], shows the interdependence between the connected domination number $\gamma_c(G)$ and the domination number $\gamma(G)$ of a connected graph G.

Observation 1.2 ([5]). For every connected graph G, it holds that $\gamma_c(G) \leq 3 \cdot \gamma(G) - 2$.

Adapting the terminology used above for vertex cover, Observation 1.2 implies that the price of connectivity for dominating set on general graphs is upper bounded by 3. The class of all paths again shows that this bound is asymptotically sharp. Zverovich [9] proved that for any graph G, it holds that $\gamma_c(H) = \gamma(H)$ for every induced subgraph H of G if and only if G is (P_5, C_5)-free, that is, if and only if G does not contain an induced subgraph isomorphic to P_5 or C_5. This implies that the price of connectivity for dominating set is exactly 1 for the class of (P_5, C_5)-free graphs. Camby and Schaudt [3] proved that $\gamma_c(G) \leq \gamma(G) + 1$ for every connected (P_6, C_6)-free graph G, and showed that this bound is best possible. They also obtained a sharp upper bound of 2 on the price of connectivity for dominating set for (P_8, C_8)-free graphs, and showed that the general upper bound of 3 is asymptotically sharp for (P_9, C_9)-free graphs.

A *feedback vertex set* of a graph G is a set F of vertices such that deleting F makes G acyclic, that is, the graph $G - F$ is a forest. The cardinalities of a minimum feedback vertex set and a minimum connected feedback vertex set of a graph G are denoted by $\mathrm{fvs}(G)$ and $\mathrm{cfvs}(G)$, respectively. For any graph class $\mathcal{G}$, we define the price of connectivity for feedback vertex set to be the worst-case ratio $\mathrm{cfvs}(G)/\mathrm{fvs}(G)$ over all connected graphs G in $\mathcal{G}$. In contrast to the aforementioned upper bounds of 2 and 3 on the price of connectivity for vertex cover and dominating set, respectively, the price of connectivity for feedback vertex is not upper bounded by a constant. Graphs consisting of two disjoint cycles that are connected to each other by an arbitrarily long path show that the price of connectivity for feedback vertex set is not even bounded by a constant for planar graphs. Interestingly, Grigoriev and Sitters [6] showed that for planar graphs of minimum degree at least 3, the price of connectivity for feedback vertex set is at most 11. This upper bound of 11 was later improved to 5 by Schweitzer and Schweitzer [8], who also showed that this bound is tight.

Our Results. We study the price of connectivity for feedback vertex set for graph classes characterized by one forbidden induced subgraph H. A graph is called *H-free* if it does not contain an induced subgraph isomorphic to H. We show that the price of connectivity for feedback vertex set is bounded by a constant on the class of H-free graphs if and only if H is a linear forest, that is, a forest of maximum degree at most 2. In fact, we obtain a more refined tetrachotomy result on the interdependence between $\mathrm{fvs}(G)$ and $\mathrm{cfvs}(G)$ for all connected H-free graphs G, depending on the structure of the graph H. In order to formally state our result, we need the following terminology. The *disjoint union $G+H$* of two vertex-disjoint graphs G and H is the graph with vertex set $V(G) \cup V(H)$ and edge set $E(G) \cup E(H)$. We write sH to denote the disjoint union of s copies of H, and P_n to denote the path on n vertices. Let H be a graph and $\mathcal{G}$ the class of H-free graphs. The class $\mathcal{G}$ is called *identical*, *additive*, or *multiplicative* if for all connected graphs G in $\mathcal{G}$, it holds that $\mathrm{cfvs}(G) = \mathrm{fvs}(G)$, $\mathrm{cfvs}(G) \leq \mathrm{fvs}(G) + c_H$ for some constant $c_H \geq 0$, or $\mathrm{cfvs}(G) \leq d_H \cdot \mathrm{fvs}(G)$ for some constant $d_H \geq 0$, respectively. Our result can now be formulated as follows.

Theorem 1.3. *Let H be a graph, and let $\mathcal{G}$ be the class of H-free graphs. Then*

(i) *$\mathcal{G}$ is multiplicative if and only if H is a linear forest;*
(ii) *$\mathcal{G}$ is additive if and only if H is an induced subgraph of $P_5 + sP_1$ or sP_3 for some $s \geq 0$;*
(iii) *$\mathcal{G}$ is identical if and only if H is an induced subgraph of P_3.*

2 The proof of Theorem 1.3

Statements (i), (ii) and (iii) in Theorem 1.3 follow from Lemmas 2.1, 2.3 and 2.4 below, respectively.

Lemma 2.1. *Let H be a graph. Then there is a constant d_H such that $\text{cfvs}(G) \leq d_H \cdot \text{fvs}(G)$ for every connected H-free graph G if and only if H is a linear forest.*

Proof. If H is a linear forest, then the statement of the lemma follows from the fact that every connected H-free graph has bounded diameter.

Before proving the reverse direction, we first introduce a family of graphs that will be used later in the proof. Let C_n denote the cycle on n vertices. For any three integers i, j, k, we define $B_{i,j,k}$ to be the graph obtained from $C_i + C_j$ by choosing a vertex x in C_i and a vertex y in C_j, and adding a path of length k between x and y; if $k = 0$, then we simply identify x and y. The graph $B_{3,3,0}$ is called the *butterfly*.

Now suppose H is not a linear forest. We distinguish two cases. Suppose H contains a cycle, and let C be a shortest cycle in H; in particular, C is an induced cycle. For any integer ℓ, the graph $B_\ell := B_{|V(C)|+1,|V(C)|+1,\ell}$ is C-free and therefore H-free. The observation that $\text{cfvs}(B_\ell) = \text{fvs}(B_\ell) + \ell - 1$ for every $\ell \geq 1$ shows that no constant d_H exists as described in the lemma. If H does not contain a cycle, then H is a forest. For any integer ℓ, the graph $B_{3,3,\ell}$ is claw-free. Since we assumed that H is not a linear forest and hence contains a claw, $B_{3,3,\ell}$ is also H-free. The observation that $\text{cfvs}(B_{3,3,\ell}) = \text{fvs}(B_{3,3,\ell}) + \ell - 1$ for every $\ell \geq 1$ completes the proof of Lemma 2.1. □

Lemma 2.2 below exhibits a structural property of sP_3-free graphs that will be used in the proof of Lemma 2.3 below. The proof of Lemma 2.2 has been omitted due to page restrictions.

Lemma 2.2. *For every integer s, there is a constant c_s such that $\text{cfvs}(G) \leq \text{fvs}(G) + c_s$ for every connected sP_3-free graph G.*

Lemma 2.3. *Let H be a graph. Then there is a constant c_H such that $\text{cfvs}(G) \leq \text{fvs}(G) + c_H$ for every connected H-free graph G if and only if H is an induced subgraph of $P_5 + sP_1$ or sP_3 for some integer s.*

Proof. First suppose H is an induced subgraph of P_5. Let G be a connected H-free graph. In particular, G is P_5-free. Hence, due to a result by Bacsó and Tuza [1], there exists a dominating set $D \subseteq V(G)$ such that D is a clique or D induces a P_3 in G. Let F be a minimum feedback vertex set of G. Note that $|D \setminus F| \leq 2$ if D is a clique and $|D \setminus F| \leq 3$ if D induces a P_3. Since D is a connected dominating set in G, the set

$F \cup D$ is a connected feedback vertex set of G of size at most $|F| + 3$. Hence, we can take $c_H = 3$.

Now suppose H is an induced subgraph of $P_5 + sP_1$ for some integer s. Let G be a connected H-free graph. If G is P_5-free, then we can take $c_H = 3$ due to the above arguments. Suppose G contains an induced path P on 5 vertices. Let S be a maximal independent set in the graph obtained from G by deleting the five vertices of P as well as all their neighbors in G. Since G is $P_5 + sP_1$-free, we know that $|S| \leq s - 1$. Note that $V(P) \cup S$ is a dominating set of G. Hence, by Observation 1.2, there is a connected dominating set D in G of size at most $3(|V(P) \cup S|) - 2 \leq 3s + 10$. Let F be a minimum feedback vertex set in G. Then $F \cup D$ is a connected feedback vertex set in G of size at most $|F| + 3s + 10$. Hence, we can take $c_H = 3s + 10$.

If H is an induced subgraph of sP_3 for some integer s, then the existence of a constant c_H as mentioned in Lemma 2.3 is guaranteed by Lemma 2.2.

It remains to show that if H is not an induced subgraph of $P_5 + sP_1$ or sP_3 for any integer s, then there is no constant c_H such that $\text{cfvs}(G) - \text{fvs}(G) + c_H$ for every connected H-free graph G. Let H be a graph that is not an induced subgraph of $P_5 + sP_1$ or sP_3 for any integer s. First suppose H is not a linear forest. Then, by Lemma 2.1, there does not exist a constant c such that $\text{cfvs}(G) \leq c \cdot \text{fvs}(G)$ for every connected H-free graph G. This implies that there cannot be a constant c_H such that $\text{cfvs}(G) \leq \text{fvs}(G) + c_H$ for every connected H-free graph G. Finally, suppose H is a linear forest. Since H is not an induced subgraph of $P_5 + sP_1$ or sP_3 for any integer s, it contains P_6 or $P_4 + P_2$ as an induced subgraph. Consequently, the class of H-free graphs is a superclass of the class of $\{P_6, P_4 + P_2\}$-free graphs. Hence, in order to complete the proof of Lemma 2.3, it suffices to show that if $\mathcal{G}$ is the class of $\{P_6, P_4 + P_2\}$-free graphs, then there exists no constant c_H such that $\text{cfvs}(G) \leq \text{fvs}(G) + c_H$ for every connected $G \in \mathcal{G}$.

For every integer $k \geq 1$, let L_k be the graph obtained from k disjoint copies of the butterfly by adding a new vertex x that is made adjacent to all vertices of degree 2. For every $k \geq 1$, the unique minimum feedback vertex set in L_k is the set $\{x, y_1, y_2, \ldots, y_k\}$, so $\text{fvs}(L_k) = k + 1$. Every minimum connected feedback vertex set in L_k contains the set $\{x, y_1, y_2, \ldots, y_k\}$, as well as exactly one additional vertex for each of the vertices y_i to make this set connected. Hence, $\text{cfvs}(L_k) = 2k + 1 = \text{fvs}(L_k) + k$. The observation that L_k is $\{P_6, P_4 + P_2\}$-free for every $k \geq 1$ implies that if $\mathcal{G}$ is the class of $\{P_6, P_4 + P_2\}$-free graphs, then there exists no constant c such that $\text{cfvs}(G) \leq \text{fvs}(G) + c$ for every connected $G \in \mathcal{G}$. □

The proof of the next lemma has been omitted due to page restrictions.

Lemma 2.4. *Let H be a graph. Then* $\mathrm{cfvs}(G) = \mathrm{fvs}(G)$ *for every connected H-free graph G if and only if H is an induced subgraph of* P_3.

References

[1] G. Bacsó and Zs. Tuza, *Dominating cliques in P_5-free graphs*, Period. Math. Hung. **21** (4) (1990), 303–308.

[2] E. Camby, J. Cardinal, S. Fiorini and O. Schaudt, *The price of connectivity for vertex cover*, manuscript, arXiv:1303.2478 (2013).

[3] E. Camby and O. Schaudt, *A note on connected dominating set in graphs without long paths and cycles*, manuscript, arXiv:1303.2868 (2013).

[4] J. Cardinal and E. Levy, *Connected vertex covers in dense graphs*, Theor. Comput. Sci. **411** (26-28) (2010), 2581–2590.

[5] P. Duchet and H. Meyniel, *On Hadwiger's number and the stability number*, Ann. Discrete Math. **13** (1982), 71–74.

[6] A. Grigoriev and R. Sitters, *Connected feedback vertex set in planar graphs*, In: WG 2009, LNCS, Vol. 5911, Springer (2010), 143–153.

[7] E. Levy, "Approximation Algorithms for Covering Problems in Dense Graphs", Ph.D. thesis. Université libre de Bruxelles, Brussels, 2009.

[8] P. Schweitzer and P. Schweitzer, *Connecting face hitting sets in planar graphs*, Inf. Process. Lett. **111** (1) (2010), 11–15.

[9] I. E. Zverovich, *Perfect connected-dominant graphs*, Discuss. Math. Graph Theory **23** (2003), 159–162.

A local flow algorithm in bounded degree networks

Endre Csóka[1]

1 Introduction

We show a deterministic local algorithm for constructing an almost maximum flow and an almost minimum cut in multisource-multitarget networks with bounded degrees. Locality means that we decide about each edge or node depending only on its constant radius neighbourhood. We show two applications of the flow algorithm, one is about how the neighborhood distributions of arbitrary bounded degree graphs can be approximated by bounded size graphs, and the other one is related to the Aldous–Lyons conjecture.

In our case, network means a graph with three kind of nodes: sources, targets and regular nodes; and we have directed edges with capacities. A local flow algorithm means a function that gets an edge and a constant-radius neighbourhood of it, including the types of vertices and the capacities of the edges; and the function outputs the amount of flow on that edge. A local flow algorithm is correct if for each network N, the amounts produced by the local algorithm on all edges provide a flow.

For typical problems, we expect from local algorithms approximate solutions only. We measure this error compared to the sum of all capacities $cap(N)$. Namely, we say that we can find an almost maximum flow in multisource-multitarget networks if for each $\varepsilon > 0$, there exists a correct local flow algorithm that for each network N, outputs a flow with size at most $\varepsilon \cdot cap(n)$ less than the size of the maximum flow.

Local algorithms are an equivalent description of constant-time dis-

[1] Mathematics Institute and DIMAP, University of Warwick. Email: e.csoka@warwick.ac.uk. This research was partially supported by TÁMOP 4.2.4. A/1-11-1-2012-0001 "National Excellence Program" project, EU–ESF, Hungary; European Research Council (grant agreement no. 306493); MTA Renyi "Lendulet" Groups and Graphs Research Group; ERC Advanced Research Grant No. 227701, KTIA-OTKA grant No. 77780.

Full paper: http://arxiv.org/pdf/1005.0513

tributed algorithms. For more about local algorithms, see the survey paper by Suomela [5]. It turned out that this is a useful tool for property testing, as well. Property testing is an analogue of the Szemerédi regularity theory, but for bounded degree graphs. Namely, we want to describe very large graphs by bounded-size structures, such that this structure describes several properties and parameters of the entire graph. Nguyen and Onak [4] proved the estimability of several parameters using randomized local algorithms. For more about this topic, see the book by Lovász [3].

2 Model and results

There is an input network $N = (G, c)$, as follows. $G = (S, R, T, \vec{E})$ is a graph with degrees bounded by d. d is a global constant throughout the paper. The vertices of G are separated into the disjoint union of the sets S (source), R (regular) and T (target). $\vec{E}$ is the set of directed edges of G, which satisfies that $(a, b) \in \vec{E} \Leftrightarrow (b, a) \in \vec{E}$. We have a capacity function $c : \vec{E} \to [0, \infty)$ of the directed edges. The total capacity on all edges is denoted by $cap(N) = \sum_{e \in \vec{E}} c(e)$. We will use the terms "graph", "path" and "edge" in the directed sense. Let $V = V(G) = V(N) = S \cup R \cup T$, $|V| = n$, $out(A) = \{(a, b) \in \vec{E} \mid a \in A, b \notin A\}$, and $out(v) = out(\{v\})$, and for an edge $e = (a, b)$, let $-e = (b, a)$ denote the edge in the opposite direction.

A function $f : \vec{E} \to \mathbb{R}$ is called a *flow* if it satisfies that $\forall e \in \vec{E}(G)$: $f(-e) = -f(e)$ and $f(e) \le c(e)$, and $\forall r \in R$: $\sum_{e \in out(r)} f(e) = 0$. The value of a flow f is $\|f\| = \sum_{e \in out(S)} f(e)$. Denote a maximum flow by $f^* = f^*(N)$.

A set $S \subseteq X \subseteq S \cup R$ is called a *cut*. The value of a cut is $\|X\| = \|X\|_N = \sum_{e \in out(X)} c(e)$. The Maximum Flow Minimum Cut Theorem [2] says that $\min_{S \subseteq X \subseteq S \cup R} \|X\| = \|f^*\|$.

The rooted r-neighborhood of a vertex v or edge e, denoted by $h_r(v) = h_r(G, v)$ and $h_r(e)$, means the (vertex- or edge-)rooted induced subnetwork of the vertices at distance at most r from v or e, rooted at v or e, respectively. The set of all possible r-neighborhoods are denoted by $\mathcal{B}(r)$ and $\mathcal{B}^{(2)}(r)$, respectively. A function $F : \mathcal{B}^{(2)}(r) \to \mathbb{R}$ is called a *local flow algorithm*, and for each network N, we define the flow on N generated by F as $F(N) = \big(e \to F(h_r(e))\big)$. Similarly, $C : \mathcal{B}(r) \to \{true, false\}$ is called a *local cut algorithm*, and for each network N, we define the cut on N generated by C as $C(N) = \Big\{v \in V(G) \Big| C\big(h_r(v)\big)\Big\}$.

Theorem 2.1. *For each $\varepsilon > 0$, there exists a local flow algorithm F that for each network N, $\|F(N)\| \geq \|f^*(N)\| - \varepsilon \cdot cap(n)$.*

Summary of proof. We start from the empty flow, and we augment on all augmenting paths in increasing order of length, and in a random order within the same length. This provides an expectedly almost maximum flow. In order to keep the locality, we skip the augmentation in some specific low probability events. At each edge, this changes the flow with low probability. Finally, we average the flow on all random orderings. □

Theorem 2.2. *For each $\varepsilon > 0$, there exists a probability distribution $\mathcal{D}$ of local cut algorithms such that for each network N, $F(N)$ is a flow, and*

$$\mathbb{E}_{C \in \mathcal{D}} \|C(N)\| \leq \|f^*(N)\| + \varepsilon \cdot cap(n).$$

Summary of proof. First, we construct the flow with the local algorithm. Then we construct a fractional cut, based on the length of the shortest path from each node to a target node, using edges with not too small free capacities. This fractional cut naturally provides a probability distribution on cuts. □

Corollary 2.3. *In the class of networks with capacities bounded by a constant, $\|f^*(N)\|/n$ is estimable. In other words, for every $\varepsilon > 0$, there exist $k, r \in \mathbb{N}$ and a function $g : \mathcal{B}(r)^k \to \mathbb{R}$ such that if the vertices $v_1, v_2, \ldots v_k$ are chosen independently with uniform distribution, then*

$$\mathbb{E}\left(\left|\frac{\|f^*(N)\|}{n} - g\big(h_r(v_1), h_r(v_2), \ldots h_r(v_n)\big)\right|\right) < \varepsilon.$$

Summary of proof. If we run the local flow algorithm on each neighborhood, then the sample proportion of source nodes, multiplied by the average flow starting from these nodes is a good approximation. □

3 Applications on neighborhood distributions

Let $\mathcal{G}$ denote the set of all graphs with degrees bounded by d. Let $H_r(G)$ denote the distribution of the r-neighborhood of a random vertex of a graph $G \in \mathcal{G}$. We call a family $\mathcal{F}$ of graphs **nice** if $G_1, G_2 \in \mathcal{F} \Rightarrow G_1 \cup G_2 \in \mathcal{F}$ and $G_1 \subseteq G \in \mathcal{F} \Rightarrow G_1 \in \mathcal{F}$ – where $\subseteq$ denotes nonempty induced subgraph – and $\emptyset \notin \mathcal{F}$. Let us denote the closure of the set of all r-neighborhood distributions in F by $D(\mathcal{F}, r) = cl\{H_r(G) \big| G \in \mathcal{F}\}$. It is easy to see that $D(\mathcal{F}, r)$ is a convex compact subset of $\mathbb{R}^{\mathcal{B}(r)}$. Therefore, $D(\mathcal{F}, r)$ is determined by its dual, as follows.

Let us identify the natural base of $\mathbb{R}^{\mathcal{B}(r)}$ by the elements of $\mathcal{B}(r)$, and let the linear extension of $w : \mathcal{B}(r) \to \mathbb{R}$ defined as the function $\tilde{w} : \mathbb{R}^{\mathcal{B}(r)} \to \mathbb{R}$, $\tilde{w}\Big(\sum_{b\in B(r)} \lambda_b b\Big) = \sum_{b\in B(r)} \lambda_b w(b)$. Let us define

$$m(\mathcal{F}, w) = \max_{P\in D(\mathcal{F},r)} \sum_{b\in\mathcal{B}(r)} w(b)P(b) = \max_{P\in D(\mathcal{F},r)} \tilde{w}(P) = \sup_{G\in\mathcal{F}} \tilde{w}\big(H_r(G)\big).$$

It is easy to see that, for a distribution P on $\mathcal{B}(r)$,

$$P \in D(\mathcal{F}, r) \;\Leftrightarrow\; \forall w : \mathcal{B}(r) \to [0,1],\; \tilde{w}(P) \le m(\mathcal{F}, w).$$

The following theorem expresses that if a graph G is distinguishable with high probability from a nice family $\mathcal{F}$, then it is distinguishable based on the constant-radius neighborhood of only one random vertex, as well.

Theorem 3.1. *Assume that $H_r(G_0) \notin D(\mathcal{F}, r)$ holds for a graph $G_0 \in \mathcal{G}$ and a nice family $\mathcal{F}$ of graphs; namely, there exists a $w : \mathcal{B}(r) \to [0,1]$ satisfying*

$$\tilde{w}\big(H_r(G_0)\big) - m(\mathcal{F}, w) \ge \delta > 0.$$

Then for all $\varepsilon > 0$, with $r' = r'(r, \varepsilon, \delta)$, there exists a subset $M \subset \mathcal{B}(r')$ and an induced subgraph G_1 of G_0 such that the following holds.

$$\begin{aligned} &\mathbb{P}\big(H_{r'}(G_1) \in M\big) > 1 - \varepsilon, \\ \forall G \in \mathcal{F}: \quad &\mathbb{P}\big(H_{r'}(G) \in M\big) < \varepsilon. \end{aligned}$$

Summary of proof. For each graph G and vertex weights $\mu_1 : V(G) \to [0,1]$ and $\alpha \in [0,1]$, we construct a network with bounded degrees and bounded edge capacities, in which a maximum flow defines a redistribution on the weights to $\mu_2 : V(G) \to [0, \alpha]$ satisfying that, under some condition, $\sum\limits_{v\in V(G)} \mu_1(v) = \sum\limits_{v\in V(G)} \mu_2(v)$. If we use our local flow algorithm on this network, then it provides a local redistribution of the weights with a small loss. Using this tool two times, from $w : \mathcal{B}(r) \to [0,1]$, we get another function $w' : \mathcal{B}(r') \to [0,1]$ such that $\tilde{w}'\big(H_{r'}(G_1)\big) > 1 - \varepsilon'$ for some $G_1 \in \mathcal{G}$, but $\tilde{w}'\big(H_{r'}(G)\big) < \varepsilon'$ for all $G \in \mathcal{F}$. Then the set $M = \big\{B \in \mathcal{B}(r') : w'(B) > 1/2\big\}$ satisfies the requirements. □

Lovász [3] asked to find, for every radius $r \in \mathbb{N}$ and error bound $\varepsilon > 0$, an explicit $n = n(r, \varepsilon) \in \mathbb{N}$ such that the r-neighborhood distribution of each graph can be ε-approximated by a graph of size at most n. Formally, $\forall G \in \mathcal{G}: \exists G' \in \mathcal{G}: \big|V(G')\big| \le n$ such that $\big\| H_r(G), H_r(G')\big\|_1 < \varepsilon$.

Alon gave a simple proof of the existence of such a function, but the proof did not provide any explicit bound. It is still open whether there exists, say, a recursive function satisfying the requirement, and also how to compute the graph G' from G. See Lovász [3] for details.

As a corollary of Theorem 3.1, we can show that if there exists an arbitrary large error bound $\lambda < 1$ such that, for all r, we can find an explicit upper bound on $n(r, \lambda)$, then it provides explicit upper bounds on $n(r, \varepsilon)$ for all $r \in \mathbb{N}$ and $\varepsilon > 0$, as well.

3.1 Connection with the Aldous–Lyons Conjecture

We only show here the definitions about the Aldous-Lyons conjecture. For details about this topic, we suggest reading [1] or [3].

Consider a probability distribution U of rooted graphs, including infinite graphs, with degrees bounded by d. Select a connected rooted graph from U, and then select a uniform random edge e from the root. We consider e as oriented away from the root. This way we get a probability distribution σ with an oriented "root edge". Let $R(\sigma)$ denote the distribution obtained by reversing the orientation of the root of a random element of σ. We say that U is a **unimodular** random graph if $\sigma = R(\sigma)$.

Let $\mathcal{U}$ denote the set of unimodular random graphs. Let $H_r(U)$ denote the distribution of the r-neighborhoods of the root of $U \in \mathcal{U}$. Let $D(\mathcal{U}, r) = cl\{H_r(U) \big| U \in \mathcal{U}\}$. It can be easily shown that every graph G with a uniform random root provides a $U \in \mathcal{U}$ with $H_r(G) = H_r(U)$, therefore, $D(\mathcal{G}, r) \subseteq D(\mathcal{U}, r)$.

The Aldous–Lyons conjecture says that every unimodular distribution on rooted connected graphs with bounded degree is the "limit" of a bounded degree graph sequence. More precisely,

Conjecture 3.2 (Aldous–Lyons). $\forall r \in \mathbb{N}$: $D(\mathcal{G}, r) = D(\mathcal{U}, r)$.

About this conjecture, the same idea as what we used in Theorem 3.1 provides the following result.

Theorem 3.3. *If the Aldous–Lyons Conjecture is false, then there exists a unimodular random graph U that can be distinguished with high probability from any graph, based on the constant-radius neighborhood of only one random vertex. In other words, for all $\varepsilon > 0$, there exists an $r \in \mathbb{N}$, a subset $M \subset \mathcal{B}(r)$ and a unimodular random graph $U \in \mathcal{U}$ that for all $G \in \mathcal{G}$, the r-neighborhood of a random vertex of G is in M with probability at most ε, but the r-neighborhood of the root of U is in M with probability at least $1 - \varepsilon$.*

References

[1] A. DAVID and L. RUSSELL, *Processes on unimodular random networks*, Electron. J. Probab. **12** (54) (2007), 1454–1508.

[2] L. R. FORD and D. R. FULKERSON, *Maximal flow through a network*, Canadian Journal of Mathematics **8** (3) (1956), 399–404.

[3] L. LOVÁSZ, "Large Networks and Graph Limits" Vol. 60, American Mathematical Soc., 2012.

[4] H. N. NGUYEN and K. ONAK, *Constant-time approximation algorithms via local improvements*, In: "Foundations of Computer Science, 2008, FOCS'08", IEEE 49th Annual IEEE Symposium on, IEEE, 2008, 327–336.

[5] J. SUOMELA, *Survey of local algorithms*, 2009.

The maximum time of 2-neighbour bootstrap percolation: algorithmic aspects

Fabrício Benevides[1], Victor Campos[1], Mitre C. Dourado[2], Rudini M. Sampaio[1] and Ana Silva[1]

Abstract. In 2-neighbourhood bootstrap percolation on a graph G, an infection spreads according to the following deterministic rule: infected vertices of G remain infected forever and in consecutive rounds healthy vertices with at least 2 already infected neighbours become infected. Percolation occurs if eventually every vertex is infected. In this paper, we are interested in calculating the maximum time $t(G)$ the process can take, in terms of the number of rounds needed to eventually infect the entire vertex set. We prove that the problem of deciding if $t(G) \geq k$ is NP-complete for: (a) fixed $k \geq 4$; (b) bipartite graphs with fixed $k \geq 7$; and (c) planar bipartite graphs. Moreover, we obtain polynomial time algorithms for (a) $k \leq 2$, (b) chordal graphs and (c) $(q, q-4)$-graphs, for every fixed q.

1 Introduction

Under r-neighbour bootstrap percolation on a graph G, the spreading rule is a threshold rule in which S_{i+1} is obtained from S_i by adding to it the vertices of G which have at least r neighbours in S_i. We say that a set S_0 percolates G (or that percolation occurs) if eventually every vertex of G becomes infected, that is, there exists a t such that $S_t = V(G)$. In that case we define $t_r(S)$ as the minimum t such that $S_t = V(G)$. And define, the *percolation time of* G as $t_r(G) = \max\{t_r(S) : S \text{ percolates } G\}$. In this paper, we shall focus on the case where $r = 2$ and in such case we omit the subscript of the functions $t_r(S)$ and $t_r(G)$.

Bootstrap percolation was introduced by Chalupa et al. [8] as a model for interacting particle systems in physics. Since then it has found applications in clustering phenomena, sandpiles, and many other areas of sta-

[1] Universidade Federal do Ceará, Fortaleza, Brazil.
Email: fabricio@mat.ufc.br, anasilva@mat.ufc.br, campos@lia.ufc.br, rudini@lia.ufc.br

[2] Universidade Federal do Rio de Janeiro, Brazil. Email: mitre@dcc.ufrj.br

This research was supported by Capes, CNPq, FUNCAP and FAPERJ

tistical physics, as well as in neural networks and computer science [10].

There are two broad classes of questions one can ask about bootstrap percolation. The first, and the most extensively studied, is what happens when the initial configuration S_0 is chosen randomly under some probability distribution? One would like to know how likely percolation is to occur, and if it does occur, how long it takes. The answer to the first of these questions is now well understood for various graphs [1,2,15].

The second broad class of questions is the one of extremal questions. For example, what is the smallest or largest size of a percolating set with a given property? Interesting cases are solved in [3–5,7,9,17–19].

Here, we consider the decision version of the problem, as stated below.

PERCOLATION TIME PROBLEM
Input: A graph G and an integer k.
Question: Is $t(G) \geq k$?

It is interesting to notice that infection problems appear in the literature under many different names and were studied by researches of various fields. The particular case in which $r = 2$ in r-neighbourhood bootstrap percolation is also a particular case of a infection problem related to convexities in graph, which are also of our interest.

A finite *convexity space* is a pair $(V, \mathcal{C})$ consisting of a finite ground set V and a set $\mathcal{C}$ of subsets of V satisfying $\emptyset, V \in \mathcal{C}$ and $\mathcal{C}$ is closed under intersection. The members of $\mathcal{C}$ are called *$\mathcal{C}$-convex sets* and the *convex hull* of a set S is the minimum convex set $H(S) \in \mathcal{C}$ containing S.

A convexity space $(V, \mathcal{C})$ is an *interval convexity* [6] if there is a so-called *interval function* $I : \binom{V}{2} \to 2^V$ such that a subset C of V belongs to $\mathcal{C}$ if and only if $I(\{x, y\}) \subseteq C$ for every two distinct elements x and y of C. With no risk of confusion, for any $S \subset V$, we also denote by $I(S)$ the union of S with $\bigcup_{x,y \in S} I(\{x, y\})$. In interval convexities, the convex hull of a set S can be computed by exhaustively applying the corresponding interval function until obtaining a convex set.

The most studied graph convexities defined by interval functions are those in which $I(\{x, y\})$ is the union of paths between x and y with some particular property. Some common examples are the P_3-convexity [12], geodetic convexity [13] and monophonic convexity [11]. We observe that the spreading rule in 2-neighbours bootstrap percolation is equivalent to $S_{i+1} = I(S_i)$ where I is the interval function which defines the P_3-convexity: $I(S)$ contains S and every vertex belonging to some path of 3 vertices whose extreme vertices are in S. It will be convenient to denote S_i by $I_i(S)$, where $I_i(S)$ is obtained by applying i times the operation I. Related to the geodetic convexity, there exists the *geodetic iteration number of a graph* [14] which is similar to our definition of $t(G)$.

2 Results

We first prove the following NP-hardness results.

Theorem 2.1. PERCOLATION TIME *is NP-complete for any fixed* $k \geq 4$. *If the graph is bipartite, it is NP-Complete for any fixed* $k \geq 7$. *It is also NP-Complete for planar bipartite graphs.*

We first make a reduction from 3-SAT. We construct a graph G as follows. For each clause C_i with literals $\ell_{i,1}$, $\ell_{i,2}$ and $\ell_{i,3}$, add to G a gadget as in Figure 2(a). For each pair of literals $\ell_{i,a}$, $\ell_{j,b}$ such that one is the negation of the other, add a vertex $y_{(i,a),(j,b)}$ adjacent to $w_{i,a}$ and $w_{j,b}$. Let Y be the set of all vertices created this way. Finally, add a vertex z adjacent to all vertices in Y and a pendant vertex z' adjacent to z. It is possible to prove that the formula is satisfiable if and only if $t(G) \geq 4$. In the case of bipartite graphs, the reduction is from 4-SAT and the construction is similar, but using the gadget in Figure 2(b). The case of planar bipartite graphs is more technical and is by a reduction from PLANAR 3-SAT using the gadget in Figure 2(c).

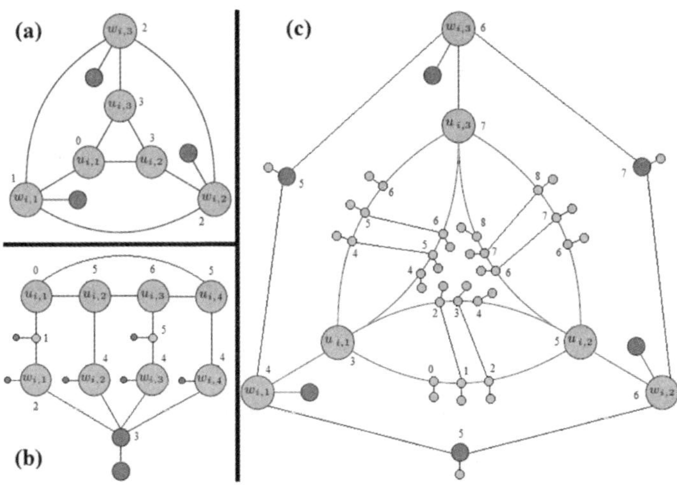

Figure 2.1. Gadget for each clause C_i.

Now we present our polynomial results. It is clear that $t(G) \geq 1$ if and only if G has a vertex of degree ≥ 2. The next result characterizes the graphs with $t(G) \geq 2$. The same question for $t(G) \geq 3$ is still open.

Theorem 2.2. *Let G be a graph. Then* $t(G) \geq 2$ *if and only if there exist* $u \in V(G)$ *and a neighbour* v *of* u *such that* $A \cup \{v\}$ *is a hull set, where* $A \subset V(G)$ *contains every vertex which is neither* u *nor a neighbour of* u.

Now we show how to determine $t(T)$ in linear time, for any tree T. Given two adjacent vertices $u, v \in V(T)$, denote by $s(u, v)$ the maximum time that u enters in the convex hull of S, among all hull sets S of the subtree of $T - v$ containing u; and by $t(u)$ the maximum time that u enters in the convex hull of S, among all hull sets S of T. Clearly, $t(T)$ equals $max_{u \in V(T)} t(u)$ and the values $s(u, v)$, $t(u)$ are given by:

$$s(u,v) = \begin{cases} 0, & \text{if } |N(u)| \leq 2, \\ 1 + \text{the second value in the non-decreasing} & \\ \text{ordering of the set } \{s(x,u) : x \in N(u) \setminus \{v\}\}, & \text{if } |N(u)| \geq 3. \end{cases}$$

$$t(u) = \begin{cases} 0, & \text{if } |N(u)| \leq 1, \\ 1 + \text{the second value in the non-decreasing} & \\ \text{ordering of the set } \{s(x,u) : x \in N(u)\}, & \text{if } |N(u)| \geq 2. \end{cases} \tag{2.1}$$

In order to compute the values $s(u, v)$ and $s(v, u)$, for each edge $uv \in E(T)$, we use a directed graph D with vertex set $V(T)$ and edges $\{(u, v), (v, u) \mid uv \in E(T)\}$. Observe that $s(u, v)$ can be computed only after all values $s(x, y)$ are known, where (x, y) is an arc of $D - v$ belonging to a directed path from some leaf of T to u. Thus, consider a partition of the arcs of D into sets $S_0, \ldots, S_{d-1}$, where d is the diameter of T, and S_i contains the arcs (u, v) such that the longest directed path of $D - v$ from some leaf to u has length i. In the beginning, we know $s(u, v)$, for all arc $(u, v) \in S_0$. Further, as long as $s(x, y)$ is known for each arc $(x, y) \in \bigcup_{j=0}^{i-1} S_j$, we can compute $s(u, v)$, for every arc $(u, v) \in S_i$, $i \in [1, d - 1]$. Therefore, we can compute all values $s(u, v)$ in linear time. With some modifications, we can adapt these ideas to obtain a polynomial time algorithm for chordal graphs.

Theorem 2.3. *If T is a tree, then $t(T)$ can be computed in linear time. Let G be a chordal graph. If G is 2-connected, then $t(G)$ can be computed in time $O(n^2 m)$; otherwise, $t(G)$ can be computed in time $O(n^2 m^2)$.*

Considering $(q, q - 4)$-graphs, q fixed, we proved that the percolation time $t(G)$ is bounded by q, and we give linear time algorithms to obtain $t(G)$. We mention that these graphs generalize the P_4-sparse graphs.

References

[1] J. Balogh, B. Bollobás, H. Duminil-Copin and R. Morris, *The sharp threshold for bootstrap percolation in all dimensions*, Trans. Amer. Math. Soc. **364** (5) (2012), 2667–2701.

[2] J. BALOGH, B. BOLLOBÁS and R. MORRIS, *Bootstrap percolation in three dimensions*, Ann. Probab. **37** (4) (2009), 1329–1380.

[3] J. BALOGH and G. PETE, *Random disease on the square grid*, Random Structures and Algorithms **13** (1998), 409–422.

[4] F. BENEVIDES and M. PRZYKUCKI, *Maximal percolation time in two-dimensional bootstrap percolation*, submitted.

[5] F. BENEVIDES and M. PRZYKUCKI, *On slowly percolating sets of minimal size in bootstrap percolation*, The Elec. J. Combinatorics, to appear.

[6] J. CALDER, *Some elementary properties of interval convexities*, J. London Math. Soc. **3** (1971), 422–428.

[7] C. CENTENO, M. DOURADO, L. PENSO, D. RAUTENBACH and J. L. SZWARCFITER, *Irreversible conversion of graphs*, Theo. Comp. Science **412** (2011).

[8] J. CHALUPA, P. L. LEATH and G. R. REICH, *Bootstrap percolation on a Bethe lattice*, J. Phys. C **12** (1) (1979), 31–35.

[9] N. CHEN, *On the Approximability of Influence in Social Networks*, SIAM J. Discrete Math. **23** (3) (2009), 1400–1415.

[10] P. A. DREYER and F. S. ROBERTS, *Irreversible k-threshold processes: Graph-theoretical threshold models of the spread of disease and of opinion*, Discrete Appl. Math. **157** (7) (2009), 1615–1627.

[11] P. DUCHET, *Convex sets in graphs. II: Minimal path convexity*, J. Comb. Theory, Ser. B **44** (1988), 307–316.

[12] P. ERDŐS, E. FRIED, A. HAJNAL and E. C. MILNER, *Some remarks on simple tournaments*, Algebra Univers. **2** (1972), 238–245.

[13] M. FARBER and R. E. JAMISON, *Convexity in graphs and hypergraphs*, SIAM J. Algebraic Discrete Methods **7** (1986), 433–444.

[14] F. HARARY and J. NIEMINEN, *Convexity in graphs*, Journal of Diferential Geometry **16** (1981), 185–190.

[15] A. E. HOLROYD, *Sharp metastability threshold for two-dimensional bootstrap percolation*, Prob. Th. Rel. Fields **125** (2003), 195–224.

[16] D. LICHTENSTEIN, *Planar Formulae and Their Uses*, SIAM J. Comput. **11** (1982), 329–343.

[17] R. MORRIS, *Minimal percolating sets in bootstrap percolation*, Electron. J. Combin. **16** (1) (2009), 20 pp.

[18] M. PRZYKUCKI, *Maximal Percolation Time in Hypercubes Under 2-Bootstrap Percolation*, The Elec. J. Combinatorics **19** (2012), P41.

[19] E. RIEDL, *Largest minimal percolating sets in hypercubes under 2-bootstrap percolation*, Electron. J. Combin. **17** (1) (2010), 13 pp.

A multipartite Hajnal-Szemerédi theorem

Peter Keevash[1] and Richard Mycroft[2]

Abstract. The celebrated Hajnal-Szemerédi theorem gives the precise minimum degree threshold that forces a graph to contain a perfect K_k-packing. Fischer's conjecture states that the analogous result holds for all multipartite graphs except for those formed by a single construction. Using recent results on perfect matchings in hypergraphs, we prove that (a generalisation of) this conjecture holds for any sufficiently large graph.

1 Introduction

The celebrated Hajnal-Szemerédi theorem [6] states that if k divides n then any graph G on n vertices with minimum degree $\delta(G) \geq (k-1)n/k$ contains a perfect K_k-packing[3]. This theorem generalised a result of Corradi and Hajnal [3], who established the case $k = 3$, and is best-possible in the sense that the theorem would not hold assuming any weaker minimum degree condition. More recently, a series of papers [1,2,8,9] determined the minimum degree thresholds which force a perfect H-packing in a graph for non-complete graphs H, culminating in the work of Kühn and Osthus [11], who essentially settled the problem by giving the best-possible such condition (up to an additive constant) for any graph H, in terms of the so-called *critical chromatic number*.

In many applications it is natural to instead consider packings in a multipartite setting, in which the analogous problem seems to be considerably more difficult. More precisely, let $V_1, \dots, V_k$ be pairwise-disjoint sets of n vertices each, and G be a k-partite graph with vertex classes $V_1, \dots, V_k$ (so G has vertex set $V_1 \cup \dots \cup V_k$ and each V_j is an inde-

[1] School of Mathematical Sciences, Queen Mary, University of London, London E1 4NS, United Kingdom. Email: p.keevash@qmul.ac.uk

[2] School of Mathematics, University of Birmingham, Birmingham B15 2TT, United Kingdom. Email: r.mycroft@bham.ac.uk

[3] A *perfect H-packing* in a graph G is a spanning collection of vertex-disjoint copies of H in G; other sources have referred to the same notion as a *perfect H-tiling* or *H-factor*.

pendent set in G). We define the *partite minimum degree* of G, denoted $\delta^*(G)$, to be the largest m such that every vertex has at least m neighbours in each part other than its own, so

$$\delta^*(G) := \min_{i\in[k]} \min_{v\in V_i} \min_{j\in[k]\setminus\{i\}} |N(v) \cap V_j|,$$

where $N(v)$ denotes the neighbourhood of v.

Fischer [5] conjectured that the natural multipartite analogue of the Hajnal-Szemerédi theorem should hold. That is, he conjectured that if $\delta^*(G) \geq (k-1)n/k$ then G must contain a perfect K_k-packing. This conjecture is straightforward for $k = 2$, as it is not hard to see that any maximal matching must be perfect. However, Magyar and Martin [13] constructed a counterexample for $k = 3$, and furthermore showed that their construction gives the only counterexample for large n. More precisely, they showed that if n is sufficiently large, G is a 3-partite graph with vertex classes each of size n and $\delta^*(G) \geq 2n/3$, then either G contains a perfect K_3-packing, or G is isomorphic to the graph $\Gamma_{n,3,3}$ defined in Construction 1 for some odd n which is divisible by 3.

The implicit conjecture behind this result (stated explicitly by Kühn and Osthus [10]) is that the only counterexamples to Fischer's original conjecture are the constructions given by the graphs $\Gamma_{n,k,k}$ defined in Construction 1 when n is odd and divisible by k. We refer to this as the modified Fischer conjecture. If k is even then n cannot be both odd and divisible by k, so the modified Fischer conjecture is the same as the original conjecture in this case. Martin and Szemerédi [15] proved that (the modified) Fischer's conjecture holds for $k = 4$. Another partial result was obtained by Csaba and Mydlarz [4], who gave a function $f(k)$ with $f(k) \to 0$ as $k \to \infty$ such that the conjecture holds for large n if one strengthens the degree assumption to $\delta^*(G) \geq (k-1)n/k + f(k)n$. However, for general k the validity of even an asymptotic version of Fischer's conjecture (*i.e.* assuming that $\delta^*(H) \geq (k-1)n/k + o(n)$) was unknown until recently, when the results described below were obtained.

2 New results

Keevash and Mycroft [7] used new results on perfect matchings in k-uniform hypergraphs[4] to deduce the following asymptotic result (which

[4] A *hypergraph* H consists of a vertex set V and an edge set E, where each edge $e \in E$ is a subset of V. The edges are not required to be the same size; if they are then we say that H is a *k-uniform hypergraph*, or *k-graph*, where k is the common size of the edges.

was also proved independently and simultaneously by Lo and Markström [12] using the 'absorbing' method.)

Theorem 2.1. *For any k and $\varepsilon > 0$ there exists n_0 such that any k-partite graph G whose vertex classes each have size $n \geq n_0$ with $\delta^*(G) \geq (k-1)n/k + \varepsilon n$ contains a perfect K_k-packing.*

An r-partite graph can only contain a K_k-packing for $r \geq k$, since otherwise we do not have even a single copy of K_k. Fischer's conjecture pertains to the case $r = k$, but it is natural to ask also for an analogous result for the case $r > k$. By a careful analysis of the extremal cases of Theorem 2.1, we can prove an exact result answering both Fischer's conjecture and also this more general question for large n. This is the following theorem, the case $r = k$ of which shows that (the modified) Fischer's conjecture holds for any sufficiently large graph. (The graph $\Gamma_{n,r,k}$ referred to in the statement is defined in Construction 1.)

Theorem 2.2. *For any $r \geq k$ there exists n_0 such that for any $n \geq n_0$ with $k \mid rn$ the following statement holds. Let G be a r-partite graph whose vertex classes each have size n such that $\delta^*(G) \geq (k-1)n/k$. Then G contains a perfect K_k-packing, unless rn/k is odd, $k \mid n$, and $G \cong \Gamma_{n,r,k}$.*

We now give the generalised version of the construction of Magyar and Martin [13] showing Fischer's original conjecture to be false.

Construction 1. *Suppose rn/k is odd and k divides n. Let V be a vertex set partitioned into parts $V_1, \dots, V_r$ of size n. Partition each V_i, $i \in [r]$ into subparts V_i^j, $j \in [k]$ of size n/k. Define a graph $\Gamma_{n,r,k}$, where for each $i, i' \in [r]$ with $i \neq i'$ and $j \in [k]$, if $j \geq 3$ then any vertex in V_i^j is adjacent to all vertices in $V_{i'}^{j'}$ with $j' \in [k] \setminus \{j\}$, and if $j = 1$ or $j = 2$ then any vertex in V_i^j is adjacent to all vertices in $V_{i'}^{j'}$ with $j' \in [k] \setminus \{3-j\}$.*

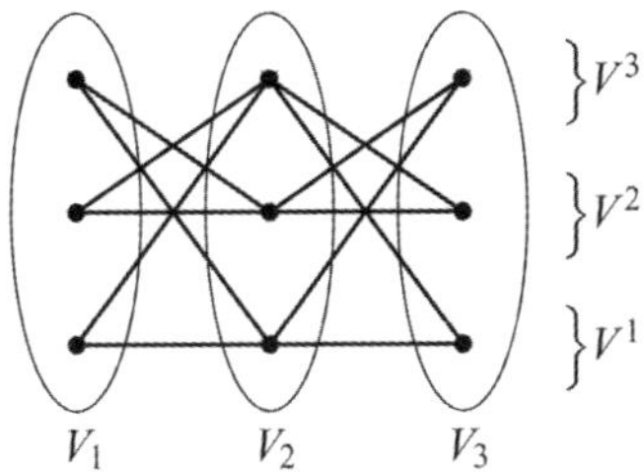

Figure 2.1. Construction 1 for the case $k = r = 3$.

Figure 2.1 shows Construction 1 for the case $k = r = 3$. To avoid complicating the diagram, edges between V_1 and V_3 are not shown: these are analogous to those between V_1 and V_2 and between V_2 and V_3. For $n = k$ this is the exact graph of the construction; for larger n we 'blow up' the graph above, replacing each vertex by a set of size n/k, and each edge by a complete bipartite graph between the corresponding sets. In general, it is helpful to picture the construction as an r by k grid, with columns corresponding to parts V_i, $i \in [r]$ and rows $V^j = \bigcup_{i\in[r]} V_i^j$, $j \in [k]$ corresponding to subparts of the same superscript. Vertices have neighbours in other rows and columns to their own, except in rows V^1 and V^2, where vertices have neighbours in other columns in their own row and other rows besides rows V^1 and V^2. Thus $\delta^*(G) = (k-1)n/k$. We claim that there is no perfect K_k-packing. For any K_k has at most one vertex in any V^j with $j \geq 3$, so at most $k-2$ vertices in $\bigcup_{j\geq 3} V^j$. Also $\left|\bigcup_{j\geq 3} \right| V^j = (k-2)rn/k$, and there are rn/k copies of K_k in a perfect packing. Thus each K_k must have $k-2$ vertices in $\bigcup_{j\geq 3} V^j$, and so 2 vertices in $V^1 \cup V^2$, which must either both lie in V^1 or both lie in V^2. However, $|V^1| = rn/k$ is odd, so V^1 cannot be perfectly covered by pairs. Thus G contains no perfect K_k-packing.

3 Rough outline of the proofs

As described above, Theorem 2.1, the asymptotic version of Fischer's conjecture, is proved by a short deduction from results on perfect matchings in uniform hypergraphs proved in [7]. Indeed, the result used gives fairly general conditions on a k-graph H which guarantee that either

(a) H contains a perfect matching, or
(b) H is close to a 'divisibility barrier', one of a family of lattice-based constructions which do not contain a perfect matching.

Given a graph G, we define the *clique k-complex* of G to be the hypergraph J on $V(G)$ whose edges are the cliques of size j in G for $1 \leq j \leq k$. Then a perfect K_k-packing in G is a perfect matching in the k-graph J_k consisting of all edges of J of size k. It is straightforward to show that if G meets the conditions of Theorem 2.1, then J_k satisfies the conditions necessary to apply the theorem from [7] described above. Furthermore, it is similarly not difficult to show that J_k is not close to a divisibility barrier, ruling out (b). So the theorem implies that (a) must hold, completing the proof of Theorem 2.1.

However, if we instead only assume that G satisfies the weaker conditions of Theorem 2.2, we can no longer deduce that J_k is not close to a divisibility barrier. Indeed, the clique k-complex of the graph $\Gamma_{n,r,k}$ constructed in Construction 1 is actually isomorphic to a divisibility barrier. On the other hand, if J_k is close to a divisibility barrier then we can obtain significant structural information regarding G. In fact, for $k \geq 3$ we find that we may partition G into two 'rows'. That is, we may find a subset U_i of each vertex class V_i of size pn/k for some $1 \leq p \leq k-1$ such that the bipartite graphs $G[U_i, V_j \setminus U_j]$ for $i \neq j$ are almost-complete. Except for a small number of 'bad' vertices, the rows $G_1 := G[\bigcup U_i]$ and $G_2 := G[\bigcup V_i \setminus U_i]$ satisfy a similar degree condition to G, but with p and $k-p$ respectively in place of k. This suggests our approach: we argue inductively to find a perfect K_p-packing in G_1 and a perfect K_{k-p}-packing in G_2. Using the fact that we have almost all edges between rows, we join each copy of K_p in the former packing to a copy of K_{k-p} in the latter packing to form a K_k-packing in G, as required.

However, for $k = 2$ there is another possibility for G for which J_k is close to a divisibility barrier. This is that G is *pair-complete*, meaning that we may choose $U_i \subseteq V_i$ of size $n/2$ for each i so that $G_1 := G[\bigcup U_i]$ and $G_2 := G[\bigcup V_i \setminus U_i]$ are almost-complete r-partite graphs, and there are very few edges in the bipartite graphs $G[U_i, V_j \setminus U_j]$. If there are in fact no edges in these bipartite graphs, and r and $n/2$ are both odd, then G cannot contain a perfect matching (*i.e.* perfect K_2-packing). This presents an obstacle to the proof strategy described above for $k \geq 3$ (since our inductive argument may fail for this reason). It transpires that we can avoid this problem by initially deleting a well-chosen small K_k-packing in G except for when G is exactly isomorphic to the graph $\Gamma_{n,r,k}$, and the theorem follows from this.

4 Future directions

As described in the introduction, the Hajnal-Szemerédi theorem on perfect K_k-packings in a graph G was followed by a sequence of papers addressing the problem of finding an H-packing in G for an arbitrary graph H. Following Theorem 2.2, it seems natural to ask for multipartite analogues of these theorems as well. In this direction, Martin and Skokan [14] recently proved an approximate multipartite version of the Alon-Yuster theorem. That is, they proved that if H is a graph with $\chi(H) \leq k$, and G is a k-partite graph with vertex classes $V_1, \ldots, V_k$ of size n which satisfies $\delta^*(G) \geq (k-1)n/k + o(n)$, then G contains a perfect H-packing. One natural question is whether this minimum degree bound can be improved to include only a constant error term. Moreover, this bound is not even asymptotically best possible for many graphs: to find the degree threshold which forces a perfect H-packing in a k-partite graph for an arbitrary k-partite graph H an analogue of the critical chromatic number seems necessary.

References

[1] N. ALON and E. FISCHER, *Refining the graph density condition for the existence of almost K-factors*, Ars Combinatorica **52** (1999), 296–208.

[2] N. ALON and R. YUSTER, *H-factors in dense graphs*, J. Combinatorial Theory, Series B **66** (1996), 269–282.

[3] K. CORRÁDI and A. HAJNAL, *On the maximal number of independent circuits in a graph*, Acta Math. Acad. Sci. Hungar. **14** (1963), 423–439.

[4] B. CSABA and M. MYDLARZ, *Approximate multipartite version of the Hajnal–Szemerédi theorem*, J. Combinatorial Theory, Series B, **102** (2012), 395–410.

[5] E. FISCHER, *Variants of the Hajnal-Szemerédi Theorem*, J. Graph Theory **31** (1999), 275–282.

[6] A. HAJNAL and E. SZEMERÉDI, *Proof of a conjecture of Erdős*, In: "Combinatorial Theory and its Applications" (Vol. 2), P. Erdős, A. Rényi and V. T. Sós (eds.), Colloq. Math. Soc. J. Bolyai 4, North-Holland, Amsterdam (1970), 601–623.

[7] P. KEEVASH and R. MYCROFT, *A geometric theory for hypergraph matching*, Mem. Amer. Math. Soc., to appear.

[8] J. KOMLÓS, *Tiling Turán theorems*, Comb. Probab. Comput. **8** (1999), 161–176.

[9] J. KOMLÓS, G. N. SÁRKÖZY and E. SZEMERÉDI, *Proof of the Alon-Yuster conjecture*, Disc. Math. **235** (2001), 255–269.

[10] D. KÜHN and D. OSTHUS, *Embedding large subgraphs into dense graphs*, "Surveys in Combinatorics 2009", Cambridge University Press, 2009, 137–167.

[11] D. KÜHN and D. OSTHUS, *The minimum degree threshold for perfect graph packings*, Combinatorica **29** (2009), 65–107.

[12] A. LO and K. MARKSTRÖM, *A multipartite version of the Hajnal-Szemerédi theorem for graphs and hypergraphs*, Comb. Probab. Comput. **22** (2013), 97–111.

[13] C. MAGYAR and R. MARTIN, *Tripartite version of the Corrádi-Hajnal theorem*, Disc. Math. **254** (2002), 289–308.

[14] R. MARTIN and J. SKOKAN, personal communication.

[15] R. MARTIN and E. SZEMERÉDI, *Quadripartite version of the Hajnal-Szemerédi theorem*, Disc. Math. **308** (2008), 4337–4360.

Directed cycle double covers: hexagon graphs

Andrea Jiménez[1], Mihyun Kang[2] and Martin Loebl[3]

Abstract. Jaeger's directed cycle double cover conjecture can be formulated as a problem of existence of special perfect matchings in a class of graphs that we call hexagon graphs. A hexagon graph can be associated with any cubic graph. We show that the hexagon graphs of cubic *bridgeless graphs* are braces that can be generated from the ladder on 8 vertices using two types of McCuaig's augmentations.

The long-standing Jaeger's directed cycle double cover conjecture [1] (DCDC conjecture in short) is broadly considered to be among the most important open problems in graph theory. A typical formulation asks whether every 2-connected graph admits a family of cycles such that one may prescribe an orientation on each cycle of the family in such a way that each edge e of the graph belongs to exactly two cycles and these cycles induce opposite orientations on e. In order to prove the DCDC conjecture, a wide variety of approaches have arisen [1,4], including a topological approach. The topological formulation of the DCDC conjecture is as follows: every cubic bridgeless graph admits an embedding in a closed Riemann surface such that every edge belongs to exactly two distinct face boundaries defined by the embedding; that is, with no dual loop.

Our results. In this work, we formulate the DCDC conjecture as a problem of existence of special perfect matchings in a class of graphs that we call *hexagon graphs.*

The class of the hexagon graphs of the cubic bridgeless graphs turns out to be a subclass of braces. A brace is a simple (*i.e.* no loops and no multiple edges), connected, bipartite graph on at least six vertices such

[1] Instituto de Matemática e Estatística, Universidade de São Paulo. Email: ajimenez@ime.usp.br.
Partially supported by CNPq (Proc. 477203/2012-4), FAPESP (Proc. 2011/19978-5) and Project USP MaCLinC/NUMEC, Brazil.

[2] Institut für Optimierung und Diskrete Mathematik, Technische Universität Graz. Email: kang@math.tugraz.at. Partially supported by the German Research Foundation (KA 2748/2-1).

[3] Department of Applied Mathematics and Institute for Theoretical Computer Science, Charles University. Email: loebl@kam.mff.cuni.cz. Partially supported by the project of the Czech Science Foundation No. P202-13-219885.

that for every pair of non-adjacent edges, there is a perfect matching containing the pair of edges. The class of braces, along with bricks, are a fundamental class of graphs in matching theory, mainly because they are building blocks of a perfect matching decomposition procedure, namely of the tight cut decomposition procedure [2]. In [3], McCuaig introduced a method for generating all braces starting from a large base set of graphs and recursively making use of 4 distinct types of operations. Our main results are:

A cubic graph G has a directed cycle double cover if and only if its hexagon graph H admits a safe perfect matching.

The hexagon graphs arising from cubic bridgeless graphs are braces that can be generated from the ladder on 8 *vertices using* 2 *simple augmentations of McCuaig* (see Figure 1).

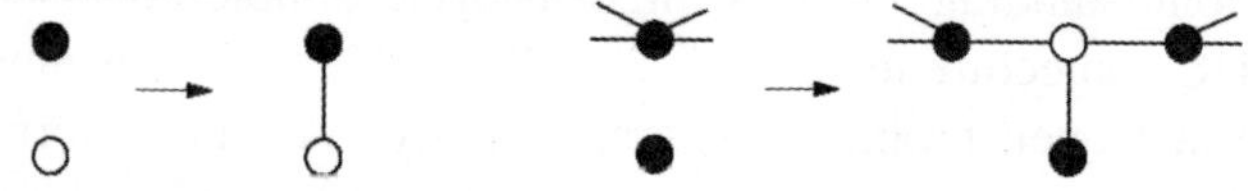

Figure 1. Simple augmentations.

Next we define the hexagon graphs and explain our results.

Hexagon graphs. We refer to the complete bipartite graph $K_{3,3}$ as a *hexagon*. For a graph G and a vertex v of G, let $N_G(v)$ denote the set of neighbors of v in G.

Let G be a cubic graph with vertex set V and edge set E. A *hexagon graph* of G is a graph H obtained from G following the next rules:

1. We replace each vertex v in V by a hexagon h_v so that for every pair $u, v \in V$, if $u \neq v$, then h_u and h_v are vertex disjoint. Let $V(H) = \{V(h_v) : v \in V\}$.
2. For each vertex $v \in V$, let $\{v_i : i \in \mathbb{Z}_6\}$ denote the vertex set of h_v and $\{v_i v_{i+1}, v_i v_{i+3} : i \in \mathbb{Z}_6\}$ its edge set. With each neighbor u of v in G, we associate an index $i_{v(u)}$ from the set $\{0, 1, 2\} \subset \mathbb{Z}_6$ so that if $N_G(v) = \{u, w, z\}$, then $i_{v(u)}, i_{v(w)}, i_{v(z)}$ are pairwise distinct.
3. (See Figure 2). Let $X = \cup_{v\in V}\{v_{2i} : i \in \mathbb{Z}_6\}$ and $Y = \cup_{v\in V}\{v_{2i+1} : i \in \mathbb{Z}_6\}$. We replace each edge uv in E by two vertex disjoint edges e_{uv}, e'_{uv} so that if both $v_{i_{v(u)}}, u_{i_{u(v)}}$ belong to either X or Y, then $e_{uv} = v_{i_{v(u)}} u_{i_{u(v)}+3}$, $e'_{uv} = v_{i_{v(u)}+3} u_{i_{u(v)}}$. Otherwise, $e_{uv} = v_{i_{v(u)}} u_{i_{u(v)}}$, $e'_{uv} = v_{i_{v(u)}+3} u_{i_{u(v)}+3}$. Let $E(H) = \{E(h_v) : v \in V\} \cup \{e_{uv}, e'_{uv} : uv \in E\}$.

We say that h_v is the hexagon of H associated with the vertex v of G and that $\{h_v : v \in V\}$ is the *set of hexagons of H*. We shall refer to the set of edges $\bigcup_{v\in V}\{v_i v_{i+3} : i \in \mathbb{Z}_6\}$ as the set of *red edges of H*, to the set

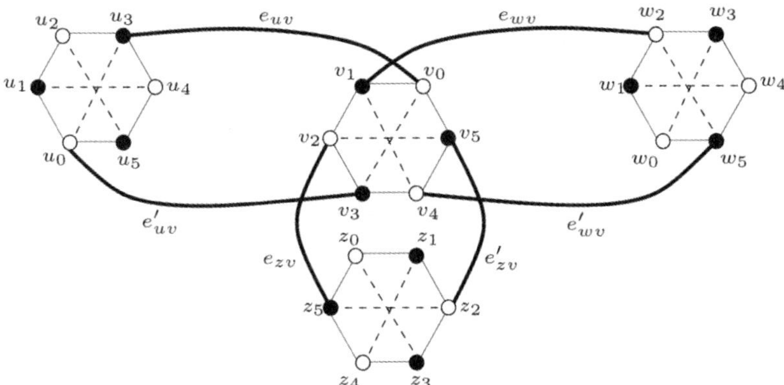

Figure 2. Red edges:= dashed lines, blue edges:= thin lines and white edges:= thick lines. Moreover, $X :=$ white vertices, $Y :=$ black vertices.

of edges $\{e_{uv}, e'_{uv} : uv \in E\}$ as the set of *white edges of* H, and finally to the set of edges $\bigcup_{v\in V}\{v_i v_{i+1} : i \in \mathbb{Z}_6\}$ as the set of *blue edges of* H (see Figure 2). Moreover, we shall say that a perfect matching of H containing only blue edges is a *blue perfect matching*.

Let G be a cubic graph and H be a hexagon graph of G. We observe two important properties: (i) H is bipartite; and (ii) if H' is another hexagon graph of G, then H and H' are isomorphic.

Embeddings and hexagon graphs. Let G be a cubic graph, H the hexagon graph of G, and W the set of white edges of H. Each blue perfect matching M of H encodes an embedding of G on a closed Riemann surface with set of face boundaries the set of subgraphs of G induced by the cycles in $M\Delta W$. The converse also holds. That is, each embedding of G on a closed Riemann surface defines a blue perfect matching M of H, where the set of subgraphs of G induced by all cycles in $M\Delta W$ coincides with the set of face boundaries of the embedding.

Finally, the embedding of G encoded by M has a dual loop if and only if there is a cycle in $M\Delta W$ that contains both end vertices of a red edge. We shall say that a blue perfect matching M is *safe* if no cycle of $M\Delta W$ contains the end vertices of a red edge. This implies the first main result stated above.

Braces and hexagon graphs. The generation of the hexagon graphs of the cubic bridgeless graphs from the ladder on 8 vertices using 2 simple augmentations of McCuaig is a hard technical result and it is contained in the full version of this write-up.

The weak form is: Let G be a cubic graph. Then the hexagon graph H of G is a brace if and only if G is bridgeless. Below we sketch the proof.

Let B, W, and R denote the set of blue, white, and red edges, respectively. Moreover, a blue edge is denoted by b, a white edge by w, and a red edge by r. Each pair of disjoint edges, $\{b, b'\}$, $\{r, r'\}$, or $\{b, r\}$, can be simply extended to a perfect matching of H.

We note that each component of $W \cup R$ is a cycle on four vertices, a *square*. Let w, w' be a pair of disjoint white edges. The edges w, w' belong to the same square of $W \cup R$, or to two different squares of $W \cup R$. In either case w, w' can be naturally extended to a perfect matching of H. Similarly, each edge of a pair w, r of disjoint white and red edge belongs to different square of $W \cup R$, and therefore it can be completed into a perfect matching of H.

Finally we consider a pair b, w of disjoint white and blue edge. If the hexagon with b does not contain an end vertex of w, then it is not difficult to extend b, w to a perfect matching of H. Hence, let h_u be the hexagon that contains b and an end vertex of w, and let h_v be the hexagon that contains the other end vertex of w. Let $b = u_i u_{i+1}$, $w = u_k v_j$, where $i, j, k \in \mathbb{Z}_6$.

If $k \notin \{i+3, i+4\}$, then b, w can be completed into a perfect matching of H that contains the edges b, w, and $u_{i+3}u_{i+4}$.

Hence, without loss of generality we can assume that $k = i + 3$. Let $e_{uv} = u_i v_{j+3}$ and $e_{uz} = u_{i+1} z_l$, where z is the neighbor of v in G such that the white edge with an end vertex u_{i+1} has an end vertex in h_z, and $l \in \mathbb{Z}_6$. Given that in G, edges uv, uz have a common end vertex u represented by hexagon h_u, edge $b = u_i u_{i+1}$ can be seen as *the transition* between uv, uz, while $u_k u_{k+1}$ can be seen as this transition reversed.

Now let G be bridgeless. We observe that two adjacent edges in a cubic bridgeless graph belong to a common cycle. Let C be such a cycle for uv, uz. The two possible orientations of C correspond to two disjoint cycles C_b, C_w in H, where $b \in C_b$ and $w \in C_w$; they contain the transition and transition reversed (between uv, uz), respectively. Let M_b be the perfect matching of C_b consisting of all blue edges and M_w be the perfect matching of C_w consisting of all white edges. In particular, $b \in M_b$ and $w \in M_w$. Since each hexagon of H is intersected by $C_b \cup C_w$ either in a pair of disjoint blue edges, or in the empty set, $M_b \cup M_w$ can be extended to a perfect matching of H.

On the other hand, if G has a bridge $e = \{u, v\}$, then let V_1 be the component of $G - e$ containing u. Any perfect matching of G extending b, w must induce a perfect matching of $\cup_{x \in V_1} h_x \setminus \{u_{i+3}\}$, but this set consists of an odd number of vertices and thus no perfect matching containing b, w can exist.

References

[1] F. JAEGER, *A survey of the cycle double cover conjecture*, In: "Annals of Discrete Mathematics 27 Cycles in Graphs", B. R. Alspach and C. D. Godsil (eds.), Vol. 115, North-Holland Mathematics Studies, North-Holland, 1985, 1–12.

[2] L. LOVÁSZ, *Matching structure and the matching lattice*, J. Comb. Theory Ser. B **43**(2) (1987), 187–222.

[3] W. MCCUAIG, *Brace generation*, Journal of Graph Theory **38**(3) (2001), 124–169.

[4] C. Q. ZHANG, "Integer Flows and Cycle Covers of Graphs", In: Department of Mathematics, University of Illinois, Urbana, Illinois, Marcel Dekker, Inc, 1997.

Finding an Odd $K_{3,3}$

Peter Whalen[1] and Robin Thomas[1]

Abstract. In their 1999 paper, "Permanents, Pfaffian orientations, and even directed curcuits", Neil Robertson, Paul Seymour, and Robin Thomas provided a good characterization for whether or not a bipartite graph has a Pfaffian orientation. Robin Thomas and Peter Whalen recently provided a shorter proof of the central result using elementary methods. The first step in this new proof is the finding of a subgraph of a brace isomorphic to a subdivision of $K_{3,3}$ in which each edge has been subdivided an even number of times. Here, we extend this result to prove that any internally 4-connected non-planar bipartite graph contains a subgraph isomorphic to a subdivision of $K_{3,3}$ in which each edge has been subdivided an even number of times.

1 Introduction

While a well-known algorithm of Edmonds [1] provides a polynomial time solution to the problem of finding a perfect matching in a graph, the problem of counting the number of perfect matchings is #P-Complete [5]. Work by Kasteleyn [2], however, has led to the notion of a Pfaffian orientation and an algorithm for computing the number of perfect matchings in graphs admitting a Pfaffian orientation.

Definition 1.1. Let G be a directed graph. Then G is *Pfaffian* if for every even cycle, C such that $G - V(C)$ contains a perfect matching, the number of edges directed in either direction of the cycle is odd. If G is an undirected graph, we say that G is *Pfaffian* if there is an orientation of the edges such that the resulting directed graph is Pfaffian. Such an orientation is called a *Pfaffian orientation.*

The question of determining whether or not a graph has a Pfaffian orientation has been solved for several cases, though remains open in gen-

[1] School of Mathematics, Georgia Institute of Technology, Atlanta, Georgia 30332-0160, USA. Email: pwhalen3@math.gatech.edu, thomas@math.gatech.edu

eral. One solved case is that of bipartite graphs. The following elegant result of Little [3] provides a characterization for this problem:

Theorem 1.2. *Let G be a bipartite graph. Then G is Pfaffian if and only if it does not contain a subgraph H and a perfect matching M such that $E(G) = E(H) \cup M$, $E(H) \cap M = \emptyset$, and H is isomorphic to a subdivision of $K_{3,3}$ in which each edge is subdivided an even number of times.*

The disadvantage to this theorem is that it does not seem to immediately give rise to a polynomial time algorithm. Instead, a result of Robertson, Seymour, and Thomas [4] provides an algorithm for determining whether a brace has a Pfaffian orientation which relies on a theorem also independently proven by McCuaig.

Recently, Robin Thomas and Peter Whalen have provided a new proof for the central theorem behind this algorithm that relies on more elementary methods. The first step of this process is to find a subgraph of G isomorphic to a subdivision of $K_{3,3}$ in which each edge is subdivided an even number of times. We refer to such a subdivision as an *odd* $K_{3,3}$ since each branch of the $K_{3,3}$ is a path of odd length. In the context of that result, it sufficed to find such a subgraph only when G is a brace.

The quality of having an odd $K_{3,3}$, however, seems more naturally tied to connectivity rather than to matching properties. So in this result, we extended our theorem to show that any internally 4-connected, non-planar bipartite graph contains a subgraph isomorphic to an odd $K_{3,3}$. This result seems natural and has practical application in the study of Pfaffian orientations.

2 Statement of the theorem

We first require several definitions

Definition 2.1. Let G be a 3-connected graph on at least 5 vertices such that for every pair of sets $X, Y \subseteq E(G)$ with $E(G) = X \cup Y$, $X \cap Y = \emptyset$, if three vertices are incident with an edge in A and an edge in B then $|A| \leq 3$ or $|B| \leq 4$. Then we say that G is *internally 4-connected.*

Definition 2.2. Let G be a graph and H a subgraph of G isomorphic to a subdivision of $K_{3,3}$. Let $v_1, v_2, \ldots, v_6$ be the degree three vertices of H and $P_1, P_2, \ldots, P_9$ be the paths in H between v_i. We then refer to H as a *hex* or a *hex of* G, the v_i as the *feet* of H, and the P_i as the *segments* of H.

Definition 2.3. Let G be a graph and H a hex of G. We refer to a segment P of H as *even* if it has even length and *odd* otherwise. We refer to H as even if all of its segments are even and similarly for odd.

The remainder of the paper deals with the proof of the following theorem:

Theorem 2.4. *Let G be an internally 4-connected non-planar bipartite graph. Then G has an odd hex.*

3 The proof technique

The sketch of the proof of the theorem is as follows. G contains a hex since G is 3-connected, non-planar, and not K_5. Choose the hex with the fewest number of even segments. If some segment of that hex is even, it has a vertex v with degree 2 with respect to the hex. Since G is 3-connected, we can use Menger's theorem to augment the segment of the hex containing v to find a path from v to elsewhere in the hex. With some analysis, we show that this (or a similar argument) will always produce a hex with more odd segments which completes the proof.

At each step in our proof, we consider a subgraph of G, H, that consists of a hex plus potentially additional paths. Our main operation is to find some piece of H that is not internally 4-connected and try to find a new path that we might be able to use to find a hex with more odd segments than our original choice.

Let the bipartition of G be (A, B). Suppose we have three internally-vertex disjoint paths P_1, P_2, P_3 meeting at a single vertex v. Let the ends of P_1, P_2, P_3 be a, b, c repectively and let $v, a \in A$ and $b, c \in B$. Let X be a subset of the vertices of G disjoint from this structure. We would like a path with one end in the union of the interior of our paths and the other in X. The core lemma that we use describes possible outcomes of this situation. Note first that in the context of this subgraph, $\{a, b, c\}$ represents a three separation, so we should be able to find another path out of this structure. By possibly rerouting the original paths, we find four outcomes, one of which is a straightforward jump from a B vertex along P_1 to an A vertex in X. The other three are similarly explicit, but somewhat more complicated. The proof of this lemma follows from Menger's theorem and induction and carefully analyzing the resulting outcomes.

To prove the main theorem given the lemma, let H be the hex of G with the fewest even segments. If at most 3 of the segments are odd, say most of the feet are in A, then choose a B-vertex on one of the even segments, v and apply Menger's theorem to find a path to one of the other segments. Choosing v as one of the feet of our new hex, we can immediately find a hex with more odd segments. When H has at least 4 odd segments, we apply a similar argument. Suppose we look at a B vertex on an $A - A$ segment and find that the path we get from Menger's theorem ends at another B vertex. Then we can apply the main lemma. In the first outcome, we find a new path that ends instead at an A-vertex. In the other outcomes, we get a large amount of structure (including parity) in a neighborhood of B. We can generally take advantage of this structure to find a new hex with more odd segments than our original.

References

[1] J. EDMONDS, *Paths, trees, and flowers*, Canad. J. Math., **17** (1965), 449–467.

[2] P. W. KASTELEYN, *The statistics of dimers on a lattice, I., The number of dimer arrangements on a quadratic lattice*, Physica **83** (1961), 1664–1672.

[3] C. H. C. LITTLE, *A characterization of convertible (0, 1)-matrices*, J. Combin. Theory Ser. B **18** (1975), 187–208.

[4] N. ROBERTSON, P. D. SEYMOUR and R. THOMAS, *Permanents, Pfaffian orientations, and even directed circuits*, Ann. Math. **150** (1999), 929–975.

[5] L. G. VALIANT, *The complexity of computing the permanent*, Theoret. Comput. Sci. **8** (1979), 189–201.

Zero-error source-channel coding with entanglement

Jop Briët[1], Harry Buhrman[1], Monique Laurent[1], Teresa Piovesan[1] and Giannicola Scarpa[1]

1 Introduction

We study a problem from zero-error information theory—a topic well-known for its rich connections to combinatorics [1, 8, 10–12, 14]—in a setting where a sender and receiver may use quantum entanglement, one of the most striking features of quantum mechanics. The problem that we consider is the classical *source-channel coding problem*, where Alice and Bob are each given an input from a random source and get access to a noisy channel through which Alice can send messages to Bob. Their goal is to minimize the average number of channel uses per source input while allowing Bob to learn Alice's inputs. Here we show that entanglement can allow for an unbounded decrease in the asymptotic rate of classical source-channel codes. We also consider the *source problem*, the case where Alice can send messages to Bob without noise. We prove a lower bound on the rate of source codes with entanglement in terms of a variant of the Lovász theta number [10, 13], a graph parameter given by a semidefinite program.

1.1 Classical source-channel coding

We briefly explain the three relevant problems from zero-error information theory and their well-known graph-theoretical characterizations.

A *discrete dual source* $\mathcal{M} = (\mathsf{X}, \mathsf{U}, P)$ consists of a finite set X, a (possibly infinite) set U and a probability distribution P over $\mathsf{X} \times \mathsf{U}$. In a dual-source instance, with probability $P(x, u)$, Alice gets an input

[1] Centrum Wiskunde & Informatica (CWI), Science Park 123, 1098 XG Amsterdam, The Netherlands. Email: j.briet@cwi.nl, Harry.Buhrman@cwi.nl, Monique.Laurent@cwi.nl, Teresa.Piovesan@cwi.nl, g.scarpa@cwi.nl

J. B. and H. B. were supported by the European Commission under the project QCS (Grant No. 255961), and G. S. was supported by Vidi grant 639.072.803 from the Netherlands Organization for Scientific Research (NWO).

$x \in \mathsf{X}$ and Bob a $u \in \mathsf{U}$. Bob needs to learn Alice's input without error by having Alice send Bob as few bits as possible. Associated to $\mathcal{M}$ is its *characteristic graph* $G = (\mathsf{X}, E)$, where $\{x, y\} \in E$ if there exists a $u \in \mathsf{U}$ such that $P(x, u) > 0$ and $P(y, u) > 0$. As observed in [14], solving the zero-error source coding problem is equivalent to finding a proper coloring of G that uses the minimum number of colors and the *Witsenhausen rate*

$$R(G) = \lim_{m\to\infty} \frac{1}{m} \log \chi(G^{\boxtimes m}) \tag{1.1}$$

is the minimum asymptotic *cost rate* (*i.e.*, the average number of bits Alice needs to send Bob per source input) of a zero-error code for $\mathcal{M}$. Here $G^{\boxtimes m}$ is the mth strong graph power [12] and log is the logarithm in base 2.

A *discrete channel* $\mathcal{N} = (\mathsf{S}, \mathsf{V}, Q)$ consists of a finite input set S, a (possibly infinite) output set V and a probability distribution $Q(\cdot|s)$ over V for each $s \in \mathsf{S}$. If Alice sends $s \in \mathsf{S}$ through the channel, then Bob receives $v \in \mathsf{V}$ with probability $Q(v|s)$. Associated to a channel is its *confusability graph* $H = (\mathsf{S}, F)$, where $\{s, t\} \in F$ if there exists a $v \in \mathsf{V}$ such that both $Q(v|s) > 0$ and $Q(v|t) > 0$. As observed in [12], $\alpha(H^{\boxtimes n})$ is the maximum number of distinct possible messages that Alice can send to Bob without error by using the channel n times. The *Shannon capacity*

$$c(H) = \lim_{n\to\infty} \frac{1}{n} \log \alpha(H^{\boxtimes n}) \tag{1.2}$$

gives the maximum asymptotic rate of a zero-error channel code.

In the *source-channel coding problem* the parties get inputs from a dual source $\mathcal{M}$ and get access to a channel $\mathcal{N}$. As observed in [11], if $\mathcal{M}$ has characteristic graph G and $\mathcal{N}$ has confusability graph H, then a zero-error coding scheme which encodes length m source-input-sequences into length n channel-input-sequences defines a homomorphism from $G^{\boxtimes m}$ to $\overline{H^{\boxtimes n}}$. The parameter

$$\eta(G, H) := \lim_{m\to\infty} \frac{1}{m} \min\left\{n \in \mathbb{N} : G^{\boxtimes m} \xrightarrow{\exists\,\text{homomorphism}} \overline{H^{\boxtimes n}}\right\}$$

gives the minimum asymptotic cost rate (*i.e.*, the minimum average number of channel uses per source input) of a zero-error code. Note that $R(G) = \eta(G, \overline{K_2})$ and $1/c(H) = \eta(K_2, H)$. In general $\eta(G, H) \leq R(G)/c(H)$ and in [11] it is shown that unbounded separations between $\eta(G, H)$ and $R(G)/c(H)$ can occur.

1.2 Source-channel coding with entanglement

We briefly introduce the model of entanglement-assisted coding, but a direct algebraic definition of the entanglement-assisted variant of $\eta(G, H)$ is given in Definition 1.1. A *state* is a complex positive semidefinite matrix with trace 1. The possible states of a pair of d-dimensional quantum systems $(\mathcal{A}, \mathcal{B})$ are the states in $\mathbb{C}^{d\times d} \otimes \mathbb{C}^{d\times d}$. Such a pair is *entangled* if its state is not a convex combination of states of the form $\rho_A \otimes \rho_B$ with $\rho_A \in \mathbb{C}^{d\times d}$ and $\rho_B \in \mathbb{C}^{d\times d}$. A *$t$-outcome measurement* is a collection $\mathsf{A} = \{A^i \in \mathbb{C}^{d\times d} : i \in [t]\}$ of positive semidefinite matrices A^i that satisfy $\sum_{i=1}^{t} A^i = I$. A measurement describes an experiment that one may perform on a d-dimensional quantum system. If $(\mathcal{A}, \mathcal{B})$ is in a state σ, and Alice performs a t-outcome measurement A on $\mathcal{A}$ and Bob performs an r-outcome measurement B on $\mathcal{B}$, then they obtain outcomes $i \in [t]$ and $j \in [r]$, respectively, with probability $\mathsf{Tr}\big((A^i \otimes B^j)\sigma\big)$.

The *entanglement-assisted* protocol for solving the source-channel coding problem is as follows:

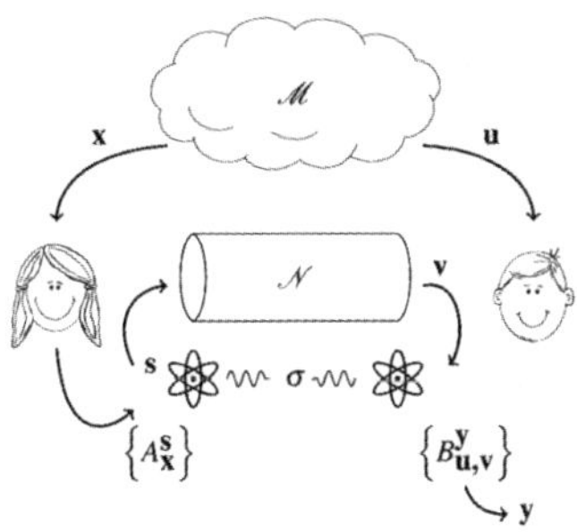

(1) Alice and Bob get inputs $\mathbf{x} \in \mathsf{X}^m$ and $\mathbf{u} \in \mathsf{U}^m$, respectively, from the source $\mathcal{M}$;
(2) Alice performs a measurement $\{A^{\mathbf{s}}_{\mathbf{x}} \in \mathbb{C}^{d\times d} : \mathbf{s} \in \mathsf{S}^n\}$ on her system $\mathcal{A}$ (the measurement may depend on $\mathbf{x}$);
(3) Alice sends the outcome $\mathbf{s}$ over $\mathcal{N}$;
(4) Bob receives an output $\mathbf{v}$ from $\mathcal{N}$;
(5) Bob performs a measurement $\{B^{\mathbf{y}}_{\mathbf{u},\mathbf{v}} \in \mathbb{C}^{d\times d} : \mathbf{y} \in \mathsf{X}^m\}$ (which may depend on $\mathbf{u}$ and $\mathbf{v}$) on his system $\mathcal{B}$;
(6) Bob obtains a measurement outcome $\mathbf{y} \in \mathsf{X}^n$.

A zero-error entanglement-assisted coding scheme satisfies that Bob's measurement outcome $\mathbf{y}$ is Alice's input $\mathbf{x}$ with probability 1. With a similar technique as in [6], we can define the entanglement variant of $\eta(G, H)$ as follows.

Definition 1.1. For graphs G, H and $m \in \mathbb{N}$, define $\eta^\star_m(G, H)$ as the minimum positive integer n such that there exist $d \in \mathbb{N}$ and $d \times d$ positive semidefinite matrices $\{\rho^{\mathbf{s}}_{\mathbf{x}} : \mathbf{x} \in V(G^{\boxtimes m}), \mathbf{s} \in V(H^{\boxtimes n})\}$ and ρ such that $\mathsf{Tr}(\rho) = 1$,

$$\sum_{\mathbf{s} \in V(H^{\boxtimes n})} \rho^{\mathbf{s}}_{\mathbf{x}} = \rho \quad \forall \mathbf{x} \in V(G^{\boxtimes m}),$$

$$\rho^{\mathbf{s}}_{\mathbf{x}} \rho^{\mathbf{t}}_{\mathbf{y}} = 0 \quad \forall \{\mathbf{x}, \mathbf{y}\} \in E(G^{\boxtimes m}), \{\mathbf{s}, \mathbf{t}\} \in V(H^{\boxtimes n}) \cup E(H^{\boxtimes n}).$$

Define $\eta^\star(G, H) = \lim_{m \to \infty} \eta^\star_m(G, H)/m$.

We regain the parameter $\eta(G, H)$ if we restrict the above matrices ρ and $\rho^{\mathbf{s}}_{\mathbf{x}}$ to be $\{0, 1\}$-valued scalars. As in the classical setting, we obtain the entangled variants of the Witsenhausen rate $R^\star(G) = \eta^\star(G, \overline{K_2})$ and Shannon capacity $1/c^\star(H) = \eta^\star(K_2, H)$. Alternatively these parameters can be defined analogously to (1.1) and (1.2) based on entangled variants of the chromatic and independence numbers $\chi^\star(G)$ and $\alpha^\star(H)$, whose definitions are similar to Definition 1.1. The parameters $\alpha^\star(H)$ and $c^\star(H)$ were first defined in [6], where it was first shown that a separation $\alpha < \alpha^\star$ is possible. It was later shown in [5, 9] that even the zero-error *capacity* can be increased with entanglement (*i.e.*, $c < c^\star$). To the best of our knowledge, neither source nor source-channel coding has been considered in the context of shared entanglement before.

2 Our results

2.1 The entangled chromatic number and Szegedy's number

Our first result gives a lower bound for the entangled chromatic number, which can be efficiently computed with semidefinite programming.

Theorem 2.1. *For every graph G, $\vartheta^+(G) \leq \chi^\star(\overline{G})$ and $\log \vartheta(G) \leq R^\star(\overline{G})$.*

Here $\vartheta(G)$ is the celebrated theta number of Lovász [10] defined by

$$\vartheta(G) = \min\{\lambda : \exists\, Z \in \mathbb{R}^{V \times V}_{\succeq 0}, Z_{u,u} = \lambda - 1 \text{ for } u \in V, Z_{u,v} = -1 \text{ for } \{u, v\} \notin E\},$$

where $\mathbb{R}^{V \times V}_{\succeq 0}$ is the space of positive semidefinite matrices, and $\vartheta^+(G) \geq \vartheta(G)$ is the variant of Szegedy [13] obtained by adding the constraint $Z_{u,v} \geq -1$ for $\{u, v\} \in E$.

Combining with results of [3, 7], we get the chain of inequalities

$$c(G) \leq c^\star(G) \leq \log \vartheta(G) \leq R^\star(\overline{G}) \leq R(\overline{G}).$$

As in the classical setting, the problem of giving stronger bounds on the entangled Witsenhausen rate and Shannon capacity is wide open.

2.2 Classical versus entangled source-channel coding rates

Our second result says that entanglement allows for an unbounded advantage in the asymptotic cost rate of a zero-error source-channel coding scheme. For this we use (as in [5]) the "*quarter orthogonality graph*" H_k (for odd k), with vertices all vectors in $\{-1, 1\}^k$ with an even number of "-1" entries and with edges the pairs with inner product -1. We also use the result of [15] showing the existence of a Hadamard matrix (*i.e.*, a matrix $A \in \{-1, 1\}^{N \times N}$ that satisfies $AA^\mathsf{T} = NI$) of size $N = 4q^2$ if q is an odd prime power with $q \equiv 1 \bmod 4$.

Theorem 2.2. *For every odd integer $k \geq 5$, we have*

$$\eta^\star(H_k, H_k) \leq \frac{\log(k+1)}{(k-1)\left(1 - \frac{4\log(k+1)}{k-3}\right)}. \tag{2.1}$$

Moreover, if p is an odd prime and $\ell \in \mathbb{N}$ such that there exists a Hadamard matrix of size $4p^\ell$ (which holds e.g. *for $p = 5$ and ℓ even) and $k = 4p^\ell - 1$, then*

$$\eta(H_k, H_k) > \frac{0.154\,k - 1}{k - 1 - \log(k+1)}. \tag{2.2}$$

The proof of the bound (2.1) uses the inequality $\eta^\star(H_k, H_k) \leq R^\star(H_k)/c^\star(H_k)$. To show $R^\star(H_k) \leq \log(k+1)$, we prove $\chi^\star(H_k) \leq k+1$ (by constructing feasible operators from a $(k+1)$-dimensional orthonormal representation of H_k) and then conclude using the sub-multiplicativity of $\chi^\star$ under strong graph powers. To show $c^\star(H_k) \geq (k-1)\left(1 - \frac{4\log(k+1)}{k-3}\right)$, we use the celebrated *quantum teleportation* scheme of [4] to exhibit an explicit protocol that achieves such capacity on any channel with confusability graph H_k. This proof-technique appears to be new in the context of zero-error entanglement-assisted communication.

To show (2.2) we use properties of the fractional chromatic number and vertex transitivity of H_k by which we lower bound $\eta(H_k, H_k)$ by lower bounding $\alpha(\overline{H_k})$ and upper bounding $\alpha(H_k)$. The lower bound uses the existence of a Hadamard matrix of size $k+1$ and the upper bound combines the linear algebra method of Alon [1] with the beautiful construction of certain polynomials in [2].

References

[1] N. Alon. *The Shannon capacity of a union*, Combinatorica **18** (3) (1998), 301–310.

[2] D. Barrington, R. Beigel and S. Rudich, *Representing Boolean functions as polynomials modulo composite numbers*, Comput. Complexity **4** (4) (1994), 367–382.

[3] S. Beigi, *Entanglement-assisted zero-error capacity is upper-bounded by the Lovász ϑ function*, Phys. Rev. A **82** (1) (2010), 010303.

[4] C. H. Bennett, G. Brassard, C. Crépeau, R. Jozsa, A. Peres and W. K. Wootters, *Teleporting an unknown quantum state via dual classical and Einstein-Podolsky-Rosen channels*, Phys. Rev. Lett. **70** (1993), 1895–1899.

[5] J. Briët, H. Buhrman and D. Gijswijt, *Violating the Shannon capacity of metric graphs with entanglement*, Proc. Nat. Acad. Sci. U.S.A., 2012.

[6] T. S. Cubitt, D. Leung, W. Matthews and A. Winter, *Improving zero-error classical communication with entanglement*, Phys. Rev. Lett. **104** (23) (2010), 230503.

[7] R. Duan, S. Severini and A. Winter, *Zero-error communication via quantum channels, noncommutative graphs, and a quantum Lovász number*, IEEE Trans. Inf. Theory **59** (2) (2013), 1164 –1174.

[8] J. Körner and A. Orlitsky, *Zero-error information theory*, IEEE Trans. Inf. Theory **44** (6) (1998), 2207–2229.

[9] D. Leung, L. Mancinska, W. Matthews, M. Ozols and A. Roy, *Entanglement can increase asymptotic rates of zero-error classical communication over classical channels*, Comm. Math. Phys. **311** (2012), 97–111.

[10] L. Lovász, *On the Shannon capacity of a graph*, IEEE Trans. Inf. Theory **25** (1) (1979), 1–7.

[11] J. Nayak, E. Tuncel and K. Rose, *Zero-error source-channel coding with side information*, IEEE Trans. Inf. Theory **52** (10) (2006), 4626–4629.

[12] C. Shannon, *The zero error capacity of a noisy channel*, IRE Trans. Inf. Theory **2** (3) (1956), 8–19.

[13] M. Szegedy, *A note on the ϑ number of Lovász and the generalized Delsarte bound*, FOCS 1994 (1994), 36–39.

[14] H. S. Witsenhausen, *The zero-error side information problem and chromatic numbers*, IEEE Trans. Inf. Theory **22** (5) (1976), 592–593.

[15] M. Xia and G. Liu, *An infinite class of supplementary difference sets and Williamson matrices*, J. Combin. Theory Ser. A **58** (2) (1991), 310–317.

Ramsey Theory

Ramsey numbers for bipartite graphs with small bandwidth

Guilherme O. Mota[1], Gábor N. Sárközy[2], Mathias Schacht[3] and Anusch Taraz[4]

Abstract. We determine asymptotically the two color Ramsey numbers for bipartite graphs with small bandwidth and constant maximum degree and the three color Ramsey numbers for *balanced* bipartite graphs with small bandwidth and constant maximum degree. In particular, we determine asymptotically the two and three color Ramsey numbers for grid graphs.

1 Introduction and Results

For graphs $G_1, G_2, ..., G_r$, the Ramsey number $R(G_1, G_2, ..., G_r)$ is the smallest positive integer n such that if the edges of a complete graph K_n are partitioned into r disjoint color classes giving r graphs $H_1, H_2, ..., H_r$, then at least one H_i $(1 \leq i \leq r)$ contains a subgraph isomorphic to G_i. The existence of such a positive integer is guaranteed by Ramsey's classical theorem. The number $R(G_1, G_2, \ldots, G_r)$ is called the Ramsey number for the graphs $G_1, G_2, \ldots, G_r$. Determining $R(G_1, G_2, \ldots, G_r)$ for general graphs appears to be a difficult problem (see *e.g.* [13]). For $r = 2$, a well-known theorem of Gerencsér and Gyárfás [7] states that $R(P_n, P_n) = \left\lfloor \frac{3n-2}{2} \right\rfloor$, where P_n denotes the path with $n \geq 2$ vertices.

[1] Instituto de Matemática e Estatística, Universidade de São Paulo, mota@ime.usp.br

[2] Computer Science Department, Worcester Polytechnic Institute, USA, gsarkozy@cs.wpi.edu; and Alfréd Rényi Institute of Mathematics, Hungarian Academy of Sciences, Budapest, Hungary

[3] Fachbereich Mathematik, Universität Hamburg, schacht@math.uni-hamburg.de

[4] Zentrum Mathematik, Technische Universität München, Germany, taraz@ma.tum.de

The first author was supported by FAPESP (2009/06294-0 and 2012/00036-2). He gratefully acknowledges the support of NUMEC/USP, Project MaCLinC/USP. The second author was supported in part by the National Science Foundation under Grant No. DMS–0968699 and by OTKA Grant K104373. The third author was supported by DFG grant SCHA 1263/4-1. The fourth author was supported in part by DFG grant TA 309/2-2. The cooperation was supported by a joint CAPES-DAAD project (415/ppp-probral/po/D08/11629, Proj. no. 333/09).

In [9] more general trees were considered. For a tree T, we write t_1 and t_2, $t_2 \geq t_1$, for the sizes of the vertex classes of T as a bipartite graph. Note that if $2t_1 \geq t_2$, then $R(T, T) \geq 2t_1 + t_2 - 1$, since the following edge-coloring of $K_{2t_1+t_2-2}$ has no monochromatic copy of T. Partition the vertices into two classes V_1 and V_2 such that $|V_1| = t_1 - 1$ and $|V_2| = t_1 + t_2 - 1$, then use color "red" for all edges inside the classes and use color "blue" for all edges between the classes. On the other hand, if $2t_1 < t_2$, a similar edge-coloring of K_{2t_2-2} with two classes both of size $t_2 - 1$ shows that $R(T, T) \geq 2t_2$. Thus, $R(T, T) \geq \max\{2t_1 + t_2, 2t_2\} - 1$. Haxell, Łuczak and Tingley proved in [9] that for a tree T with maximum degree $o(t_2)$, this lower bound is the asymptotically correct value of $R(T, T)$.

We try to extend this to bipartite graphs with small bandwidth (although with a more restrictive maximum degree condition). A graph is said to have *bandwidth* at most b, if there exists a labelling of the vertices by numbers $1, \ldots, n$ such that for every edge $\{i, j\}$ of the graph we have $|i - j| \leq b$. We will focus on the following class of bipartite graphs.

Definition 1.1. A bipartite graph H is called a (β, Δ)-graph if it has bandwidth at most $\beta|V(H)|$ and maximum degree at most Δ. Furthermore, we say that H is a balanced (β, Δ)-graph if it has a legal 2-coloring $\chi : V(H) \to [2]$ such that $1 - \beta \leq |\chi^{-1}(1)|/|\chi^{-1}(2)| \leq 1 + \beta$.

For example, all bounded degree planar graphs G are $(\beta, \Delta(G))$-graphs for any $\beta > 0$ [3]. Our first theorem is an analogue of the result in [9] for (β, Δ)-graphs.

Theorem 1.2. *For every $\gamma > 0$ and natural number Δ, there exist a constant $\beta > 0$ and natural number n_0 such that for every (β, Δ)-graph H on $n \geq n_0$ vertices with a legal 2-coloring $\chi : V(H) \to [2]$ where $t_1 = |\chi^{-1}(1)|$ and $t_2 = |\chi^{-1}(2)|$, $t_1 \leq t_2$, we have*

$$R(H, H) \leq (1 + \gamma) \max\{2t_1 + t_2, 2t_2\}.$$

For more recent results on the Ramsey number of graphs of higher chromatic number and sublinear bandwidth, we refer the reader to the recent paper by Allen, Brightwell and Skokan [1].

For $r \geq 3$ less is known about Ramsey numbers. Proving a conjecture of Faudree and Schelp [5], it was shown in [8] that for sufficiently large n $R(P_n, P_n, P_n) = 2n - 1$, for odd n and $R(P_n, P_n, P_n) = 2n - 2$, for even n. Asymptotically this was also proved independently by Figaj and Łuczak [6]. Benevides and Skokan proved [2] that $R(C_n, C_n, C_n) = 2n$ for sufficiently large even n. In our second theorem we extend these results (asymptotically) to balanced (β, Δ)-graphs.

Theorem 1.3. *For every $\gamma > 0$ and natural number Δ, there exist a constant $\beta > 0$ and natural number n_0 such that for every balanced (β, Δ)-graph H on $n \geq n_0$ vertices we have $R(H, H, H) \leq (2+\gamma)n$.*

In particular, Theorems 1.2 and 1.3 determine asymptotically the two and three color Ramsey numbers for grid graphs.

We conclude this section with a few words about the proof method for our main theorems. The proof of Theorem 1.2 combines ideas from [9] and [4], while the proof of Theorem 1.3 follows a similar approach as in [6], again, together with the result in [4]. Since the strategies for both theorems are close to each other, we focus on the proof of Theorem 1.3, for which we present an outline in the next section. Details can be found in [12].

2 Sketch of the proof of Theorem 1.3

Here we will sketch the main ideas of our proof. The proof relies on the regularity method for graphs and we refer the reader to the survey [10] for related notation and definitions.

The first part of the proof follows the same pattern as the proof by Figaj and Łuczak [6] for the case where H is a path. Namely, we apply a multicolored variant of Szemerédi's Regularity Lemma [14] to the 3-colored complete graph K_N with $N = (2+\gamma)n$ and get a partition with a very dense reduced graph. The edges of the reduced graph inherit the majority color of the respective pair. Applying Lemma 8 from [6] gives us a monochromatic tree T in the reduced graph that contains a matching M covering almost half of the vertices.

Switching back from the reduced graph to the colored complete graph, we denote by G_T the subgraph of K_N whose vertices are contained in the clusters represented by the vertices of T and whose edges run inside the pairs represented by the edges of T and have the same color as the edges of T. Thus G_T is a monochromatic subgraph of K_N whose regular pairs are arranged in a structure mirroring that of T and all have density at least $1/3$. Finally, we localize almost spanning super-regular subgraphs in the pairs in G_T represented by edges in M and denote the subgraph formed by the union of these pairs by $G_M \subset G_T$.

To understand the motivation for the second part, recall that our overall goal is to embed H into G_T. Notice that G_M has in fact enough vertices to accomodate all of H. Indeed, most of the vertices of H will be mapped to G_M, and we will only have to use parts from $G_T \setminus G_M$ because we may need to connect the various parts of H embedded into G_M. Let us explain this more precisely by assuming for the moment that H is just a path. Let

m be an integer which is just a bit smaller than the size of the clusters in G_T (that we assume to be all of the same size). Applying the Blow-up Lemma by Komlós, Sárközy and Szemerédi [11], we then embed the first m vertices from each color class of H into the super-regular pair represented by the first matching edge in M.

To be able to 'reach' the next super-regular pair in G_M, where we can embed the next $m+m$ vertices of H, we need to make use of the fact that the vertices representing these two pairs are connected by a path in T. This path translates into a sequence of regular pairs in G_T, into each of which we embed an intermediate edge of H, thereby 'walking' towards the next super-regular pair in G_M. In this way, we only use few edges of the regular pairs in $G_T \setminus G_M$, thus keeping them regular all the way through, and leaving a bit of space in the super-regular pairs in G_M, in case we need to walk through them later again.

The task for the second part of our proof is to restructure our balanced (β, Δ)-graph H in such a way that it behaves like the path in the embedding approach described before. Here two major problems occur:

- Suppose for example that H is a graph consisting of a path whose vertices are labelled by $1, \dots, n$, with some additional edges between vertices whose label differ by at most βn and have different parity (because H is bipartite). For such a graph 'making the connections' as above is now more difficult. Suppose, for instance, that we have a chain of regular pairs (V_i, V_{i+1}) in G_T for $i = 1, \dots, 4$ and want to use it to 'walk' with H from V_1 to V_5. We cannot simply assign vertex 1 to V_1, then 2 to V_2 and so on up to 5 to V_5, because maybe $\{2, 5\}$ forms an edge in H but (V_2, V_5) is not a regular pair in G_T.
  The solution to this problem is to walk more slowly: assign vertex 1 to V_1, then with the vertices $2, 3, \dots, \beta n+1$ alternate between V_2 and V_3, the next βn vertices continue the zig-zag pattern between V_3 and V_4, and finally we send the last vertex to V_5. What does this buy us? Consider, *e.g.*, the final vertex y that got mapped to V_5. Due to the bandwidth condition, all its potential neighbours were embedded in V_3 or V_4, and due to the parity condition, they must all lie in V_4. This is good, because we have a regular pair (V_4, V_5).
- The second problem that we have to face is as follows. By definition, H has a 2-coloring of its vertices that uses both colors similarly often *in total*, but this does not have to be true *locally* – among the first $m+m$ vertices of H, there could be far more vertices of color 1 than of color 2, which means that our approach to embed them into a super-regular pair with two classes of the same size would fail.
  The solution to this problem is to re-balance H. We use an ordering

of H with bandwidth at most βn and cut H into small blocks of size ξn, where $\beta n \ll \xi n \ll m$. Then it is not hard to see that we can obtain a new ordering of the vertices of H by changing the order in which the blocks appear, so that in every interval of blocks summing to roughly m consecutive vertices of H the two colors are balanced up to $2\xi n$ vertices. We can now assign the blocks forming these intervals to super-regular pairs in G_M in such a way that they there represent a balanced 2-coloring and can therefore be embedded via the Blow-up Lemma into the super-regular pair.

Both these problems can appear at the same time, but one can combine these two solutions. Hence H can indeed be embedded into G_T similarly to the example of the path example given above, which finishes the proof.

References

[1] P. ALLEN, G. BRIGHTWELL and J. SKOKAN, *Ramsey-goodness – and otherwise*, Combinatorica, to appear.

[2] F. S. BENEVIDES and J. SKOKAN, *The 3-colored Ramsey number of even cycles*, J. Combin. Theory Ser. B **99** (4) (2009), 690–708.

[3] J. BÖTTCHER, K. P. PRUESSMANN, A. TARAZ and A. WÜRFL, *Bandwidth, expansion, treewidth, separators and universality for bounded-degree graphs*, European J. Combin. **31** (2010), no. 5, 1217–1227.

[4] J. BÖTTCHER, M. SCHACHT and A. TARAZ, *Proof of the bandwidth conjecture of Bollobás and Komlós*, Math. Ann. **343** (2009), no. 1, 175–205.

[5] R. J. FAUDREE and R. H. SCHELP, *Path Ramsey numbers in multicolorings*, J. Combinatorial Theory Ser. B **19** (1975), no. 2, 150–160.

[6] A. FIGAJ and T. ŁUCZAK, *The Ramsey number for a triple of long even cycles*, J. Combin. Theory Ser. B **97** (2007), no. 4, 584–596.

[7] L. GERENCSÉR and A. GYÁRFÁS, *On Ramsey-type problems*, Ann. Univ. Sci. Budapest. Eötvös Sect. Math. **10** (1967), 167–170.

[8] A. GYÁRFÁS, M. RUSZINKÓ, G. N. SÁRKÖZY and E. SZEMERÉDI, *Three-color Ramsey numbers for paths*, Combinatorica **27** (2007), no. 1, 35–69.

[9] P. E. HAXELL, T. ŁUCZAK and P. W. TINGLEY, *Ramsey numbers for trees of small maximum degree*, Combinatorica **22** (2002), no. 2, 287–320, Special issue: Paul Erdős and his mathematics.

[10] J. KOMLÓS and M. SIMONOVITS, *Szemerédi's regularity lemma and its applications in graph theory*, Combinatorics, Paul Erdős is

eighty, Vol. 2 (Keszthely, 1993), Bolyai Soc. Math. Stud., vol. 2, János Bolyai Math. Soc., Budapest, 1996, 295–352.

[11] J. KOMLÓS, G. N. SÁRKÖZY and E. SZEMERÉDI, *Blow-up lemma*, Combinatorica **17** (1997), no. 1, 109–123.

[12] G. O. MOTA, G. N. SÁRKÖZY, M. SCHACHT and A. TARAZ, *Ramsey numbers for bipartite graphs with small bandwidth*, manuscript.

[13] S. P. RADZISZOWSKI, *Small Ramsey numbers*, Electron. J. Combin. **1** (1994 (Lastest update: 2011)), Dynamic Survey 1, 84 pp.

[14] E. SZEMERÉDI, *Regular partitions of graphs*, Problèmes combinatoires et théorie des graphes (Colloq. Internat. CNRS, Univ. Orsay, Orsay, 1976), Colloq. Internat. CNRS, Vol. 260, CNRS, Paris, 1978, 399–401.

Polynomial bounds on geometric Ramsey numbers of ladder graphs

Josef Cibulka[1], Pu Gao[2], Marek Krčál[1], Tomáš Valla[3] and Pavel Valtr[1]

Abstract. We prove that the geometric Ramsey numbers of the ladder graph on $2n$ vertices are bounded by $O(n^3)$ and $O(n^{10})$, in the convex and general case, respectively. We also prove polynomial upper bounds of geometric Ramsey numbers of pathwidth-2 outerplanar triangulations in both convex and general cases.

1 Introduction and basic definitions

A finite set $P \subset \mathbb{R}^2$ of points is in a *general position* if no three points of P are collinear. The *complete geometric graph on* P, denoted by K_P, is the complete graph with vertex set P, whose edges are drawn as the straight-line segments between pairs of points of P.

The set of points P is in *convex position* if P is the set of vertices of a convex polygon. If P is in convex position, we say that K_P is a *convex complete geometric graph.*

Károlyi, Pach and Tóth [4] introduced the concept of Ramsey numbers for geometric graphs as follows. Given a graph G, the *geometric Ramsey number* of G, denoted by $R_g(G)$, is the smallest integer n such that every complete geometric graph K_P on n vertices with edges arbitrarily coloured by two colours contains a monochromatic non-crossing copy of G. The *convex geometric Ramsey number* of G, $R_c(G)$, is defined the same way except that K_P is restricted to the convex complete geometric

[1] Department of Applied Mathematics, Charles University, Faculty of Mathematics and Physics, Malostranské nám. 25, 118 00 Praha 1, Czech Republic. Email: cibulka@kam.mff.cuni.cz, krcal@kam.mff.cuni.cz

[2] Max-Planck-Institut für Informatik, Saarbrücken, Saarland, Germany.
Email: janegao@mpi-inf.mpg.de

[3] Czech Technical University, Faculty of Information Technology, Prague, Czech Republic. Email: tomas.valla@fit.cvut.cz

This research was started at the 2nd Emléktábla Workshop held in Gyöngyöstarján, January 24–27, 2011. Research was supported by the project CE-ITI (GAČR P202/12/G061) of the Czech Science Foundation and by the grant SVV-2012-265313 (Discrete Methods and Algorithms). Josef Cibulka and Pavel Valtr were also supported by the project no. 52410 of the Grant Agency of Charles University. Pu Gao was supported by the Humboldt Foundation and is currently affiliated with University of Toronto.

graph. A graph G is said to be *outerplanar* if G can be drawn in the plane without any edge crossing and with all vertices of G incident to the unbounded face. Apparently, the numbers $R_g(G)$ and $R_c(G)$ are finite only if G is outerplanar, and it follows immediately from the definitions that $R_c(G) \leq R_g(G)$ for every outerplanar graph G.

The Ramsey numbers of outerplanar graphs, as well as of all planar graphs, are bounded by a function linear in the number of vertices by a result of Chen and Schelp [2]. However, the corresponding geometric Ramsey numbers can be larger and it remains open whether there is a general polynomial bound for all outerplanar graphs. By a simple constructive proof, it is easy to see that for any G on n vertices that contains a Hamilton cycle, $R_c(G) \geq (n-1)^2 + 1$. It has also been proved that, for any cycle C_n on n vertices, $R_g(C_n) \leq 2(n-2)(n-1) + 2$ [5], and this upper bound is only known to be tight when $n \in \{3, 4\}$. Károlyi *et al.* [5] found the exact value $R_c(P_n) = 2n - 3$ and the upper bound $R_g(P_n) \in O(n^{3/2})$, where P_n is a path on $n > 2$ vertices. The bounds $2n - 3 \leq R_g(P_n) \leq O(n^{3/2})$ remain the best known bounds on the geometric Ramsey number of paths.

In this extended abstract, we contribute to this subject by showing polynomial upper bounds on the geometric Ramsey numbers of the ladder graphs, see Definition 2.1, and their generalisation. In Section 2, we show that the geometric Ramsey numbers of the ladder graph on $2n$ vertices are bounded by $O(n^3)$ and $O(n^{10})$ in the convex and general case, respectively. In Section 3, we generalise the polynomial upper bounds to the class of all subgraphs of pathwidth-2 outerplanar triangulations, see Definition 3.1. These bounds are $20n^7$ and $O(n^{22})$ in the convex and general case, respectively

We note here that all colourings in this extended abstract, unless specified, refer to edge colourings. As a convention, in any 2-colouring, we assume that the colours used are blue and red.

We abbreviate the set $\{1, 2, \ldots, k\}$ with $[k]$ and $\{l, l+1, \ldots, k\}$ with $[l, k]$. We write $(x_i)_{i=1}^k$ for the sequence $x_1, x_2, \ldots, x_k$. The sequence of vertices $(v_i)_{i=1}^{\ell+1}$ is a *path of colour c and length ℓ* if every pair $\{v_i v_{i+1}\}$, $i \in [\ell]$ is an edge and has colour c. A sequence $(A_i)_{i=1}^k$ is said to be a *partition* of A if A_i are pairwise disjoint and $\cup_{i=1}^k A_i = A$.

2 Ladder graphs

The ladder graphs are defined as follows.

Definition 2.1. For any integer $n \geq 1$, the ladder graph on $2n$ vertices, denoted by L_{2n}, is the graph composed of two paths $(u_i)_{i=1}^n$ and $(v_i)_{i=1}^n$, together with the set of edges $\{u_i v_i : i \in [n]\}$.

In the following two subsections, we prove upper bounds on $R_c(L_{2n})$ and $R_g(L_{2n})$. Both proofs use the following lemma due to Gritzmann *et al.* [3].

Lemma 2.2 (Gritzmann *et al.* 1991 [3]). *Let G be an outerplanar graph on n vertices and let P be a set of n points in general position. Then K_P contains a non-crossing copy of G.*

2.1 Convex position

In this section, let C denote a set of $32n^3$ points in convex position. That is, C is the set of vertices of some convex polygon. We label the vertices $v_1, v_2, \ldots, v_{|C|}$ in the clockwise order starting at an arbitrarily chosen vertex v_1. We write $v_i \prec v_j$ if and only if $i < j$. Let $A, B \subset C$. We say that A precedes B and write $A \prec B$ if and only if for every $u \in A$ and every $v \in B$, $u \prec v$. Notice that if $A \prec B$, then the sets A and B can be separated by a line.

For a pair of disjoint vertex sets (L, R), $L \subset C$, $R \subset C$, the complete bipartite graph on (L, R), denoted by $K_{L,R}$, is the set of edges $\{u, v\}$, where $u \in L$ and $v \in R$. A complete bipartite graph $K_{L,R}$ is said to be *well-split* if $L \prec R$ or $R \prec L$. A well-split $K_{m,n}$ is a well-split $K_{L,R}$, for some L and R such that $|L| = m$, $|R| = n$.

The following lemmas are used frequently throughout the proofs.

Lemma 2.3. *If a 2-colouring of K_C contains a monochromatic well-split $K_{2n^2,2n^2}$, then it contains a monochromatic non-crossing copy of L_{2n}.*

Lemma 2.4. *Let N be a positive integer. Let G be the complete graph on a set A of points in general position and let $(A_i)_{i=1}^n$ be a partition of A with $|A_i| \geq N$ for every $i \in [n]$. Then for any 2-colouring of the edges of G, either there is a red path $(u_i)_{i=1}^n$ with $u_i \in A_i$ for each $i \in [n]$ or for some $i \in [n-1]$ there exists a blue $K_{B_i,B_{i+1}}$ with $B_i \subseteq A_i$, $B_{i+1} \subseteq A_{i+1}$ and $\min\{|B_i|, |B_{i+1}|\} \geq N/2$.*

Using Lemma 2.3 and 2.4, we are able to prove the following theorem.

Theorem 2.5. *For every $n \geq 1$, $R_c(L_{2n}) \leq 32n^3$.*

2.2 General geometric position

Definition 2.6. Two sets of points A and B in the plane are *mutually avoiding* if $|A|, |B| \geq 2$ and no line subtended by a pair of points in A intersects the convex hull of B, and vice versa.

A simple example of a pair of mutually avoiding sets are sets A and B such that $A \cup B$ is in convex position and A and B can be separated by a straight line. Observe that for any mutually avoiding pair (A, B), every point in A "sees" all the vertices in B in the same order and vice versa. That is, there are unique total orders $u_1 \prec u_2 \prec \cdots \prec u_{|A|}$ of the points in A and $v_1 \prec v_2 \prec \cdots \prec v_{|B|}$ of the the points in B such that every point in B "sees" $u_1, \ldots, u_{|A|}$ consecutively in a clockwise order before seeing any vertex in B, whereas every point in A "sees" $v_1, \ldots, v_{|B|}$ consecutively in a counterclockwise order before seeing any vertex in A. A path $(p_i)_{i=1}^{\ell}$ in either A or B is an *increasing path* if $p_1 \prec p_2 \prec \cdots \prec p_\ell$.

For any two sets of vertices A_1, A_2 both contained in A (or B), we write $A_1 \prec A_2$ if and only if for every $u \in A_1$ and $v \in A_2$, $u \prec v$. An embedding of the complete bipartite graph $K_{m,n}$ is said to be *well-split* if the two sets of points representing the two vertex parts are mutually avoiding.

The following proposition follows from the definition of a pair of mutually avoiding sets.

Proposition 2.7. *Assume A and B are mutually avoiding. Let $P_u = (x_i)_{i=1}^n$ be an increasing path in A and let $P_v = (y_i)_{i=1}^n$ be an increasing path in B. Then the ladder graph composed of the paths P_u and P_v and edges $\{\{x_i, y_i\} : i \in [n]\}$ is non-crossing.*

Given a set of points in general position, the following theorem guarantees the existence of two mutually avoiding subsets, each with reasonably large sizes.

Theorem 2.8 (Aronov *et al.* 1994 [1]). *Let A' and B' be two sets of points separated by a line, each of size $6n^2$. Then there exist mutually avoiding subsets $A \subset A'$ and $B \subset B'$ such that A and B are both of size n.*

By Lemma 2.2 and Proposition 2.7, we have the following generalisation of Lemma 2.3.

Lemma 2.9. *If a 2-colouring of K_P contains a monochromatic well-split $K_{2n^2,2n^2}$, then it contains a monochromatic non-crossing L_{2n}.*

A complete geometric bipartite graph $K_{L,R}$ is said to be *separable* if L and R can be separated by a line. Notice that if $L \cup R$ is in convex position, then $K_{L,R}$ is separable if and only if it is well-split. Obviously, every complete bipartite geometric graph $K_{L,R}$ contains a separable complete bipartite graph with parts of sizes $|L|/2$ and $|R|/2$. However, all complete bipartite geometric graphs that we encounter in subsequent proofs are separable, so we state the following corollary of Theorem 2.8 and Lemma 2.9 for separable complete bipartite graphs only.

Corollary 2.10. *Every 2-colouring of K_P containing a monochromatic separable $K_{24n^4,24n^4}$ contains a monochromatic non-crossing copy of L_{2n}.*

We conclude by the resulting theorem.

Theorem 2.11. *The geometric Ramsey number of the ladder graph L_{2n} satisfies $R_g(L_{2n}) = O(n^{10})$.*

3 Generalisation to pathwidth-2 outerplanar triangulations

An *outerplanar triangulation* G is a planar graph that can be drawn in the plane in such a way that the outer face is incident with all the vertices of G and every other face is incident with exactly three vertices.

The pathwidth of a graph was first defined by Robertson and Seymour [6] as follows. A *path decomposition* of a graph G is a sequence $(G_i)_{i=1}^m$ of subgraphs of G such that each edge of G is in at least one of G_i and for every vertex v of G, the set of graphs G_i containing v forms a contiguous subsequence of $(G_i)_{i=1}^m$. The *pathwidth* of a graph G is the smallest k such that G has a path decomposition in which every G_i has at most $k+1$ vertices. Let $pw(G)$ denote the pathwidth of G.

We now give an equivalent definition of pathwidth-2 outerplanar triangulations that will be used in the later proofs.

Definition 3.1. Let $PW_2(n)$ be the class of outerplanar triangulations G whose vertices can be decomposed into two disjoint sets $V_u \cup V_v = V(G)$ such that the subgraphs induced by the two sets, $P_u = G[V_u]$ and $P_v = G[V_v]$, are paths.

We then state the following theorem.

Theorem 3.2. *For any $G \subseteq G' \in PW_2(n)$, $R_c(G) \leq 20n^7$ and $R_g(G) \leq O(n^{22})$.*

References

[1] B. ARONOV, P. ERDŐS, W. GODDARD, D. J. KLEITMAN, M. KLUGERMAN, J. PACH and L. J. SCHULMAN, *Crossing families*, Combinatorica **14** (2) (1994), 127–134.

[2] G. T. CHEN and R. H. SCHELP, *Graphs with linearly bounded Ramsey numbers*, Journal of Combinatorial Theory, Series B **57** (1) (1993), 138–149.

[3] P. Gritzmann, B. Mohar, J. Pach and R. Pollack, *Embedding a planar triangulation with vertices at specified points*, Am. Math. Monthly **98** (1991), 165–166 (Solution to problem E3341).

[4] G. Károlyi, J. Pach and G. Tóth, *Ramsey-type results for geometric graphs, I*, Discrete & Computational Geometry **18** (3) (1997), 247–255.

[5] G. Károlyi, J. Pach, G. Tóth and P. Valtr, *Ramsey-type results for geometric graphs, II*, Discrete & Computational Geometry **20** (3) (1998), 375–388.

[6] N. Robertson and P. D. Seymour, *Graph minors. I. Excluding a forest*, Journal of Combinatorial Theory, Series B **35** (1) (1983), 39–61.

Geometry and Surfaces

Arrangements of pseudocircles and circles

Ross J. Kang[1] and Tobias Müller[1]

Abstract. An arrangement of pseudocircles is a finite collection of Jordan curves in the plane with the additional properties that **(i)** no three of the curves meet in a point; **(ii)** every two curves meet in at most two points; and **(iii)** if two curves meet in a point p, then they cross at p.

We say that two arrangements $\mathscr{C} = (c_1, \ldots, c_n)$, $\mathscr{D} = (d_1, \ldots, d_n)$ are equivalent if there is a homeomorphism φ of the plane onto itself such that $\varphi[c_i] = d_i$ for all $1 \leq i \leq n$. Linhart and Ortner (2005) gave an example of an arrangement of five pseudocircles that is not equivalent to an arrangement of circles, and conjectured that every arrangement of at most four pseudocircles is equivalent to an arrangement of circles. We prove their conjecture.

We consider two related recognition problems. The first is the problem of deciding, given a pseudocircle arrangement, whether it is equivalent to an arrangement of circles. The second is deciding, given a pseudocircle arrangement, whether it is equivalent to an arrangement of convex pseudocircles. We prove that both problems are NP-hard, answering questions of Bultena, Grünbaum and Ruskey (1998) and of Linhart and Ortner (2008).

We also give an example of a collection of convex pseudocircles with the property that its intersection graph (*i.e.* the graph with one vertex for each pseudocircle and an edge between two vertices if and only if the corresponding pseudocircles intersect) cannot be realized as the intersection graph of a collection of circles. This disproves a folklore conjecture communicated to us by Pyatkin.

1 Introduction

An *arrangement of pseudocircles* is a finite list $\mathscr{C} = (c_1, \ldots, c_n)$ of Jordan curves in the plane satisfying the following three conditions:

1. no three curves intersect in a point;
2. every two curves intersect in at most two points; and
3. if two curves meet in a point p, then they cross at p.

[1] Utrecht University, Utrecht, the Netherlands. E-mail: ross.kang@gmail.com, t.muller@uu.nl

Part of the work in this paper was done while this author was at Durham University, supported by EPSRC grant EP/G066604/1. He is now supported by a VENI grant from Netherlands Organisation for Scientific Research (NWO).
Part of the work in this paper was done while this author was supported by a VENI grant from Netherlands Organisation for Scientific Research (NWO).

We will say that two arrangements $\mathscr{C}=(c_1,\ldots,c_n)$ and $\mathscr{D}=(d_1,\ldots,d_n)$ are *equivalent* if there exists a homeomorphism φ from the plane onto itself with the property that $\varphi[c_i]=d_i$ for all $1\le i\le n$.

Naturally, an arrangement of pseudocircles $\mathscr{C}=(c_1,\ldots,c_n)$ is called an *arrangement of circles* if each c_i is a circle. We will say that an arrangement of pseudocircles is *circleable* if it is equivalent to an arrangement of circles. In this work, we are interested in a number of problems which are related to the question of whether a given pseudocircle arrangement is circleable. Unfortunately, due to page limitations, we are unable to provide any full proofs for the results given below; however, we give indications of the proof methodology used.

ACKNOWLEDGEMENTS. We thank Artem Pyatkin for bringing to our attention the folkore conjecture described in Section 4 below as well as for helpful pointers to the literature.

2 A smallest circleable arrangement

A natural question is whether every arrangement of pseudocircles is circleable. This question was studied before by Linhart and Ortner [6], who showed that the pseudocircle arrangement in Figure 2.1 is not circleable.

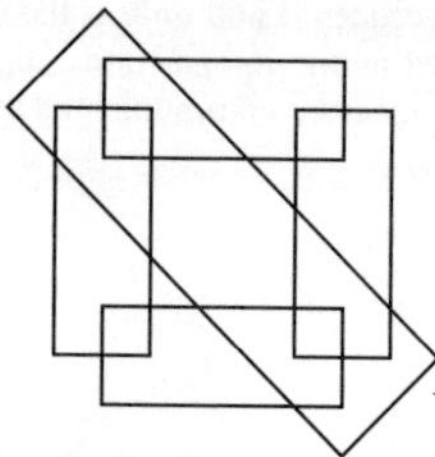

Figure 2.1. An arrangement of five pseudocircles that is not circleable.

They also conjectured that this is a minimal example of a non-circleable pseudocircle arrangement. We confirm their conjecture.

Theorem 2.1. *Every arrangement of at most four pseudocircles is equivalent to an arrangement of circles.*

The proof of this theorem necessitates an involved case analysis which is thankfully shortened by the use of circle inversions in the extended complex plane. Theorem 2.1 provides a natural analogue of a celebrated result of Goodman and Pollack [5], who showed that every arrangement of up to eight pseudolines is equivalent to an arrangement of lines. Prior to this an example was known of an arrangement of nine pseudolines not equivalent to an arrangement of lines.

3 Circleability and convexibility complexity

In this section, we shall discuss the computation complexity of some pseudocircle recognition problems. Before we do so, we must appropriately define the input format.

Given an arbitrary list $\mathscr{C} = (c_1, \ldots, c_n)$ of Jordan curves in the plane, the *arrangement multigraph* for $\mathscr{C}$ is defined as follows: the vertices are all the intersection points among the curves, while the edges are all the maximal curve segments that have no intersection points. This multigraph is already naturally endowed with an embedding in the plane, which we may specify combinatorially for instance by a rotation system (*i.e.* each vertex v is assigned a cyclic permutation of the edges incident with v). Because of conditions **(i)** and **(iii)** of a pseudocircle arrangement, we can straightforwardly derive from the embedded multigraph the information of which edge belongs to which curve of $\mathscr{C}$. Therefore, we may assume that the input to the computational problems below are given as the rotation system of a multigraph.

Given that not all pseudocircle arrangements are circleable, one might wonder if there is an efficient characterization of circleable arrangements. We write CIRCLEABILITY for the computational problem of deciding, given a list $\mathscr{C}$ of Jordan curves in the plane, whether $\mathscr{C}$ can be realized as an oriented circle arrangement. We indeed consider a restricted version of this problem, CONVEX CIRCLEABILITY, on the input of arrangements of convex pseudocircles. By using a corollary of a deep result of Mnëv [8] on the recognition of stretchable pseudoline arrangements (see also Shor [10]), we show the following.

Theorem 3.1. *CONVEX CIRCLEABILITY is NP-hard.*

It is natural also to consider the recognition problem for arrangements of convex pseudocircles. We say that an arrangement of pseudocircles is *convexible* if it is equivalent to an arrangement of convex pseudocircles. We write PSEUDOCIRCLE CONVEXIBILITY for the problem of deciding, given a list $\mathscr{C}$ of Jordan curves in the plane, whether $\mathscr{C}$ can be realized as an arrangement of convex pseudocircles. Bultena, Grünbaum and Ruskey [1] have asked about the complexity of this computational problem. Later, Linhart and Ortner [7] asked the weaker question of whether there exists a pseudocircle arrangement that is not convexible. Again using the result of Mnëv, we answer both of these questions.

Theorem 3.2. *PSEUDOCIRCLE CONVEXIBILITY is NP-hard.*

4 A folklore conjecture

If $\mathscr{A} = (A_1, \ldots, A_n)$ is a list of sets, then the *intersection graph* of $\mathscr{A}$ is the graph $G = (V, E)$ with vertex set $V = \{1, \ldots, n\}$ and an edge $ij \in E$ if and only if $A_i \cap A_j \neq \emptyset$. A folklore conjecture that was communicated to us by Artem Pyatkin [9] states that every intersection graph of a list of convex curves is also the intersection graph of a list of circles. (We do not use the word "arrangement" here because we do not necessarily want to impose the restrictions 1–3 above.) This conjecture was apparently inspired by the work of Dobrynin and Mel′nikov [2–4] on the chromatic number of "arrangement graphs" of Jordan curves in the plane (*i.e.* graphs whose vertices are the intersection points of the curves and whose edges are the curve segments between these intersection points). To get a feel for the conjecture observe for instance that, while all the pseudocircles of the arrangement in Figure 2.1 are convex curves and the arrangement is not equivalent to any arrangement of circles, one can easily construct a family of five circles in the plane with the same intersection graph. We are however able to produce a counterexample to the folklore conjecture by placing additional pseudocircles as in Figure 4.1.

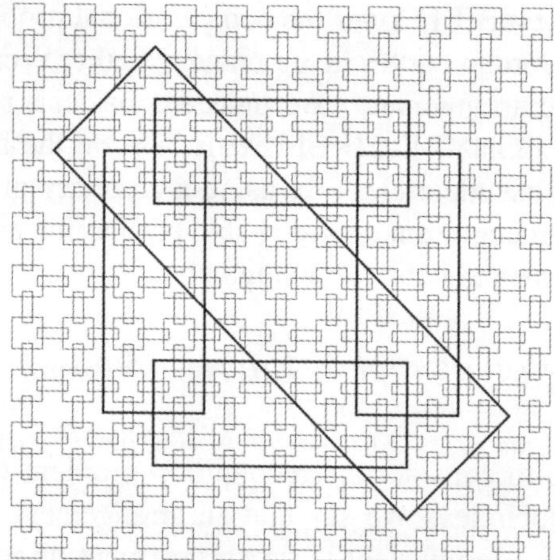

Figure 4.1. A collection of convex curves such that no collection of circles defines the same intersection graph.

Theorem 4.1. *The intersection graph of the convex pseudocircles in Figure* 4.1 *cannot be realized as the intersection graph of a list of circles.*

The idea for this additional "gridding" of smaller pseudocircles is that it ensures that, if the resultant intersection graph were realizable as the intersection graph of a list of circles, then the original (Linhart-Ortner) arrangement of pseudocircles would be circleable, a contradiction.

References

[1] B. BULTENA, B. GRÜNBAUM and F. RUSKEY, *Convex drawings of intersecting families of simple closed curves*, In: "11th Canadian Conference on Computational Geometry", 1998, 18–21.

[2] A. A. DOBRYNIN and L. S. MEL'NIKOV, *Counterexamples to Grötzsch-Sachs-Koester's conjecture*, Discrete Math. **306** (6) (2006), 591–594.

[3] A. A. DOBRYNIN and L. S. MEL'NIKOV, *Infinite families of 4-chromatic Grötzsch-Sachs graphs*, J. Graph Theory **59** (4) (2008), 279–292.

[4] A. A. DOBRYNIN and L. S. MEL'NIKOV, *4-chromatic edge critical Grötzsch-Sachs graphs*, Discrete Math. **309** (8) (2009), 2564–2566.

[5] J. E. GOODMAN and R. POLLACK, *Proof of Grünbaum's conjecture on the stretchability of certain arrangements of pseudolines*, J. Combin. Theory Ser. A **29**(3) (1980), 385–390.

[6] J. LINHART and R. ORTNER, *An arrangement of pseudocircles not realizable with circles*, Beiträge Algebra Geom. **46**(2) (2005), 351–356.

[7] J. LINHART and R. ORTNER, *A note on convex realizability of arrangements of pseudocircles*, Geombinatorics **18** (2) (2008), 66–71.

[8] N. E. MNËV, *The universality theorems on the classification problem of configuration varieties and convex polytopes varieties*, In: "Topology and Geometry—Rohlin Seminar", volume 1346 of Lecture Notes in Math., Springer, Berlin, 1988, 527–543.

[9] A. V. Pyatkin. Personal communication, December 2011.

[10] P. W. SHOR, *Stretchability of pseudolines is NP-hard*, In: "Applied Geometry and Discrete Mathematics", volume 4 of DIMACS Ser. Discrete Math. Theoret. Comput. Sci., Amer. Math. Soc., Providence, RI, 1991, 531–554.

Extended abstract for structure results for multiple tilings in 3D

Nick Gravin[1], Mihail N. Kolountzakis[2], Sinai Robins[1]
and Dmitry Shiryaev[1]

Abstract. We study multiple tilings of 3-dimensional Euclidean space by a convex body. In a multiple tiling, a convex body P is translated with a discrete multiset Λ in such a way that each point of $\mathbb{R}^d$ gets covered exactly k times, except perhaps the translated copies of the boundary of P. It is known that all possible multiple tilers in $\mathbb{R}^3$ are zonotopes. In $\mathbb{R}^2$ it was known by the work of M. Kolountzakis [9] that, unless P is a parallelogram, the multiset of translation vectors Λ must be a finite union of translated lattices (also known as quasi periodic sets). In that work [9] the author asked whether the same quasi-periodic structure on the translation vectors would be true in $\mathbb{R}^3$. Here we prove that this conclusion is indeed true for $\mathbb{R}^3$.

Namely, we show that if P is a convex multiple tiler in $\mathbb{R}^3$, with a discrete multiset Λ of translation vectors, then Λ has to be a finite union of translated lattices, unless P belongs to a special class of zonotopes. This exceptional class consists of two-flat zonotopes P, defined by the Minkowski sum of two 2-dimensional symmetric polygons in $\mathbb{R}^3$, one of which may degenerate into a single line segment. It turns out that rational two-flat zonotopes admit a multiple tiling with an aperiodic (non-quasi-periodic) set of translation vectors Λ. We note that it may be quite difficult to offer a visualization of these 3-dimensional non-quasi-periodic tilings, and that we discovered them by using Fourier methods.

The study of multiple tilings of Euclidean space began in 1936, when the famous Minkowski facet-to-facet conjecture [15] for classical tilings was extended to the setting of k-tilings with the unit cube, by Furtwängler [3]. Minkowski's facet-to-facet conjecture states that for any *lattice* tiling of $\mathbb{R}^d$ by translations of the unit cube, there exist at least two translated cubes that share a facet (face of co-dimension 1). The conjecture was strengthened by Furtwängler [3] who conjectured the same conclusion for any *multiple* lattice tiling.

[1] Division of Mathematical Sciences, Nanyang Technological University SPMS, MAS-03-01, 21 Nanyang Link, Singapore 637371. Email: ngravin@pmail.ntu.edu.sg, rsinai@ntu.edu.sg, shir0010@ntu.edu.sg

[2] Department of Mathematics, University of Crete, Knossos Ave., GR-714 09, Iraklio, Greece. Email: kolount@math.uoc.gr

To define a multiple tiling, suppose we translate a convex body P with a discrete multiset Λ, in such a way that each point of $\mathbb{R}^d$ gets covered exactly k times, except perhaps the translated copies of the boundary of P. We then call such a body a k-tiler, and such an action has been given the following names in the literature: a **k-tiling**, a **tiling at level** k, a **tiling with multiplicity** k, and sometimes simply a **multiple tiling**. We may use any of these synonyms here, and we immediately point out, for polytopes P, a trivial but useful algebraic equivalence for a tiling at level k:

$$\sum_{\lambda \in \Lambda} \mathbf{1}_{P+\lambda}(x) = k, \tag{1}$$

for almost all $x \in \mathbb{R}^d$, where $\mathbf{1}_P$ is the indicator function of the polytope P.

Furtwängler's conjecture was disproved by Hajós [7] for dimension larger than 3 and for $k \geq 9$ while Furtwängler himself [3] proved it for dimension at most 3. Hajós [8] also proved Minkowski's conjecture in all dimensions. The ideas of Furtwängler were subsequently extended (but still restricted to cubes) by the important work of Perron [16], Robinson [17], Szabó [21], Gordon [4] and Lagarias and Shor [11]. These authors showed that for some levels k and dimensions d and under the lattice assumption as well as not, a facet-to-facet conclusion for k-tilings is true in $\mathbb{R}^d$, while for most values of k and d it is false.

It was known to Bolle [2] that in $\mathbb{R}^2$, every k-tiling convex polytope has to be a centrally symmetric polygon, and using combinatorial methods Bolle [2] gave a characterization for all polygons in $\mathbb{R}^2$ that admit a k-tiling with a *lattice* Λ of translation vectors. Kolountzakis [10] proved that if a convex polygon P tiles $\mathbb{R}^2$ multiply with *any* discrete multiset Λ, then Λ must be a *finite union of two-dimensional lattices*. The ingredients of Kolountzakis' proof include the idempotent theorem for the Fourier transform of a measure. Roughly speaking, the idempotent theorem of Meyer [14] tells us that if the square of the Fourier transform of a measure is itself, then the support of the measure is contained in a finite union of lattices.

A multiple tiling is called **quasi-periodic** if its multiset of discrete translation vectors Λ is a finite union of translated lattices, not necessarily all of the same dimension.

Theorem (Kolountzakis, 2002 [9]). *Suppose that K is a symmetric convex polygon which is not a parallelogram. Then K admits only quasi-periodic multiple tilings if any.*

In this work, we extend this result to $\mathbb{R}^3$, and we also find a fascinating class of polytopes analogous to the parallelogram of the theorem above. To describe this class, we first recall the definition of a **zonotope**, which is the Minkowski sum of a finite number of line segments. In other words, a zonotope equals a translate of $[-\mathbf{v}_1, \mathbf{v}_1] + \cdots + [-\mathbf{v}_N, \mathbf{v}_N]$, for some positive integer N and vectors $\mathbf{v}_1, \ldots, \mathbf{v}_N \in \mathbb{R}^d$. A zonotope may equivalently be defined as the projection of some l-dimensional cube. A third equivalent condition is that for a d-dimensional zonotope, all of its k-dimensional faces are centrally symmetric, for $1 \leq k \leq d$. For example, the zonotopes in $\mathbb{R}^2$ are the centrally symmetric polygons.

We shall say that a polytope $P \subseteq \mathbb{R}^3$ is a **two-flat zonotope** if P is the Minkowski sum of $n + m$ line segments which lie in the union of two different two-dimensional subspaces H_1 and H_2. In other words, H_1 contains n of the segments and H_2 contains m of the segments (if one of the segments belongs to both H_1 and H_2 we list it twice, once for each plane). Equivalently, P may be thought of as the Minkowski sum of two 2-dimensional symmetric polygons one of which may degenerate into a single line segment.

It turns out that if P is a rational two-flat zonotope, then P admits a k-tiling with a non-quasi-periodic set of translation vectors Λ. For some of the classical study of 1-tilings, and their interesting connections to zonotopes, the reader may refer to the work of [12, 13, 19, 23], and [1]. Here we find it very useful to use the intuitive language of distributions [18, 20] in order to think – and indeed discover – facts about k-tilings. To that end we introduce the distribution (which is locally a measure)

$$\delta_\Lambda := \sum_{\lambda \in \Lambda} \delta_\lambda, \tag{2}$$

where δ_λ is the Dirac delta function at $\lambda \in \mathbb{R}^d$. To develop some intuition, we may check formally that

$$\delta_\Lambda * \mathbf{1}_P = \sum_{\lambda \in \Lambda} \delta_\lambda * \mathbf{1}_P = \sum_{\lambda \in \Lambda} \mathbf{1}_{P+\lambda},$$

so that from the first definition (1) of k-tiling, we see that a polytope P is a k-tiler if and only if

$$\delta_\Lambda * \mathbf{1}_P = k. \tag{3}$$

Suppose the polytope P tiles multiply with the translates $\Lambda \subseteq \mathbb{R}^d$. We will need to understand some basic facts about how the Λ points are distributed.

For any symmetric polytope P, and any face $F \subset P$, we define F^- to be the face of P symmetric to F with respect to P's center of symmetry. We call F^- the **opposite face** of F. We use the standard convention of boldfacing all vectors, to differentiate between $\mathbf{v}$ and v, for example. We furthermore use the convention that $[\mathbf{e}]$ denotes the 1-dimensional line segment from 0 to the endpoint of the vector $\mathbf{e}$. Whenever it is clear from context, we will also write $[e]$ to denote the same line segment - for example, in the case that e denotes an edge of a polytope.

Suppose $P \in \mathbb{R}^3$ is a zonotope (symmetric polytope with symmetric facets). A collection of four edges of P is called a **4-legged-frame** if whenever e is one of the edges then there exist two vectors $\boldsymbol{\tau}_1$ and $\boldsymbol{\tau}_2$ such that the four edges are

$$[e],\ [e] + \boldsymbol{\tau}_1,\ [e] + \boldsymbol{\tau}_2 \text{ and } [e] + \boldsymbol{\tau}_1 + \boldsymbol{\tau}_2,$$

and such that the edges $[e]$ and $[e] + \boldsymbol{\tau}_1$ belong to the same face of P and the edges $[e] + \boldsymbol{\tau}_2$ and $[e] + \boldsymbol{\tau}_1 + \boldsymbol{\tau}_2$ belong to the opposite face.

With the notion of 4-legged frames, we can introduce the following so-called **intersection property**, which plays an important role in the proof of the main theorem.

Suppose P is a k-tiler with a discrete multiset Λ, in $\mathbb{R}^3$. We say that the intersection property holds, if

$$\bigcap_{e,\tau_1,\tau_2} \left(\mathbf{e}^{\perp} \cup \boldsymbol{\tau}_1^{\perp} \cup \boldsymbol{\tau}_2^{\perp}\right) = \{0\}, \tag{4}$$

where the intersection above is taken over all sets of 4-legged frames of P.

Recently, a structure theorem for convex k-tilers in $\mathbb{R}^d$ was found, and is as follows.

Theorem (Gravin, Robins, Shiryaev 2012 [6]). *If a convex polytope k-tiles $\mathbb{R}^d$ by translations, then it is centrally symmetric and its facets are centrally symmetric.*

This theorem generalizes the theorem for 1-tilers by Minkowski [15]. One-tiler case was extensively studied in the past, and the complete characterization for 1-tilers was given independently by Venkov [22] and McMullen [13].

It follows immediately from the latter theorem that a k-tiler $P \subset \mathbb{R}^3$ is necessarily a zonotope. The following theorem extends the result of Kolountzakis [9] from $\mathbb{R}^2$ to $\mathbb{R}^3$, providing a structure theorem for multiple tilings by polytopes in three dimensions.

The main result here is the following theorem [5]:

Main Theorem. *Suppose a polytope P k-tiles $\mathbb{R}^3$ with a discrete multiset Λ, and suppose that P is not a two-flat zonotope. Then Λ is a finite union of translated lattices.*

Although the proof is involved, one of the main ideas is to compute the Fourier transform of any 4-legged frame of a polytope, and show that its zeros form a certain countable union of hyperplanes. Another main idea is to show that if the intersection property holds for P, then δ_Λ has discrete support.

We also show that each rational two-flat zonotope admits a very peculiar non-quasi-periodic k-tiling. We note that it may be quite difficult to offer a visualization of these 3-dimensional non-quasi-periodic tilings, and that we discovered them by using Fourier methods.

Open Questions

The proof of the main theorem does not directly generalize to dimensions higher that 3, since k-tilers in these dimensions are no longer all zonotopes. So it might require some new ideas and methods to deal with the higher dimension case, and we propose this as a primary direction for future work.

It would also be very helpful to generalize Bolle's characterization of 2-dimensional lattice k-tilers to higher dimensions.

Another important topic of future research would be to generalize the Venkov-McMullen characterization [13] from 1-tilers in $\mathbb{R}^n$ to k-tilers. It is already established that any k-tiler in $\mathbb{R}^n$ is centrally symmetric and has centrally symmetric facets, and it is reasonable to conjecture that it is enough to add one more condition on co-dimension 2 facets to get a complete characterization.

Finally, it would be interesting to prove or disprove the following conjecture: if a polytope has a quasi-periodic multiple tiling, then it also has a tiling with just one lattice.

References

[1] A. D. ALEXANDROV, "Convex Polyhedra", Springer Monographs in Mathematics. Springer-Verlag, Berlin, 2005. Translated from the 1950 Russian edition by N. S. Dairbekov, S. S. Kutateladze and A. B. Sossinsky, With comments and bibliography by V. A. Zalgaller and appendices by L. A. Shor and Yu. A. Volkov.

[2] U. BOLLE, *On multiple tiles in E^2*, In: "Intuitive geometry (Szeged, 1991)", volume 63 of Colloq. Math. Soc. János Bolyai, North-Holland, Amsterdam, 1994, 39–43.

[3] P. FURTWÄNGLER, *Über Gitter konstanter Dichte*, Monatsh. Math. Phys. **43** (1) (1936), 281–288.

[4] B. GORDON, *Multiple tilings of Euclidean space by unit cubes*, Comput. Math. Appl. **39** (11) (2000), 49–53. Sol Golomb's 60th Birthday Symposium (Oxnard, CA, 1992).

[5] N. GRAVIN, M. N. KOLOUNTZAKIS, S. ROBINS and D. SHIRYAEV, *Structure results for multiple tilings in 3d*, submitted.

[6] N. GRAVIN, S. ROBINS and D. SHIRYAEV, *Translational tilings by a polytope, with multiplicity* Combinatorica **32** (6) (2012), 629–648.

[7] G. HAJÓS, *Többméretu terek befedése kockaráccsal*, Mat. Fiz. Lapok **45** (1938), 171–190.

[8] G. HAJÓS, *Über einfache und mehrfache Bedeckung des n-dimensionalen Raumes mit einem Würfelgitter*, Math. Z. **47** (1941), 427–467.

[9] M. KOLOUNTZAKIS, *On the structure of multiple translational tilings by polygonal regions*, Discrete & Computational Geometry **23** (4) (2000), 537–553.

[10] M. N. KOLOUNTZAKIS, *The study of translational tiling with Fourier analysis*, In: "Fourier Analysis and Convexity", Appl. Numer. Harmon. Anal., Birkhäuser Boston, Boston, MA, 2004, 131–187.

[11] J. LAGARIAS and P. SHOR, *Keller's cube-tiling conjecture is false in high dimensions*, Bulletin, new series, of the American Mathematical Society **27** (2) (1992), 279–283.

[12] P. MCMULLEN, *Space tiling zonotopes*, Mathematika **22** (2) (1975), 202–211.

[13] P. MCMULLEN, *Convex bodies which tile space by translation* Mathematika **27** (01) (1980), 113–121.

[14] Y. MEYER "Nombres de Pisot, nombres de Salem, et analyse harmonique", volume 117. Springer-Verlag, 1970.

[15] H. MINKOWSKI, "Diophantische approximationen: Eine einführung in die zahlentheorie", volume 2. BG Teubner, 1907.

[16] O. PERRON, *Über lückenlose Ausfüllung des n-dimensionalen Raumes durch kongruente Würfel*, Mathematische Zeitschrift **46** (1) (1940), 1–26.

[17] R. M. ROBINSON, *Multiple tilings of n-dimensional space by unit cubes*, Math. Z. **166** (3) (1979), 225–264.

[18] W. RUDIN, "Functional Analysis" McGraw-Hill, New York, 1973.

[19] G. C. SHEPHARD, *Space-filling zonotopes*, Mathematika **21** (1974), 261–269.

[20] R. STRICHARTZ, "A Guide to Distribution Theory and Fourier Transforms", World Scientific Pub Co Inc, 2003.

[21] S. SZABÓ, *Multiple tilings by cubes with no shared faces*, Aequationes Mathematicae **25** (1) (1982), 83–89.

[22] B. VENKOV, *On a class of Euclidean polyhedra*, Vestnik Leningrad Univ. Ser. Mat. Fiz. Him **9** (1954), 11–31.

[23] G. M. ZIEGLER, "Lectures on Polytopes", volume 152 of *Graduate Texts in Mathematics*. Springer-Verlag, New York, 1995.

On the nonexistence of k-reptile simplices in $\mathbb{R}^3$ and $\mathbb{R}^4$

Jan Kynčl[1] and Zuzana Safernová[1]

Abstract. A d-dimensional simplex S is called a *k-reptile* (or a *k-reptile simplex*) if it can be tiled without overlaps by k simplices with disjoint interiors that are all mutually congruent and similar to S. For $d = 2$, triangular k-reptiles exist for many values of k and they have been completely characterized by Snover, Waiveris, and Williams. On the other hand, the only k-reptile simplices that are known for $d \geq 3$, have $k = m^d$, where m is a positive integer. We substantially simplify the proof by Matoušek and the second author that for $d = 3$, k-reptile tetrahedra can exist only for $k = m^3$. We also prove a weaker analogue of this result for $d = 4$ by showing that four-dimensional k-reptile simplices can exist only for $k = m^2$.

1 Introduction

A closed set $X \subset \mathbb{R}^d$ with nonempty interior is called a *k-reptile* (or a *k-reptile set*) if there are sets $X_1, X_2, \ldots, X_k$ with disjoint interiors and with $X = X_1 \cup X_2 \cup \cdots \cup X_k$ that are all mutually congruent and similar to X. Such sets have been studied in connection with fractals and also with crystallography and tilings of $\mathbb{R}^d$ [3,6,8].

It easy to see that whenever S is a d-dimensional k-reptile simplex, then all of $\mathbb{R}^d$ can be tiled by congruent copies of S: indeed, using the tiling of S by its smaller copies $S_1, \ldots, S_k$ as a pattern, one can inductively tile larger and larger similar copies of S. On the other hand, not all space-filling simplices must be k-reptiles for some $k \geq 2$.

Clearly, every triangle tiles $\mathbb{R}^2$. Moreover, every triangle T is a k-reptile for $k = m^2$, since T can be tiled in a regular way with m^2 congruent tiles, each positively or negatively homothetic to T. See *e.g.* Snover *et al.* [16] for an illustration.

[1] Department of Applied Mathematics and Institute for Theoretical Computer Science, Charles University, Faculty of Mathematics and Physics, Malostranské nám. 25, 118 00 Praha 1, Czech Republic; kyncl@kam.mff.cuni.cz, zuzka@kam.mff.cuni.cz

The authors were supported by the project CE-ITI (GACR P202/12/G061) of the Czech Science Foundation, by the grant SVV-2012-265313 (Discrete Models and Algorithms) and by project GAUK 52410. The research was partly conducted during the Special Semester on Discrete and Computational Geometry at École Polytechnique Féderale de Lausanne, organized and supported by the CIB (Centre Interfacultaire Bernoulli) and the SNSF (Swiss National Science Foundation).

The question of characterizing the tetrahedra that tile $\mathbb{R}^3$ is still open and apparently rather difficult. The first systematic study of space-filling tetrahedra was made by Sommerville. Sommerville [17] discovered a list of exactly four tilings (up to isometry and rescaling), but he assumed that all tiles are *properly congruent* (that is, congruent by an orientation-preserving isometry) and meet face-to-face. Edmonds [5] noticed a gap in Sommerville's proof and by completing the analysis, he confirmed that Sommerville's classification of proper, face-to-face tilings is complete. In the non-proper and non face-to-face situations there are infinite families of non-similar tetrahedral tilers. Goldberg [7] described three such families, obtained by partitioning a triangular prism. In fact, Goldberg's first family was found by Sommerville [17] before, but he selected only special cases with a certain symmetry. Goldberg [7] noticed that even the general case admits a proper tiling of $\mathbb{R}^3$. Goldberg's first family also coincides with the family of simplices found by Hill [11], whose aim was to classify *rectifiable* simplices, that is, simplices that can be cut by straight cuts into finitely many pieces that can be rearranged to form a cube. The simplices in Goldberg's second and third families are obtained from the simplices in the first family by splitting into two congruent halves. According to Senechal's survey [14], no other space-filling tetrahedra than those described by Sommerville and Goldberg are known.

For general d, Debrunner [4] constructed $\lfloor d/2 \rfloor + 2$ one-parameter families and a finite number of additional special types of d-dimensional simplices that tile $\mathbb{R}^d$. Smith [15] generalized Goldberg's construction and using Debrunner's ideas, he obtained $(\lfloor d/2 \rfloor + 2)\phi(d)/2$ one-parameter families of space-filling d-dimensional simplices; here $\phi(d)$ is the Euler's totient function. It is not known whether for some d there is an acute space-filling simplex or a two-parameter family of space-filling simplices [15].

In recent years the subject of tilings has received a certain impulse from computer graphics and other computer applications. In fact, our original motivation for studying simplices that are k-reptiles comes from a problem of probabilistic marking of Internet packets for IP traceback [1, 2]. See [12] for a brief summary of the ideas of this method. For this application, it would be interesting to find a d-dimensional simplex that is a k-reptile with k as small as possible.

For dimension 2 there are several possible types of k-reptile triangles, and they have been completely classified by Snover *et al.* [16]. In particular, k-reptile triangles exist for all k of the form $a^2 + b^2$, a^2 or $3a^2$ for arbitrary integers a, b. In contrast, for $d \geq 3$, reptile simplices seem to be much more rare. The only known constructions of higher-dimensional k-reptile simplices have $k = m^d$. The best known examples are the *Hill sim-*

plices (or the *Hadwiger–Hill simplices*) [4,9,11]. A d-dimensional Hill simplex is the convex hull of vectors $0, b_1, b_1+b_2, \ldots, b_1+\cdots+b_d$, where $b_1, b_2, \ldots, b_d$ are vectors of equal length such that the angle between every two of them is the same and lies in the interval $(0, \frac{\pi}{2} + \arcsin \frac{1}{d-1})$.

Concerning nonexistence of k-reptile simplices in dimension $d \geq 3$, Hertel [10] proved that a 3-dimensional simplex is an m^3-reptile using a "standard" way of dissection (which we will not define here) if and only if it is a Hill simplex. He conjectured that Hill simplices are the only 3-dimensional reptile simplices. Herman Haverkort recently pointed us to an example of a k-reptile tetrahedron which is not Hill, which contradicts Hertel's conjecture. In fact, except for the one-parameter family of Hill tetrahedra, three other space-filling tetrahedra described by Sommerville [17] and Goldberg [7] are also k-reptiles for every $k = m^3$. The simplices and their tiling are based on the barycentric subdivision of the cube. The construction can be naturally extended to find similar examples of d-dimensional k-reptile simplices for $d \geq 4$ and $k = m^d$. Matoušek [12] showed that there are no 2-reptile simplices of dimension 3 or larger. For dimension $d = 3$ Matoušek and the second author [13] proved the following theorem.

Theorem 1.1 ([13]). *In $\mathbb{R}^3$, k-reptile simplices (tetrahedra) exist only for k of the form m^3, where m is a positive integer.*

We give a new, simple proof of Theorem 1.1 in Section 3.

Matoušek and the second author [13] conjectured that a d-dimensional k-reptile simplex can exist only for k of the form m^d for some positive integer m. We prove a weaker version of this conjecture for four-dimensional simplices.

Theorem 1.2. *Four-dimensional k-reptile simplices can exist only for k of the form m^2, where m is a positive integer.*

Four-dimensional Hill simplices are examples of k-reptile simplices for $k = m^4$. Whether there exists a four-dimensional m^2-reptile simplex for m non-square remains an open question.

2 Angles in simplices and Coxeter diagrams

Given a d-dimensional simplex S with vertices $v_0, \ldots, v_d$, let F_i be the facet opposite to v_i. A *dihedral angle* $\beta_{i,j}$ of S is the internal angle of the facets F_i and F_j, that meet at the $(d-2)$-face $F_i \cap F_j$.

The *Coxeter diagram* of S is a graph $c(S)$ with labeled edges such that the vertices of $c(S)$ represent the facets of S and for every pair of facets F_i and F_j, there is an edge $e_{i,j}$ labeled by the dihedral angle $\beta_{i,j}$.

The most important tool we use is Debrunner's lemma [4, Lemma 1], which connects the symmetries of a d-simplex with the symmetries of its Coxeter diagram (which represents the "arrangement" of the dihedral angles). This lemma allows us to substantially simplify the proof of Theorem 1.1 and enables us to step up by one dimension and prove Theorem 1.2, which seemed unmanageable before.

Lemma 2.1 (Debrunner's lemma [4]). *Let S be a d-dimensional simplex. The symmetries of S are in one-to-one correspondence with the symmetries of its Coxeter diagram $c(S)$ in the following sense: each symmetry φ of S induces a symmetry Φ of $c(S)$ so that $\varphi(v_i) = v_j \Leftrightarrow \Phi(F_i) = F_j$, and vice versa.*

3 A simple proof of Theorem 1.1

We proceed as in the original proof, but instead of using the theory of scissor congruence, Jahnel's theorem about values of rational angles and Fiedler's theorem, wc only use Debrunner's lemma (Lemma 2.1).

Assume for contradiction that S is a k-reptile tetrahedron where k is not a third power of a positive integer. A dihedral angle α is called *indivisible* if it cannot be written as a linear combination of other dihedral angles in S with nonnegative integer coefficients. Call the edges of S (and of $c(S)$) with dihedral angle α the *α-edges*.

The following lemmas are proved in [13].

Lemma 3.1 ([13, Lemma 3.1]). *If α is an indivisible dihedral angle in S, then the α-edges of S have at least three different lengths.*

Lemma 3.2 ([13, Lemma 3.3]). *One of the following two possibilities occur:*

(i) *All the dihedral angles of S are integer multiples of the minimal dihedral angle α, which has the form $\frac{\pi}{n}$ for an integer $n \geq 3$. If two α-edges meet at a vertex v of S, then the third edge incident to v has dihedral angle $\pi - \alpha$.*
(ii) *There are exactly two distinct dihedral angles β_1 and β_2, each of them occurring three times in S.*

First we exclude case (ii) of Lemma 3.2. The two triples of edges with dihedral angles β_1 and β_2 form either a triangle and a claw, or two paths of length three. In both cases, for each $i \in \{1, 2\}$, the Coxeter diagram of S has at least one nontrivial symmetry which swaps two distinct edges with label β_i. By Debrunner's lemma, the corresponding symmetry of S swaps two distinct edges with dihedral angle β_i, which thus have the same length. But then the edges with dihedral angle β_i have at most 2 different lengths and this contradicts Lemma 3.1, since the smaller of the two angles β_1, β_2 is indivisible.

Now we exclude case (i). Since there are at least three α-edges in S, there is a vertex v of S where two α-edges meet. Let $\beta = \pi - \alpha$. We distinguish several cases depending on the subgraph H_α of $c(S)$ formed by the α-edges.

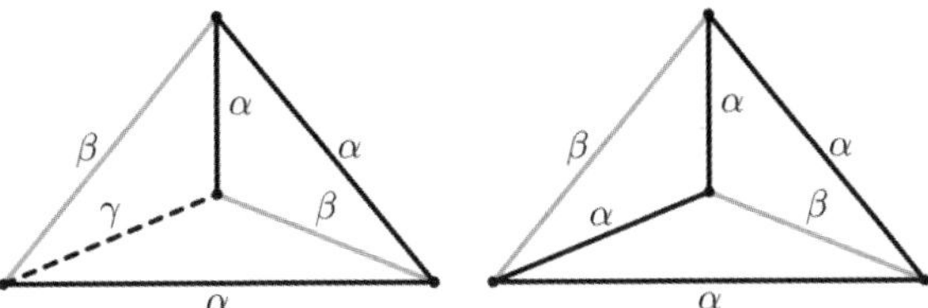

Figure 3.1. The α-edges form a path (left) or a four-cycle (right) in $c(S)$.

- H_α contains three edges incident to a common vertex (which correspond to a triangle in S). Then all the other edges must have the angle β and we get the configuration with three-fold symmetry, which we excluded in case (ii).
- H_α contains a triangle (the corresponding edges in S meet at a single vertex). Then $\beta = \alpha$, and thus $\alpha = \frac{\pi}{2}$, which contradicts Lemma 3.2 (i).
- H_α is a path of length three (this corresponds to a path in S, too). Then two edges have the angle $\beta > \alpha$ and the remaining edge has some angle $\gamma \neq \alpha$. See Figure. 3.1 (left). The resulting Coxeter diagram has a nontrivial involution swapping two α-edges. By Debrunner's lemma, this contradicts Lemma 3.1.
- It remains to deal with the case where H_α is a four-cycle (which corresponds to a four-cycle in S). In this case the remaining two edges have dihedral angle β, so the Coxeter diagram has a dihedral symmetry group D_4 acting transitively on the α-edges. By Debrunner's lemma, all the α-edges have the same length. This again contradicts Lemma 3.1.

We obtained a contradiction in each of the cases, hence the proof of Theorem 1.1 is finished.

References

[1] M. ADLER, *Tradeoffs in probabilistic packet marking for IP traceback*, Proc. 34th Annu. ACM Symposium on Theory of Computing (2002), 407–418.

[2] M. ADLER, J. EDMONDS and J. MATOUŠEK, *Towards asymptotic optimality in probabilistic packet marking*, Proc. 37th Annu. ACM Symposium on Theory of Computing (2005), 450–459.
[3] C. BANDT, *Self-similar sets. V. Integer matrices and fractal tilings of* $\mathbf{R}^n$, Proc. Amer. Math. Soc. **112**(2) (1991), 549–562.
[4] H. E. DEBRUNNER, *Tiling Euclidean d-space with congruent simplexes*, Discrete geometry and convexity (New York, 1982), vol. 440 of Ann. New York Acad. Sci., New York Acad. Sci., New York (1985) 230–261.
[5] A. L. EDMONDS, *Sommerville's missing tetrahedra*, Discrete Comput. Geom. **37** (2) (2007), 287–296.
[6] G. GELBRICH, *Crystallographic reptiles*, Geom. Dedicata **51** (3) (1994), 235–256.
[7] M. GOLDBERG, *Three infinite families of tetrahedral space-fillers*, J. Comb. Theory, Ser. A **16** (1974), 348–354.
[8] S. W. GOLOMB, *Replicating figures in the plane*, Mathematical Gazette **48** (1964), 403–412.
[9] H. HADWIGER, *Hillsche Hypertetraeder* (in German), Gaz. Mat. (Lisboa) **12** (50) (1951), 47–48.
[10] E. HERTEL, *Self-similar simplices*, Beiträge Algebra Geom. **41** (2) (2000), 589–595.
[11] M. J. M. HILL, *Determination of the volumes of certain species of tetrahedra without employment of the method of limits*, Proc. London Math. Soc. **27** (1896), 39–52.
[12] J. MATOUŠEK, *Nonexistence of 2-reptile simplices*, Discrete and Computational Geometry: Japanese Conference, JCDCG 2004, Lecture Notes in Computer Science 3742, Springer, Berlin (2005) 151–160, erratum at http://kam.mff.cuni.cz/ matousek/no2r-err.pdf.
[13] J. MATOUŠEK and Z. SAFERNOVÁ, *On the nonexistence of k-reptile tetrahedra*, Discrete Comput. Geom. **46** (3) (2011), 599–609.
[14] M. SENECHAL, *Which tetrahedra fill space?*, Math. Magazine **54** (5) (1981), 227–243.
[15] W. D. SMITH, *Pythagorean triples, rational angles, and space-filling simplices* (2003), manuscript.
[16] S. L. SNOVER, C. WAIVERIS and J. K. WILLIAMS, *Rep-tiling for triangles*, Discrete Math. **91** (2) (1991), 193–200.
[17] D. M. Y. SOMMERVILLE, *Division of space by congruent triangles and tetrahedra*, Proc. Roy. Soc. Edinburgh **4** (1923), 85–116.

Homogeneous selections from hyperplanes

Imre Bárány[1] and János Pach[2]

Abstract. Given $d+1$ hyperplanes $h_1, \dots, h_{d+1}$ in general position in $\mathbb{R}^d$, let $\Delta(h_1, \dots, h_{d+1})$ denote the unique bounded simplex enclosed by them. There exists a constant $c(d) > 0$ such that for any finite families $H_1, \dots, H_{d+1}$ of hyperplanes in $\mathbb{R}^d$, there are subfamilies $H_i^* \subset H_i$ with $|H_i^*| \ge c(d)|H_i|$ and a point $p \in \mathbb{R}^d$ with the property that $p \in \Delta(h_1, \dots, h_{d+1})$ for all $h_i \in H_i^*$.

1 The main result

Throughout this paper, let $H_1, \dots, H_{d+1}$ be finite families of hyperplanes in $\mathbb{R}^d$ in general position. That is, we assume that (1) no element of $\cup_{i=1}^{d+1} H_i$ passes through the origin, (2) any d elements have precisely one point in common, and (3) no $d+1$ of them have a nonempty intersection. A transversal to these families is an ordered $(d+1)$-tuple $h = (h_1, \dots, h_{d+1}) \in \prod_{i=1}^{d+1} H_i$, where $h_i \in H_i$ for every i.

Given hyperplanes $h_1, \dots, h_{d+1} \subset \mathbb{R}^d$ in general position in $\mathbb{R}^d$, there is a unique simplex denoted by $\Delta = \Delta(h_1, \dots, h_{d+1})$ whose boundary is contained in $\cup_1^{d+1} h_i$. For simpler writing we let $[n]$ stand for the set $\{1, 2, \dots, n\}$. Our main result is the following.

Theorem 1.1. *For any $d \ge 1$, there is a constant $c(d) > 0$ with the following property. Given finite families $H_1, \dots, H_{d+1}$ of hyperplanes in $\mathbb{R}^d$ in general position, there are subfamilies $H_i^* \subset H_i$ with $|H_i^*| \ge c(d)|H_i|$ for $i = 1, \dots, d+1$ and a point $p \in \mathbb{R}^d$ such that p is contained in $\Delta(h)$ for every transversal $h \in \prod_{i=1}^{d+1} H_i$.*

[1] Rényi Institute, Budapest, Hungary. Email: barany.imre@renyi.mta.hu

[2] EPFL, Lausanne, Switzerland. Email: janos.pach@epfl.ch

Research of the first author was partially supported by ERC Advanced Research Grant no 267165 (DISCONV), and by Hungarian National Research Grant K 83767. The second author was partially supported by Hungarian Science Foundation EuroGIGA Grant OTKA NN 102029, by Swiss National Science Foundation Grants 200021-137574 and 200020-144531, and by NSF grant CCF-08-30272. Both authors are grateful to R. Radoičić for his valuable suggestions.

It follows from the general position assumption that the simplices $\Delta(h)$ in Theorem 1.1 also have an interior point in common.

It will be convenient to use the language of hypergraphs. Let $\mathcal{H} = \mathcal{H}(H_1, \dots, H_{d+1})$ be the complete $(d+1)$-partite hypergraph with vertex classes $H_1, \dots, H_{d+1}$. We refer to $\mathcal{H}$ as the *hyperplane hypergraph*, or *h-hypergraph* associated with the hyperplane families $H_1, \dots, H_{d+1}$. The hyperedges of $\mathcal{H}$ are the transversals of the families $H_1, \dots, H_{d+1}$. Our main result can now be reformulated as follows.

Theorem 1.2. *For every positive integer d, there is a constant $c(d) > 0$ with the following property. Every complete $(d+1)$-partite h-hypergraph $\mathcal{H}(H_1, \dots, H_{d+1})$ contains a complete $(d+1)$-partite h-subhypergraph $\mathcal{H}^*(H_1^*, \dots, H_{d+1}^*)$ such that $|H_i^*| \geq c(d)|H_i|$ for all $i \in [d+1]$ and $\bigcap_{h \in \mathcal{H}^*} \Delta(h) \neq \emptyset$.*

In some sense, our theorem extends the following recent and beautiful result of Karasev [5].

Theorem 1.3. [5] *Assume r is a prime power and $t \geq 2r - 1$. Let $\mathcal{H}$ be a complete $(d+1)$-partite h-hypergraph with partition classes of size t. Then there are vertex-disjoint hyperedges (transversals) $h^1, \dots, h^r$ of $\mathcal{H}$ such that $\bigcap_{j=1}^{r} \Delta(h^j) \neq \emptyset$.*

Two hyperedges (transversals) h and h' of $\mathcal{H}$ are vertex-disjoint if h_i and h'_i are distinct for each i.

Our Theorem 1.1 implies a weaker version of Karasev's theorem. Namely, the same conclusion holds with arbitrary r and $t \geq r/c(d)$. Since $c(d)$ will turn out to be doubly exponential in d, our result is quantitatively much weaker than the bound $t \geq 4r$ that follows from Karasev's theorem for any r.

Theorem 1.3 is a kind of dual to Tverberg's famous theorem [8]. In the same sense, our result is dual to the homogeneous point selection theorem of Pach [7] (see also [6]), which guarantees the existence of an absolute constant $c_d > 0$ with the following property. Let $X_1, \dots, X_{d+1}$ be finite sets of points in general position in $\mathbb{R}^d$ with $|X_i| = n$ for every i. Then there exist subsets $X_i^* \subset X_i$ of size at least $c_d n$ for every $i \in [d+1]$ and a point $p \in \mathbb{R}^d$ such that $p \in \operatorname{conv}\{x_1, \dots, x_{d+1}\}$ for all transversals $(x_1, \dots, x_{d+1}) \in \prod_{i+1}^{d+1} X_i^*$. Here the assumption that the sets X_i are of the same size can be removed (see *e.g.* [4]).

To establish Theorem 1.2, we need some preparation. Let h be an edge of the h-hypergraph $\mathcal{H} = \mathcal{H}(H_1, \dots, H_{d+1})$. The simplex $\Delta(h)$ is the convex hull of the points

$$v_i = \bigcap_{j \neq i} h_j, \quad i \in [d+1].$$

Similarly, for $h' \in \mathcal{H}$ the vertices of $\triangle(h')$ are $v'_1, \ldots, v'_{d+1}$. Here v_i (and v'_i) is the vertex of $\triangle(h)$ (and $\triangle(h')$) opposite to the facet contained in h_i (and h'_i, respectively). The *edges* h and h' are said to be of the same type if, for each $i \in [d+1]$, the vertices v_i and v'_i are not separated by either of the hyperplanes h_i and h'_i. We say that the h-hypergraph $\mathcal{H}$ is homogeneous if every pair of its edges is of the same type.

The heart of the proof of Theorem 1.2 is the following "same type lemma" for hyperplanes (*cf.* [2])

Lemma 1.4. *For any $d \geq 1$, there exists a constant $b(d) > 0$ with the following property. Every complete $(d+1)$-partite h-hypergraph $\mathcal{H}(H_1, \ldots, H_{d+1})$ contains a complete $(d+1)$-partite subhypergraph $\mathcal{H}^*(H_1^*, \ldots, H_{d+1}^*)$ with $|H_i^*| \geq b(d)|H_i|$ for all $i \in [d+1]$ which is homogeneous.*

In this extended abstract we show how our main result, Theorem 1.1 follows from the same type lemma for hyperplanes. The proof of the latter appears in the full version of the paper. Actually, there are two separate proofs there. The first one, which provides a better estimate for the value of the constant $b(d)$, uses duality and is based on the original same type lemma (for points), [2] (see also [6]). The second proof is shorter, and it utilizes a far reaching generalization of the same type lemma to semialgebraic relations of several variables, found by Fox, Gromov, Lafforgue, Naor, and Pach [4], see also Bukh and Hubard [3] for a quantitative form. The same result for binary semialgebraic relations was first established by Alon, Pach, Pinchasi, Radoičić, and Sharir [1].

2 Proof of Theorem 1.2

Here we deduce Theorem 1.2 from Lemma 1.4.

Let $\mathcal{H}^*$ denote the complete $(d+1)$-partite subhypergraph of $\mathcal{H}$ whose existence is guaranteed by the lemma. For a fixed $h \in \mathcal{H}^*$ let h_i^+ denote the half-space bounded by h_i that contains vertex v_i of $\triangle(h)$, for $i \in [d+1]$. The lemma implies that, for every hyperedge $k = (k_1, \ldots, k_{d+1}) \in \mathcal{H}^*$ and for every i, the half-space h_i^+ contains the vertex u_i of $\triangle(k)$ opposite to hyperplane k_i. To prove the theorem, it suffices to establish the following claim.

$$\bigcap_{h \in \mathcal{H}^*} \triangle(h) \neq \emptyset.$$

For $h = (h_1, \ldots, h_{d+1}) \in \mathcal{H}^*$, let $\rho(h)$ denote the distance between h_1 and $v_1 = \cap_2^{d+1} h_j$, and let $h' \in \mathcal{H}^*$ be the edge for which $\rho(h)$ is minimal. By the general position assumption, we have $\rho(h') > 0$. Set $v' = \cap_2^{d+1} h'_j$. We show that $v' \in \Delta(h)$ for every $h \in \mathcal{H}^*$, which implies the claim. To see this, we have to verify that $v' \in h_i^+$ for every $h \in \mathcal{H}^*$ and for every i.

This is trivial for $i = 1$. Suppose that $i \geq 2$. By symmetry, we may assume that $i = d + 1$. We have to show that $v' \in h_{d+1}^+$ for every $h_{d+1} \in H_{d+1}^*$.

Assume to the contrary that $v' \notin h_{d+1}^+$ for some $h_{d+1} \in H_{d+1}^*$. Setting $k = (h'_1, \ldots, h'_d, h_{d+1})$, we clearly have $k \in \mathcal{H}^*$. The simplices $\Delta(k)$ and $\Delta(h')$ share the vertex $v_{d+1} = \cap_1^d h'_i$. As $v_{d+1} \in h_{d+1}^+$, by the construction, $v' \notin h_{d+1}^+$ implies that h_{d+1} intersects the segment $[v_{d+1}, v']$ in a point u in its relative interior, see Figure 1. On the other hand, we know that $u = \cap_2^{d+1} k_i$ is the vertex of $\Delta(k)$ opposite to $h_1 = k_1$. Thus, u is closer to $h_1 = k_1$ than v' is. Therefore, we obtain that $\rho(k) < \rho(h')$, contradicting the definition of h'. □

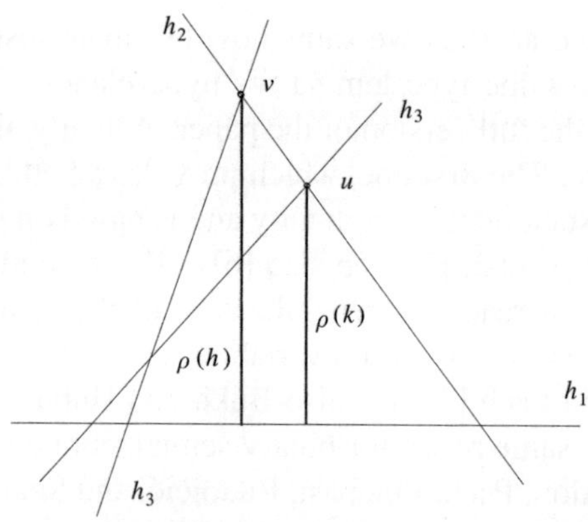

Figure 2.1. Illustration for Theorem 1.2, $d = 2$.

It follows from the above proof that Theorems 1.2 and 1.1 hold with $c(d) = b(d)$.

References

[1] N. ALON, J. PACH, R. PINCHASI, M. RADOIČIĆ and M. SHARIR, *Crossing patterns of semi-algebraic sets*, J. Combin. Theory Ser. A **111** (2005), 310–326.

[2] I. BÁRÁNY and P. VALTR, *Positive fraction Erdős–Szekeres theorem*, Discrete Comput. Geometry **19** (1998), 335–342.

[3] B. BUKH and A. HUBARD, *Space crossing numbers*, In: "Symposium on Computational Geometry", ACM Press, 2011, 163–170.
[4] J. FOX, M. GROMOV, V. LAFFORGUE, A. NAOR and J. PACH, *Overlap properties of geometric expanders*, Journal für die reine und angewandte Mathematik (Crelle's Journal), in press.
[5] R. N. KARASEV, *Dual central point theorems and their generalizations*, Math. Sbornik **199** (2008), 1459–1479.
[6] J. MATOUŠEK, "Lectures on Discrete Geometry", Spinger, Heidelberg, 2002.
[7] J. PACH, *A Tverberg-type result on multicolored simplices*, Computational Geometry: Theory and Appls **1** (1998), 71–76.
[8] H. TVERBERG *A generalization of Radon's theorem*, J. London Math. Soc. **41** (1966), 123–128.

Conic theta functions and their relations to theta functions

Amanda Folsom[1], Winfried Kohnen[2] and Sinai Robins[3]

Abstract. It is natural to ask when the spherical volume defined by the intersection of a sphere at the apex of an integer polyhedral cone is rational. We use number theoretic methods to study a new class of polyhedral functions called conic theta functions, which are closely related to classical theta functions. We show that if K is a Weyl chamber for any finite reflection group, then its conic theta function lies in a graded ring of classical theta functions and in this sense is 'almost' modular. It is then natural to ask whether or not the conic theta functions are themselves modular, and we prove that (generally) they are not. In other words, we uncover some connections between the class of integer polyhedral cones that have a rational solid angle, and the class of conic theta functions that are almost modular.

The present investigations arose from an interest in studying the volume of a spherical polytope, also known as a solid angle, and extending the so-called Gram relations by use of conic theta functions [9]. We study the relationship between volumes of spherical polytopes, and 'almost' modular conic theta functions associated to them, by considering some connections between the following two apriori different problems:

Problem 1. Which lattice polyhedral cones K give rise to spherical polytopes with a rational volume?

Problem 2. Analyzing a certain conic theta function Φ_K attached to a polyhedral cone K, how 'close' is Φ_K to being modular?

We first recall some of the basic definitions from the combinatorial geometry of cones and the theory of modular forms. Suppose we are

[1] Yale University, Mathematics Department, P.O. Box 208283, New Haven, Connecticut 06520-8283, USA. Email: amanda.folsom@yale.edu

[2] Ruprecht-Karls-Universität Heidelberg, Mathematisches Institut, Im Neuenheimer Feld 288, 69120 Heidelberg, Deutschland. Email: winfried@mathi.uni-heidelberg.de

[3] Division of Mathematical Sciences, Nanyang Technological University, SPMS-MAS-03-01, 21 Nanyang Link, Singapore 637371. Email: rsinai@ntu.edu.sg

given a d-dimensional (simple) **polyhedral cone**, defined by

$$K := \left\{ \sum_{j=1}^{d} \lambda_j w_j \mid \text{all } \lambda_j \geq 0 \right\},$$

where the edges of the cone are some fixed set of d linearly independent vectors $w_j \in \mathbb{R}^d$. Such a cone K is called a **pointed cone**, and in the present work every cone has the origin as its vertex. One important special case of a polyhedral cone is the **positive orthant**, defined by $K_0 := \{(x_1, \ldots, x_d) \in \mathbb{R}^d \mid each\ x_j \geq 0\} := \mathbb{R}^d_{\geq 0}$.
For each pointed cone K, we define its **conic theta function** by:

$$\Phi_{K,\mathcal{L}}(\tau) := \sum_{m \in \mathcal{L} \cap K} e^{\pi i \tau ||m||^2}, \tag{1}$$

where τ lies in the complex upper half plane $\mathbb{H} := \{\tau := x + iy \mid x \in \mathbb{R}, y \in \mathbb{R}^+\} \subset \mathbb{C}$, and where $\mathcal{L}$ is a rank d lattice in $\mathbb{R}^d$. If $\mathcal{L}$ is clear from the context, we will only write Φ_K. The conic theta function $\Phi_{K,\mathcal{L}}(\tau)$ given in (1) is reminiscent of the modular theta function

$$\theta(\tau) := \sum_{n \in \mathbb{Z}} e^{\pi i \tau n^2}, \tag{2}$$

a statement that we will make more precise in what follows. The function $\theta(\tau)$ is a classical example of a **modular form**. Loosely speaking, a holomorphic modular form of integer weight k on a suitable subgroup $\Gamma \subseteq \mathrm{SL}_2(\mathbb{Z})$ is any holomorphic function $f : \mathbb{H} \to \mathbb{C}$ satisfying $f(\gamma\tau) = (c\tau + d)^k f(\tau)$ for all $\gamma := \left(\begin{smallmatrix} a & b \\ c & d \end{smallmatrix}\right) \in \Gamma$, as well as a suitable growth condition in the cusps of Γ. It is well known that the **modular group** $\mathrm{SL}_2(\mathbb{Z})$ acts on $\mathbb{H}$ by fractional linear transformation $\left(\begin{smallmatrix} a & b \\ c & d \end{smallmatrix}\right)\tau := \frac{a\tau+b}{c\tau+d}$, and so modular forms can be thought of as complex analytic functions that obey a certain symmetry with respect to this action.

To see where the conic theta function naturally comes from, we let S^{d-1} be the unit sphere centered at the origin. We define the **solid angle** ω_K at the vertex of K by:

$$\omega_K := \frac{\mathrm{vol}\left(K \cap S^{d-1}\right)}{\mathrm{vol}\left(S^{d-1}\right)}. \tag{3}$$

In other words, ω_K is the normalized volume of a $(d-1)$-dimensional spherical polytope. With this normalization, we note that $0 \leq \omega_K \leq 1$, and that in two dimensions we have $\omega_K = \theta/2\pi$, where θ is the usual 2-dimensional angle, measured in radians, at the vertex of the 2-dimensional cone K. It is an elementary fact that

$$\omega_K = \int_K e^{-\pi ||x||^2} dx, \tag{4}$$

We note that when K is replaced by all of Euclidean space, the integral (4) becomes $\int_{\mathbb{R}^d} e^{-\pi||x||^2} dx = 1$, confirming that we do indeed have the proper normalization $0 \le \omega_K \le 1$. For more information about rational pointed cones, and connections between discrete volumes of polytopes and local spherical angle contributions, the reader may consult [11] and [3]. The papers [7, 8], provide further background for solid angles and their relations.

Thus, the foregoing discussion shows that a strong motivation for defining the conic theta function $\Phi_K(\tau)$ is that it is essentially a discrete, Riemann sum approximation to the integral definition of the volume ω_K of a spherical polytope, as defined by (4). We will make precise sense of this intuition in the following section, which will be used later to consider carefully the putative expansion of $\Phi_K(\tau)$ at the cusp $\tau = 0$.

Conic theta functions are also related to the representation numbers of quadratic forms, a link which we explicate here. From the above definitions, and using the lattice $\mathcal{L} := A(\mathbb{Z}^d)$, it is immediate that

$$\Phi_{K,\mathcal{L}}(\tau) = \sum_{m \in \mathcal{L} \cap K} e^{\pi i \tau ||m||^2} = \sum_{m \in \mathbb{Z}^d \cap K} e^{\pi i \tau (m^t (A^t A) m)} = \sum_{k=0}^{\infty} a(k) q^{k/2}, \tag{5}$$

where $q = e^{2\pi i \tau}$, and $a(k) := \#\{m \in \mathbb{Z}^d \mid m^t(A^t A)m = k, \ and \ k \in K\}$. This combinatorial interpretation of the Fourier coefficients tells us that the k'th Fourier coefficient is the number of ways to represent the integer k by the quadratic form $m^t(A^t A)m$, while m is simultaneously constrained to satisfy the finite system of linear inequalities defined by the polyhedral cone K.

Another very recent analysis of cones from a different perspective takes place in [5]. The authors of [5] define certain zeta functions attached to polyhedral cones and analyze conical zeta values as a geometric generalization of the celebrated multiple zeta values.

Here, we endeavor to show that the conic theta function $\Phi_K(\tau)$ is never a modular form. However, when K is a Weyl chamber of a finite reflection group, we show that Φ_K nevertheless belongs to a certain graded ring of theta functions (see Theorem 5 below), and it is in this sense that Φ_K is 'almost' modular. We next give the simplest family of examples of conic theta functions, arising from the positive orthant in each dimension.

Example 3. For the conic theta function of the positive orthant $K_0 := \mathbb{R}^d_{\geq 0}$ (and $\mathcal{L} = \mathbb{Z}^d$), we claim that

$$\Phi_{K_0}(\tau) = \frac{1}{2^d}\left(\theta(\tau) + 1\right)^d, \tag{6}$$

where $\theta(\tau) := \sum_{n\in\mathbb{Z}} e^{\pi i \tau n^2}$, the classical weight 1/2 modular form. In particular,

$$\Phi_{K_0}(\tau) = \frac{1}{2^d}\sum_{k=0}^{d}\binom{d}{k}\theta^k(\tau), \tag{7}$$

a linear combination of modular forms of distinct weights, with nonzero coefficients, and hence Φ_{K_0} is not a modular form.

To see (7), we begin with the case in which the dimension of K_0 equals 1, so that here $K_0 := \mathbb{R}_{\geq 0}$. We note that $1 + \theta(\tau) = 1 + \sum_{n\in\mathbb{Z}} e^{\pi i \tau n^2} = 2 + 2\sum_{n\geq 1} e^{\pi i \tau n^2}$. Therefore, $1 + \theta(\tau) = 2\Phi_{\mathbb{R}_{\geq 0}}(\tau)$. Finally, the relation $\Phi_{K_0}(\tau) = \Phi^d_{\mathbb{R}_{\geq 0}}(\tau)$ gives us the desired expansion (6) above. □

We notice that the positive orthant possesses a lot of symmetry, so it is natural to ask if other cones with less symmetry might not be modular, and for which cones K might we get a phenomenon similar in spirit to the example above, in the sense that Φ_K could be written as a linear combination of "classical" theta functions attached to lower-dimensional lattices.

We also note that, almost by definition, every cone K which is a Weyl chamber necessarily has a *rational* solid angle $\omega_K = \frac{1}{|W|}$, because the Weyl group W tiles $\mathbb{R}^d$ with isometric copies of the cone K. For each even integral lattice $\mathcal{L}$, we define its usual theta function by:

$$\Theta_{\mathcal{L}}(\tau) := \sum_{n\in\mathcal{L}} e^{\pi i \tau ||n||^2},$$

where τ lies in the upper half plane H. Next, we quote the standard fact that when $\mathcal{L}$ is an even integral lattice, the theta function $\Theta_{\mathcal{L}}(\tau)$ turns out to be a modular form, of weight $\frac{d}{2}$ and level N, where N is the smallest positive integer M such that $M\left(A^t A\right)^{-1}$ is also even integral.

Example 4. Consider the 2-dimensional root system defined by

$$S := \{(1,1), (-1,1), (1,-1), (-1,-1), (2,0), (-2,0), (0,2), (0,-2)\},$$

so that we have the root lattice $\mathcal{L}_{root} := \{m(1,1) + n(2,0) \mid m, n \in \mathbb{Z}\}$. One fundamental domain for this group action on $\mathbb{R}^2$ is the polyhedral cone K whose edge vectors are the roots $(1,1)$ and $(2,0)$, and whose (solid) angle is $\omega = \frac{1}{8}$. Here, the conic theta function is

$$\Phi_{K,\mathcal{L}_{root}}(\tau) = \sum_{m\geq 0, n\geq 0} e^{\pi i \tau ||m(2,0)+n(1,1)||^2} \tag{8}$$

$$= \sum_{m\geq 0, n\geq 0} e^{\pi i \tau (4m^2+4mn+2n^2)}. \tag{9}$$

We arrive at the following representation of $\Phi_{K,\mathcal{L}_{root}}(\tau)$ as a nontrivial rational linear combination of classical theta functions:

$$\Phi_{K,\mathcal{L}_{root}}(\tau) = \frac{3}{8} + \frac{1}{4}\sum_{k\in\mathbb{Z}} e^{\pi i \tau (2k^2)} + \frac{1}{4}\sum_{k\in\mathbb{Z}} e^{\pi i \tau (4k^2)} + \frac{1}{8}\sum_{(m,n)\in\mathbb{Z}^2} e^{\pi i \tau (4m^2+4mn+n^2)}. \tag{10}$$

Therefore we see that for this example, $\Phi_{K,\mathcal{L}_{root}}(\tau)$ is a nontrivial rational linear combination of theta functions of different weights. In particular, $\Phi_{K,\mathcal{L}_{root}}(\tau)$ is not modular. □

We define R to be the ring of all finite, rational linear combinations of theta functions $\Theta_{\mathcal{L}}$, for any d-dimensional even integral lattice $\mathcal{L} \subset \mathbb{R}^d$, varying over all dimensions d. The ring R has a natural grading, namely it is graded by the weight $k = \frac{d}{2}$ of the relevant theta functions $\Theta_{\mathcal{L}}$, for each rank d lattice $\mathcal{L} \subset \mathbb{R}^d$. Equivalently, we may also grade R by the dimension d of the lattices $\mathcal{L} \subset \mathbb{R}^d$, as d varies over the positive integers. When a conic theta function lies in R, it is 'almost modular'. The two main results here are the following.

Theorem 5. *If the polyhedral cone K is the Weyl chamber of a finite reflection group W, then the conic theta function $\Phi_{K,2\mathcal{L}_{root}}(\tau)$ is in the graded ring R.*

On the other hand, we also have the following result.

Theorem 6. *Suppose that the polyhedral cone $K \subset \mathbb{R}^d$ has the solid angle ω_K at it vertex, located at the origin, and that $\mathcal{L} := A(\mathbb{Z}^d)$ is an even integral lattice of full rank. If $\frac{\omega_K}{|\det A|}$ is irrational, then $\Phi_{K,\mathcal{L}}(\tau)$ is not a modular form of weight k on any congruence subgroup, and for any $k \in \frac{1}{2}\mathbb{Z}$, $k \geq \frac{1}{2}$.*

The tool that we use for the proof of Theorem 6 is the "q-expansion principle", due to Deligne and Rapoport [2, Théorème 3.9, p.304], which tells us that if an integer weight modular form f has *rational* Fourier coefficients at the cusp $i\infty$, then the Fourier expansion of f at all other cusps must also have rational coefficients.

References

[1] K. Bringmann and A. Folsom, *Almost harmonic Maass forms and Kac-Wakimoto characters*, J. Reine Angew. Math. (Crelle's J.), to appear, (2013).

[2] P. Deligne and M. Rapoport, *Les schemas de modules de courbes elliptiques. (French) Modular functions of one variable*, II (Proc. Internat. Summer School, Univ. Antwerp, Antwerp, (1972), Lecture Notes in Math. **349**, Springer, Berlin, (1973), 143–316.

[3] D. DeSario and S. Robins, *Generalized solid-angle theory for real polytopes*, The Quarterly Journal of Mathematics **62** (4) (2011), 1003–1015.

[4] L. C. Grove and C. T. Benson, "Finite Reflection Groups", Graduate Texts in Mathematics, Vol. **99**, Originally published by Bogden & Quigley, (1971), 2nd ed. Corr. 2nd printing, (1985), 1–156.

[5] Li Guo, Sylvie Paycha and Bin Zhang, *Conical zeta values and their double subdivision relations*, Jan. 2013, arXiv:1301.3370 [math.NT].

[6] W. Kohnen, *On certain generalized modular forms*, Functiones et Approximatio **43** (2010), 23–29.

[7] P. McMullen, *Non-linear angle-sum relations for polyhedral cones and polytopes*, Math. Proc. Cambridge Philos. Soc. **78** (1975), 247–261.

[8] M. A. Perles and G. C. Shephard, *Angle sums of convex polytopes*, Math. Scand. **21** (1967), 199–218.

[9] S. Robins, *An extension of the Gram relations, using conic theta functions*, preprint.

[10] S. D. Schmoll, "Eine Charakterisierung von Spitzenformen", Diploma Thesis, unpublished, Heidelberg (2011).

[11] R. P. Stanley, *Decompositions of rational convex polytopes*, Ann. Discrete Math. **6** (1980), 333–342.

[12] D. Zagier, "The Dilogarithm Function, in Frontiers in Number Theory", Physics and Geometry II, Springer (2007), 3-65.

The Carathéodory number of the P_3 convexity of chordal graphs

Erika M.M. Coelho[1], Mitre C. Dourado[2], Dieter Rautenbach[3] and Jayme L. Szwarcfiter[4]

Abstract. If S is a set of vertices of a graph G, then the convex hull of S in the P_3-convexity of G is the smallest set $H_G(S)$ of vertices of G that contains S such that no vertex in $V(G) \setminus H_G(S)$ has at least two neighbors in S. The Carathéodory number of the P_3 convexity of G is the smallest integer c such that for every set S of vertices of G and every vertex u in $H_G(S)$, there is a set $F \subseteq S$ with $|F| \leq c$ and $u \in H_G(F)$. We describe a polynomial time algorithm to determine the Carathéodory number of the P_3 convexity of a chordal graph.

1 Introduction

Graph convexities are a well studied topic. For a finite, simple, and undirected graph G with vertex set $V(G)$, a *graph convexity* on $V(G)$ is a collection $\mathcal{C}$ of subsets of $V(G)$ such that

- $\emptyset, V(G) \in \mathcal{C}$ and
- $\mathcal{C}$ is closed under intersections.

The sets in $\mathcal{C}$ are called *convex sets* and the *convex hull* in $\mathcal{C}$ of a set S of vertices of G is the smallest set $H_{\mathcal{C}}(S)$ in $\mathcal{C}$ containing S.

Several well known graph convexities $\mathcal{C}$ are defined using some set $\mathcal{P}$ of paths of the underlying graph G. In this case, a subset S of vertices of G is convex, that is, belongs to $\mathcal{C}$, if for every path P in $\mathcal{P}$ whose endvertices belong to S also every vertex of P belongs to S. When $\mathcal{P}$ is the set of all shortest paths in G, this leads to the *geodetic convexity*

[1] Instituto de Informática, Universidade Federal de Goiás, Goiás, Brazil.
Email: erikamorais@inf.ufg.br

[2] Instituto de Matemática, Universidade Federal do Rio de Janeiro, Rio de Janeiro, Brazil. Email: mitre@dcc.ufrj.br

[3] Institut für Optimierung und Operations Research, Universität Ulm, Ulm, Germany. Email: dieter.rautenbach@uni-ulm.de

[4] Instituto de Matemática and COPPE, Universidade Federal do Rio de Janeiro, Instituto Nacional de Metrologia, Qualidade e Tecnologia, Rio de Janeiro, Brazil. Email: jayme@nce.ufrj.br

We gratefully acknowledge financial support by the CAPES/DAAD Probral project “Cycles, Convexity, and Searching in Graphs” (PPP Project ID 56102978).

[2, 8, 12, 14]. The *monophonic convexity* is defined by considering as $\mathcal{P}$ the set of all induced paths of G [9, 10, 13]. The set of all paths of G leads to the *all path convexity* [7]. Similarly, if $\mathcal{P}$ is the set of all triangle paths in G, then $\mathcal{C}$ is the *triangle path convexity* [6]. Here we consider the P_3 *convexity* of G, which is defined when $\mathcal{P}$ is the set of all paths of length two. The P_3 convexity was first considered for directed graphs [11, 15–17]. For undirected graphs, the P_3 convexity was studied in [1,4,5].

A famous result about convex sets in $\mathbb{R}^d$ is *Carathéodory's theorem* [3]. It states that every point u in the convex hull of a set $S \subseteq \mathbb{R}^d$ lies in the convex hull of a subset F of S of order at most $d + 1$. Let G be a graph and let $\mathcal{C}$ be a graph convexity on $V(G)$. The *Carathéodory number* of $\mathcal{C}$ is the smallest integer c such that for every set S of vertices of G and every vertex u in $H_{\mathcal{C}}(S)$, there is a set $F \subseteq S$ with $|F| \leq c$ and $u \in H_{\mathcal{C}}(F)$. A set S of vertices of G is a *Carathéodory set* of $\mathcal{C}$ if the set $\partial H_{\mathcal{C}}(S)$ defined as $H_{\mathcal{C}}(S) \setminus \bigcup_{u \in S} H_{\mathcal{C}}(S \setminus \{u\})$ is not empty. This notion allows an alternative definition of the *Carathéodory number* of $\mathcal{C}$ as the largest cardinality of a Carathéodory set of $\mathcal{C}$.

The Carathéodory number was determined for several graph convexities. The Carathéodory number of the monophonic convexity of a graph G is 1 if G is complete and 2 otherwise [10]. The Carathéodory number of the triangle path convexity of G is 2 whenever G has at least one edge [6]. It is known that the maximum Carathéodory number of the P_3 convexity of a multipartite tournament is 3 [16]. Some general results concerning the Carathéodory number of the P_3 convexity are shown in [1]. On the one hand, [1] contains efficient algorithms to determine the Carathéodory number of the P_3 convexity of trees and, more generally, block graphs. On the other hand, it is NP-hard to determine the Carathéodory number of the P_3 convexity of bipartite graphs [1].

In the present extended abstract we exclusively study the Carathéodory number of P_3 convexities of graphs. Since a graph G uniquely determines its P_3 convexity $\mathcal{C}$, we speak of a Carathéodory set of G and the Carathéodory number $c(G)$ of G. Furthermore, we write $H_G(S)$ and $\partial H_G(S)$ instead of $H_{\mathcal{C}}(S)$ and $\partial H_{\mathcal{C}}(S)$, respectively.

Our result is a polynomial time algorithm for the computation of the Carathéodory number of a chordal graph, which extends results from [1].

2 Results

The following result from [5] plays a central role in our approach.

Theorem 2.1 (Centeno *et al.* [5]). *If u and v are two vertices at distance at most 2 in a 2-connected chordal graph G, then $H_G(\{u, v\}) = V(G)$.*

Let G be a connected chordal graph. Let r be a vertex of G. Let the graph G' arise from G by adding edges between all pairs of vertices of G that lie in the same block of G. Let T be the breadth first search tree of G' rooted in r.

For a vertex u of G, let $V(u)$ denote the set of vertices of G that are either u or a descendant of u in T. By the definition of T, we have that $V(r) = V(G)$ and if u is a vertex of G that is distinct from r, then $V(u)$ is the union of $\{u\}$ and all vertex sets of components of G that do not contain r. Furthermore, the set of children of u in T is the set of all vertices of G that belong to $V(u)$ and lie in a common block of G with u. Note that a vertex of G that is distinct from r is a leaf of T if and only if it is no cut vertex of G.

In order to determine the Carathéodory number of G, we consider the following three values for every vertex u of G:

- $c_{(G,r)}(u)$ is the maximum cardinality of a set S with
  - $S \subseteq V(u)$ and
  - $u \in \partial H_{G[V(u)]}(S)$.
- $c'_{(G,r)}(u)$ is the maximum cardinality of a set S with
  - $S \subseteq V(u)$,
  - $u \notin H_{G[V(u)]}(S)$,
  - $|H_{G[V(u)]}(S) \cap N_G(u)| = 1$, and
  - $H_{G[V(u)]}(S) \cap N_G(u) \subseteq \partial H_{G[V(u)]}(S)$.
- $c''_{(G,r)}(u)$ is the maximum cardinality of a set S with
  - $S \subseteq V(u)$,
  - $u \in H_{G[V(u)]}(S)$, and
  - $H_{G[V(u)]}(S) \cap N_G[u] \subseteq \partial H_{G[V(u)]}(S)$.

Let $S_{(G,r)}(u)$, $S'_{(G,r)}(u)$, and $S''_{(G,r)}(u)$ denote sets of maximum cardinality satisfying the conditions in the above definitions of $c_{(G,r)}(u)$, $c'_{(G,r)}(u)$, and $c''_{(G,r)}(u)$, respectively, that is, for instance, $c_{(G,r)}(u) = |S_{(G,r)}(u)|$. Note that if u is a leaf of T, then no set as in the definition of $c'_{(G,r)}(u)$ exists. In this case, let $c'_{(G,r)}(u) = -\infty$ and let $S'_{(G,r)}(u)$ be undefined.

The following lemma describes recursions for $c_{(G,r)}(u)$, $c'_{(G,r)}(u)$, and $c''_{(G,r)}(u)$, which allow their efficient recursive computation.

Lemma 2.2. *Let G, r, and T be as above. Let u be a vertex of G.*

(a) *$c_{(G,r)}(u)$ is the maximum of the following values:*

   (i) 1.

   (ii) *$c_{(G,r)}(v) + c_{(G,r)}(w)$, where v and w are children of u in T and* $\mathrm{dist}_G(v, w) = 2$.

   (iii) *$c_{(G,r)}(v) + c'_{(G,r)}(w)$, where v and w are children of u in T and* $\mathrm{dist}_G(v, w) = 1$.

   (iv) *$c''_{(G,r)}(v) + c''_{(G,r)}(w)$, where v and w are children of u in T and* $\mathrm{dist}_G(v, w) = 1$.

(b) *$c'_{(G,r)}(u)$ is the maximum of the following values:*

   (i) $-\infty$.

   (ii) *$c_{(G,r)}(u')$, where u' is a child of u in T that is a neighbor of u in G.*

(c) *$c''_{(G,r)}(u)$ is the maximum of the following values:*

   (i) 1.

   (ii) *$c_{(G,r)}(v) + c_{(G,r)}(w)$, where v and w are children of u in T such that v and w are no neighbors of u in G and* $\mathrm{dist}_G(v, w) = 2$.

   (iii) *$c_{(G,r)}(v) + c'_{(G,r)}(w)$, where v and w are children of u in T such that v is no neighbor of u in G and* $\mathrm{dist}_G(v, w) = 1$.

   (iv) $c''_{(G,r)}(v) + c''_{(G,r)}(w)$, where v and w are children of u in T such that v and w are no neighbors of u in G and $\mathrm{dist}_G(v, w) = 1$.

Theorem 2.3. *The Carathéodory number of a chordal graph can be determined in polynomial time.*

Proof. Let G be a chordal graph. Since the Carathéodory number of G is the maximum of the Carathéodory numbers of the components of G, we may assume that G is connected. Since $c(G) = \max\{c_{(G,r)}(r) : r \in V(G)\}$, it suffices to argue that, for every vertex r of G, the value $c_{(G,r)}(r)$ can be determined in polynomial time. In fact, using the recursions given in Lemma 2.2, it is possible to determine $c_{(G,r)}(r)$ in polynomial time calculating the values $c_{(G,r)}(u)$, $c'_{(G,r)}(u)$, and $c''_{(G,r)}(u)$ for all vertices u of G in a bottom up fashion along the corresponding breadth first search tree T. This completes the proof. □

While the running time of the algorithm described in the proof of Theorem 2.3 is obviously polynomial, it is possible to reduce it below some immediate estimates, because many values are essentially calculated several times, that is, there are many triples (r, s, u) of vertices of G with $c_{(G,r)}(u) = c_{(G,s)}(u)$, $c'_{(G,r)}(u) = c'_{(G,s)}(u)$, and $c''_{(G,r)}(u) = c''_{(G,s)}(u)$.

References

[1] R. M. BARBOSA, E. M. M. COELHO, M. C. DOURADO, D. RAUTENBACH and J. L. SZWARCFITER, *On the Carathéodory number for the convexity of paths of order three*, SIAM J. Discrete Math. **26** (2012), 929–939.

[2] J. CÁCERES, C. HERNANDO, M. MORA, I. M. PELAYO, M. L. PUERTAS and C. SEARA, *On geodetic sets formed by boundary vertices*, Discrete Math. **306** (2006), 188–198.

[3] C. CARATHÉODORY, *Über den Variabilitätsbereich der Fourierschen Konstanten von positiven harmonischen Funktionen*, Rend. Circ. Mat. Palermo **32** (1911), 193–217.

[4] C. C. CENTENO, S. DANTAS, M. C. DOURADO, D. RAUTENBACH and J. L. SZWARCFITER, *Convex Partitions of Graphs induced by Paths of Order Three*, Discrete Mathematics and Theoretical Computer Science **12** (2010), 175–184.

[5] C. C. CENTENO, M. C. DOURADO, L. D. PENSO, D. RAUTENBACH and J. L. SZWARCFITER, *Irreversible conversion of graphs*, Theoretical Computer Science 412 (2011), 3693–3700.

[6] M. CHANGAT and J. MATHEW, *On triangle path convexity in graphs*, Discrete Math. **206** (1999), 91–95.

[7] M. CHANGAT, S. KLAVŽAR and H. M. MULDER, *The all-paths transit function of a graph*, Czech. Math. J. **51** (126) (2001), 439–448.

[8] M. C. DOURADO, F. PROTTI, D. RAUTENBACH and J. L. SZWARCFITER, *On the hull number of triangle-free graphs*, SIAM J. Discrete Math. **23** (2010), 2163–2172.

[9] M. C. DOURADO, F. PROTTI and J. L. SZWARCFITER, *Complexity results related to monophonic convexity*, Discrete Appl. Math. **158** (2010), 1269–1274.

[10] P. DUCHET, *Convex sets in graphs II: Minimal path convexity*, J. Combin. Theory, Ser. B **44** (1988), 307–316.

[11] P. ERDŐS, E. FRIED, A. HAJNAL and E.C. MILNER, *Some remarks on simple tournaments*, Algebra Univers. **2** (1972), 238–245.

[12] M. G. EVERETT and S. B. SEIDMAN, *The hull number of a graph*, Discrete Math. **57** (1985), 217–223.

[13] M. FARBER and R. E. JAMISON, *Convexity in graphs and hypergraphs*, SIAM J. Algebraic Discrete Methods **7** (1986), 433–444.

[14] M. FARBER and R. E. JAMISON, *On local convexity in graphs*, Discrete Math. **66** (1987), 231-247.

[15] J. W. MOON, *Embedding tournaments in simple tournaments*, Discrete Math. **2** (1972), 389-395.

[16] D. B. Parker, R. F. Westhoff and M. J. Wolf, *On two-path convexity in multipartite tournaments*, European J. Combin. **29** (2008), 641–651.

[17] J. C. Varlet, *Convexity in tournaments*, Bull. Soc. R. Sci. Liège **45** (1976), 570–586.

Locally-maximal embeddings of graphs in orientable surfaces

Michal Kotrbčík[1] and Martin Škoviera[1]

1 Introduction

The set of all orientable cellular embeddings of a graph has the intrinsic structure of adjacency between embeddings based on elementary operations on rotation schemes. Several types of elementary operations were considered in the past, usually in proofs of interpolation theorems: moving a single arc within its local rotation, moving both ends of an edge in the respective local rotations, and interchanging two arcs in the local rotation at a given vertex, see [2, 6, 7, 10]. We call these operations *rotation moves*. Each type of a rotation move gives rise to the structure of a *stratified graph* on the set of all embeddings of a given graph. Stratified graphs were studied by Gross, Rieper, and Tucker [5, 6, 8], although they were implicit already in the works of Duke [2] and Stahl [10]. Very little is known about stratified graphs in general, although their structure is crucial for understanding the entire system of all embeddings of a given graph. In the present paper we focus on embeddings that correspond to local maxima in stratified graphs. We call a cellular embedding of a graph into an orientable surface *locally maximal* if its genus cannot be raised by moving a single arc within its local rotation. Somewhat surprisingly, the concept of a locally-maximal embedding does not depend on which type of a move is taken as a basis for the stratified graph, indicating its important position in the hierarchy of graph embeddings between the minimum genus and the maximum genus.

The main results of this paper are (1) a characterisation of locally-maximal embeddings, (2) analysis of their relationship to the minimum and the maximum genus of a graph, and (3) a simple greedy 2-approxi-

[1] Department of Computer Science, Faculty of Mathematics, Physics and Informatics, Comenius University, 842 48 Bratislava, Slovakia.
Email: kotrbcik@dcs.fmph.uniba.sk, skoviera@dcs.fmph.uniba.sk

This research was partially supported by APVV-0223-10, VEGA 1/1005/12, UK 513/2013, and by the EUROCORES Programme EUROGIGA (project GReGAS) of the European Science Foundation, under the contract APVV-ESF-EC-0009-10.

mation algorithm for maximum genus based on the notion of a locally-maximal embedding. Full proofs and additional results can be found in the forthcoming papers.

2 Fundamentals

The fundamental property of rotation moves is that any rotation move changes the number of faces of an embedding by -2, 0, or $+2$, changing the genus by -1, 0, or 1. This property is well known and was used in several proofs of the interpolation theorem for the genus range of a graph.

The following result asserts that whenever a vertex v is incident with at least three faces, there is a rotation move at v that merges three faces into one, thus raising the genus. In this regard, all three types of rotation moves display a similar behaviour.

Theorem 2.1. *If a vertex v is incident with at least three faces of an embedding Π, then there exists a move of an arc at v that merges three faces at v into one while leaving all other faces of Π intact. There also exists an interchange of two arcs at v that merges three faces at v into one and leaves all other faces intact.*

A particularly useful corollary of our proof of Theorem 2.1 is that moving any arc lying on the boundary of two distinct faces into a corner belonging to a third face merges these three faces into one.

The next result is the cornerstone of our theory of locally-maximal embeddings as it shows that the concept of a locally-maximal embedding is independent on which type of a move is chosen as its basis. At the same time, it provides a characterisation of locally-maximal embeddings in terms of the multiplicity of vertex-face incidences.

Theorem 2.2. *For any orientable embedding Π of a connected graph G the following statements are equivalent.*

(i) *The embedding Π is locally maximal.*
(ii) *The genus of Π cannot be raised by interchanging two arcs in a local rotation.*
(iii) *The genus of Π cannot be raised by moving any edge in the rotation of Π.*
(iv) *Every vertex of G is incident with at most two faces of Π.*

A natural question about locally-maximal embeddings concerns their distribution within the embedding range. We therefore define the *locally-maximal genus* of a graph G, $\gamma_L(G)$, as the minimum among the genera of all locally-maximal embeddings of G. For a graph G we further denote by $\mu(G)$ the largest number of pairwise disjoint circuits contained in G; this number is sometimes known as the *cycle packing number* of G. We also define the *reduced Betti number* $\beta'(G)$ by setting $\beta'(G) = \beta(G) - \mu(G)$; note that the reduced Betti number is nonnegative since $\mu(G) \leq \beta(G)$. The importance of these two invariants is explained by the following theorem.

Theorem 2.3. *The maximum number of faces in a locally-maximal embedding of a graph G does not exceed $\mu(G) + 1$ or $\mu(G)$, depending on whether $\beta'(G)$ is even or odd, respectively.*

The previous result enables us to prove that the relationship between the Betti number, the reduced Betti number, and the genus parameters of a graph is governed by the following inequalities.

Theorem 2.4. *The following inequalities hold for every connected graph G:*

(i) $\gamma(G) \leq \beta'(G)/2 \leq \gamma_L(G) \leq \gamma_M(G) \leq \beta(G)/2$
(ii) $\beta'(G)/2 \leq \gamma_L(G) \leq \gamma_M(G) \leq \beta'(G)$
(iii) $\gamma_M(G)/2 \leq \gamma_L(G) \leq \gamma_M(G)$

Graphs G for which $\gamma_M(G) = \lfloor \beta(G)/2 \rfloor$ are known as *upper-embeddable* graphs. With this analogy in mind we define a graph G to be *lower-embeddable* if $\gamma_L(G) = \lceil \beta'(G)/2 \rceil$. As with upper-embeddable graphs, many important classes of graphs are lower-embeddable.

Theorem 2.5. *All graphs in the following classes are lower embeddable: complete graphs K_n for all $n \geq 1$, complete bipartite graphs $K_{m,n}$ for all $m, n \geq 1$, complete equipartite graphs $K_{n,\dots,n}$ for all $n \geq 1$, and hypercubes Q_n for all $n \geq 1$.*

A connected graph is called a *cycle-tree* if any two of its cycles are vertex disjoint. It turns out that planar locally-maximal embeddings admit a simple characterisation: A connected graph G has a planar locally-maximal embedding if and only if G is a cycle-tree.

3 Constructions

The proof of the next theorem is based on the well-known edge-addition technique of raising the genus by adding a pair of adjacent edges; for technical details of the method see for example [9], [3], or [1].

We need the following definitions to state and prove our result. A cycle-tree graph is called a *k-cycle-tree* if it contains exactly k cycles. A component of a graph is called *even* if it has even number of edges.

A face of an embedding is *spanning* if it is incident with each vertex of the graph. The idea of a spanning face was used in [1] to construct embeddings with genus $\gamma_M(G) - 1$;

Theorem 3.1. *Let G be a connected graph. If G has a spanning k-cycle-tree S such that $G - E(S)$ has only even components, then G has a locally-maximal embedding with $k + 1$ faces. In particular, $\gamma_L(G) \leq (\beta(G) - k)/2$.*

Sketch of a proof. Since each component of $G - E(S)$ is even, $G - E(S)$ has a partition $\mathcal{P}$ into pairs of adjacent edges (see for example [9, Lemma 4]). Let us arrange the pairs from $\mathcal{P}$ into a linear order, and let $\{e_i, f_i\}$ be the i-th pair. Consider the graphs

$$\begin{aligned} G_0 &= S, \\ G_i &= G_{i-1} \cup \{e_i, f_i\} \quad \text{for } i \geq 1. \end{aligned}$$

Assuming that the number of pairs in $\mathcal{P}$ is n, we get $G_n = G$. The proof is finished by using the edge-addition technique and employing induction to prove that for each $i \in \{0, 1, \ldots, n\}$ the graph G_i has a locally-maximal embedding with $i + 1$ faces, at least one of them being spanning. □

Although Theorem 3.1 cannot be used to construct all locally-maximal embeddings, it is often useful in determining the locally-maximal genus provided that one can obtain good lower bounds. In particular, Theorem 3.1 can be used to prove lower-embeddability in Theorem 2.5, where lower bounds follow from Theorem 2.3. The idea is as follows. Every lower-embeddable graph G satisfies $\gamma_L(G) = \lceil \beta'(G)/2 \rceil$. Therefore, it suffices to construct a spanning cycle-tree S of G with $\mu(G)$ or $\mu(G) - 1$ cycles such that $G - E(S)$ consists of even components. In many cases, the graph $G - E(S)$ is connected. This method is in its nature similar to the one used in the proofs of upper-embeddability, which are often carried out by finding a connected cotree of the given graph.

4 Algorithms

Let us start with Greedy-Max-Genus Algorithm described in Figure 4.1. The algorithm repeatedly increases the genus by employing suitable rotation moves. By part (iv) of Theorem 2.2, the genus of an embedding can be raised by a rotation move at a vertex v if and only if v is incident with at least three faces. Note that testing whether a vertex is incident with at least three faces, as well as finding and performing a rotation move increasing the genus if such a move exists, can be easily done in polynomial time. It follows that the running time of Greedy-Max-Genus Algorithm is polynomial. Moreover, by Theorem 2.2, the embedding output by Greedy-Max-Genus Algorithm is locally maximal. Since the genus of any locally-maximal embedding is at least $\gamma_L(G)$, and $\gamma_L(G) \geq \gamma_M(G)/2$ by Theorem 2.4, we obtain the following theorem.

Greedy-Max-Genus Algorithm	
Input:	A connected graph G
Output:	An embedding of G and its genus
1:	randomly choose an embedding Π of G
2:	**while** there is a rotation move increasing the genus of Π
3:	apply one of such rotation moves to Π
4:	output Π and the genus of Π

Figure 4.1. 2-approximation algorithm for maximum genus.

Theorem 4.1. *The Greedy-Max-Genus Algorithm from Figure 4.1 is a polynomial-time 2-approximation algorithm for maximum genus. Furthermore, the embeddings output by Greedy-Max-Genus Algorithm are precisely the locally-maximal embeddings.*

Part (iii) of Theorem 2.4 enables us to efficiently approximate also the locally-maximal genus using a polynomial-time algorithm computing the maximum genus of an arbitrary graph by Glukhov [4], or Furst et al. [3].

Theorem 4.2. *There is a polynomial-time 2-approximation algorithm for locally-maximal genus.* □

References

[1] J. CHEN, S. P. KANCHI and A. KANEVSKY, *On the complexity of graph embeddings*, extended abstract, In: " Algorithms and Data Structures", F.D. et al. (eds.), Lecture Notes in Comp. Sci., vol. 709, Springer-Verlag, Berlin (1993), 234–245.

[2] R. A. DUKE, *The genus, regional number, and the Betti number of a graph*, Canad. J. Math. **18** (1966), 817–822.

[3] M. L. FURST, J. L. GROSS and L. A. MCGEOCH, *Finding a maximum-genus embedding*. J. Assoc. Comput. Mach. **35** (1988), 523–534.

[4] A. D. GLUKHOV, *A contribution to the theory of maximum genus of a graph*, In: "Structure and Topological Properties of Graphs", Inst. Mat. Akad. Nauk Ukrain. SSR, Kiev (1981), 15–29. In Russian.

[5] J. L. GROSS and R. G. RIEPER, *Local extrema in genus-stratified graphs*, J. Graph Theory **15** (1991), 159–171.

[6] J. L. GROSS and T. W. TUCKER *Local maxima in graded graphs of embeddings*, Ann. NY Acad. Sci. **319** (1979), 254–257.

[7] J. L. GROSS and T. W. TUCKER, "Topological Graph Theory", Wiley, 1987.

[8] J. L. GROSS and T. W. TUCKER, *Stratified graphs for imbedding systems*, Discrete Math. **143**(1-3) (1995), 71–85.

[9] M. JUNGERMAN, *A characrterization of upper embeddable graphs*, Trans. Amer. Math. Soc. **241** (1978), 401–406.

[10] S. STAHL, *A counting theorem for topological graph theory*, In: "Theory and Applications of Graphs", Lecture Notes in Math., vol. 642, Springer-Verlag (1978), 534–544.

A characterization of triangulations of closed surfaces

Jorge Arocha[1], Javier Bracho[1], Natalia García-Colín[1] and Isabel Hubard[1]

Abstract. In this paper we prove that a finite triangulation of a connected closed surface is completely determined by its intersection matrix. The *intersection matrix* of a finite triangulation, K, is defined as $M_K = (\dim(s_i \cap s_j))_{0\leq i,0\leq j}^{n-1}$, where $K_2 = \{s_0, \ldots s_{n-1}\}$ is a labelling of the triangles of K.

1 Introduction

Within the theory of convex polytopes, the study of the combinatorial equivalence of k-skeleta of pairs polytopes which are not equivalent themselves has been of interest, this phenomena is referred to in the literature as ambiguity [3].

It is well known that for $k \geq \lfloor \frac{d}{2} \rfloor$ the k-skeleton of a convex polytope is not dimensionally ambiguous, this is, it defines the entire structure of its underlying d-polytope. However for $k < \lfloor \frac{d}{2} \rfloor$ the question is much more intricate.

One of the most interesting results in this direction is the solution to Perle's conjecture by P. Blind and R. Mani [1] and, separately, by G. Kalai [2] which states that the 1-skeleta of convex simple d-polytopes define their entire combinatorial structure. Or, on its dual version, that the dual graph (facet adjacency graph) of a convex simplicial d-polytope determines its entire combinatorial structure.

2 Motivation & contribution

Allured by Perles' conjecture, we decided to explore the extent to which an adequate combination of combinatorial and topological assumptions

[1] Instituto de Matemáticas, Universidad Nacional Autónoma de México.
Email: arocha@matem.unam.mx, roli@matem.unam.mx, natalia.garciacolin@im.unam.mx, isahubard@im.unam.mx

Research supported by PAPIIT-México IN112511 y CONACyT 166951.

would prove as powerful for characterising certain simplicial complexes. The purpose of this work is to present our first result, product of this exploration.

For topological assumption we will, in this instance, ask for the simplicial complex of study to be a connected closed surface. As for combinatorial assumption, one might be tempted to choose only to have the information provided by its dual graph. However, the dual graph of a triangulation of a closed surface does not provide enough information to characterise it, as there are some dual graphs to triangulations which have been shown in [4] to have combinatorically different polyhedral embeddings.

Therefore, we will need to strengthen the combinatorial hypothesis. In order to do so we will introduce the concept of an intersection preserving mapping of simplices of a simplicial complex.

Definition 2.1. A bijective mapping $f : K_d \to K'_d$ between the sets of d-simplices of two simplicial complexes, K and K', is an intersection preserving mapping if for every pair of simplices $s, t \in K_d$ $\dim(s \cap t) = \dim(f(s) \cap f(t))$.

Throughout this paper we will use the notation K_l to refer to the set of l-dimensional simplices of the complex K. Additionally, we will define two particular triangulations of the projective plane, which are of interest for this work.

Definition 2.2. We define a 10-triangle triangulation of the projective plane, $T\mathbb{P}_{10}$, as the triangulation whose triangles have the vertex sets $(s_i)_0 = \{a_{i \bmod 5}, a_{i+1 \bmod 5}, x\}$, and $(r_i)_0 = \{a_{i \bmod 5}, a_{i+1 \bmod 5}, a_{i-2 \bmod 5}\}$ for $0 \le i \le 4$.

Definition 2.3. We define a 12-triangle triangulation of the projective plane, $T\mathbb{P}_{12}$, as the triangulation whose triangles have the vertex sets $(s_i)_0 = \{a_{i \bmod 6}, a_{i+1 \bmod 6}, x\}$, for $0 \le i \le 5$ and $(r_i)_0 = \{a_{i \bmod 6}, a_{i+1 \bmod 6}, a_{i+4 \bmod 6}\}$ for $0 \le i \le 4$ even, and $(r_i)_0 = \{a_{i \bmod 6}, a_{i+1 \bmod 6}, a_{i+3 \bmod 6}\}$ for $0 \le i \le 4$ odd.

We now use the aforementioned definitions to state the main result:

Theorem 2.4. *Let $\|K\|$ and $\|K'\|$ be geometric realizations of finite triangulations which are homeomorphic to connected closed surfaces, and let $f : K_2 \to K'_2$ be an intersection preserving mapping, then one of the following three statements holds:*

(1) *f can be extended into a bijective simplicial mapping between K and K'*

(2) *f cannot be extended into a simplicial mapping between K and K', but both $\|K\|$ and $\|K'\|$ are $T\mathbb{P}_{10}$*
(3) *f cannot be extended into a simplicial mapping between K and K', but both $\|K\|$ and $\|K'\|$ are $T\mathbb{P}_{12}$.*

Consider the *intersection matrix*, $M_K = (\dim(s_i \cap s_j))_{0\leq i,0\leq j}^{n-1}$, of a finite triangulation, K, where $K_2 = \{s_0, \dots s_{n-1}\}$ is a labelling of the triangles of K then, in the spirit of Perles' conjecture, we can state the previous theorem as;

Corollary 2.5. *A finite triangulation of a connected closed surface is completely determined by its intersection matrix.*

3 Preliminaries

One of the peculiarities of triangulations of a closed surface is that the neighbourhood of every vertex is a disk. Furthermore, the triangles incident to any vertex of such surface form the simplest of triangulations of a disk, namely an n-gon whose vertices are all linked by an edge to a central vertex in the centre of the n-gon. We start off by analysing the intersection patterns of such a structure.

Definition 3.1. An n-shell is the abstract triangulation ${}_n\Pi^2$ such that ${}_n\Pi_2^2 = \{s_0, s_1, \dots, s_{n-1}\}$, $\dim(s_i \cap s_{i+1\ modn}) = 1$, $\dim(s_i \cap s_j) = 0$ for $|i - j| \geq 2$ with $i, j \in \{0, \dots, n-1\}$.

We will now focus on studying what other structures can have an intersection pattern equal to that of a triangulated disk.

Lemma 3.2. *The vertex sets of the triangles in an n-shell, with $n \geq 3$, ${}_n\Pi_2^2 = \{s_0, \dots s_{n-1}\}$ can only take one of the following three types*

(1) $(s_i)_0 = \{a_{i \mod n}, a_{i+1 \mod n}, x\}$ *for all* $0 \leq i \leq n$, *for any* n*;*

(2) $(s_0)_0 = \{a_0, a_2, a_1\}$, $(s_1)_0 = \{a_1, a_3, a_2\}$, $(s_2)_0 = \{a_2, a_4, a_3\}$, $(s_3)_0 = \{a_3, a_0, a_4\}$, *and* $(s_4)_0 = \{a_4, a_1, a_0\}$, *when* $n = 5$*; or*

(3) $(s_0)_0 = \{a_0, a_1, a_2\}$, $(s_1)_0 = \{a_1, a_2, a_4\}$, $(s_2)_0 = \{a_2, a_3, a_4\}$, $(s_3)_0 = \{a_3, a_0, a_4\}$, $(s_4)_0 = \{a_0, a_5, a_4\}$, *and* $(s_5)_0 = \{a_5, a_2, a_0\}$, *when* $n = 6$.

The proof of the lemma above consists of several parts and follows largely by a detailed analysis of the combinatorial structure of n-cycles of triangles.

It is easy to see that geometric realisations of the three types of triangulations associated to puzzles of n-shells are an n-triangulation of a disk, a 5-triangulation of a Möbius band and a 6-triangulation of a Möbius band, respectively.

4 Proof of the theorem

Proof. For each vertex $x \in K_0$ let ${}_{n_x}\Pi$ be the n_x-cycle around x, by hypothesis $\|{}_{n_x}\Pi\|$ is necessarily a disk.

(1) If, for all $x \in K_0$, $\|f({}_{n_x}\Pi)\|$ is also a disk, then the mapping $h : K_0 \to K_0'$ such that $h(x) = \bigcap_{i=1}^{n_x} f(s_i)$ is a bijective simplicial mapping.

Assume then that there is a vertex $x \in K_0$ such that $\|f({}_{n_x}\Pi)\|$ is not a disk.

(2) Suppose $\|f({}_{n_x}\Pi)\|$ is the 5-triangulation of the Möbius band described in Lemma 3.2.

Let ${}_{n_x}\Pi_2 = \{s_0, s_1, s_2, s_3, s_4\}$, where $(s_i)_0 = \{a_{i \bmod 5}, a_{i+1 \bmod 5}, x\}$ and $f({}_{n_x}\Pi)_2 = \{s_0', s_1', s_2', s_3', s_4'\}$ where $(s_0')_0 = \{a_0', a_2', a_1'\}$, $(s_1')_0 = \{a_1', a_3', a_2'\}$, $(s_2')_0 = \{a_2', a_4', a_3'\}$, $(s_3')_0 = \{a_3', a_0', a_4'\}$, and $(s_4')_0 = \{a_4', a_1', a_0'\}$.

Given that K' is also a closed surface, then each of the simplices $s_0', s_1', s_2', s_3', s_4'$ has got a triangle adjacent to its remaining free edge. Let r_i' be the simplices such that $\dim(r_i' \cap s_i') = 1$, then $(r_0')_0 = \{a_0', a_2', x_0'\}$, $(r_1')_0 = \{a_1', a_3', x_1'\}$, $(r_2')_0 = \{a_2', a_4', x_2'\}$, $(r_3')_0 = \{a_3', a_0', x_3'\}$, and $(r_4')_0 = \{a_4', a_1', x_4'\}$. This is $(r_i')_0 = \{a'_{i \bmod 5}, a'_{i+2 \bmod 5}, x_i'\}$. It follows that, $\dim(r_i' \cap s_j') \geq 0$ for all $i \neq j$.

Note that the interior of each of the edges $\{a'_{i \bmod 5}, a'_{i+1 \bmod 5}\}$ is in the interior of the Möbius band, thus this edges cannot be repeated in any further simplex in the complex.

This implies that $x_i \notin \{a_0', a_1', a_2', a_3', a_4'\}$, because, if this was the case, at least one of the edges $\{a'_{i \bmod 5}, a'_{i+1 \bmod 5}\}$ would belong to $(r_i')_1$. Then, $\dim(r_i' \cap s_j') = 0$

Let $r_i = f^{-1}(r_i')$, then $\dim(r_i \cap s_i) = 1$ and $\dim(r_i \cap s_j) = 0$ for all $i \neq j$. As $(s_i)_0 = \{a_{i \bmod 5}, a_{i+1 \bmod 5}, x\}$ then $(r_i)_0 = \{a_{i \bmod 5}, a_{i+1 \bmod 5}, x_{i \bmod 5}\}$.

Here $\dim(r_{i \bmod 5} \cap s_{i+1 \bmod 5}) \geq 0$ and $\dim(r_{i \bmod 5} \cap s_{i-1 \bmod 5}) \geq 0$ trivially, hence $a_{i-1 \bmod 5}, a_{i+2 \bmod 5} \notin (r_i)_0$. However, for $\dim(r_{i \bmod 5} \cap s_{i+2 \bmod 5}) = 0$ and $\dim(r_{i \bmod 5} \cap s_{i-2 \bmod 5}) = 0$ to be accomplished, necessarily $x_i = a_{i+3 \bmod 5} = a_{i-2 \bmod 5}$.

That is, $(r_i)_0 = \{a_{i \bmod 5}, a_{i+1 \bmod 5}, a_{i-2 \bmod 5}\}$, hence the simplicial complex asociated to $\bigcup_{i=0}^{4} r_i$ is a 5-triangulation of a Móbius band, where $\dim(r_i \cap r_{i+2 \bmod 5}) = \dim(r_i \cap r_{i-2 \bmod 5}) = 1$, and as necessarily $K_2 = \bigcup_{i=0}^{4} r_i \cup \bigcup_{i=0}^{4} s_i$ and the geometric simplicial complexes associated to $\bigcup_{i=0}^{4} r_i$ and $\bigcup_{i=0}^{4} s_i$ are a Möbius band and a disk, respectively, then $\|K\|$ is equal to $T\mathbb{P}_{10}$.

The above also implies that $\dim(r_i' \cap r_{i+2 \bmod 5}') = \dim(r_i' \cap r_{i-2 \bmod 5}') = 1$ then $v' = v_i'$ for all $i = 1, \ldots 4$, so that $K_2' = \bigcup_{i=0}^{4} r_i' \cup \bigcup_{i=0}^{4} s_i'$, hence $\|K'\|$ is also equal to $T\mathbb{P}_{10}$.

(3) Suppose $\|f({}_{n_x}\Pi)\|$ is the 6-triangulation of the Möbius band described in Lemma 3.2.

Let ${}_{n_x}\Pi_2 = \{s_0, s_1, s_2, s_3, s_4, s_5\}$, where $(s_i)_0 = \{a_{i \bmod 6}, a_{i+1 \bmod 6}, x\}$ and $f({}_{n_x}\Pi)_2 = \{s_0', s_1', s_2', s_3', s_4', s_5'\}$ where $(s_0')_0 = \{a_0', a_1', a_2'\}$, $(s_1')_0 = \{a_1', a_2', a_4'\}$, $(s_2')_0 = \{a_2', a_3', a_4'\}$, $(s_3')_0 = \{a_3', a_0', a_4'\}$, $(s_4')_0 = \{a_0', a_5', a_4'\}$, and $(s_5')_0 = \{a_5', a_2', a_0'\}$.

As $\|K'\|$ is a closed surface, then each of the simplices $\{s_0', s_1', s_2', s_3', s_4', s_5'\}$ has got a triangle adjacent to its remaining free edge. Let r_i' be the simplices such that $\dim(r_i' \cap s_i') = 1$, then $(r_0')_0 = \{a_0', a_1', x_0'\}$, $(r_1')_0 = \{a_1', a_4', x_1'\}$, $(r_2')_0 = \{a_2', a_3', x_2'\}$, $(r_3')_0 = \{a_3', a_0', x_3'\}$, $(r_4')_0 = \{a_4', a_5', x_4'\}$, and $(r_5')_0 = \{a_2', a_5', x_5'\}$.

Here it follows that, $\dim(r_i' \cap s_j') \geq 0$ for all $i \neq j$, except for the pairs $i = 0$ and $j = 2$, $i = 1$ and $j = 5$, $i = 2$ and $j = 4$, $i = 3$ and $j = 1$, $i = 4$ and $j = 0$ and $i = 5$ and $j = 3$; for these exceptions the intersection might be empty.

The above implies that, if $r_i = f^{-1}(r_i')$, then $\dim(r_i \cap s_i) = 1$ and $\dim(r_i \cap s_j) \geq 0$ for all $i \neq j$, except for the pairs $i = 0$ and $j = 2$, $i = 1$ and $j = 5$, $i = 2$ and $j = 4$, $i = 3$ and $j = 1$, $i = 4$ and $j = 0$, and $i = 5$ and $j = 3$; for these exceptions the intersection might be empty.

As $(s_i)_0 = \{a_{i \bmod 6}, a_{i+1 \bmod 6}, x\}$ then the vertex sets of the r_i's are $(r_i)_0 = \{a_{i \bmod 6}, a_{i+1 \bmod 6}, x_{i \bmod 6}\}$.

Note that $x_0' \notin \{a_0', a_1', a_2', a_4', a_5'\}$ as the edges $\{a_0', a_2\}$, $\{a_0', a_4\}$, $\{a_0', a_5\}$, $\{a_0', a_4\}$ are edges whose interior is in the interior of the Möbius band. Thus we might have $v_0' = a_3'$, however if that was the case $\dim(r_0' \cap r_3') = 1$, and $\dim(r_0 \cap r_3) = 1$, but this is not possible. Then necessarily $x_0' \notin \{a_0', a_1', a_2', a_3', a_4', a_5'\}$.

Using an argument analogous to the one in the previous case, we deduce that for each i, $x_i' \notin \{a_0', a_1', a_2', a_3', a_4', a_5'\}$; so that $\dim(r_i' \cap s_j') = 0$ for all $i \neq j$, except for the pairs $i = 0$ and $j = 2$, $i = 1$ and $j = 5$, $i = 2$ and $j = 4$, $i = 3$ and $j = 1$, $i = 4$ and $j = 0$, and $i = 5$ and $j = 3$, for which the intersection is empty.

The above implies $\dim(r_i \cap s_i) = 1$ and $\dim(r_i \cap s_j) = 0$ for all $i \neq j$, except for the pairs $i = 0$ and $j = 2$, $i = 1$ and $j = 5$, $i = 2$ and $j = 4$, $i = 3$ and $j = 1$, $i = 4$ and $j = 0$, and $i = 5$ and $j = 3$, for which the intersection is empty. Hence, in order to accomplish the intersection dimensions indicated by the puzzle necessarily, $x_0 = x_1 = a_4$, $x_2 =$

$x_3 = a_0$, $x_4 = x_5 = a_2$, thus; $r_0 = \{a_0, a_1, a_4\}$, $r_1 = \{a_1, a_2, a_4\}$, $r_2 = \{a_2, a_3, a_0\}$, $r_3 = \{a_3, a_4, a_0\}$, $r_4 = \{a_4, a_5, a_2\}$, and $r_5 = \{a_0, a_5, a_2\}$.

Therefore, the simplicial complex associated to $\bigcup_{i=0}^{5} r_i$ is a 6-triangulation of a Möbius band and $K_2 = \bigcup_{i=0}^{5} s_i \cup \bigcup_{i=0}^{5} r_i$, so that $\|K\|$ is equal to $T\mathbb{P}_{12}$.

The implication for K' is that $\dim(r'_0 \cap r'_1) = 1$, $\dim(r'_1 \cap r'_4) = 1$, $\dim(r'_4 \cap r'_5) = 1$, $\dim(r'_5 \cap r'_2) = 1$, $\dim(r'_2 \cap r'_3) = 1$, $\dim(r'_3 \cap r'_0) = 1$, which in turn implies $v' = v'_i$ for all $i \in \{0, \ldots 5\}$ and, further, $K'_2 = \bigcup_{i=0}^{5} s'_i \cup \bigcup_{i=0}^{5} r'_i$, so that K' is also equal to $T\mathbb{P}_{12}$. □

References

[1] R. BLIND and P. MANI-LEVITSKA, *Puzzles and polytope isomorphisms*, Aequationes Mathematicae **34** (1987), n. 3-4, 287–297.

[2] G. KALAI, *A simple way to tell a simple polytope from its graph*, Journal of combinatorial theory, Series A. **49** (1988), n. 2, 381–383.

[3] B. GRÜNBAUM, "Convex Polytopes" Springer, 1967.

[4] B. MOHAR and A. VODOPIVEC *On Polyhedral Embeddings of Cubic Graphs*, Comb. Probab. Comput. **15** (2006), n. 6, 877–893.

Quasi-perfect linear codes from singular plane cubics

Massimo Giulietti[1]

Abstract. We present some recently obtained constructions of quasi-perfect linear codes with small density arising from plane cubic curves defined over finite fields.

1 Introduction

Galois Geometry, that is the theory of combinatorial objects embedded in projective spaces over finite fields, is well known to be rich of nice algebraic, combinatorial and group theoretic aspects that have also found wide and relevant applications in Coding Theory and Cryptography; see *e.g.* the monography [5]. In this context an important role is played by plane arcs and their generalizations - especially complete caps, saturating sets and arcs in higher dimensions - since their code theoretic counterparts are distinguished types of error-correcting and covering linear codes. In this extended abstract we present some recent results on small complete caps and quasi-perfect linear codes, obtained in few joint works with N. Anbar, D. Bartoli and I. Platoni, mostly unpublished.

Let $\mathbb{F}_q$ be the finite field with q elements and let $\mathbf{C}$ be an $[n, k, d]_q$-code, *i.e.*, a q-ary linear code of length n, dimension k and minimum distance d. The covering radius of $\mathbf{C}$ is the minimum integer $R(\mathbf{C})$ such that for any vector $\mathbf{v} \in \mathbb{F}_q^n$ there exists $\mathbf{x} \in C$ with $d(\mathbf{v}, \mathbf{x}) \leq R(\mathbf{C})$. An $[n, k, d]_q$-code with covering radius R is denoted by $[n, k, d]_q R$. Let t be the integer part of $(d-1)/2$. Clearly, $R(\mathbf{C}) \geq t$ holds and when equality is attained the code $\mathbf{C}$ is said to be perfect. As there are only finitely many classes of linear perfect codes, of particular interest are those codes $\mathbf{C}$ with $R(\mathbf{C}) = t + 1$, called quasi-perfect codes. One of the parameters characterizing the covering quality of an $[n, k, d]_q R$-code $\mathbf{C}$ is its covering density $\mu(\mathbf{C})$, introduced in [3] as the average number of codewords at distance less than or equal to R from a vector in $\mathbb{F}_q^n$. The covering

[1] Dipartimento di Matematica e Informatica, Università degli Studi di Perugia, Italia. Email: giuliet@dmi.unipg.it

density $\mu(\mathbf{C})$ is always greater than or equal to 1, and equality holds precisely when $\mathbf{C}$ is perfect. Among codes with the same codimension s and covering radius R, the shortest ones have the best covering density. This explains why the problem of determining the minimal length n for which there exists an $[n, n-s, d]_q R$-code with given s, q, d and R, has been broadly investigated. Throughout, such minimal length will be denoted as $l(s, R, q)_d$.

Here we will restrict our attention to codes with covering radius $R = 2$ and $d = 4$, *i.e.* quasi-perfect linear codes that are both 1-error correcting and 3-error detecting. Interestingly, such codes have a nice geometrical counterpart: the columns of a parity check matrix of an $[n, n-s, 4]_q 2$-code can be considered as points of a complete cap of size n in the finite projective space $PG(s-1, q)$. In particular, $l(s, 2, q)_4$ coincides with the minimum size of a complete cap in $PG(s-1, q)$. This makes it possible to use methods from both Galois Geometries and Algebraic Geometry in order to investigate covering-radius-2 codes with small density. Here, we are going to discuss some recently obtained upper bounds on the minimum size of a complete cap which are valid for arbitrarily large values of q. The key tool is the construction of complete caps in higher dimensional spaces from singular plane cubic curves defined over $\mathbb{F}_q$.

2 Complete caps from bicovering arcs

An n-cap in an (affine or projective) Galois space over $\mathbb{F}_q$ is a set of n points no three of which are collinear. An n-cap is said to be complete if it is not contained in an $(n+1)$-cap. A plane n-cap is also called an n-arc. Let $t\,(AG\,(N, q))$ be the size of the smallest complete cap in the Galois affine space $AG(N, q)$ of dimension N over $\mathbb{F}_q$. Since the affine space $AG(N, q)$ is embedded in the projective space $PG(N, q)$, a complete cap in $AG(N, q)$ can be viewed as a cap in $PG(N, q)$, whose completeness can be achieved by adding some extra-points at the hyperplane at infinity. Therefore, the following relation holds.

Proposition 2.1. *Let $M(N, q)$ denote the maximal size of a complete cap in $PG(N-1, q)$. Then*

$$l(N+1, 2, q)_4 \leq t(AG(N, q)) + M(N, q).$$

In particular, $l(5, 2, q)_4 \leq t(AG(4, q)) + q^2 + 1$.

The trivial lower bound for $t(AG(N, q))$ is $\sqrt{2}q^{\frac{N-1}{2}}$. General constructions of complete caps whose size is close to this lower bound are only known for q even and N odd. When N is even, complete caps of size of the same order of magnitude as $cq^{N/2}$, with c a constant independent of q, are known to exist for both the odd and the even order case.

Small complete caps in dimensions $N \equiv 0 \pmod 4$ can be obtained from plane arcs via the product method for caps: let $q' = q^{\frac{N-2}{2}}$ and fix a basis of $\mathbb{F}_{q'}$ as a linear space over $\mathbb{F}_q$; identify points in $AG(N, q)$ with vectors of $\mathbb{F}_{q'} \times \mathbb{F}_{q'} \times \mathbb{F}_q \times \mathbb{F}_q$; for an arc A in $AG(2, q)$, let

$$K_A = \{(\alpha, \alpha^2, u, v) \in AG(N, q) \mid \alpha \in \mathbb{F}_{q'}, (u, v) \in A\};$$

then the set K_A is a cap in $AG(N, q)$. For q odd, the completeness of the cap K_A depends on the bicovering properties of A in $AG(2, q)$; see Theorem 2.3 below. According to Segre [6], given three pairwise distinct points P, P_1, P_2 on a line ℓ in $AG(2, q)$, P is external or internal to the segment P_1P_2 depending on whether $(x-x_1)(x-x_2)$ is a non-zero square or a non-square in $\mathbb{F}_q$, where x, x_1 and x_2 are the coordinates of P, P_1 and P_2 with respect to any affine frame of ℓ.

Definition 2.2. Let A be a complete arc in $AG(2, q)$. A point $P \in AG(2, q) \setminus A$ is said to be bicovered with respect to A if there exist $P_1, P_2, P_3, P_4 \in A$ such that P is both external to the segment P_1P_2 and internal to the segment P_3P_4. If every $P \in AG(2, q) \setminus A$ is bicovered by A, then A is said to be a bicovering arc.

Theorem 2.3 ([4]). *Let A be a bicovering n-arc in $AG(2, q)$; then K_A is a complete cap in $AG(N, q)$.*

3 Small complete caps from cubic curves

From now on we assume that the characteristic of $\mathbb{F}_q$ is $p > 3$. Let $\mathcal{X}$ be an irreducible plane cubic curve defined over $\mathbb{F}_q$, and consider the set G of the non-singular $\mathbb{F}_q$-rational points of $\mathcal{X}$. As it is well known, for any point O of G it is possible to give a group structure to G, by defining a binary operation $\boxplus$ in such a way that $(G, \boxplus)$ is an abelian group with neutral element $O \in G$. The point O is usually chosen as an inflection point. One of the main properties of this operation is that three distinct points in G are collinear if and only if their sum is the neutral element in G. Then it is easy to see that for a subgroup H of G of index m, with $(3, m) = 1$, and a point P in $G \setminus H$, the coset $K = H \boxplus P$ is an arc.

In order to investigate the covering properties of such an arc, we recall a general method, due to Segre and Lombardo-Radice, that uses Hasse-Weil's Theorem to prove the completeness of arcs contained in conic or cubic curves. This method is based on the following idea for proving that the secants of K cover a generic point P off the curve $\mathcal{X}$: (1) write K in an algebraically parametrized form; in the case of cubic curves $\mathcal{X}$, this can be easily done when $\mathcal{X}$ is singular; (2) construct an algebraic curve $\mathcal{C}_P$, defined over $\mathbb{F}_q$, describing the collinearity of two points of K and P; (3) show that $\mathcal{C}_P$ is absolutely irreducible or has at least an absolutely irreducible component defined over $\mathbb{F}_q$; (4) apply the Hasse-Weil bound to guarantee the existence of a suitable $\mathbb{F}_q$-rational point of $\mathcal{C}_P$ (or of its irreducible component): this is sufficient to deduce the collinearity between P and two points in K. Finally, in order to obtain the completeness, it might be necessary to extend the arc K with some points on $\mathcal{X}$.

By Theorem 2.3, bicovering arcs in affine planes are a powerfool tool to construct small complete caps in $AG(N,q)$ with q odd and $N \equiv 0 \pmod 4$. However, to establish whether a complete arc is bicovering can be a difficult task. So far, three different types of irreducible plane cubic curves have been investigated in order to prove the bicovering properties of the associated arcs: non-singular, cuspidal and nodal. The non-singular (or elliptic) case was investigated in [1].

Theorem 3.1 ([1]). *Let q be odd, and let m be a prime divisor of $q-1$, with $7 < m < \frac{1}{8}\sqrt[4]{q}$. Assume that the cyclic group of order m admits a maximal-3-independent subset of size s. Then for any positive integer $N \equiv 0 \pmod 4$,*

$$t(AG(N,q)) \le s \cdot q^{\frac{N-2}{2}} \cdot \left(\left\lfloor \frac{q - 2\sqrt{q} + 1}{m} \right\rfloor + 31 \right). \tag{3.1}$$

It has been noticed in [7] that in the cyclic group of order m there exists a maximal 3-independent subset of size $s \le (m+1)/3$. For specific values of m, the upper bound on $t(AG(N,q))$ can be improved, as there exist maximal-3-independent subsets of the cyclic group of order m of size significantly less than $m/3$ (see [1, Table 1]).

The case of a cubic with a cuspidal rational singularity and a rational inflection point is the object of the preprint [2].

Theorem 3.2 ([2]). *Let $q = p^h$ with $p > 3$ a prime, $h > 8$. Let $N \equiv 0 \pmod 4$, $N \ge 4$. Let t_h be the integer in $\{1,\dots,4\}$ such that $t_h \equiv h \pmod 4$. Assume that $p^{t_h} > 144$. Then*

$$t(AG(N,q)) \le 2pq^{\frac{N}{2}-\frac{1}{8}}.$$

We now present some new results on cubics with both a rational node and a rational inflection point, which for infinite q's improve both Theorems 3.1 and 3.2.

Theorem 3.3. *Let m be an odd divisor of $q-1$ such that $(3,m) = 1$ and*

$$q + 1 - (12m^2 - 8m + 2)\sqrt{q} \ge 8m^2 + 8m + 1. \tag{3.2}$$

Assume that the cyclic group of order m admits a maximal 3-independent subset of size s. Then there exists a bicovering arc in $AG(2,q)$ of size $\frac{s(q-1)}{m}$ contained in the cubic curve with equation $XY = (X-1)^3$.

Theorem 3.3 can be used together with Theorem 2.3 in order to construct small complete caps in affine spaces. Note that (3.2) holds whenever $m \leq \frac{\sqrt[4]{q}}{3.5}$.

Corollary 3.4. *Let m be an odd divisor of $q-1$ such that $(3,m)=1$ and $m \leq \frac{\sqrt[4]{q}}{3.5}$. Assume that the cyclic group of order m admits a maximal 3-independent subset of size s. Then for $N \equiv 0 \pmod 4$, $N \geq 4$,*

$$t(AG(N,q)) \leq \frac{s(q-1)}{m} q^{\frac{N-2}{2}}.$$

In the case where a group $\mathcal{G}$ is the direct product of two groups $\mathcal{G}_1 \times \mathcal{G}_2$ of order at least 4, neither of which elementary 3-abelian, there exists a maximal 3-independent subset of $\mathcal{G}$ of size less than or equal to $(\#\mathcal{G}_1) + (\#\mathcal{G}_2)$. Then the following holds.

Theorem 3.5. *Let m be an odd divisor of $q-1$ such that $(3,m)=1$ and $m \leq \frac{\sqrt[4]{q}}{3.5}$. Assume that $m = m_1 m_2$ with $(m_1, m_2) = 1$ and $m_1, m_2 \geq 4$. Then for $N \equiv 0 \pmod 4$, $N \geq 4$*

$$t(AG(N,q)) \leq \frac{(m_1+m_2)(q-1)}{m_1 m_2} q^{\frac{N-2}{2}}$$

Corollary 3.6. *Let $q = \bar{q}^8$ for an odd prime power $\bar{q}$. Let $m = (\bar{q}^2 - 1)/(2^h 3^k)$, where $2^h \geq 4$ is the highest power of 2 which divides $\bar{q}^2 - 1$, and similarly $3^k \geq 3$ is the highest power of 3 which divides $\bar{q}^2 - 1$. Then for $N \equiv 0 \pmod 4$, $N \geq 4$*

$$t(AG(N,q)) \leq (2^{h_2} + 2^{h_1} 3^k) q^{\frac{N}{2} - \frac{1}{8}}.$$

with $h_1 + h_2 = h$.

Results on complete arcs contained in cubics with an isolated double point have not appeared in the literature so far. This case is currently under investigation by the authors of [2].

References

[1] N. Anbar and M. Giulietti, *Bicovering arcs and complete caps from elliptic curves*, J. Algebraic Combin., published online 25/10/2012, DOI: 10.1007/s10801-012-0407-8.

[2] N. ANBAR, D. BARTOLI, M. GIULIETTI and I. PLATONI, *Small complete caps from singular cubics*, preprint.

[3] G. D. COHEN, A. C. LOBSTEIN and N. J. A. SLOANE, *Further results on the covering radius of codes*, IEEE Trans. Inform. Theory **32** (5) (1986), 680–694.

[4] M. GIULIETTI, *Small complete caps in Galois affine spaces*, J. Algebraic Combin. **25** (2) (2007), 149–168.

[5] J. W. P. HIRSCHFELD, "Projective Geometries over Finite Fields", Oxford Mathematical Monographs. The Clarendon Press Oxford University Press, New York, second edition, 1998.

[6] B. SEGRE, *Proprietà elementari relative ai segmenti ed alle coniche sopra un campo qualsiasi ed una congettura di Seppo Ilkka per il caso dei campi di Galois*, Ann. Mat. Pura Appl. (4) **96** (1972), 289–337.

[7] J. F. VOLOCH, *On the completeness of certain plane arcs. II*, European J. Combin. **11** (5) (1990), 491–496.

Boxicity and cubicity of product graphs

L. Sunil Chandran[1], Wilfried Imrich[2], Rogers Mathew[3]
and Deepak Rajendraprasad[4]

Abstract. The *boxicity* (*cubicity*) of a graph G is the minimum natural number k such that G can be represented as an intersection graph of axis-parallel rectangular boxes (axis-parallel unit cubes) in $\mathbb{R}^k$. In this article, we give estimates on the boxicity and the cubicity of *Cartesian*, *strong* and *direct products* of graphs in terms of invariants of the component graphs. In particular, we study the growth, as a function of d, of the boxicity and the cubicity of the d-th power of a graph with respect to the three products. Among others, we show a surprising result that the boxicity and the cubicity of the d-th Cartesian power of any given finite graph is in $O(\log d / \log\log d)$ and $\Theta(d/\log d)$, respectively. On the other hand, we show that there cannot exist any sublinear bound on the growth of the boxicity of powers of a general graph with respect to strong and direct products.

1 Introduction

Throughout this discussion, a k-*box* is the Cartesian product of k closed intervals on the real line $\mathbb{R}$, and a k-*cube* is the Cartesian product of k closed unit length intervals on $\mathbb{R}$. Hence both are subsets of $\mathbb{R}^k$ with edges parallel to one of the coordinate axes. All the graphs considered here are finite, undirected and simple.

Definition 1.1 (Boxicity, Cubicity). A k-*box representation* (k-*cube representation*) of a graph G is a function f that maps each vertex of G to a k-box (k-cube) such that for any two distinct vertices u and v of G, the pair uv is an edge in G if and only if the boxes $f(u)$ and $f(v)$ have a non-empty intersection. The *boxicity* (*cubicity*) of a graph G, denoted by boxicity(G) (cubicity(G)), is the smallest natural number k such that G has a k-box (k-cube) representation.

[1] Department of Computer Science and Automation, Indian Institute of Science, Bangalore, India - 560012. Email: sunil@csa.iisc.ernet.in

[2] Department Mathematics and Information Technology, Montanuniversität Leoben, Austria. Email: imrich@unileoben.ac.at

[3] Department of Mathematics and Statistics, Dalhousie University, Halifax, Canada - B3H 3J5. Supported by an AARMS Postdoctoral Fellowship. Email: rogersm@mathstat.dal.ca

[4] Department of Computer Science and Automation, Indian Institute of Science, Bangalore, India - 560012. Supported by Microsoft Research India PhD Fellowship. Email: deepakr@csa.iisc.ernet.in

It follows from the above definition that complete graphs have boxicity and cubicity 0 and interval graphs (unit interval graphs) are precisely the graphs with boxicity (cubicity) at most 1. The concepts of boxicity and cubicity were introduced by F.S. Roberts in 1969 [9]. He showed that every graph on n vertices has an $\lfloor n/2 \rfloor$-box and a $\lfloor 2n/3 \rfloor$-cube representation.

Given two graphs G_1 and G_2 with respective box representations f_1 and f_2, let G denote the graph on the vertex set $V(G_1) \times V(G_2)$ whose box representation is a function f defined by $f((v_1, v_2)) = f_1(v_1) \times f_2(v_2)$. It is not difficult to see that G is the usual strong product of G_1 and G_2 (cf. Definition 1.2). Hence it follows that the boxicity (cubicity) of G is at most the sum of the boxicities (cubicities) of G_1 and G_2. The interesting question here is: *can it be smaller?* We show that *it can be smaller* in general. But in the case when G_1 and G_2 have at least one universal vertex each, we show that that the boxicity (cubicity) of G is equal to the sum of the boxicities (cubicities) of G_1 and G_2 (Theorem 2.1).

Definition 1.2 (Graph products). The *strong product*, the *Cartesian product* and the *direct product* of two graphs G_1 and G_2, denoted respectively by $G_1 \boxtimes G_2$, $G_1 \square G_2$ and $G_1 \times G_2$, are graphs on the vertex set $V(G_1) \times V(G_2)$ with the following edge sets:

$$
\begin{aligned}
E(G_1 \boxtimes G_2) &= \{(u_1, u_2)(v_1, v_2) : (u_1 = v_1 \text{ or } u_1v_1 \in E(G_1)) \text{ and} \\
&\qquad (u_2 = v_2 \text{ or } u_2v_2 \in E(G_2))\}, \\
E(G_1 \square G_2) &= \{(u_1, u_2)(v_1, v_2) : (u_1 = v_1, u_2v_2 \in E(G_2)) \text{ or} \\
&\qquad (u_1v_1 \in E(G_1), u_2 = v_2)\}, \\
E(G_1 \times G_2) &= \{(u_1, u_2)(v_1, v_2) : u_1v_1 \in E(G_1) \text{ and } u_2v_2 \in E(G_2)\}.
\end{aligned}
$$

The *d-th strong power*, *Cartesian power* and *direct power* of a graph G with respect to each of these products, that is, the respective product of d copies of G, are denoted by $G^{\boxtimes d}$, $G^{\square d}$ and $G^{\times d}$, respectively. Please refer to [7] to know more about graph products.

Unlike the case in strong product, the boxicity (cubicity) of the Cartesian and direct products can have a boxicity (cubicity) larger than the sum of the individual boxicities (cubicities). For example, while the complete graph on n vertices K_n has boxicity 0, we show that the Cartesian product of two copies of K_n has boxicity at least $\log n$ and the direct product of two copies of K_n has boxicity at least $n-2$. In this note, we give estimates on boxicity and cubicity of Cartesian and direct products in terms of the boxicities (cubicities) and chromatic number of the component graphs. This answers a question raised by Douglas B. West in 2009 [10].

We also study the growth, as a function of d, of the boxicity and the cubicity of the d-th power of a graph with respect to these three products. Among others, we show a surprising result that the boxicity and the cubicity of the d-th Cartesian power of any given finite graph is in $O(\log d/\log\log d)$ and $\Theta(d/\log d)$, respectively (Corollary 2.7). To get this result, we had to obtain non-trivial estimates on boxicity and cubicity of hypercubes and Hamming graphs and a bound on boxicity and cubicity of the Cartesian product which does not involve the sum of the boxicities or cubicities of the component graphs.

The results are summarised in the next section after a brief note on notations. The proofs and figures are included in the full version of the paper [3].

1.1 Notational note

The vertex set and edge set of a graph G are denoted, respectively, by $V(G)$ and $E(G)$. A pair of distinct vertices u and v is denoted at times by uv instead of $\{u, v\}$ in order to avoid clutter. A vertex in a graph is *universal* if it is adjacent to every other vertex in the graph. If S is a subset of vertices of a graph G, the subgraph of G induced on the vertex set S is denoted by $G[S]$. If A and B are sets, then $A \triangle B$ denotes their symmetric difference and $A \times B$ denotes their Cartesian product. The set $\{1, \ldots, n\}$ is denoted by $[n]$. All logarithms mentioned are to the base 2.

2 Our Results

2.1 Strong products

Theorem 2.1. *Let G_i, $i \in [d]$, be graphs with* $\text{boxicity}(G_i) = b_i$ *and* $\text{cubicity}(G_i) = c_i$. *Then*

$$\max_{i=1}^{d} b_i \leq \text{boxicity}(\boxtimes_{i=1}^{d} G_i) \leq \sum_{i=1}^{d} b_i, \text{ and}$$
$$\max_{i=1}^{d} c_i \leq \text{cubicity}(\boxtimes_{i=1}^{d} G_i) \leq \sum_{i=1}^{d} c_i.$$

Furthermore, if each G_i, $i \in [d]$ has a universal vertex, then the second inequality in both the above chains is tight.

If we consider the strong product of a 4-cycle C_4 with a path on 3 vertices P_3, we get an example where the upper bound in Theorem 2.1 is not tight.

Corollary 2.2. *For any given graph G,* $\text{boxicity}(G^{\boxtimes d})$ *and* cubicity $(G^{\boxtimes d})$ *are in $O(d)$ and there exist graphs for which they are in $\Omega(d)$.*

2.2 Cartesian products

We show two different upper bounds on the boxicity and cubicity of Cartesian products. The first and the easier result bounds from above the boxicity (cubicity) of a Cartesian product in terms of the boxicity (cubicity) of the corresponding strong product and the boxicity (cubicity) of a Hamming graph whose size is determined by the chromatic number of the component graphs. The second bound is in terms of the maximum cubicity among the component graphs and the boxicity (cubicity) of a Hamming graph whose size is determined by the sizes of the component graphs. The second bound is much more useful to study the growth of boxicity and cubicity of higher Cartesian powers since the first term remains a constant.

Theorem 2.3. *For graphs $G_1, \ldots, G_d$,*

$$\text{boxicity}(\square_{i=1}^d G_i) \leq \text{boxicity}(\boxtimes_{i=1}^d G_i) + \text{boxicity}(\square_{i=1}^d K_{\chi_i}) \text{ and}$$
$$\text{cubicity}(\square_{i=1}^d G_i) \leq \text{cubicity}(\boxtimes_{i=1}^d G_i) + \text{cubicity}(\square_{i=1}^d K_{\chi_i})$$

where χ_i denotes the chromatic number of G_i, $i \in [d]$.

When $G_i = K_q$ for every $i \in [d]$, $G = \boxtimes_{i=1}^d G_i$ is a complete graph on q^d vertices and hence has boxicity and cubicity 0. In this case it is easy to see that both the bounds in Theorem 2.3 are tight.

Theorem 2.4. *For graphs $G_1, \ldots, G_d$, with $|V(G_i)| = q_i$ and* cubicity$(G_i) = c_i$*, for each $i \in [d]$,*

$$\text{boxicity}(\square_{i=1}^d G_i) \leq \max_{i\in[d]} c_i + \text{boxicity}(\square_{i=1}^d K_{q_i}), \text{ and}$$
$$\text{cubicity}(\square_{i=1}^d G_i) \leq \max_{i\in[d]} c_i + \text{cubicity}(\square_{i=1}^d K_{q_i}).$$

In wake of the two results above, it becomes important to have a good upper bound on the boxicity and the cubicity of Hamming graphs. The *Hamming graph* K_q^d is the Cartesian product of d copies of a complete graph on q vertices. We call the K_2^d the *d-dimensional hypercube*.

The cubicity of hypercubes is known to be in $\Theta\left(\frac{d}{\log d}\right)$. The lower bound is due to Chandran, Mannino and Oriolo [4] and the upper bound is due to Chandran and Sivadasan [6]. But we do not have such tight estimates on the boxicity of hypercubes. The only explicitly known upper bound is one of $O(d/\log d)$ which follows from the bound on cubicity since boxicity is bounded above by cubicity for all graphs. The only non-trivial lower bound is one of $\frac{1}{2}(\lceil \log\log d\rceil + 1)$ due to Chandran, Mathew and Sivadasan [5].

We make use of a non-trivial upper bound shown by Kostochka on the dimension of the partially ordered set (poset) formed by two neighbouring levels of a Boolean lattice [8] and a connection between boxicity and poset dimension established by Adiga, Bhowmick and Chandran in [1] to obtain the following result.

Theorem 2.5. *Let b_d be the largest dimension possible of a poset formed by two adjacent levels of a Boolean lattice over a universe of d elements. Then*

$$\tfrac{1}{2}b_d \leq \text{boxicity}(K_2^d) \leq 3b_d.$$

Furthermore, $\text{boxicity}(K_2^d) \leq 12 \log d/\log\log d$.

Below, we extend the results on hypercubes to Hamming graphs.

Theorem 2.6. *Let K_q^d be the d-dimensional Hamming graph on the alphabet $[q]$ and let K_2^d be the d-dimensional hypercube. Then for $d \geq 2$,*

$$\begin{aligned} \log q &\leq \text{boxicity}(K_q^d) \leq \lceil 10 \log q \rceil \,\text{boxicity}(K_2^d), \text{ and} \\ \log q &\leq \text{cubicity}(K_q^d) \leq \lceil 10 \log q \rceil \,\text{cubicity}(K_2^d). \end{aligned}$$

Theorem 2.4, along with the bounds on boxicity and cubicity of Hamming graphs, gives the following corollary which is the main result in this article. The lower bound on the order of growth is due to the presence of K_2^d as an induced subgraph in the d-the Cartesian power of any non-trivial graph.

Corollary 2.7. *For any given graph G with at least one edge,*

$$\begin{aligned} \text{boxicity}(G^{\square d}) &\in O\left(\log d/\log\log d\right) \cap \Omega\left(\log\log d\right), \text{ and} \\ \text{cubicity}(G^{\square d}) &\in \Theta\left(d/\log d\right). \end{aligned}$$

2.3 Direct products

Theorem 2.8. *For graphs $G_1, \ldots, G_d$,*

$$\begin{aligned} \text{boxicity}(\times_{i=1}^{d} G_i) &\leq \text{boxicity}(\boxtimes_{i=1}^{d} G_i) + \text{boxicity}(\times_{i=1}^{d} K_{\chi_i}) \text{ and} \\ \text{cubicity}(\times_{i=1}^{d} G_i) &\leq \text{cubicity}(\boxtimes_{i=1}^{d} G_i) + \text{cubicity}(\times_{i=1}^{d} K_{\chi_i}) \end{aligned}$$

where χ_i denotes the chromatic number of G_i, $i \in [d]$.

In the wake of Theorem 2.8, it is useful to estimate the boxicity and the cubicity of the direct product of complete graphs. Before stating our result on the same, we would like to discuss a few special cases. If $G = \times_{i=1}^{d} K_2$ then G is a perfect matching on 2^d vertices and hence has boxicity and cubicity equal to 1. If $G = K_q \times K_2$, then it is isomorphic to a graph obtained by removing a perfect matching from the complete bipartite graph with q vertices on each part. This is known as the *crown graph* and its boxicity is known to be $\lceil q/2 \rceil$ [2].

Theorem 2.9. *Let $q_i \geq 2$ for each $i \in [d]$. Then,*

$$\frac{1}{2}\sum_{i=1}^{d}(q_i - 2) \leq \text{boxicity}\left(\times_{i=1}^{d} K_{q_i}\right) \leq \sum_{i=1}^{d} q_i, \text{ and}$$
$$\frac{1}{2}\sum_{i=1}^{d}(q_i - 2) \leq \text{cubicity}\left(\times_{i=1}^{d} K_{q_i}\right) \leq \sum_{i=1}^{d} q_i \log(n/q_i),$$

where $n = \Pi_{i=1}^{d} q_i$ is the number of vertices in $\times_{i=1}^{d} K_{q_i}$.

Corollary 2.10. *For graphs $G_1, \ldots, G_d$,*

$$\text{boxicity}(\times_{i=1}^{d} G_i) \leq \sum_{i=1}^{d}(\text{boxicity}(G_i) + \chi(G_i)).$$

Corollary 2.11. *For any given graph G, $boxicity(G^{\times d})$ is in $O(d)$ and there exist graphs for which it is in $\Omega(d)$.*

References

[1] A. Adiga, D. Bhowmick and L. Sunil Chandran, *Boxicity and poset dimension*, In: "COCOON", 2010, 3–12.

[2] L. Sunil Chandran, A. Das and Chintan D. Shah, *Cubicity, boxicity, and vertex cover*, Discrete Mathematics **309** (8) (2009), 2488–2496.

[3] L. Sunil Chandran, W. Imrich, R. Mathew and D. Rajendraprasad, *Boxicity and cubicity of product graphs*, arXiv preprint arXiv:1305.5233, 2013.

[4] L. Sunil Chandran, C. Mannino and G. Orialo, *On the cubicity of certain graphs*, Information Processing Letters **94** (2005), 113–118.

[5] L. Sunil Chandran, R. Mathew and N. Sivadasan, *Boxicity of line graphs*, Discrete Mathematics **311** (21) (2011), 2359–2367.

[6] L. Sunil Chandran and N. Sivadasan, *The cubicity of hypercube graphs*, Discrete Mathematics **308** (23) (2008), 5795–5800.

[7] R. Hammack, W. Imrich and S. Klavžar, "Handbook of Product Graphs", CRC press, 2011.

[8] AV Kostochka, *The dimension of neighboring levels of the boolean lattice*, Order, **14** (3) (1997), 267–268.

[9] F. S. Roberts, "Recent Progresses in Combinatorics", chapter On the boxicity and cubicity of a graph, Academic Press, New York, 1969, 301–310.

[10] Douglas B. West, *Boxicity and maximum degree*, http://www.math.uiuc.edu/~west/regs/boxdeg.html, 2008. Accessed: Jan 12, 2013.

Planarity

Planar graphs with $\Delta \geq 8$ are $(\Delta + 1)$-edge-choosable

Marthe Bonamy[1]

Abstract. We consider the problem of *list edge coloring* for planar graphs. Edge coloring is the problem of coloring the edges while ensuring that two edges that are incident receive different colors. A graph is k-edge-choosable if for any assignment of k colors to every edge, there is an edge coloring such that the color of every edge belongs to its color assignment. Vizing conjectured in 1965 that every graph is $(\Delta + 1)$-edge-choosable. In 1990, Borodin solved the conjecture for planar graphs with maximum degree $\Delta \geq 9$, and asked whether the bound could be lowered to 8. We prove here that planar graphs with $\Delta \geq 8$ are $(\Delta + 1)$-edge-choosable.

1 Introduction

We consider simple graphs. A *k-edge-coloring* of a graph G is a coloring of the edges of G with k colors such that two edges that are incident receive distinct colors. We denote by $\chi'(G)$ the smallest k such that G admits a k-edge-coloring. Let $\Delta(G)$ be the maximum degree of G. Since incident edges have to receive distinct colors in an edge coloring, every graph G satisfies $\chi'(G) \geq \Delta(G)$. A trivial upper-bound on $\chi'(G)$ is $2\Delta(G) - 1$, which can can be greatly improved, as follows.

Theorem 1.1 (Vizing [12]). *Every graph G satisfies $\Delta(G) \leq \chi'(G) \leq \Delta(G) + 1$.*

Vizing [13] proved that $\chi'(G) = \Delta(G)$ for every planar graph G with $\Delta(G) \geq 8$. He gave examples of planar graphs with $\Delta(G) = 4, 5$ that are not $\Delta(G)$-edge-colorable, and conjectured that no such graph exists for $\Delta(G) = 6, 7$. This remains open for $\Delta(G) = 6$, but the case $\Delta(G) = 7$ was solved by Sanders and Zhao [10], as follows.

Theorem 1.2 (Sanders and Zhao [10]). *Every planar graph G with $\Delta(G) \geq 7$ satisfies $\chi'(G) = \Delta(G)$.*

An extension of the problem of edge coloring is the *list edge coloring* problem, defined as follows. For any $L : E \rightarrow \mathcal{P}(\mathbb{N})$ list assignment of colors to the edges of a graph $G = (V, E)$, the graph G is *L-edge-colorable* if there exists an edge coloring of G such that the color of

[1] LIRMM, Université Montpellier 2. Email: marthe.bonamy@lirmm.fr. Work supported by the ANR Grant EGOS (2012-2015) 12 JS02 002 01.

every edge $e \in E$ belongs to $L(e)$. A graph $G = (V, E)$ is said to be *list k-edge-colorable* (or *k-edge-choosable*) if G is L-edge-colorable for any list assignment L such that $|L(e)| \geq k$ for any edge $e \in E$. We denote by $\chi'_\ell(G)$ the smallest k such that G is k-edge-choosable.

One can note that edge coloring is a special case of list edge coloring, where all the lists are equal. Thus $\chi'(G) \leq \chi'_\ell(G)$. This inequality is in fact conjectured to be an equality (see [8] for more information).

Conjecture 1.3 (List Coloring Conjecture). Every graph G satisfies $\chi'(G) = \chi'_\ell(G)$.

The conjecture is still widely open. Some partial results were however obtained in the special case of planar graphs: for example, the conjecture is true for planar graphs of maximum degree at least 12, as follows.

Theorem 1.4 (Borodin *et al.* [6]). *Every planar graph G with $\Delta(G) \geq 12$ satisfies $\chi'_\ell(G) = \Delta(G)$.*

There is still a large gap with the lower bound of 7 that should hold by Theorem 1.2 if Conjecture 1.3 were true.

Using Vizing's theorem, the List Coloring Conjecture can be weakened into Conjecture 1.5.

Conjecture 1.5 (Vizing [14]). Every graph G satisfies $\chi'_\ell(G) \leq \Delta(G)+1$.

Conjecture 1.5 has been actively studied in the case of planar graphs with some restrictions on cycles (see for example [11, 15, 16]), and was settled by Borodin [4] for planar graphs of maximum degree at least 9 (a simpler proof was later found by Cohen and Havet [7]).

Theorem 1.6 (Borodin [4]). *Every planar graph G with $\Delta(G) \geq 9$ satisfies $\chi'_\ell(G) \leq \Delta(G) + 1$.*

Here we prove the following theorem.

Theorem 1.7. *Every planar graph G with $\Delta(G) \leq 8$ satisfies $\chi'_\ell(G) \leq 9$.*

This improves Theorem 1.6 and settles Conjecture 1.5 for planar graphs of maximum degree 8.

Corollary 1.8. *Every planar graph G with $\Delta(G) \geq 8$ satisfies $\chi'_\ell(G) \leq \Delta(G) + 1$.*

This answers Problem 5.9 in a survey by Borodin [5]. For small values of Δ, Theorem 1.7 implies that every planar graph G with $5 \leq \Delta(G) \leq 7$ is also 9-edge-choosable. To our knowledge, this was not known. It is however known that planar graphs with $\Delta(G) \leq 4$ are $(\Delta(G) + 1)$-edge-choosable [9, 14].

2 Method

The discharging method was introduced in the beginning of the 20th century. It has been used to prove the celebrated Four Color Theorem ([1] and [2]).

We prove Theorem 1.7 using a discharging method, as follows. A graph is *minimal* for a property if it satisfies this property but none of its proper subgraphs does. The first step is to consider a minimal counter-example G (*i.e.* a graph G such that $\Delta(G) \leq 8$ and $\chi'_\ell(G) > 9$, whose every proper subgraph is 9-edge-choosable), and prove it cannot contain some configurations. We assume by contradiction that G contains one of the configurations. We consider a particular subgraph H of G. For any list assignment L on the edges of G, with $|L(e)| \geq 9$ for every edge e, we L-edge-color H by minimality. We show how to extend the L-edge-coloring of H to G, a contradiction.

The second step is to prove that a connected planar graph on at least two vertices with $\Delta \leq 8$ that does not contain any of these configurations does not satisfy Euler's Formula. To that purpose, we consider a planar embedding of the graph. We assign to each vertex its degree minus six as a weight, and to each face two times its degree minus six. We apply discharging rules to redistribute weights along the graph with conservation of the total weight. As some configurations are forbidden, we can prove that after application of the discharging rules, every vertex and every face has a non-negative final weight. This implies that $\sum_v (d(v)-6) + \sum_f (2d(f)-6) = 2\times|E(G)| - 6\times|V(G)| + 4\times|E(G)| - 6\times|F(G)| \geq 0$, a contradiction with Euler's Formula that $|E| - |V| - |F| = -2$. Hence a minimal counter-example cannot exist.

The complete proof was omitted due to space limitations. The proof requires eleven forbidden configurations and eleven discharging rules, which we do not present here because they rely on additionnal definitions. The full proof can be found in [3].

3 Conclusion

The key idea in the proof lies in some recoloring arguments using directed graphs. It allowed us to deal with configurations that would not yield under usual techniques, and thus to improve Theorem 1.6. Though this simple argument does not seem to be enough to prove Conjecture 1.5 for $\Delta = 7$, it might be interesting to try to improve similarly Theorem 1.4.

Note that the proof could easily be adapted to prove that planar graphs with $\Delta \geq 8$ are $(\Delta + 1)$-edge-choosable. This would however be of little interest considering the simple proof for $\Delta \geq 9$ presented in [7].

Conjecture 1.5 remains open for $\Delta = 5, 6$ and 7. It might be interesting to weaken the conjecture and ask whether all planar graphs are $(\Delta + 2)$-edge-choosable. This is true for planar graphs with $\Delta \geq 7$ by Theorems 1.6 and 1.7. What about planar graphs with $\Delta = 6$?

References

[1] K. APPEL and W. HAKEN, *Every map is four colorable: Part 1, Discharging*, Illinois J. Math. **21** (1977), 422–490.

[2] Appel, K., W. Haken and J. Koch, *Every map is four colorable: Part 2, Reducibility*, Illinois J. Math. **21** (1977), pp. 491–567.

[3] M. BONAMY, *Planar graphs with maximum degree D at least 8 are (D+1)-edge-choosable*, preprint, http://arxiv.org/abs/1303.4025

[4] O. V. BORODIN, *Generalization of a theorem of Kotzig and a prescribed coloring of the edges of planar graphs*, Mathematical Notes **6** (1991), 1186–1190.

[5] O. V. BORODIN, *Colorings of plane graphs: A survey*, Discrete Mathematics **313** (2013), 517–539

[6] O. V. BORODIN, A. V. KOSTOCHKA and D. R. WOODALL, *List Edge and List Total Colourings of Multigraphs*, Journal of Combinatorial Theory, Series B **71** (2) (1997), 184–204.

[7] N. COHEN and F. HAVET, *Planar graphs with maximum degree $\Delta \geq 9$ are $(\Delta + 1)$-edge-choosable - A short proof*, Discrete Mathematics **310** (21) (2010), 3049–3051.

[8] T. JENSEN and B. TOFT, "Graph Coloring Problems", Wiley Interscience, New York (1995).

[9] M. JUVAN, B. MOHAR and R. ŠKREKOVSKI, *Graphs of degree 4 are 5-edge-choosable*, Journal of Graph Theory **32** (1999), 250–264.

[10] D. SANDERS and Y. ZHAO, *Planar graphs of maximum degree seven are class I*, Journal of Combinatorial Theory, Series B **83 (2)** (2001), 201–221.

[11] Y. SHEN, G. ZHENG, W. HE and Y. ZHAO, *Structural properties and edge choosability of planar graphs without 4-cycles*, Discrete Mathematics **308** (2008), 5789–5794.

[12] V. G. VIZING, *On an estimate of the chromatic class of a p-graph (in russian)*, Diskret. Analiz **3** (1964), 25–30.

[13] V. G. VIZING,, *Critical graphs with given chromatic index (in russian)*, Diskret. Analiz **5** (1965), 9–17.

[14] V. G. VIZING, *Colouring the vertices of a graph with prescribed colours (in russian)*, Diskret. Analiz **29** (1976), 3–10.

[15] F. W. WANG and K. W. LIH, *Choosability and Edge Choosability of Planar Graphs without Five Cycles*, Applied Mathematics Letters **15** (2002), 561–565.

[16] L. ZHANG and B. WU, *Edge choosability of planar graphs without small cycles*, Discrete Mathematics **283** (2004), 289–293.

Planar emulators conjecture is nearly true for cubic graphs

Martin Derka[1] and Petr Hliněný[1]

Abstract. We prove that a cubic nonprojective graph cannot have a finite planar emulator, unless one of two very special cases happen (in which the answer is open). This shows that Fellows' planar emulator conjecture, disproved for general graphs by Rieck and Yamashita in 2008, is nearly true on cubic graphs, and might very well be true there definitely.

1 Introduction

A graph G has a finite *planar emulator* H if H is a planar graph and there is a graph homomorphism $\varphi : V(H) \to V(G)$ where φ is locally surjective, *i.e.* for every vertex $v \in V(H)$, the neighbours of v in H are mapped surjectively onto the neighbours of $\varphi(v)$ in G. We also say that such a G is planar-emulable. If we insist on φ being locally bijective, we get a *planar cover*.

The concept of planar emulators was proposed in 1985 by M. Fellows [5], and it tightly relates (although of independent origin) to the better known *planar cover conjecture* of Negami [10]. Fellows also raised the main question: What is the class of graphs with finite planar emulators? Soon later he conjectured that the class of planar-emulable graphs coincides with the class of graphs with finite planar covers (conjectured to be the class of projective graphs by Negami [10]—still open nowadays). This was later restated as follows:

Conjecture 1.1 (M. Fellows, falsified in 2008). A connected graph has a finite planar emulator if and only if it embeds in the projective plane.

For two decades the research focus was exclusively on Negami's conjecture and no substantial new results on planar emulators had

[1] Faculty of Informatics, Masaryk University Brno, Czech Republic. Email: xderka@fi.muni.cz, hlineny@fi.muni.cz

Supported by the research centre Institute for Theoretical Computer Science (CE-ITI); Czech Science foundation project No. P202/12/G061.

been presented until 2008, when emulators for two nonprojective graphs were given by Rieck and Yamashita [12], effectively disproving Conjecture 1.1.

Planar emulable nonprojective graphs. Following Rieck and Yamashita, Chimani et al [2] constructed finite planar emulators of all the minor minimal obstructions for the projective plane except those which have been shown non-planar-emulable already by Fellows ($K_{3,5}$ and "two disjoint k-graphs" cases, Def. 2.1), and $K_{4,4} - e$. The graph $K_{4,4} - e$ is thus the only forbidden minor for the projective plane where the existence of a finite planar emulator remains open. Even though we do not have a definite replacement for falsified Conjecture 1.1 yet, the results obtained so far [2, 4] suggest that, vaguely speaking, up to some trivial operations ("planar expansions"), there is only a finite family of nonprojective planar-emulable graphs. A result like that would nicely correspond with the current state-of-art [9] of Negami's conjecture.

While characterization of planar-emulable graphs has proven itself to be difficult in general, significant progress can be made in a special case. Negami's conjecture has been confirmed in the case of cubic graphs in [11], and the same readily follows from [9]. Here we prove:

Definition 1.2. A *planar expansion* of a graph G is a graph which results from G by repeatedly adding a planar graph sharing one vertex with G, or by replacing an edge or a cubic vertex with a connected planar graph with its attachments (two or three, resp.) on the outer face.

Theorem 1.3. *If a cubic nonprojective graph H has a finite planar emulator, then H is a planar expansion of one of two minimal cubic nonprojective graphs shown in Figure* 1.1.

A computerized search for possible counterexamples to Conjecture 1.1, carried out so far [4], shows that a nonprojective planar-emulable graph G cannot be cubic, unless G contains a minor isomorphic to $\mathcal{E}_2$, $K_{4,5} - 4K_2$, or a member of the so called "$K_7 - C_4$ family". Our new approach, Theorem 1.3, dismisses the former two possibilities completely and strongly restricts the latter one.

2 Cubic planar-emulable graphs

The purpose of this section is to prove Theorem 1.3. In order to do so, we need to introduce some basic related concepts.

Definition 2.1. Graph G is said to *contain two disjoint k-graphs* if there exist two vertex-disjoint subgraphs $J_1, J_2 \subseteq G$ such that, for $i = 1, 2$, the graph J_i is isomorphic to a subdivision of K_4 or $K_{2,3}$, the subgraph

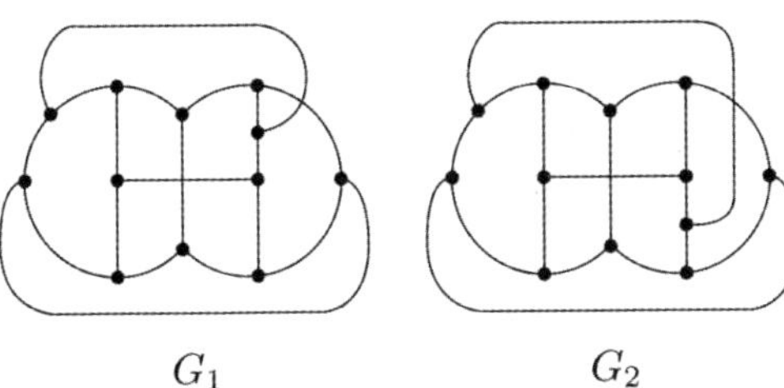

Figure 1.1. Two (out of six in total) cubic irreducible obstructions for the projective plane [6]. Although these graphs result by splitting nonprojective graphs for which we have finite planar emulators [2] (namely $K_7 - C_4$ and its "relatives"), it is still open whether they are planar-emulable.

$G - V(J_i)$ is connected and adjacent to J_i, and contracting in G all the vertices of $V(G) \setminus V(J_i)$ into one results in a nonplanar graph.

Proposition 2.2 (Fellows, unpublished).

a) *The class of planar-emulable graphs is closed under taking minors.*
b) *If G is projective and connected, then G has a finite planar emulator in form of its finite planar cover.*
c) *If G contains two disjoint k-graphs or a $K_{3,5}$ minor, then G is not planar-emulable.*
d) *G is planar-emulable if, and only if, so is any planar expansion of G.*

Proof of Theorem 1.3. Glover and Huneke [6] characterized the cubic graphs with projective embedding by giving a set $\mathcal{I}$ of six cubic graphs such that; if H is a cubic graph that does not embed in the projective plane, then H contains a graph $G \in \mathcal{I}$ as a topological minor.

Let us point out that four out of the six graphs in $\mathcal{I}$ contain two disjoint k-graphs, and so only the remaining two— $G_1 \in \mathcal{I}$ and $G_2 \in \mathcal{I}$ of Figure 1.1, can potentially be planar-emulable. Hence the cubic graph H in Theorem 1.3 contains one of G_1, G_2 as a topological minor. In other words, there is a subgraph $G' \subseteq H$ being a subdivision of a cubic $G \in \{G_1, G_2\}$.

We call a *bridge of G' in H* any connected component B of $H - V(G')$ together will all the incident edges. In a degenerate case, B might consist just of one edge from $E(H) \setminus E(G')$ with both ends in G'. We would like, for simplicity, to speak about positions of bridges with respect to the underlying cubic graph G: Such a bridge B connects to vertices u of G' which subdivide edges f of G —this is due to the cubic degree bound, and we (with neglectable abuse of terminology) say that B *attaches to* this edge f in G itself.

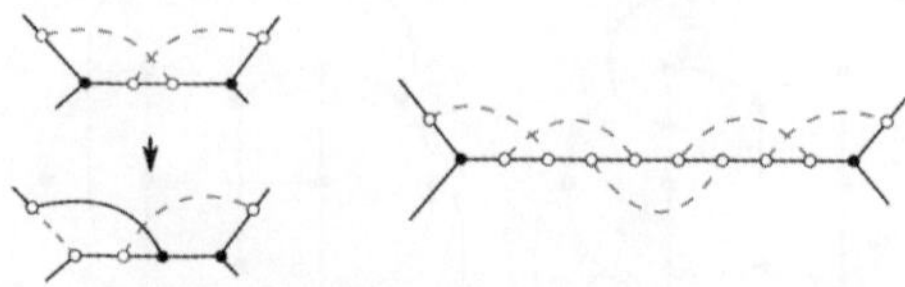

Figure 2.1. Illustration for Lemma 2.3. The trivial bridge on the left takes over the role of a branch vertex of G in the subdivision, resulting in existence of a nontrivial bridge. The other case shows when the transitive closure of declared attachment becomes important.

A bridge B is *nontrivial* if B attaches to some two nonadjacent edges of G, and B is *trivial* otherwise. For a trivial bridge B; either B attaches to only one edge in G, and we say *exclusively*, or all thc edges to which B attaches in G have a vertex w in common and we say that B *attaches to* this w.

We divide the rest of the proof into two main cases; that either some bridge of G' in H is nontrivial or all such bridges are trivial. In the "all-trivial" case one more technical condition has to be observed: Suppose B_1, B_2 are bridges such that B_1 attaches to w and B_2 attaches to an edge f incident to w in G (perhaps B_2 exclusively to f). On the path P_f which replaces (subdivides) f in G', suppose that B_2 connects to some vertex which is closer to w on P_f than some other vertex to which B_1 connects to. Then we *declare that* B_2 *attaches to* w, too. This is well defined because of the following (Figure 2.1):

Claim 2.3. Let $G' \subseteq H$ be a subdivision of G where G, H are cubic graphs. Suppose that all bridges of G' in H are trivial, and that a bridge B_0 attaches (is declared to) both to w_1 and w_2, where $w_1 w_2 \in E(G)$. Then there is $G'' \subseteq H$ which is a subdivision of G, too, and a nontrivial bridge of G'' in H exists.

In the described situation, we call B_0 a *conflicting* bridge. We then continue with the following claim obtained by routine examination of (collections of) trivial bridges in view of Definition 2.1 (Figure 2.2).

Claim 2.4. Let $G' \subseteq H$ be a subdivision of G where G, H are cubic nonprojective graphs and G does not contain two disjoint k-graphs. Suppose that all bridges of G' in H are trivial, and no one is conflicting (cf. Lemma 2.3). Then H does not contain two disjoint k-graphs if, and only if, H is a planar expansion of G.

After that, we use an exhaustive computerized enumeration of nontrivial bridges to conclude the following. We would like to point out that

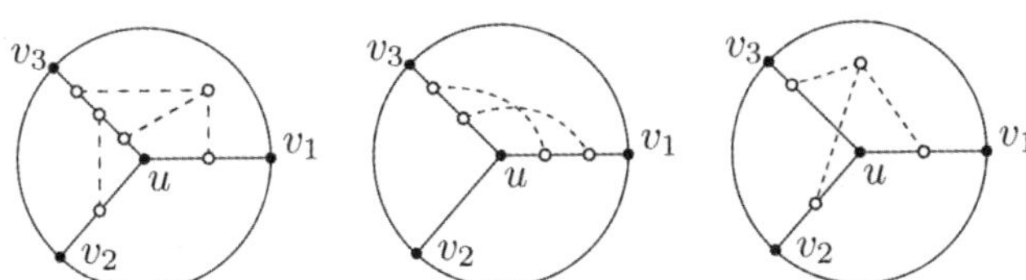

Figure 2.2. Illustration of three collections of trivial bridges that attach to a cubic vertex u. The first collection gives a planar expansion, while the other two are the "minimal" non-planar-expansion cases.

due to necessity of the $K_{3,5}$ case, there is likely no simple handwritten argument summarizing the cases similarly as done in Lemma 2.4.

Claim 2.5. Let $G' \subseteq H$ be a subdivision of G where G, H are cubic nonprojective graphs. If there exists a nontrivial bridge of G' in H, then H contains two disjoint k-graphs or a $K_{3,5}$ minor.

3 Conclusions

We identified two graphs (Figure 1.1), for which existence of finite planar emulator now becomes extremely interesting. We would like to point out that similarity of these two graphs suggest that if one has a finite planar emulator, so does the other one. If we however elaborate on this idea and attempt to "unify" the graphs in the form of a common supergraph, we have to use a nontrivial bridge. Perhaps, this provides a clue that these two graphs should not be planar-emulable. Thus, providing an answer for any of these two graphs would bring a better insight to the problem of planar emulations not only for the cubic case, but also in general.

References

[1] D. ARCHDEACON, *A Kuratowski Theorem for the Projective Plane*, J. Graph Theory **5** (1981), 243–246.

[2] M. CHIMANI, M. DERKA, P. HLINĚNÝ and M. KLUSÁČEK, *How Not to Characterize Planar-emulable Graphs*, Advances in Applied Mathematics **50** (2013), 46–68.

[3] M. DERKA, "Planar Graph Emulators: Fellows' Conjecture", Bc. Thesis, Masaryk University, Brno, 2010.

[4] M. DERKA, *Towards Finite Characterization of Planar-emulable Non-projective Graphs*, Congressus Numerantium **207** (2011), 33–68.

[5] M. FELLOWS, "Encoding Graphs in Graphs", Ph.D. Dissertation, Univ. of California, San Diego, 1985.

[6] H. GLOVER andJ. P. HUNEKE, *Cubic Irreducible Graphs for the Projective Plane*, Discrete Mathematics **13** (1975), 341–355.

[7] P. HLINĚNÝ, "Planar Covers of Graphs: Negami's Conjecture", Ph.D. Dissertation, Georgia Institute of Technology, Atlanta, 1999.

[8] P. HLINĚNÝ, *20 Years of Negami's Planar Cover Conjecture*, Graphs and Combinatorics **26** (2010), 525–536.

[9] P. HLINĚNÝ and R. THOMAS, *On possible counterexamples to Negami's planar cover conjecture*, J. of Graph Theory **46** (2004), 183–206.

[10] S. NEGAMI, *Enumeration of projective-planar embeddings of graphs*, Discrete Math. **62** (1986), 299–306.

[11] S. NEGAMI and T. WATANABE, *Planar cover conjecture for* 3-*regular graphs*, Journal of the Faculty of Education and Human Sciences, Yokohama National University **4** (2002), 73–76.

[12] Y. RIECK and Y. YAMASHITA, *Finite planar emulators for* $K_{4,5} - 4K_2$ *and* $K_{1,2,2,2}$ *and Fellows' conjecture*, European Journal of Combinatorics **31** (2010), 903–907.

Random planar graphs with minimum degree two and three

Marc Noy[1] and Lander Ramos[1]

1 Main results

The main goal of this paper is to enumerate planar graphs subject to a condition on the minimum degree δ. Asking for $\delta \geq 1$ is not very interesting, since a random planar graph only contains a constant number of isolated vertices. The condition $\delta \geq 2$ is directly related to the concept of the core of a graph. Given a connected graph $\mathcal{G}$, its *core* (also called 2-core in the literature) is the maximum subgraph $\mathcal{C}$ with minimum degree at least two. The core $\mathcal{C}$ is obtained from $\mathcal{G}$ by repeatedly removing vertices of degree one. Conversely, $\mathcal{G}$ is obtained by attaching rooted trees at the vertices of $\mathcal{C}$. The *kernel* of $\mathcal{G}$ is obtained by replacing each maximal path of vertices of degree two in $\mathcal{C}$ by a single edge. The kernel has minimum degree at least three, and $\mathcal{C}$ can be recovered from $\mathcal{K}$ by replacing edges by paths. As shown in the figure below, the kernel may have loops and multiple edges, which must be taken into account since our goal is to analyze simple graphs. Another difficulty is that when replacing loops and multiple edge by paths the same graph can be produced several times. To this end we weight multigraphs appropriately according to the number of loops and edges of each multiplicity. We remark that the concepts of core and kernel of a graph are instrumental in the theory of random graphs [3,5].

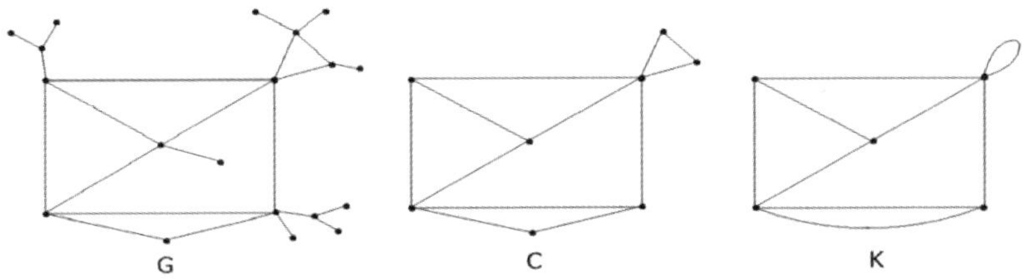

[1] Department of Applied Mathematics, Universitat Politècnica de Catalunya, Barcelona. Email: marc.noy@upc.edu, landertxu@gmail.com

Notice that $\mathcal{G}$ is planar if and only if $\mathcal{C}$ is planar, if and only if $\mathcal{K}$ is planar. It is convenient to introduce the following definitions: a *2-graph* is a connected graph with minimum degree at least two, and a *3-graph* is a connected graph with minimum degree at least three. In order to enumerate planar 2- and 3-graphs, we use generating functions. From now on all graphs are labelled and generating functions are of the exponential type. Let c_n, h_n and k_n be, respectively, the number of planar connected graphs, 2-graphs and 3-graphs with n vertices, and let

$$C(x) = \sum c_n \frac{x^n}{n!}, \qquad H(x) = \sum h_n \frac{x^n}{n!}, \qquad K(x) = \sum k_n \frac{x^n}{n!}$$

be the associated generating functions. Also, let $t_n = n^{n-1}$ be the number of (labelled) rooted trees with n vertices and let $T(x) = \sum t_n x^n/n!$. The decomposition of a connected graph into its core and the attached trees implies the following equation

$$C(x) = H(T(x)) + U(x), \tag{1.1}$$

where $U(x) = T(x) - T(x)^2/2$ is the generating function of unrooted trees. Since $C(x)$ is known completely [2], we have access to $H(x)$. Since $T(x) = xe^{T(x)}$, we can invert the above relation and obtain

$$H(x) = C(xe^{-x}) - x + \frac{x^2}{2}.$$

The equation defining $K(x)$ is more involved and requires the bivariate generating function

$$C(x, y) = \sum c_{n,k} y^k \frac{x^n}{n!},$$

where $c_{n,k}$ is the number of connected planar graphs with n vertices and k edges. We can express $K(x)$ in terms of $C(x, y)$ as

$$K(x) = C(A(x), B(x)) + E(x), \tag{1.2}$$

where $A(x)$, $B(x)$, $E(x)$ are certain elementary functions.

The knowledge of $C(x)$ as the solution of a system of functional-differential equations [2] leads to the following expression:

$$c_n \sim \kappa n^{-7/2} \gamma^n n!,$$

where $\kappa \approx 0.4261 \cdot 10^{-5}$ and $\gamma \approx 27.2269$ are computable constants. Analyzing the enriched generating function $C(x, y)$ it is possible to obtain results on the number of edges and other basic parameters in random planar graphs. Our main goal is to extend these results to planar 2-graphs and 3-graphs.

Using Equations (1.1) and (1.2) we obtain precise asymptotic estimates for the number of planar 2- and 3-graphs.

Theorem 1.1. *The numbers of planar 2- and 3-graphs are asymptotically*

$$h_n \sim \kappa_2 n^{-7/2} \gamma_2^n n!, \qquad \gamma_2 \approx 26.2076,$$

$$k_n \sim \kappa_3 n^{-7/2} \gamma_3^n n!, \qquad \gamma_2 \approx 21.3102.$$

As is natural to expect, h_n and k_n are exponentially smaller than c_n. Also, the number of 2-connected planar graphs is known to be asymptotically $c \cdot n^{-7/2}(26.1849)^n n!$ (see [2]). This is consistent with the estimate for h_n, since a 2-connected has minimum degree at least two.

By enriching Equations (1.1) and (1.2) taking into account the number of edges, we prove the following.

Theorem 1.2. *The number of edges in random planar 2-graphs and 3-graphs are both asymptotically normal with linear expectation and variance. The respective expected values are asymptotically $\mu_2 n$ and $\mu_3 n$, where $\mu_2 \approx 2.2614$ and $\mu_3 \approx 2.3227$.*

The number of edges in connected planar graphs was shown to be [2] asymptotically normal with expectation $\sim \mu n$, where $\mu \approx 2.2133$. This conforms to our intuition that increasing the minimum degree also increases the expected number of edges.

It is also of interest to analyze the size X_n of the core in a random connected planar graph, and the size Y_n of the kernel in a random planar 2-graph.

Theorem 1.3. *The variables X_n and Y_n are asymptotically normal with linear expectation and variance and*

$$\mathbf{E}X_n \sim \lambda_2 n, \qquad \lambda_2 \approx 0.9626,$$

$$\mathbf{E}Y_n \sim \lambda_3 n, \qquad \lambda_3 \approx 0.8259.$$

We remark that the value of λ_2 has been recently found by McDiarmid [4] using alternative methods. Also, we remark that the expected size of the largest block (2-connected component) in random connected planar graphs is asymptotically $0.9598n$. Again this is consistent since the largest block is contained in the core.

The picture is completed by analyzing the size of the trees attached to the core.

Theorem 1.4. *The number of trees with k vertices attached to the core of a random planar connected graph is asymptotically normal with linear expectation and variance. The expected value is asymptotically*

$$C\frac{k^{k-1}}{k!}\rho^k n,$$

where $C > 0$ is a constant $\rho \approx 0.03673$ corresponding to the radius of convergence of $C(x)$.

For k large, the previous quantity grows like

$$\frac{C}{\sqrt{2\pi}} \cdot k^{-3/2}(\rho e)^k n.$$

This quantity is negligible if $k \gg \log(n)/(\log(1/\rho e))$. Using the first and second moment method, it can be shown that the size of the largest tree attached to the core is in fact asymptotically

$$\frac{\log(n)}{\log(1/\rho e)}.$$

Our last result concerns the distribution of the vertex degrees in random planar 2-graphs and 3-graphs. It was proved in [1] that the probability that a random vertex has degree $k \geq 1$ in a random planar tends to a constant $d(k)$. The probability generating function

$$p(w) = \sum_{k\geq 1} d(k)w^k$$

is explicit although very complicated [1]. He prove a similar result for planar 2- and 3-graphs.

Theorem 1.5. *For each fixed $k \geq 2$ the probability that a random vertex has degree k in a random planar 2-graph tends to a constant $d_H(k)$, and for each fixed $k \geq 3$ the probability that a random vertex has degree k in a random planar 3-graph tends to a constant $d_K(k)$. Moreover $\sum_{k\geq 2} d_H(k) = \sum_{k\geq 3} d_K(k) = 1$, and the probability generating functions*

$$p_H(w) = \sum_{k\geq 2} d_H(k)w^k, \qquad p_K(w) = \sum_{k\geq 3} d_K(k)w^k$$

are computable in terms the probability generating function $p(w)$ of connected planar graphs (and other constants related to the enumeration of planar graphs).

We show that almost all planar 2-graphs have a vertex of degree two, and almost all planar 3-graphs have a vertex of degree three. Hence asymptotically all our results hold for planar graphs with minimum degree exactly two and three, respectively. In addition, all the results for connected planar graphs extend easily to arbitrary planar graphs.

2 Results for maps

We also find analogous results for planar maps. They are simpler to derive and serve as a preparation for the results on planar graphs, while at the same time they are new and interesting by themselves. A planar map is a connected planar multigraph (loops and multiple edges are allowed) embedded in the plane up to homeomorphism. A map is rooted if one of the edges is distinguished and given a direction. In this way a rooted map has a root edge and a root vertex (the tail of the root edge). A rooted map has no automorphisms, in the sense that every vertex, edge and face is distinguishable.

The enumeration of rooted planar maps was started by Tutte in his seminal paper [6]. If m_n is the number of maps with n edges, then

$$m_n = \frac{2 \cdot 3^n}{(n+2)(n+1)} \binom{2n}{n}, \qquad n \geq 0.$$

The generating function $M(z) = \sum_{n \geq 0} m_n z^n$ is equal to

$$M(z) = \frac{18z - 1 + (1-12z)^{3/2}}{54z^2}. \tag{2.1}$$

Either from the explicit formula or from the expression for $M(z)$ and singularity analysis, it follows that

$$m_n \sim \frac{2}{\sqrt{\pi}} n^{-5/2} 12^n.$$

If $m_{n,k}$ is the number of maps with n edges and degree of the root face equal to k, then $M(z,u) = \sum m_{n,k} u^k z^n$ satisfies the equation

$$M(z,u) = 1 + zu^2 M(z,u)^2 + uz \frac{uM(z,u) - M(z,1)}{u-1}. \tag{2.2}$$

By duality, $M(z,u)$ is also the generating function of maps in which u marks the degree of the root vertex.

The core and the kernel of a map are defined as for connected graphs. The situation is simpler for kernels since loops and multiple edges are allowed. We define a 2-map as a map with minimum degree at least two, and a 3-map as a map with minimum degree at least three.

Theorem 2.1. *Let h_n and k_n be, respectively, the number of 2-maps and 3-maps with n edges. The associated generating functions $H(z)$ and $K(z)$ are given by*

$$H(x) = \frac{1-x}{1+x} M\left(\frac{x}{(1+x)^2}\right) + x - 1,$$

$$K(x) = \frac{H\left(\frac{x}{1+x}\right) - x}{1+x}.$$

The following estimates hold:

$$h_n \sim \kappa_2 n^{-5/2}(5+2\sqrt{6})^n, \qquad k_n \sim \kappa_3 n^{-5/2}(4+2\sqrt{6})^n,$$

where

$$\kappa_2 = \frac{2}{\sqrt{\pi}}\left(\frac{2}{3}\right)^{5/4}, \qquad \kappa_3 = \frac{2}{\sqrt{\pi}}\left(4 - 4\sqrt{\frac{2}{3}}\right)^{5/2}.$$

The same estimates hold for maps with minimum degree exactly two and three, respectively.

We also prove a limit law for the size of the core and the kernel in random maps.

Theorem 2.2. *The size X_n of the core of a random map with n edges, and the size Y_n of the kernel of a random 2-map with n edges are asymptotically Gaussian with*

$$\mathbf{E}X_n \sim \frac{\sqrt{6}}{3}n, \qquad \mathbf{Var}(X_n) \sim \frac{n}{6},$$

$$\mathbf{E}Y_n \sim (2\sqrt{6}-4)n, \qquad \mathbf{Var}(Y_n) \sim (18\sqrt{6}-44)n.$$

Furthermore, the size Z_n of the kernel of a random map with n edges is also Gaussian with

$$\mathbf{E}Z_n \sim \left(4 - \frac{4\sqrt{6}}{3}\right)n, \qquad \mathbf{Var}(Z_n) \sim \left(\frac{128}{3} - \frac{52}{3}\sqrt{6}\right)n.$$

We also analyze the size of the trees attached to the core of a random map.

Theorem 2.3. *Let $X_{n,k}$ count trees with k edges attached to the core of a random map with n edges. Then $X_{n,k}$ is asymptotically normal and*

$$\mathbf{E}X_{n,k} \sim \alpha_k n$$

where

$$\alpha_k = \left(4 + \frac{5}{3}\sqrt{6}\right)\frac{1}{k+1}\binom{2k}{k}\left(\frac{1}{12}\right)^k.$$

By the same argument as in the previous section, the size of the largest tree attached to the core is asymptotically $\log n / \log 3$.

Our last result in this section deals with the distribution of the degree of the root vertex in 2-maps and 3-maps. The limiting probability $p_M(K)$ that a random map has a root vertex (or face) of degree k exists for all $k \geq 1$. The probability generating function of the distribution is known to be

$$p_M(u) = \sum p_M(k)u^k = \frac{u\sqrt{3}}{\sqrt{(2+u)\,(6-5\,u)^3}}.$$

The tail of the distribution is asymptotically

$$p_M(k) \sim \frac{1}{2\sqrt{10\pi}}k^{1/2}\left(\frac{5}{6}\right)^k.$$

We obtain analogous results for 2-maps and 3-maps.

Theorem 2.4. *Let $p_M(u)$ be as before, and let $p_H(u)$ and $p_K(u)$ be the probability generating functions for the distribution of the root degree in 2-maps and 3-maps, respectively. Then we have*

$$p_H(u) = \frac{p_M\left(\dfrac{w(1+\sigma)}{1+w\sigma}\right)\dfrac{1+\sigma}{1+w\sigma} - w\sigma}{1-\sigma},$$

$$p_K(u) = \frac{p_H(u) - \sigma u^2}{1-\sigma}.$$

Furthermore, the limiting probabilities that the degree of the root vertex is equal to k exists, both for 2-maps and 3-maps, and are asymptotically

$$p_H(k) \sim \frac{1}{8}\sqrt{\frac{3(1-\sigma)}{\pi}}k^{1/2}\left(\sqrt{\frac{2}{3}}\right)^k,$$

$$p_K(k) \sim \frac{1}{8}\sqrt{\frac{3}{(1-\sigma)\pi}}k^{1/2}\left(\sqrt{\frac{2}{3}}\right)^k.$$

References

[1] M. DRMOTA, O. GIMÉNEZ and M. NOY, *Degree distribution in random planar graphs*, J. Combin. Theory Ser. A **118** (2011), 2102–2130.

[2] O. GIMÉNEZ and M. NOY, *Asymptotic enumeration and limit laws of planar graphs*, J. Amer. Math. Soc. **22** (2009), 309–329.

[3] S. JANSON, D. E. KNUTH, T. ŁUCZAK and B. PITTEL, *The birth of the giant component*, Random Structures Algorithms **4** (1993), 233–358.

[4] C. MCDIARMID, *Random graphs from a weighted minor-closed class*, arXiv:1210.2701.

[5] M. NOY, V. RAVELOMANANA and J. RUÉ, *On the probability of planarity of a random graph near the critical point*, Proc. Amer. Math. Soc., to appear.

[6] W. T. TUTTE, *A census of planar maps*, Canad. J. Math. **15** (1963), 249-271.

Degenerated induced subgraphs of planar graphs

Robert Lukot'ka[1], Ján Mazák[1] and Xuding Zhu[2]

Abstract. A graph G is *k-degenerated* if it can be deleted by subsequent removals of vertices of degree k or less. We survey known results on the size of maximal k-degenerated induced subgraph in a planar graph. In addition, we sketch the proof that every planar graph of order n has a 4-degenerated induced subgraph of order at least $8/9 \cdot n$. We also show that in every planar graph with at least 7 vertices, deleting a suitable vertex allows us to subsequently remove at least 6 more vertices of degree four or less.

1 Introduction

A graph G is *k-degenerated* if every subgraph of G has a vertex of degree k or less. Equivalently, a graph is k-degenerated if we can delete the whole graph by subsequently removing vertices of degree at most k. The reverse of this sequence of removed vertices can be used to colour (or even list-colour) G with $k + 1$ colours in a greedy fashion. Graph degeneracy is therefore a natural bound on both chromatic number and list chromatic number. In fact, for some problems graph degeneracy provides the best known bounds on the choice number [3].

Every planar graph has a vertex of degree at most 5. Since a subgraph of a planar graph is planar, it also has a vertex of degree at most 5. Therefore, every planar graph is 5-degenerated. If $k < 5$, we can still choose at least some of the vertices of G, and if these vertices induce a

[1] Trnava University in Trnava, Slovakia. Email: robert.lukotka@truni.sk, jan.mazak@truni.sk

[2] Zhejiang Normal University, China. Email: xudingzhu@gmail.com

The first author acknowledges partial support from the research grant 7/TU/13 and from the APVV grant ESF-EC-0009-10 within the EUROCORES Programme EUROGIGA (project GReGAS) of the European Science Foundation.
The work of the second author leading to this invention has received funding from the European Research Council under the European Union's Seventh Framework Programme (FP7/2007-2013)/ERC grant agreement no. 259385.
The third author acknowledges partial support from the research grants NSF No. 11171730 and ZJNSF No. Z6110786.

k-degenerated graph, then they can be greedily coloured by $k+1$ colours. Thus an interesting question is how large k-degenerated subgraph can be guaranteed in a planar graph G.

Without the restriction to planar graphs Alon, Kahn, and Seymour [2] determined exactly how large k-degenerated induced subgraph one can guarantee depending only on the degree sequence of G. For planar graphs the only settled case is $k = 0$, due to four colour theorem. Except for $k = 4$ all other values of k were examined. Most attention is devoted to the case $k = 1$. The Albertson-Berman conjecture [1] asserts that every planar graph has an induced forest with at least half of the vertices. The best known bound, guaranteeing a forest of size at least $2/5 \cdot |V(G)|$, is implied by the fact that planar graphs are acyclic 5-colourable [4].

We do not know of any results on maximum 4-degenerated induced subgraphs of planar graphs. A likely reason is that such a bound is not interesting for list-colouring applications: Thomassen [5] proved that every planar graph is 5-choosable. The rest of this paper focuses on degeneracy 4.

We define two operations for vertex removal: deletion and collection. To *delete* a vertex v, we remove v and its incident edges from the graph. To *collect* a vertex v is the same as to delete v, but to be able to collect v we require v to be of degree at most 4. The collected vertices induce a 4-degenerated subgraph. Vertices that are deleted or collected are collectively called *removed*.

Our main results are the following two theorems.

Theorem 1.1. *In every planar graph G we can delete at most* $1/9$ *of its vertices in such a way that we can collect all the remaining ones.*

Theorem 1.2. *In every planar graph with at least* 7 *vertices we can delete a vertex in such a way that we can subsequently collect at least* 6 *vertices.*

These results are probably not the best possible. The worst example known to us is the icosahedron from which we need to delete one vertex out of twelve to be able to collect the remaining eleven. We believe that this is the worst case possible.

Conjecture 1.3. In each planar graph G we can delete at most $1/12$ of its vertices in such a way that we can collect all the remaining ones.

Conjecture 1.4. In each planar graph with at least 12 vertices we can delete a vertex in such a way that we can subsequently collect at least 11 vertices.

2 Sketch of the proof

To prove Theorem 1.1 we incorporate the degrees of the vertices of G to the statement; let

$$\Phi(G) = \sum_{v \in V(G)} (\deg(v) - 5). \tag{2.1}$$

We prove the following theorem by induction on the number of vertices of G. The function $\mathrm{tc}(G)$ stands for the number of tree components of G.

Theorem 2.1. *If G is a planar graph, then there is a set $S \subset V(G)$ with at most $\Gamma(G) = |V(G)|/12 + 1/36 \cdot \Phi(G) + 1/18 \cdot \mathrm{tc}(G)$ vertices such that if we delete S we can subsequently collect all the vertices of G.*

Since $\Phi(G) + 2\,\mathrm{tc}(G) \leq |V(G)|$, Theorem 2.1 implies Theorem 1.1. Theorem 1.2 can be proved alongside Theorem 2.1.

We prove Theorem 2.1 by a discharging procedure. Let G be a minimal counterexample to Theorem 2.1 (with respect to the number of vertices); it can be easily shown that G is connected with minimal degree 5. We embed G into the plane (for now, we let the embedding be arbitrary).

Type (t)	Degree (deg)	Min. number of non-tr. faces ($n_{\sqcup}$)	Max. number of V_5 neigh. (n_5)	Maximal charge (mc)
10a	10+	0	3	1
10b	10+	0	∞	1/2
9a	9	1	3	1
9b	9	0	2	1
9c	9	0	3 if consecutive	9/10, 1, 9/10
9d	9	0	9	1/2
8a	8	0	1	1
8b	8	1	2	1
8c	8	2	3 if consecutive	9/10, 1, 9/10
8d	8	0	2	9/10
8e	8	0	8	1/2
7a	7	0	1	4/5
7b	7	1	2	13/20
7c	7	0	2	2/5
7d	7	0	7	1/3
6a	6	1	1	2/5
6b	6	0	6	0

Table 1. Maximal charges that can be send to a vertex.

Each vertex of degree at least 6 is assigned a certain *type* according to Table 1. If w is of degree d, is contained on at least $n_{\sqcup}$ non-triangular faces, and has at most n_5 neighbours from V_5, then w can be of type t. If w can have more than one type, then the type of w is the type that occurs first in the table. Let vw be an edge such that v is of degree 5 and w is of degree at least 6. For every such edge we define the maximal charge $\mathrm{mc}(v, w)$ that v can send to w. This maximal charge is given in the last column of Table 1.

We start by assigning initial charges to the vertices and faces of G. Each vertex v of degree d receives charge $6 - d$ and each face of length ℓ receives charge $2(3 - \ell)$. According to Euler's theorem, the total initial charge is equal to 12. In the discharging procedure, we redistribute the charges between vertices and faces in a certain way such that no charge is created or lost. The discharging procedure consists of the following three steps.

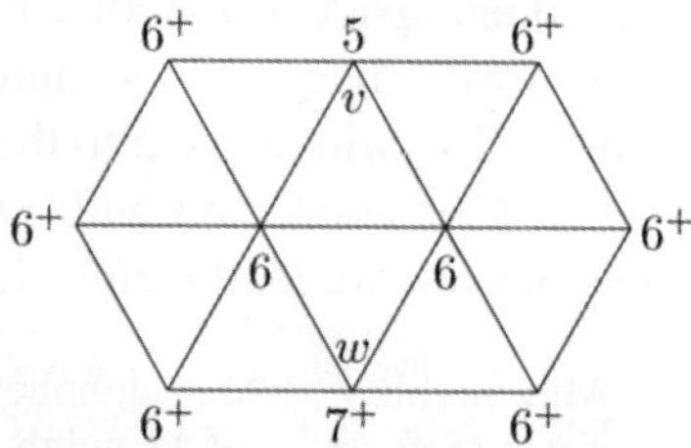

Figure 2.1. Distance discharging (d^+ denotes a vertex of degree at least d).

Step 1: Discharging to faces. For each vertex v and for each non-triangular face that contains v, send $1/2$ from v to f, except for two cases: if v is of degree 6, then send $2/5$; and if v is not of degree 6 but both its neighbours on f are of degree 6, then send $3/5$.

Step 2: Distance discharging. In every subgraph of G isomorphic to the configuration in Figure 2.1 send $1/5$ from vertex v to vertex w (vertices are denoted as in Figure 2.1; the depicted vertices are pairwise distinct; the numbers indicate degrees).

Step 3: Final discharging of the vertices of degree five. For each vertex v of degree 5 carry out the following procedure. Order the neighbours w of v which have degree at least 6 according to the value of $\mathrm{mc}(v, w)$ starting with the largest value; let $w_1, w_2, \ldots$ be the resulting ordering. If the value of $\mathrm{mc}(v, w)$ is the same for two neighbours of v, then we order them arbitrarily. For $i = 1, 2, \ldots,$ send $\max\{\mathrm{mc}(v, w_i), \mathrm{ch}_a(v)\}$ from v to w_i, where $\mathrm{ch}_a(v)$ denotes the current charge of v.

Since no face can have positive final charge, there exists a vertex with positive charge. A very technical examination allows us to show that if we start with a certain embedding of G, then there is a vertex with positive final charge not contained in any C_3-cut or C_4-cut of G.

We examine v together with its neighbourhood; this gives rise to a number of cases, each of them leading to a contradiction. We demonstrate this analysis on Configuration 7-3 shown in Figure 2.2 (the numbers in the right part of the figure indicate the degrees of the vertices shown in the left part). Since G contains no C_3-cut, the triangles depicted in Figure 2.2 are faces and contain no vertices inside. The vertex v is of type $7d$.

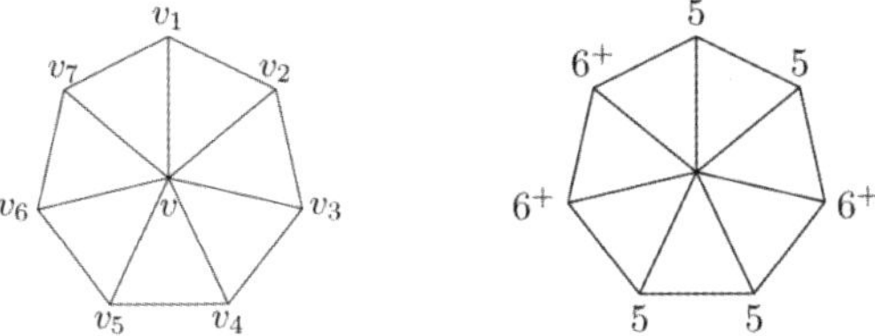

Figure 2.2. Configuration 7-3.

Configuration 7-3: Suppose first that the vertex v_7 has degree at most 7. Delete v_6. Then we can collect v_5, v_4, v, v_1, v_2, and v_7. We get a new graph G' smaller than G, so there is a set S' in G' with at most $\Gamma(G')$ vertices such that if we delete S', we can collect the rest of G'. If $\Gamma(G) - \Gamma(G') \geq 1$, we can extend the set S' by v_6 and obtain a contradiction with the fact that G is a smallest counterexample to Theorem 2.1.

We want show that $\Gamma(G) - \Gamma(G') \geq 1$. The hardest part is typically to compute $\Phi(G) - \Phi(G')$. Among the vertices removed from G, four vertices have degree at least 5, two vertices have degree at least 6, and one vertex has degree at least 7. After removing these vertices from the sum (2.1) the value of Φ decreases by at least 4. Moreover, all the neighbours of the removed vertices have smaller degree in G' than in G. From the fact that v is in no C_3-cuts we know that there is no extra edge between the neighbours of v; all such edges are shown in Figure 2.2. This decreases Φ further by at least 17. Therefore $\Phi(G) - \Phi(G') \geq 21$. It can be shown that no new tree components are created, so together $\Gamma(G) - \Gamma(G') \geq 42/36 \geq 1$.

We are left with the case where v_7 has degree at least 8. If v_7 has another neighbour w of degree 5 besides v_1, then we can delete v_7 and collect v_1, v_2, v, v_4, v_5, and w (again we need to check that Γ decreases by at least 1). Otherwise, v_7 has only one neighbour of degree 5, the vertex v_1. According to Table 1, $\mathrm{mc}(v_1, v_7) = 1$, so v_1 discharges nothing into v in Step 3 and the final charge of v is at most 0. This contradicts the fact that v has positive charge after the discharging.

References

[1] M. O. ALBERTSON and D. M. BERMAN, *A conjecture on planar graphs*, In: "Graph Theory and Related Topics" J. A. Bondy and U. S. R. Murty, (eds.), Academic Press, 1979, 357.

[2] N. ALON, J. KAHN and P. D. SEYMOUR, *Large induced degenerate subgraphs*, Graphs and Combinatorics **3** (1987), 203–211.

[3] J. BARÁT, G. JORET and D. R. WOOD, *Disproof of the list Hadwiger Conjecture*, Electron. J. Combin. **18** (2011), #P232.

[4] O. V. BORODIN, *On acyclic colorings of planar graphs*, Discrete Math. **25** (1979), 211–236.

[5] C. THOMASSEN, *Every planar graph is 5-choosable*, Journal of Comb. Theory Ser B **62** (1994), 180–181.

Strong chromatic index of planar graphs with large girth

Mickaël Montassier[1], Arnaud Pêcher[2] and André Raspaud[2]

Abstract. Let $\Delta \geq 4$ be an integer. We prove that every planar graph with maximum degree Δ and girth at least $10\Delta+46$ is strong $(2\Delta-1)$-edge-colorable, that is best possible (in terms of number of colors) as soon as G contains two adjacent vertices of degree Δ. This improves [2] when $\Delta \geq 6$.

1 Introduction

A *strong k-edge-coloring* of a graph G is a mapping from $E(G)$ to $\{1, 2, \ldots, k\}$ such that every two adjacent edges or two edges adjacent to a same edge receive two distinct colors. The *strong chromatic index* of G, denoted by $\chi'_s(G)$, is the smallest integer k such that G admits a strong k-edge-coloring.

Strong edge-colorability was introduced by Fouquet and Jolivet [7, 8] and was used to solve the frequency assignment problem in some radio networks.

An obvious upper bound on $\chi'_s(G)$ (given by a greedy coloring) is $2\Delta(\Delta-1)+1$ where Δ is the maximum degree of G. The following conjecture was posed by Erdős and Nešetřil [4,5] and revised by Faudree, Schelp, Gyárfás and Tuza [6]:

Conjecture 1.1 (See [4–6]). If G is a graph with maximum degree Δ, then

$$\chi'_s(G) \leq \frac{5}{4}\Delta^2 \text{ if } \Delta \text{ is even and } \frac{1}{4}(5\Delta^2 - 2\Delta + 1) \text{ otherwise.}$$

Moreover, they gave examples of graphs whose strong chromatic indices reach the upper bounds.

[1] University of Montpellier 2, CNRS-LIRMM, UMR5506. Email: mickael.montassier@lirmm.fr

[2] LaBRI - University of Bordeaux. Email: arnaud.pecher@labri.fr, andre.raspaud@labri.fr

This research is supported by ANR/NSC contract ANR-09-blan-0373-01 and NSC99-2923-M-110-001-MY3.

The strong chromatic index was studied for different families of graphs, as cycles, trees, d-dimensional cubes, chordal graphs, Kneser graphs, see [10]. For complexity issues, see [9, 10].

Faudree, Schelp, Gyárfás exhibited, for every integer $\Delta \geq 2$, a planar graph with maximum degree Δ and strong chromatic index $4\Delta - 4$. They established the following upper bound:

Theorem 1.2 ([6]). *Planar graphs with maximum degree* Δ *are strong* $(4\Delta + 4)$*-edge-colorable.*

The purpose of this paper is to prove that if the girth is large enough, then the upper bound can be strengthened to $2\Delta - 1$, which is best possible as soon as G contains two adjacent vertices of degree Δ. A first attempt was done Borodin and Ivanova [2] who proved: *every planar graph with maximum degree* Δ *is strong* $(2\Delta - 1)$*-edge-colorable if its girth is at least* $40 \left\lfloor \frac{\Delta}{2} \right\rfloor + 1$. Here we improved the girth condition for every $\Delta \geq 6$:

Theorem 1.3. *Let* $\mathcal{F}_\Delta$ *be the family of planar graphs with maximum degree at most* Δ. *Every graph of* $\mathcal{F}_\Delta$ *with girth at least* $10\Delta + 46$ *admits a strong* $(2\Delta - 1)$*-edge-coloring when* $\Delta \geq 4$.

Next section is devoted to the outline of the proof of Theorem 1.3. For the full version of this paper, see [3].

2 On planar graphs with large girth

A *walk* in a graph is a sequence of edges where two consecutive edges are adjacent. Throughout the paper, by *path* we mean a walk where every two consecutive edges are distinct. So a vertex or an edge can appear more than once in a path. By *cycle* we mean a closed path (the first and last edges of the sequence are adjacent).

The proof of Theorem 1.3 is based on some properties of *odd graphs*. Let n be an integer; the odd graph O_n may be defined as follows:

- the vertices are the $(n - 1)$-subsets of $\{1, 2, \ldots, 2n - 1\}$.
- two vertices are adjacent if and only if the corresponding subsets are disjoint.

The odd graph O_n is n-regular and distance transitive. Moreover its odd-girth is $2n - 1$ and its even-girth is 6 [1]. We will use the notation $S(x)$ to denote the subset assigned to the vertex x in O_n. Also we can label every edge xy by the label $\{1, \ldots, 2n - 1\} \setminus (S(x) \cup S(y))$. Remark that the obtained edge-labeling is a strong edge-coloring. As example, O_3

(the Petersen graph) is depicted in Fig 2.1. To prove Theorem 1.3, we establish that there is a path of length exactly $2(n-1)$ between every pair of vertices (not necessarily distinct) in the odd graph O_n ($n \geq 4$).

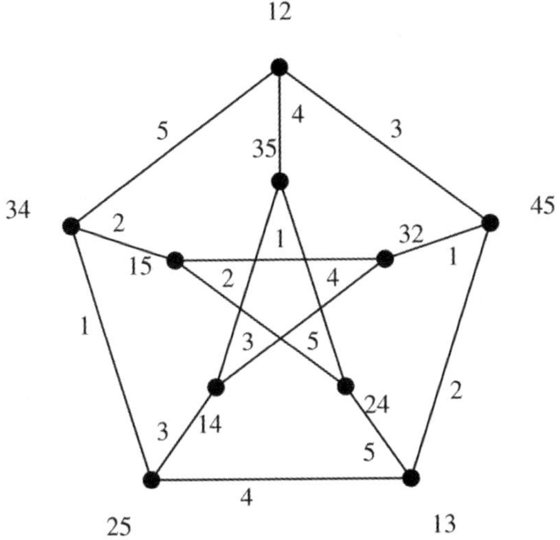

Figure 2.1. The odd graph O_3 and its edge labeling.

In the following, we consider the case $\Delta \geq 4$.

Let $G \in \mathcal{F}_\Delta$ be a counterexample to Theorem 1.3 with the minimum order. Clearly G is connected and has minimum degree at least 1.

(1) *G does not contain a vertex v adjacent to $d(v)-1$ vertices of degree* 1.

By the way of contradiction, suppose G contains such a vertex v. Let u be a vertex of degree 1 adjacent to v. By minimality of G, $G' = G - u$ admits a strong $(2\Delta - 1)$-edge-coloring. By a simple counting argument, it is easy to see that we can extend the coloring to uv, a contradiction.

Consider now $H = G - \{v : v \in G, d_G(v) = 1\}$.

(2) *The minimum degree of H is at least* 2 *(by* (1)*). Graph H is planar and has the same girth as G.*

The following observation is well-known ([11], Lemma 5):

(3) *Every planar graph with minimum degree at least* 2 *and girth at least* $5d + 1$ *contains a path consisting of d consecutive vertices of degree* 2.

Let $d = 2\Delta + 9$. It follows from the assumption on the girth, (2) and (3) that H contains a path $v_0 v_1 v_2 \ldots v_{d+1}$ in which every vertex v_i for $1 \leq i \leq d$ has degree 2. In G, the path $v_1 \ldots v_d$ is an induced path and every v_i ($1 \leq i \leq d$) may be adjacent to some vertices of degree 1, by definition of H and (1).

Now, consider G' obtained from G by

- removing all the pendant vertices adjacent to $v_1 \dots v_d$
- removing the vertices v_2 to v_{d-1}.

By minimality of G, G' admits a strong $(2\Delta - 1)$-edge coloring ϕ. Our aim is to extend ϕ to G and get a contradiction.

Let $c_\phi(u)$ be the set of colors of the edges incident to u. We can assume that $|c_\phi(v_0)| = |c_\phi(v_{d+1})| = \Delta$ (by adding vertices of degree 1 adjacent to v_0 and v_{d+1} in G' as $2\Delta < d$ and so $|V(G')| < |V(G)|$). Let $x = \phi(v_0v_1)$ and $y = \phi(v_dv_{d+1})$. For a set S of colors, define $\overline{S} = \{1, \dots, 2\Delta - 1\} \setminus S$.

Extending ϕ to G is equivalent to find a special path P in the odd graph O_Δ. This path P must have the following properties:

P1. its length is $d + 1$. Let $P = u_0u_1 \dots u_{d+1}$;
P2. u_0 is the vertex of O_Δ such that $S(u_0) = \overline{c_\phi(v_0)}$;
P3. u_{d+1} is the vertex of O_Δ such that $S(u_{d+1}) = \overline{c_\phi(v_{d+1})}$;
P4. the edge u_0u_1 is labeled with x;
P5. the edge u_du_{d+1} is labeled with y.

Informally speaking, this path may be seen as a mapping of $v_0 \dots v_{d+1}$ into O_Δ. If such a path exists, then one can extend ϕ to G by coloring the edges incident to v_i with colors of $\overline{S(u_i)}$; the edge v_iv_{i+1} is colored with the label of the edge u_iu_{i+1} in O_Δ . In the following, if P is a path, then we will denote by $||P||$ the length (number of edges) of the path P.

As O_n is distance transitive and its even-girth is 6 [1], we have:

(4) *Let xyz be a simple path of length 2 of O_n with $n \geq 3$. Then xyz is contained in a cycle of length 6.*
(5) *Let x be a vertex of O_n with $n \geq 3$. Then x is contained in a cycle of length $2k$ for any integer $k \geq 3$.*

The core of the proof is the following lemma (whose proof is omitted in this extended abstract):

Lemma 2.1. *Let u and v be two (not necessarily distinct) vertices of O_n with $n \geq 4$. There exists a simple path linking u and v of length exactly $2(n - 1)$.*

We are now able to exhibit the path P linking u_0 and u_{d+1}. Let $P_s = u_0s_1 \dots s_{2(\Delta-1)-1}u_{d+1}$ be a path linking u_0 and u_{d+1} of length $2(\Delta - 1)$

in O_Δ. Let u_1 be the neighbor of u_0 so that the edge u_0u_1 is labeled with x. Let u_d be the neighbor of u_{d+1} so that the edge u_du_{d+1} is labeled with y. As $\Delta \geq 3$, let t be a neighbor of u_0 distinct from u_1 and s_1, and let w be a neighbor of u_{d+1} distinct from u_d and $s_{2(\Delta-1)-1}$. Finally, by (4), let C_1 be a 6-cycle containing tu_0u_1 and let C_2 be a 6-cycle containing $wu_{d+1}u_d$.

1. We first start from u_0 making a loop around C_1 going through first u_1. Hence P2 and P4 are satisfied.
2. We then leave u_0 to u_{d+1} going through P_s.
3. Finally we make a loop around C_2 going through first w. Hence P3 and P5 are satisfied.

Finally observe that $||P|| = 6+||P_s||+6 = 2(\Delta-1)+12 = 2\Delta+10 = d+1$ as required by P1.

References

[1] N. Biggs, *Some odd graph theory*, Annals New York Academy of Sciences (1979), 71–81.

[2] O.V. Borodin and A. O. Ivanova, *Precise upper bound for the strong edge chromatic number of sparse planar graphs*, Discussiones Mathematicae Graph Theory, 2013, to appear.

[3] G. Chang, M. Montassier, A. Pêcher and A. Raspaud, *Strong chromatic index of planar graphs with large girth*, submitted.

[4] P. Erdős, *Problems and results in combinatorial analysis and graph theory*, Discrete Math. **72** (1988), 81–92.

[5] P. Erdős and J. Nešetřil, *Problem*, In: "Irregularities of Partitions", G. Halász and V. T. Sós (eds.), Springer, Berlin, 1989, 162–163.

[6] R.J. Faudree, A. Gyárfas, R. H. Schelp and Zs. Tuza, *The strong chromatic index of graphs*, Ars Combinatoria **29B** (1990), 205–211.

[7] J. L. Fouquet and J. L. Jolivet, *Strong edge-coloring of graphs and applications to multi-k-gons*, Ars Combinatoria **16** (1983), 141–150.

[8] J. L. Fouquet and J. L. Jolivet, *Strong edge-coloring of cubic planar graphs*, In: "Progress in Graph Theory" (Waterloo 1982), 1984, 247–264.

[9] H. Hocquard, P. Ochem and P. Valicov, *Strong edge coloring and induced matchings*, LaBRI Research Report http://hal.archives-ouvertes.fr/hal-00609454_v1/+, 2011.

[10] M. MAHDIAN, "The Strong Chromatic Index of Graphs", Master Thesis, University of Toronto, Canada, 2000.
[11] J. NEŠETŘIL, A. RASPAUD and É. SOPENA, *Colorings and girth of oriented planar graphs*, Discrete Mathematics **165-166** (1997), 519–530.

On homomorphisms of planar signed graphs to signed projective cubes

Reza Naserasr[1], Edita Rollová[2] and Éric Sopena[3]

Abstract. We conjecture that every planar signed bipartite graph of unbalanced girth $2g$ admits a homomorphism to the signed projective cube of dimension $2g-1$. Our main result is to show that for a given g, this conjecture is equivalent to the corresponding case ($k = 2g$) of a conjecture of Seymour claiming that every planar k-regular multigraph with no odd edge-cut of less than k edges is k-edge-colorable.

1 Preliminaries

It is a classic result of Tait from 1890 that the Four-Color Theorem (Conjecture at that time) is equivalent to the statement that every cubic bridgeless planar graph is 3-edge-colorable. It is easily observed that if a k-regular multigraph is k-edge-colorable, then the number of edges with exactly one end in X, assuming $|X|$ is odd, is at least k. Seymour conjectured in 1975 that for planar graphs the converse is also true:

Conjecture 1.1 (Seymour [6]). Every k-regular planar multigraph with no odd edge-cut of less than k edges is k-edge-colorable.

In this work we focus on a direct extension of the Four-Color Theorem using homomorphisms of signed graphs and its relation with Seymour's conjecture. The theory of homomorphisms of signed graphs includes in particular the theory of graph homomorphisms. Here we introduce the

1 CNRS, LRI, UMR8623, Univ. Paris-Sud 11, F-91405 Orsay Cedex, France. Email: reza@lri.fr

2 Department of Mathematics, Faculty of Applied Sciences, University of West Bohemia, Univerzitní 22, 306 14 Plzeň, Czech Republic. Email: rollova@kma.zcu.cz

3 Univ. Bordeaux, LaBRI, UMR5800, F-33400 Talence, France, CNRS, LaBRI, UMR5800, F-33400 Talence, France. Email: eric.sopena@labri.fr

We would like to acknowledge support from CNRS (France) through the PEPS project HOGRASI and partial support by APVV, Project 0223-10 (Slovakia).

basic notations but we refer to [4], to [5] and to references mentioned there for more details.

Given a graph G, a *signature* on G is a mapping that assigns to each edge of G either a positive or a negative sign. A signature is normally denoted by the set Σ of negative edges. Given a signature Σ on a graph G, *resigning* at a vertex v is to change the sign of each edge incident with v. Two signatures Σ_1 and Σ_2 on G are *equivalent* if one can be obtained from the other by a sequence of resignings or, equivalently, by changing the signs of the edges of an edge-cut. A *signed graph*, denoted (G, Σ), is a graph together with a class of equivalent signatures where Σ could be any member of the class.

An *unbalanced cycle* in a signed graph (G, Σ) is a cycle having an odd number of negative edges. Note that this is independent of the choice of a representative signature. Furthermore, the notion of unbalanced cycle is, in some sense, an extension of the classic notion of odd cycle, as a cycle of $(G, E(G))$ is unbalanced if and only if it is an odd cycle. The *unbalanced-girth* of (G, Σ) is then the shortest length of an unbalanced cycle of (G, Σ).

Theorem 1.2 (Zaslavsky [7]). *Two signatures Σ_1 and Σ_2 on a graph G are equivalent if and only if they induce the same set of unbalanced cycles.*

A cycle that is not unbalanced, i.e., a cycle that has an even number of negative edges (possibly none) is called *balanced.*

An important subclass of signed graphs, called *consistent signed graphs*, is the class of signed graphs whose balanced cycles are all of even length and whose unbalanced cycles are all of the same parity. This class itself consists of two parts. When all the unbalanced cycles are odd, then the set of unbalanced cycles of (G, Σ) is exactly the set of odd cycles of G, thus in this case, by Theorem 1.2, $E(G)$ is a signature. Such a signed graph will then be called an *odd signed graph.* When all balanced and unbalanced cycles are even, the graph G must be bipartite, and Σ can be any subset of edges. Such a signed graph will be called an *even signed graph.*

Given two graphs G and H, a *homomorphism* of G to H is a mapping $\phi : V(G) \to V(H)$ such that if $xy \in E(G)$ then $\phi(x)\phi(y) \in E(H)$. Given two signed graphs (G_1, Σ_1) and (G_2, Σ_2) we say that (G_1, Σ_1) admits a *signed homomorphism*, or a homomorphism for short, to (G_2, Σ_2) if there are signatures Σ_1' and Σ_2' equivalent to Σ_1 and Σ_2, respectively, and a homomorphism φ of G_1 to G_2 such that φ also preserves the signs of edges given by Σ_1' and Σ_2'. The binary relation $(G_1, \Sigma_1) \to (G_2, \Sigma_2)$, which denotes the existence of a homomorphism of (G_1, Σ_1) to (G_2, Σ_2), is an associative one.

Using the definition of *signed projective cube* from the later paragraph, the following conjecture is the main concern of this work:

Conjecture 1.3. Every consistent planar signed graph of unbalanced girth k admits a homomorphism to the signed projective cube of dimension $k - 1$.

In an unpublished manuscript [1], B. Guenin, after introducing the notion of signed graph homomorphisms, proposed that the larger class of consistent signed graphs with an additional property (having no $(K_5, E(K_5))$-signed-minor) satisfies the same conclusion. He has shown relations between his conjecture and several other conjectures.

The case $k = 3$ of Conjecture 1.3 is the Four-Color Theorem. For odd values of k it turns out that all the edges of both sides are negative, and thus the problem is reduced to a graph homomorphism problem. Therefore, by [3], for each such k this conjecture is equivalent to Conjecture 1.1 with the corresponding value of k. In this work we prove the analog for even values of k and thus we examine only the following case of the conjecture.

Conjecture 1.3 (even case). Every planar even signed graph of unbalanced girth at least $2g$ admits a homomorphism to $\mathcal{SPC}_{2g-1}$.

We now introduce signed projective cubes and some of their properties.

The projective cube of dimension d, denoted $\mathcal{PC}_d$, is the Cayley graph $(\mathbb{Z}_2^d, \{e_1, e_2, \dots, e_d\} \cup \{J\})$ where e_i is the vector of $\mathbb{Z}_2^d$ with the i-th coordinate being 1 and other coordinates being 0 and $J = (1, 1, 1, \dots, 1)$. Let $\mathcal{J}$ be the set of edges corresponding to J. We define the *signed projective cube* of dimension d, denoted $\mathcal{SPC}_d$, to be the signed graph $(\mathcal{PC}_d, \mathcal{J})$.

Theorem 1.4 ([5]). *All balanced cycles of $\mathcal{SPC}_d$ are of even length, all unbalanced cycles of $\mathcal{SPC}_d$ are of the same parity, and the unbalanced girth of $\mathcal{SPC}_d$ is $d + 1$. Furthermore, for each unbalanced cycle UC of $\mathcal{SPC}_d$ and for each $x \in \{e_1, e_2, \dots, e_d\} \cup \{J\}$, there is an odd number of edges of UC which correspond to x.*

Using Theorem 1.4 and Theorem 1.2 we get that the signed projective cube $\mathcal{SPC}_{2d}$ (resp. $\mathcal{SPC}_{2d-1}$) is an odd (resp. even) signed graph and thus if a signed graph (G, Σ) admits a homomorphism to $\mathcal{SPC}_{2d}$ (resp. $\mathcal{SPC}_{2d-1}$), we conclude that (G, Σ) must also be odd (resp. even). Thus, in general, consistent signed graphs are the only graphs that can map to signed projective cubes. The following theorem shows that the problem of finding a mapping of a consistent signed graph to a signed projective cube is equivalent to a packing problem.

Theorem 1.5 ([1,5]). *An even (resp. odd) signed graph admits a homomorphism to $\mathcal{SPC}_{2d-1}$ (resp. $\mathcal{SPC}_{2d}$) if and only if it admits at least $2d-1$ (resp. $2d$) edge disjoint signatures.*

2 An extension of the Four-Color Theorem

A key lemma in the study of homomorphism properties of a planar graph is the folding lemma of Klostermeyer and Zhang [2]. By considering unbalanced cycles instead of odd cycles in the lemma, using the ideas of Section 4 in [2] we will get the same result for the class of planar even signed graphs.

Lemma 2.1 (Folding lemma). [5] *Let G be a planar even signed graph of unbalanced girth at least g. If $C = v_0 \cdots v_{r-1} v_0$ is a facial cycle of G with $r \neq g$, then there is an integer $i \in \{0, \ldots, r-1\}$ such that the graph G' obtained from G by identifying v_{i-1} and v_{i+1} (subscripts are taken modulo r) is still of unbalanced girth g.*

By repeated application of this lemma we conclude that:

Corollary 2.2. *Given a planar even signed graph (G, Σ) of unbalanced girth at least g, there is a homomorphic image (H, Σ_1) of (G, Σ) such that: (i) H is planar, (ii) (H, Σ_1) is an even signed graph, (iii) (H, Σ_1) is of unbalanced girth g and (iv) every facial cycle of (H, Σ_1) is an unbalanced cycle of length g.*

We are now ready to prove the main theorem of the paper.

Theorem 2.3. *The following two statements are equivalent:*

(i) *Every planar $2g$-regular multigraph with no odd edge-cut of less than $2g$ edges is $2g$-edge-colorable.*
(ii) *Every planar even signed graph of unbalanced girth at least $2g$ admits a homomorphism to $\mathcal{SPC}_{2g-1}$.*

Proof. First assume that every planar even signed graph of unbalanced girth at least $2g$ admits a homomorphism to SPC_{2g-1} and let G be a planar $2g$-regular multigraph with no odd edge-cut of less than $2g$ edges. Using Tutte's matching theorem we can easily verify that G admits a perfect matching. Let M be a perfect matching of G. Let G^D be the dual of G with respect to some embedding of G on the plane. Since G is $2g$-regular, G^D is clearly bipartite. Let M^D be the edges in G^D corresponding to the edges of M. It is now easy to check that (G^D, M^D) is a planar even signed graph of unbalanced girth $2g$. Therefore, by our main assumption, (G^D, M^D) admits a homomorphism to $\mathcal{SPC}_{2g-1}$. This mapping induces a $2g$-edge-coloring on G^D (not necessarily a proper edge-coloring) using colors $e_1, \ldots, e_{2g-1}, J$. By Theorem 1.4 every unbalanced cycle has received exactly $2g$ different colors. In particular each face of G^D, which is an unbalanced cycle of length $2g$, has received all $2g$ colors. Thus reassigning these colors to their corresponding edges in G will result in a proper $2g$-edge-coloring of G.

Now we assume that every planar $2g$-regular multigraph with no odd edge-cut of less than $2g$ edges is (properly) $2g$-edge-colorable. Let (G, Σ) be a plane even signed graph of unbalanced girth $2g$. We would like to prove that this signed graph admits a homomorphism to $\mathcal{SPC}_{2g-1}$. By Corollary 2.2 we may assume that each face of (G, Σ) is an unbalanced cycle of length exactly $2g$. Let G^D be the dual of G with respect to its embedding on the plane. Obviously G^D is a $2g$-regular multigraph, furthermore it is easy to check that G^D has no odd edge-cut of strictly less than $2g$ edges (this is the dual of having unbalanced girth at least $2g$). Thus, by our main assumption, G^D is $2g$-edge colorable. Let M_i be one of the color classes, which, therefore, is a perfect matching. Let Σ_i be the edges of G corresponding to the edges of G^D in M_i. We claim that Σ_i is equivalent to Σ. This is the case because in both (G, Σ_i) and (G, Σ) each face is an unbalanced cycle, and any other cycle is unbalanced if and only if it bounds an odd number of faces. That means that the sets of unbalanced cycles in both signatures are the same and the claim follows by Theorem 1.2. To complete the proof note that we have partitioned edges of G into $2g$ sets Σ_i each being a signature of (G, Σ). Thus, by Theorem 1.5, (G, Σ) admits a homomorphism to $\mathcal{SPC}_{2g-1}$. □

Note that if G is a simple bipartite graph, then the unbalanced girth of (G, Σ) is at least 4. Furthermore, note that $\mathcal{SPC}_3$ is isomorphic to $(K_{4,4}, M)$ where M is a perfect matching of $K_{4,4}$. Therefore:

Corollary 2.4. *Every planar even signed graph admits a homomorphism to* $(K_{4,4}, M)$.

Using Theorem 6.2 of [4] it follows that this corollary is stronger than the Four-Color Theorem. This fact, in the edge-coloring formulation, was already proved by P. Seymour [6].

References

[1] B. GUENIN, *Packing odd circuit covers: A conjecture*, manuscript.

[2] W. KLOSTERMEYER and C. Q. ZHANG, $(2+\epsilon)$*-coloring of planar graphs with large odd girth*, J. Graph Theory **33** (2) (2000), 109–119.

[3] R. NASERASR, *Homomorphisms and edge-colourings of planar graphs*, J. Combin. Theory Ser. B **97** (3) (2007), 394–400.

[4] R. NASERASR, E. ROLLOVÁ and É. SOPENA, *Homomorphisms of signed graphs*, submitted.

[5] R. NASERASR, E. ROLLOVÁ and É. SOPENA, *Homomorphisms of planar signed graphs to signed projective cubes*, manuscript.

[6] P. SEYMOUR, “Matroids, Hypergraphs and the Max.-Flow Min.-Cut Theorem”, D. Phil. Thesis, Oxford (1975), page 34.

[7] T. ZASLAVSKY, *Signed graphs*, Discrete Applied Math. **4** (1) (1982), 47–74.

Classification of k-nets embedded in a plane

Gábor Korchmáros[1]

Abstract. We present some recent, partly unpublished, results on k-nets embedded in a projective plane $PG(2, \mathbb{K})$ defined over a field $\mathbb{K}$ of any characteristic $p \geq 0$, obtained in collaboration with G.P. Nagy and N. Pace.

1 Introduction

A general problem in finite geometry is to determine geometric structures, such as graphs, designs and incidence geometries, which can be embedded in a projective plane. Here we deal with k-nets embedded in a projective plane $PG(2, \mathbb{K})$ defined over a field $\mathbb{K}$ of any characteristic $p \geq 0$. They are line configurations in $PG(2, \mathbb{K})$ consisting of k pairwise disjoint line-sets, called components, such that any two lines from distinct families are concurrent with exactly one line from each component. The size of each component of a k-net is the same, the order of the k-net. The concept of a k-net arose in classical Differential geometry, and there is a long history about finite k-nets in Combinatorics, especially for $k = 3$, related to affine planes, latin squares, loops and strictly transitive permutation sets. In recent years a strong motivation for investigation of k-nets embedded in $PG(2, \mathbb{K})$ came from Algebraic geometry and Resonance theory see [2, 8, 9, 13, 14].

2 k-nets embedded in $PG(2, \mathbb{K})$

The Stipins-Yuzvinsky theorem states that no embedded k-net for $k \geq 5$ exists when $p = 0$; see [11, 14]. Our present investigation of k-nets embedded in $PG(2, \mathbb{K})$ includes groundfields $\mathbb{K}$ of positive characteristic

[1] Dipartimento di Matematica e Informatica, Università degli Studi della Basilicata, Italia.
Email: gabor.korchmaros@unibas.it

p, and as a matter of fact, many examples. This phenomena is not unexpected since $PG(2,\mathbb{K})$ with $\mathbb{K}$ of characteristic $p > 0$ contains an affine subplane $AG(2,\mathbb{F}_p)$ of order p from which k-nets for $3 \le k \le p+1$ arise taking k parallel line classes as components. Similarly, if $PG(2,\mathbb{K})$ also contains an affine subplane $AG(2,\mathbb{F}_{p^h})$, in particular if $\mathbb{K} = \mathbb{F}_q$ with $q = p^r$ and $h|r$, then k-nets of order p^h for $3 \le k \le p^h + 1$ exist in $PG(2,\mathbb{K})$. Actually, more families of k-nets embedded in $PG(2,\mathbb{F}_q)$ when $q = p^r$ with $r \ge 3$ arise from Lunardon's work; see [5,6]. On the other hand, no 5-net of order n with $p > n$ is known to exist. This suggests that for sufficiently large p compared with n, the Stipins-Yuzvinsky theorem remains valid in $PG(2,\mathbb{K})$. Our main result [4] in this direction proves it:

Theorem 2.1. *If $p > 3^{\varphi(n^2-n)}$ where φ is the classical Euler function then no k-net with $k \ge 5$ is embedded in $PG(2,\mathbb{K})$.*

Our approach also works in zero characteristic and provides a new proof for the Stipins-Yuzvinksy theorem. A key idea in our proof is to consider the cross-ratio of four concurrent lines from different components of a 4-net. We prove that the cross-ratio remains constant when the four lines vary without changing component. In other words, every 4-net in $PG(2,\mathbb{K})$ has constant cross-ratio. In zero characteristic, and in characteristic p with $p > 3^{\varphi(n^2-n)}$, the constant cross-ratio is restricted to two values only, namely to the roots of the polynomial $X^2 - X + 1$. From this, the non-existence of k-nets for $k \ge 5$ easily follows both in zero characteristic and in characteristic p with $p > 3^{\varphi(n^2-n)}$. It should be noted that without a suitable hypothesis on n with respect to p, the constant cross-ratio of a 4-net may assume many different values, even for finite fields.

In the complex plane, there is known only one 4-net up to projectivity; see [11–14]. This 4-net, called the classical 4-net, has order 3 and it exists since $PG(2,\mathbb{C})$ contains an affine subplane $AG(2,\mathbb{F}_3)$ of order 3, unique up to projectivity, and the four parallel line classes of $AG(2,\mathbb{F}_3)$ are the components of a 4-net in $PG(2,\mathbb{K})$. It has been conjectured that the classical 4-net is the only 4-net embedded in $PG(2,\mathbb{C})$.

3 3-nets embedded in $PG(2,\mathbb{K})$

There are known plenty of 3-nets embedded in $PG(2,\mathbb{K})$. One infinite family arises from plane cubic curves. More precisely, let C be a plane (possible reducible) cubic curve equipped with its abelian group $(G,+)$ defined on the set of the nonsingular points of C, take three distinct cosets $H+a,\ H+b,\ H+c$ of a subgroup H of G of order n, such that $a+b+c=0$. Then, in the dual plane, the lines corresponding to the points in these three cosets are the component of a 3-net. Such embedded 3-nets in $PG(2,\mathbb{K})$ are called *algebraic*. The isotopy class of quasigroups coordinatizing an algebraic 3-net contains a group isomorphic to H. Another infinite family arise from tetrahedrons of $PG(3,\mathbb{K})$ by projection, and called of *tetrahedron type*. In the dual plane, the components $\Lambda_1, \Lambda_2, \Lambda_3$ of a tetrahedron type embedded 3-net lie on the six sides (diagonals) of a non-degenerate quadrangle such a way that $\Lambda_i = \Delta_i \cup \Gamma_i$ with Δ_i and Γ_i lying on opposite sides, for $i=1,2,3$. More precisely, $(\Lambda_1, \Lambda_2, \Lambda_3)$ can be lifted to the fundamental tetrahedron of $PG(3,\mathbb{K})$ so that the projection π from the point $P_0=(1,1,1,1)$ on the plane $X_4=0$ returns $(\Lambda_1,\Lambda_2,\Lambda_3)$. For this purpose, it is enough to define the sets lying on the edges of the fundamental tetrahedron:

$$\begin{array}{ll}\Gamma'_1=\{(\xi,0,1,0)|\xi\in L_1\}, & \Gamma'_2=\{(0,\eta,1,0)|\eta\in L_2\},\\ \Gamma'_3=\{(1,-\zeta,0,0)|\zeta\in L_3\}, & \Delta'_1=\{(0,\alpha-1,0,-1)|\alpha\in M_1\},\\ \Delta'_2=\{(\beta-1,0,0,-1)|\beta\in M_2\}, & \Delta'_3=\{(0,0,\gamma-1,-1)|\gamma\in M_3\},\end{array}$$

and observe that $\pi(\Gamma'_i)=\Gamma_i$ and $\pi(\Delta'_i)=\Delta_i$ for $i=1,2,3$. Moreover, a triple (P_1,P_2,P_3) of points with $P_i\in\Gamma_i\cup\Delta_i$ consists of collinear points if and only if if their projection does. Hence, $(\Gamma'_1\cup\Gamma'_2, \Gamma'_3\cup\Delta'_1, \Delta'_2\cup\Delta'_3)$ can be viewed as a "spatial" dual 3-net realizing the same group H. Clearly, $(\Gamma'_1\cup\Gamma'_2, \Gamma'_3\cup\Delta'_1, \Delta'_2\cup\Delta'_3)$ is contained in the sides of the fundamental tetrahedron. We claim that these sides minus the vertices form an infinite spatial dual 3-net realizing the dihedral group $2.\mathbb{K}^*$. To prove this, parametrize the points as follows.

$$\begin{array}{rl}\Sigma_1 &= \{x_1=(x,0,1,0),\ (\varepsilon x)_1=(0,1,0,x)\mid x\in\mathbb{K}^*\},\\ \Sigma_2 &= \{y_2=(1,y,0,0),\ (\varepsilon y)_2=(0,0,1,y)\mid y\in\mathbb{K}^*\},\\ \Sigma_3 &= \{z_3=(0,-z,1,0),\ (\varepsilon z)_3=(1,0,0,-z)\mid z\in\mathbb{K}^*\}.\end{array}\tag{3.1}$$

Then,

$$\begin{array}{rl}x_1, y_2, z_3 \text{ are collinear} &\Leftrightarrow z=xy,\\ (\varepsilon x)_1, y_2, (\varepsilon z)_3 \text{ are collinear} &\Leftrightarrow z=xy \Leftrightarrow \varepsilon z=(\varepsilon x)y,\\ x_1, (\varepsilon y)_2, (\varepsilon z)_3 \text{ are collinear} &\Leftrightarrow z=x^{-1}y \Leftrightarrow \varepsilon z=x(\varepsilon y),\\ (\varepsilon x)_1, (\varepsilon y)_2, z_3 \text{ are collinear} &\Leftrightarrow z=x^{-1}y \Leftrightarrow z=(\varepsilon x)(\varepsilon y).\end{array}$$

Thus, $(\Gamma'_1\cup\Gamma'_2, \Gamma'_3\cup\Delta'_1, \Delta'_2\cup\Delta'_3)$ is a dual 3-subnet of $(\Sigma_1,\Sigma_2,\Sigma_3)$ and H is a subgroup of the dihedral group $2.\mathbb{K}^*$. As H is not cyclic but it has a cyclic subgroup of index 2, we conclude that H is itself dihedral.

The isotopy class of quasigroups coordinatizing a tetrahedron type 3-net is a dihedral group. We also know a sporadic example; namely the

Urzua 3-net of order 8 coordinatized by the quaternion group of order 8; see [12].

In [3] we are dealt with 3-nets embedded in $PG(2, \mathbb{K})$ which are coordinatized by groups. Our main result is a complete classification for $p = 0$:

Theorem 3.1. *In the projective plane $PG(2, \mathbb{K})$ defined over an algebraically closed field $\mathbb{K}$ of characteristic $p \geq 0$, let Λ be am embedded 3-net of order $n \geq 4$ coordinatized by a group G. If either $p = 0$ or $p > n$ then one of the following holds.*

(I) *G is either cyclic or the direct product of two cyclic groups, and Λ is algebraic.*
(II) *G is dihedral and Λ is of tetrahedron type.*
(III) *G is the quaternion group of order* 8, *and Λ is the Urzúa* 3*-net of order* 8..
(IV) *G has order* 12 *and is isomorphic to* Alt_4.
(V) *G has order* 24 *and is isomorphic to* Sym_4.
(VI) *G has order* 60 *and is isomorphic to* Alt_5.

A computer aided exhaustive search shows that if $p = 0$ then (IV) (and hence (V), (VI)) does not occur, see [7]. Theorem 3.1 shows that every realizable finite group can act in $PG(2, \mathbb{K})$ as a projectivity group. This confirms Yuzvinsky's conjecture for $p = 0$.

A combinatorial characterization of algebraic 3-nets contained in a reducible plane cubic is given in [1]:

Theorem 3.2. *Let $p > n$ or $p = 0$. If a component of a dual 3-net is contained in a pencil, then the other two components consist of lines tangents to a unique conic, and the converse also holds.*

The proofs of Theorems 3.1 and 3.2 use several results on collineation groups of $PG(2, \mathbb{K})$ together with the classification of subgroups of $PGL(2, \mathbb{K})$ due to Dickson. Furthermore, the proof of Theorem 3.2 also uses the "Rédei polynomial approach" of Szőnyi for the study of blocking-sets in $PG(2, q)$; see [10].

References

[1] A. Blokhuis, G. Korchmáros and F. Mazzocca, *On the structure of 3-nets embedded in a projective plane*, J. Combin. Theory Ser-A. **118** (2011), 1228–1238.

[2] M. Falk and S. Yuzvinsky, *Multinets, resonance varieties, and pencils of plane curves*, Compos. Math. **143** (2007), 1069–1088.

[3] G. KORCHMÁROS, G. P. NAGY and N. PACE, *3-nets realizing a group in a projective plane*, preprint.
[4] G. KORCHMÁROS, G. P. NAGY and N. PACE, *k-nets embedded in a projective plane over a field*, preprint.
[5] G. LUNARDON, *Normal spreads*, Geom. Dedicata **75** (1999), 245–261.
[6] G. LUNARDON, *Linear k-blocking sets*, Combinatorica **21** (2001), 571–581.
[7] G. P. NAGY and N. PACE, *On small 3-nets embedded in a projective plane over a field*, J. Combin. Theory Ser-A, to appear.
[8] Á. MIGUEL and M. BUZUNÁRIZ, *A description of the resonance variety of a line combinatorics via combinatorial pencils*, Graphs and Combinatorics **25** (2009), 469–488.
[9] J.V. PEREIRA and S. YUZVINSKY, *Completely reducible hypersurfaces in a pencil*, Adv. Math. **219** (2008), 672–688.
[10] T. SZŐNYI, *Around Rédei's theorem*, Discrete Math. **208179** (1999), 557175.
[11] J. STIPINS, *Old and new examples of k-nets in* $\mathbb{P}^2$, math.AG/0701046.
[12] G. URZÚA, *On line arrangements with applications to 3-nets*, Adv. Geom. **10** (2010), 287–310.
[13] S. YUZVINSKY, *Realization of finite abelian groups by nets in* $\mathbb{P}^2$, Compos. Math. **140** (2004), 1614–1624.
[14] S. YUZVINSKY, *A new bound on the number of special fibers in a pencil of curves*, Proc. Amer. Math. Soc. **137** (2009), 1641–1648.

An improved lower bound on the maximum number of non-crossing spanning trees

Clemens Huemer[1] and Anna de Mier[2]

Abstract. We address the problem of counting geometric graphs on point sets. Using analytic combinatorics we show that the so-called double chain point configuration of N points has $\Omega^*(12.31^N)$ non-crossing spanning trees and $\Omega^*(13.40^N)$ non-crossing forests. This improves the previous lower bounds on the maximum number of non-crossing spanning trees and of non-crossing forests among all sets of N points in general position given by Dumitrescu, Schulz, Sheffer and Tóth in 2011. A new upper bound of $O^*(22.12^N)$ for the number of non-crossing spanning trees of the double chain is also obtained.

1 Introduction

A geometric graph on a point set S (throughout, S has no three collinear points) is a graph with vertex set S and whose edges are straight-line segments with endpoints in S. A geometric graph is called *non-crossing* (nc-for short) if no two edges intersect except at common endpoints. Counting nc-geometric graphs is a prominent problem in combinatorial geometry. In [7] it is proved that no set of N points has more than $O^*(141.07^N)$ nc-spanning trees.[3] The maximum number of nc-spanning trees (among all sets of N points) is very likely much smaller. The point set with most nc-spanning trees known so far is the so-called *double chain*. The double chain of $N = 2n$ points consists of two sets of n points each, one forming a convex chain (the lower chain) and one forming a concave chain (the upper chain). Furthermore, each straight-line defined by two points from the upper chain leaves all the points from the lower chain on the same side, and reversely; see Figure 2.1. Counting nc-geometric graphs

[1] Universitat Politècnica de Catalunya, Barcelona, Spain. Email: clemens.huemer@upc.edu. Supported by Projects MTM2012-30951, DGR2009-SGR1040, and EuroGIGA, CRP ComPoSe: grant EUI-EURC-2011-4306.

[2] Universitat Politècnica de Catalunya, Barcelona, Spain. Email: anna.de.mier@upc.edu. Supported by Projects MTM2011-24097 and DGR2009-SGR1040.

[3] We use the O^*-, Θ^*-, and Ω^*-notation to describe the asymptotic growth of the number of geometric graphs as a function of the number N of points, neglecting polynomial factors.

on the double chain was initiated by García et al. [6] who proved that it has $\Theta^*(8^N)$ triangulations, $\Omega^*(9.35^N)$ nc-spanning trees and $\Omega^*(4.64^N)$ nc-polygonizations, where the latter bound also is the current best lower bound on the maximum number of nc-polygonizations. The lower bound for the number of nc-spanning trees of the double chain was subsequently improved to $\Omega^*(10.42^N)$ [2] and to $\Omega^*(12.0026^N)$ [3]. We further improve this bound to $\Omega^*(12.31^N)$ and also show a new lower bound of $\Omega^*(13.40^N)$ for the maximum number of nc-forests among all sets of N points. The methods we use are analytic combinatorics and singularity analysis, with the same spirit and techniques as in [4]. We finally provide a new upper bound of $O^*(22.12^N)$ for the number of nc-spanning trees of the double chain.

2 The lower bound

Theorem 2.1. *The double chain on N points has $\Omega^*(12.31^N)$ non-crossing spanning trees.*

In this section we describe a family of trees that gives the desired bound. Our construction depends on some parameters that need to be maximized later.

For any spanning tree of the double chain, the vertices on the upper and lower chains induce two forests F_U and F_L on a set of n points in convex position; there are also some edges with one endpoint on each chain (the interior edges). The first restriction is that we consider trees where only one vertex in each component of F_U is incident to interior edges; this vertex will be called the *mark* of the component and we call F_U a *marked forest.*

Of the several interior edges that are incident to a mark v, let e_v be the rightmost one. The set of edges $M_1 = \{e_v : v \text{ is a mark}\}$ induces a forest; this forest is uniquely determined by the leftmost edge in each component (assuming the set of marks is known). The set of these edges is denoted M_2.

Now we pose two further restrictions on our trees. First, no edge e_v is incident with an isolated vertex of F_L. Second, for each component C of F_L that is not incident to an edge e_v, there is a unique edge joining this component to a mark m_C in F_U, and this edge has as endpoint in F_L the leftmost vertex v_C in C. Moreover, m_C is as to the right as possible.

See Figure 2.1 for an example of such a tree.

We claim that to uniquely determine such a tree, it is enough to give:

s1) A marked nc-forest F_U with at least k components on a set of n points in convex position,

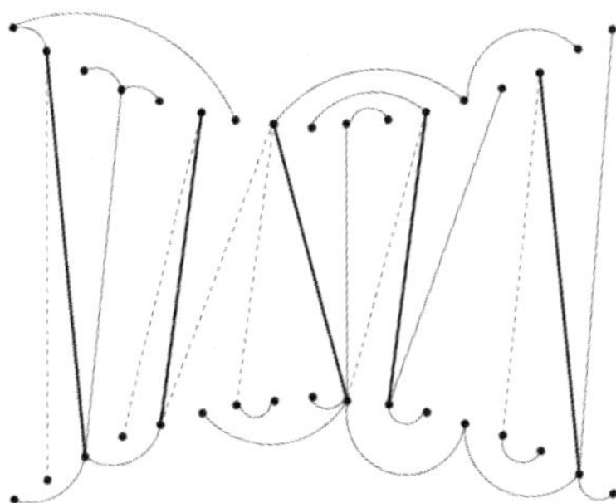

Figure 2.1. A spanning tree of the double chain. The bold edges are the edges of M_2 and the thin continuous interior edges are the other edges of M_1. The edges in F_U and F_L are drawn as arcs.

s2) a subset of k of the marks in F_U,
s3) a nc-forest F_L on a set of n points in convex position, and
s4) a subset M_L of k vertices in F_L such that none of these vertices induces a component of F_L.

Indeed, it suffices to do the following:

t1) Match the k marks from s2) with the k vertices in F_L (corresponding to the edges of M_2);
t2) join the other marks in F_U to the leftmost visible vertex in M_L (corresponding to the edges of $M_1 \backslash M_2$);
t3) for each component of F_L that has no vertex in M_L, take its leftmost vertex and join it to the rightmost visible mark;
t4) if the result is not connected, let $C_1, \ldots, C_r$ be its connected components, from left to right as encountered in the upper chain; for $i \geq 2$, join the leftmost mark in C_i to the rightmost visible vertex of M_L in C_{i-1}.

It can be shown using the methods of Section 3 that the number of nc-forests on n points in convex position where each component has a mark is $\Theta^*(9.5816\ldots^n)$, and that for n sufficiently large, at least 40% of them have $0.2237n$ or more components each. However, we choose F_U in two steps, as it actually gives us more choice. First we select ℓ vertices out of the n in the upper chain, and mark all of them, and then choose a marked forest F'_U on $n - \ell$ vertices, such that all components in this forest have at least two vertices.

To determine the forest F_L in $s3$) and $s4$) above, we choose m of the n vertices of the lower chain that will be isolated, and then choose a forest without isolated vertices on the remaining $n - m$ points.

The following estimates will be proved in Section 3.

Proposition 2.2.

(i) *The number of marked non-crossing forests on n points in convex position such that no component is an isolated vertex is* $cn^{-3/2}\omega_U^n(1+O(1/n))$*, where* $\omega_U = 8.5816\ldots$ *and c is a constant. Moreover, for n large enough at least 40% of these forests have* $0.1332n$ *or more components.*

(ii) *The number of non-crossing forests on n points in convex position such that no component is an isolated vertex is* $dn^{-3/2}\omega_L^n(1+O(1/n))$*, where* $\omega_L = 7.2246\ldots$ *and d is a constant.*

Setting $k = \beta n$, $\ell = \gamma n$ and $m = \alpha n$, and ignoring subexponential terms, we get the following lower bound on the number of nc-spanning trees of the double chain:

$$\binom{n}{\gamma n} 8.5816^{n-\gamma n} \binom{0.1332(n-\gamma n)+\gamma n}{\beta n} \binom{n}{\alpha n} 7.2246^{n-\alpha n} \binom{n-\alpha n}{\beta n}.$$

Note that when choosing the $k = \beta n$ marks from the upper forest F_U we are actually not using all the available marks, but only the ones that come from isolated vertices and the first $0.1332(n - \gamma n)$ marks of the marked forest F'_U.

Using the binary entropy function $H(x) = -x\log_2(x) - (1-x)\log_2(1-x)$, we can estimate a binomial coefficient as $\binom{\epsilon n}{\delta n} \approx 2^{\epsilon H(\frac{\delta}{\epsilon})n}$.

The values $\alpha = 0.09$, $\beta = 0.256$, $\gamma = 0.258$ give $\Omega^*(2^{7.244139N/2}) = \Omega^*(12.3126^N)$ nc-spanning trees on the double chain on N vertices, thus proving Theorem 2.1.

Using a similar construction we can also prove a lower bound on the number of nc-forests of the double chain. The details are omitted due to lack of space and will appear elsewhere.

Theorem 2.3. *The double chain on N points has* $\Omega^*(13.40^N)$ *non-crossing forests.*

The bound of $\Theta^*(9.5816\ldots^n)$ marked forests on n points in convex position implies a bound on the number of nc-spanning trees of the *single chain.* This point set has triangular convex hull and all but one point p of the set form a convex chain; also see [1]. To count the number of nc-spanning trees of this set, connect each mark of a tree in a forest on the convex chain to p.

Corollary 2.4. *The single chain on N points has* $\Theta^*(9.5816\ldots^N)$ *non-crossing spanning trees.*

3 Non-crossing forests of points in convex position

In this section we use generating functions and the techniques of analytic combinatorics to prove Proposition 2.2.

Consider a set of n points in convex position, labelled counterclockwise $p_1, \ldots, p_n$; the vertex p_1 is called the root vertex. A systematic study of several classes of non-crossing graphs with generating functions was undertaken by Flajolet and Noy in [4]; the results in this section are an extension of theirs using the same techniques (see also the book [5] by Flajolet and Sedgewick).

Let $\mathcal{T}$ be a set of nc-trees and let $\mathcal{F}_{\mathcal{T}}$ be the set of those nc-forests such that its connected components belong to $\mathcal{T}$ (by taking as the root of each component the vertex with smallest label and relabelling the other vertices suitably). Let $T(z)$ and $F_T(z)$ be the corresponding generating functions, that is,

$$T(z) = \sum_{n \geq 1} t_n z^n, \qquad F_T(z) = \sum_{n \geq 0} f_n z^n,$$

where t_n and f_n denote the number of n-vertex graphs in $\mathcal{T}$ and $\mathcal{F}_{\mathcal{T}}$, respectively. For technical convenience, we set $f_0 = 1$ (but $t_0 = 0$).

We have the following key relation

$$F_T(z) = 1 + T(zF_T(z)). \tag{3.1}$$

The combinatorial explanation is as follows (see Figure 3.1). Given a forest in $\mathcal{F}_{\mathcal{T}}$, let t_1 be the connected component that contains the root vertex; this component is of course a tree of $\mathcal{T}$. Now, the vertices (if any) that lie strictly between any two consecutive vertices of t_1 induce a nc-forest, which belongs to $\mathcal{F}_{\mathcal{T}}$. The substitution $zF_t(z)$ in $T(z)$ reflects this (where the term z corresponds to a vertex of t_1). Thus, if an expression or an equation for $T(z)$ is known, we immediately get from equation (3.1) an equation for $F_T(z)$.

It is well-known that the generating function $T(z)$ for nc-trees satisfies

$$T(z)^3 - zT(z) + z^2 = 0. \tag{3.2}$$

Let $\mathcal{T}_1$ be the class of nc-trees with more than one vertex; clearly $T_1(z) = T(z) - z$. From equations (3.2) and (3.1) it follows that $Y = F_{T_1}(z)$ satisfies

$$(1+z)^3 Y^3 - (3z^2 + 7z + 3)Y^2 + (4z + 3)Y - 1 = 0. \tag{3.3}$$

Now let $\mathcal{T}_1^*$ be the class of marked nc-trees; then $T_1^*(z) = zT'(z) - z$. By differentiating equation (3.2) and eliminating $T(z)$ we get an equation for

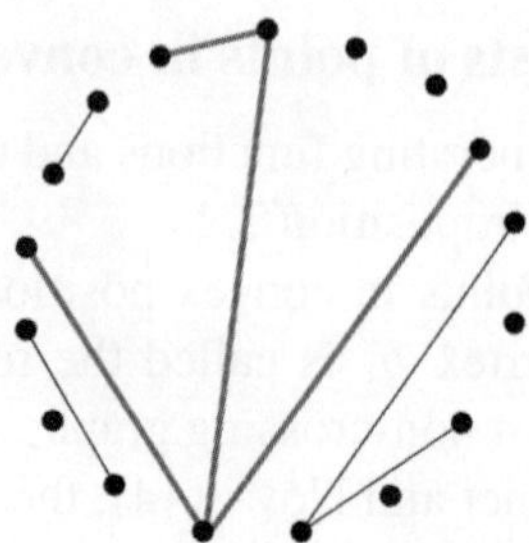

Figure 3.1. A nc-forest decomposes as a nc-tree (in grey) with a (possibly empty) nc-forest between any two of its consecutive vertices.

$T'(z)$, namely,

$$(27z^2 - 4z)T'(z)^3 + (1 - 6z)T'(z) - 1 + 8z = 0,$$

from which an equation for $T_1^*(z)$ follows immediately. Again from (3.1) we obtain an equation for $Y = F_{T_1^*}(z)$:

$$\begin{aligned}27z(1+z)^3Y^4 - (83z^3 + 180z^2 + 93z + 4)Y^3 \\ + (99z^2 + 106z + 12)Y^2 - (12 + 40z)Y + 4 = 0.\end{aligned} \tag{3.4}$$

Once an algebraic equation for $F_T(z)$ is known, it is routine to obtain an asymptotic estimate of the coefficients of $F_T(z)$. The method we apply is the one described in [4, Section 4] or, more generally, in Lemma VII.3 and Corollary VI.1 of [5]. The main idea is that the singularity of $F_T(z)$ with smallest modulus determines the asymptotic behaviour of the coefficients of $F_T(z)$. More concretely, if ρ is this singularity, and under certain conditions that in our case are immediate to check, then $[z^n]F_T(z) = \gamma n^{-3/2}\rho^{-n}(1 + \mathcal{O}(1/n))$, for some constant γ. To find ρ we use the fact that it must be one of the roots of the discriminant of the equation satisfied by $F_T(z)$, as explained in [5, Section VII.7.1]. Carrying out the calculations for equations (3.3) and (3.4), we obtain the values given in item (ii) and in the first part of item (i) in Proposition 2.2.

As for the number of components, we consider the bivariate generating function

$$F_T(z) = \sum_{n\geq 0} f_{n,k} z^n w^k,$$

where $f_{n,k}$ stands for the number of forests in $\mathcal{F}_T$ with n vertices and k components. It is easy to see that equation (3.1) becomes

$$F_T(z, w) = 1 + wT(zF_T(z, w)). \tag{3.5}$$

Let $X_{n,k} = [z^n w^k]F_T(z, w)/[z^n]F_T(z)$ and let μ_n be the mean of $X_{n,k}$. Then $\mu_n \sim \kappa n$, where κ is obtained as follows. Let $\rho(w)$ be the dominant singularity of $F_T(z, w)$ (it thus satisfies $\rho(1) = \rho$). Then $\kappa = -\rho'(1)/\rho$. Moreover, $X_{n,k}$ converges in law to a Gaussian law. (See [4, Section 5] and [5, Section IX.7.3] for details.) This implies that for each positive ε, $1/2-\varepsilon$ of the forests in $\mathcal{F}_T$ with n vertices have at least $\mu_n n$ components, for sufficiently large n.

To finish the proof of item (i) of Proposition 2.2, we use equation (3.5) to find an equation satisfied by $F_1^*(z, w)$ and then carry out the necessary calculations.

4 The upper bound

Theorem 4.1. *The double chain on N points has $O^*(22.12^N)$ non-crossing spanning trees.*

We only give a sketch of the proof. Recall that for any nc-spanning tree of the double chain, the vertices on the upper and lower chains induce two forests F_U and F_L on a set of n points in convex position and a forest F_I formed by interior edges, *i.e.*, edges with one endpoint on each chain. The product of the numbers of forests F_U, F_L and F_I gives an upper bound on the number of nc-spanning trees of the double chain. We partition the set of nc-spanning trees into $2n - 1$ classes, according to the number of edges of F_I. For asymptotic counting, it is sufficient to only consider the one class of nc-spanning trees, with k edges in forest F_I, that contains most spanning trees. For a fixed value of k, we can show that the number of nc-spanning trees of the double chain is maximized when both F_U and F_L are forests with $\frac{k}{2}$ components. We also show that the number of forests F_I with k edges is

$$\approx \sum_{\ell=1}^{\min\{k,n\}} \binom{n}{\ell}^2 \binom{2n-3-\ell}{k-\ell} = \sum_{\ell=1}^{\min\{k,n\}} g(\ell).$$

Then the value $\ell = \ell(k)$ that maximizes this sum is determined. Using the formula for $F(n, c)$, the number of nc-forests with c components on a set of n points in convex position [4], we determine the value k that maximizes $g(\ell(k))(F(n, k/2))^2$, which leads to the claimed bound $O^*(22.12^N)$.

References

[1] O. AICHHOLZER, D. ORDEN, F. SANTOS and B. SPECKMANN, *On the number of pseudo-triangulations of certain point sets*. Journal of Combinatorial Theory Series A **115** (2) (2008), 254–278.

[2] A. DUMITRESCU, *On two lower bound constructions*, Proc. 11th Canadian Conference on Computational Geometry, Vancouver, British Columbia, Canada, 1999, 111-114.

[3] A. DUMITRESCU, A. SCHULZ, A. SHEFFER and C. D. TÓTH, *Bounds on the maximum multiplicity of some common geometric graphs*, STACS, pp. 637-648, 2011, http://arxiv.org/pdf/1012.5664v2.pdf.

[4] P. FLAJOLET and M. NOY, *Analytic combinatorics of non-crossing configurations*, Discrete Mathematics **204** (1999), 203–229.

[5] P. FLAJOLET and R. SEDGEWICK, "Analytic combinatorics", Cambridge U. Press, Cambridge, 2009.

[6] A. GARCÍA, M. NOY and J. TEJEL, *Lower bounds on the number of crossing-free subgraphs of K_n*, Computational Geometry: Theory and Applications **16** (2000), 211-221.

[7] M. HOFFMANN, A. SCHULZ, M. SHARIR, A. SHEFFER, C.D. TÓTH and E. WELZL, "Counting Plane Graphs: Flippability and its Applications", Thirty Essays on Geometric Graph Theory, Springer, 2012.

On the structure of graphs with large minimum bisection

Cristina G. Fernandes[1], Tina Janne Schmidt[2] and Anusch Taraz[2]

Abstract. Bounded degree trees and bounded degree planar graphs on n vertices are known to admit bisections of width $\mathcal{O}(\log n)$ and $\mathcal{O}(\sqrt{n})$, respectively. We investigate the structure of graphs that meet this bound. In particular, we show that such a tree must have diameter $\mathcal{O}(n/\log n)$ and such a planar graph must have tree width $\Omega(\sqrt{n})$. To show the result for trees, we derive an inequality that relates the width of a minimum bisection with the diameter of a tree.

1 Introduction and results

A *bisection* of a graph G is a partition of its vertex set into two sets L and R of sizes differing by at most one. The *width* of a bisection is defined to be the number of edges with one vertex in L and one vertex in R, and the minimum width of a bisection in G is denoted by $\mathrm{MinBis}(G)$. Determining a bisection of minimum width is a famous optimization problem that is (unlike the Minimum Cut Problem) known to be NP-hard [4].

In this paper, we will concentrate on minimum bisections of trees and planar graphs. Jansen et al. showed that dynamic programming gives an algorithm with running time $\mathcal{O}(2^t n^3)$ for an arbitrary graph on n vertices when a tree decomposition of width t is provided as input [5]. Thus, the problem becomes polynomially tractable for graphs of constant tree

[1] Instituto de Matemática e Estatística, Universidade de São Paulo, Brazil. Email: cris@ime.usp.br

[2] Zentrum Mathematik, Technische Universität München, Germany. Email: schmidtt@ma.tum.de, taraz@ma.tum.de

The first author was partially supported by CNPq Proc. 308523/2012-1 and 477203/2012-4, and Project MaCLinC of NUMEC/USP. The second author gratefully acknowledges the support by the Evangelische Studienwerk Villigst e.V. and was partially supported by TopMath, the graduate program of the Elite Network of Bavaria and the graduate center of TUM Graduate School. The third author was supported by DFG grant TA 309/2-2.
The cooperation of the three authors was supported by a joint CAPES-DAAD project (415/ppp-probral/po/D08/11629, Proj. no. 333/09).

width. On the other hand, it is open whether finding a minimum bisection for a planar graph is in P or is NP-hard. Currently, the best known approximation algorithm achieves an approximation ratio of $\mathcal{O}(\log n)$ for arbitrary graphs on n vertices [6], and nothing better has been established for planar graphs. Further, Berman and Karpinski showed that the Minimum Bisection Problem restricted to 3-regular graphs is as hard to approximate as its general version [3]. Moreover, Arora et al. presented a PTAS for finding a minimum bisection in graphs with a linear minimum degree [1]. Thus, we focus on graphs that have small (in fact, constant) maximum degree and still a large minimum bisection width. If T is a tree on n vertices and maximum degree d, one can show that owing to the existence of a *separating vertex* (*i.e.* a vertex whose removal leaves no connected component of size greater than $n/2$), we always have $\text{MinBis}(T) \leq d \cdot \log_2 n$. Similarly, using the Planar Separator Theorem [2], this idea can be generalized to give an $\mathcal{O}(\sqrt{n})$ upper bound on $\text{MinBis}(G)$ for planar graphs G with bounded maximum degree. Both bounds are tight up to a constant factor as one can show that the perfect ternary tree on n vertices has minimum bisection width $\Omega(\log n)$ and the square grid on n vertices has minimum bisection width $\Omega(\sqrt{n})$.

Our aim is to investigate the structure of bounded degree graphs that have a large minimum bisection, in other words, trees T with $\text{MinBis}(T) = \Omega(\log n)$ and planar graphs G with $\text{MinBis}(G) = \Omega(\sqrt{n})$. For example, we will show that in a bounded degree tree with minimum bisection of order $\Omega(\log n)$, the length of any path must be bounded by $\mathcal{O}(n/\log n)$. More generally, we establish the following inequality:

Theorem 1.1. *Let T be a tree on n vertices and denote by $\Delta(T)$ its maximum degree and by* $\text{diam}(T)$ *its diameter. Then*

$$\text{MinBis}(T) \leq \frac{8\Delta(T)n}{\text{diam}(T)}.$$

In the case of planar graphs, we can no longer use the diameter to control the minimum bisection. For example, a graph consisting of a square grid on $\frac{3}{4}n$ vertices connected to a path on $\frac{n}{4}$ vertices has linear diameter but does not allow a bisection of constant width. However, we can show that a planar graph with a minimum bisection width close to the upper bound $\mathcal{O}(\sqrt{n})$ must indeed be far away from a tree-like structure.

Theorem 1.2. *For every $d \in \mathbb{N}$ and every $c > 0$, there is a $\gamma = \gamma(c) > 0$ such that for all planar graphs G on n vertices with $\Delta(G) \leq d$, we have*

$$\text{MinBis}(G) \geq c\sqrt{n} \qquad \Rightarrow \qquad \text{tw}(G) \geq \gamma\sqrt{n}.$$

Using Theorem 6.2 of Robertson, Seymour, and Thomas in [7], this immediately implies that there is a constant $\gamma' = \gamma'(c) > 0$ such that every such planar graph contains a square grid on $\gamma' n$ vertices as minor.

2 Trees

Although the statement in Theorem 1.1 looks like an elementary inequality, its proof is somewhat lengthy and we need to introduce a few more definitions to be able to sketch it. First, we define the *relative diameter* of a graph to be

$$\operatorname{diam}^*(G) := \frac{1}{|V(G)|} \sum_{G'\text{: component of } G} (\operatorname{diam}(G') + 1).$$

Observe that, in a tree, there is only one component to consider and thus the relative diameter of a tree denotes the fraction of the vertices in a longest path of the tree, but we will need this parameter also for graphs that may not be connected.

Moreover, we need to take a more general approach and consider partitions where the size of the classes can be specified by an input parameter m. Furthermore, we denote by $e_G(V_1, \ldots, V_k)$ the number of edges in G that have their vertices in two different sets $V_i \neq V_j$ and define $[n] = \{1, 2, 3, \ldots, n\}$ for $n \in \mathbb{N}$. The following theorem is the driving engine for the proof of Theorem 1.1.

Theorem 2.1. *For all trees T on $n \geq 3$ vertices and for all $m \in [n]$, the vertex set of T can be partitioned into three classes $L \cup R \cup S$ such that one of the following two options occurs:*

(i) $S = \emptyset$ *and* $|L| = m$ *and* $e_T(L, R, S) \leq 2$.
(ii) $S \neq \emptyset$ *and* $|L| \leq m \leq |L| + |S|$ *and* $e_T(L, R, S) \leq \frac{2}{\operatorname{diam}^*(T)} \Delta(T)$ *and* $\operatorname{diam}^*(T[S]) \geq 2 \operatorname{diam}^*(T)$.

This result states that we can either find a partition into two sets L and R with exactly the right cardinality by cutting very few edges, or there is a partition with an additional set S, such that the set L is smaller and the set $L \cup S$ is larger than the required size m, as well as the additional feature that the relative diameter of $T[S]$ is at least twice as large as that of T. Using Theorem 2.1 recursively for the graph $T[S]$, the relative diameter can therefore be doubled in each round, until it exceeds $1/2$, at which point Option *(ii)* in Theorem 2.1 is no longer feasible, which will then prove Theorem 1.1.

We conclude this section with a few words about how to prove Theorem 2.1. Consider a longest path P in T and denote by x_0 and y_0 its first and last vertex. For each vertex $z \in V(P)$, let T_z be the component of $T - E(P)$ that contains z and call z the root of T_z. We label the vertices of T with $1, 2, \ldots, n$ so that x_0 receives label 1; for each $z \in V(P)$, the vertices of T_z receive consecutive labels and z receives the largest label among those; for all $z, z' \in V(P)$ with $z \neq z'$, if x_0 is closer to z than to z', then the label of z is smaller than the label of z'. Given this labeling, we now define for every $x \in [n]$ the vertex $f(x) := x + m$. If for some vertex $x \in V(P)$ the vertex $f(x)$ lies also in P, then we are done by choosing $L := \{x+1, \ldots, x+m\}$ and $R := [n] \setminus L$, which satisfies all requirements of Option *(i)* in Theorem 2.1. Otherwise, one can show that there exists a vertex $z \in V(P)$ such that all vertices of P that are mapped into $V(T_z)$ and the vertex sets of all trees that are mapped completely into T_z by f form a set S such that the condition on $\text{diam}^*(T[S])$ is satisfied. Furthermore, this vertex z has the property that the set L, which consists of the vertex sets of all trees whose roots have labels strictly between $z - m$ and z and of some additional vertices from T_z, satisfies the remaining conditions of Option *(ii)* in Theorem 2.1.

Figure 2.1. Construction of S and $L := L_1 \cup L_2$.

3 Planar graphs

To sketch the proof for Theorem 1.2, consider its contrapositive: for all $d \in \mathbb{N}$ and for all $c > 0$, we choose $\gamma > 0$ so small that

$$\gamma \left(2 \log_2 \frac{3}{\gamma\sqrt{2}} + 7 \right) d \leq c.$$

We claim that then, for every planar graph G on n vertices with $\Delta(G) \leq d$, the following holds:

$$\text{tw}(G) + 1 \leq \gamma\sqrt{n} \quad \Rightarrow \quad \text{MinBis}(G) \leq c\sqrt{n}. \tag{3.1}$$

To explain how we find a bisection of sufficiently small width, let us assume that, for some $0 < \delta < 1$, we can find δ-*separators* of a given size in the graph G and its subgraphs, *i.e.* a vertex subset whose removal leaves connected components of size at most δn. Having removed such a separator S' from the graph, we can assign the vertices of all but one component of $G - S'$ to L and R, so that $|L| \leq \lceil \frac{n}{2} \rceil$ and $|R| \leq \lfloor \frac{n}{2} \rfloor$. We then continue recursively with the remaining component, which has size at most δn. At the end, we denote by S the union of the various separators and distribute the vertices in S to the sets L and R in such a way that L and R form a bisection of G. It is easy to see that $e_G(L, R) \leq \Delta(G) \cdot |S|$ and it only remains to find a bound on $|S|$.

In each round of a first phase, we use a cluster from an optimal tree decomposition of G that can serve as a $1/2$-separator and, by assumption, has size at most $\gamma\sqrt{n}$. Denote by n_i the number of vertices after the i-th round. This first phase stops when $\gamma\sqrt{n} > \frac{3}{\sqrt{2}}\sqrt{n_i}$. Due to $n_i \leq n/2^{i-1}$, it is easy to see that the index i^*, where this happens for the first time, can be bounded by a constant depending only on γ.

After the i^*-th round, we switch to $2/3$-separators guaranteed by the Planar Separator Theorem [2], which will have size at most $\frac{3}{\sqrt{2}}\sqrt{n_i}$. During this second phase, $n_i \leq (2/3)^{i-i^*} n_{i^*}$ holds and thus the number of vertices collected in S during the second phase can be bounded from above by

$$\sum_{i=i^*}^{\infty} \frac{3}{\sqrt{2}}\sqrt{n_i} \leq \frac{3}{\sqrt{2}}\sqrt{n_{i^*}} \sum_{i=0}^{\infty} \left(\sqrt{\frac{2}{3}}\right)^i \leq \alpha\sqrt{n_{i^*}}, \quad \text{with } \alpha = \frac{9}{\sqrt{2}} + 3\sqrt{3}.$$

Summing up, we have $|S| \leq i^*\gamma\sqrt{n} + \alpha\sqrt{n_{i^*}}$ in total. Now, computing an upper bound on i^* and on n_{i^*} will give the desired bound.

4 Concluding remarks and open questions

In Theorem 1.1 and Theorem 1.2 we have established necessary conditions for trees and planar graphs to have large minimum bisection width, namely a small diameter and a large tree width, respectively. In both cases, these conditions are not sufficient, but it would be interesting to find additional conditions that give characterizations of graphs in certain classes with large minimum bisection.

References

[1] S. Arora, D. Karger and M. Karpinski, *Polynomial time approximation schemes for dense instances of NP-hard problems*, Journal of Computer and System Sciences **58** (1) (1999), 193–210.

[2] N. ALON, P. SEYMOUR and R. THOMAS, *Planar separators*, SIAM Journal on Discrete Mathematics **7** (2) (1994), 184–193.

[3] P. BERMAN and M. KARPINSKI, *Approximation hardness of bounded degree MIN-CSP and MIN-BISECTION*, In: "Automata, Languages and Programming", vol. 2380 of LNCS, Springer, Berlin, 2002, 623–632.

[4] M. GAREY, D. JOHNSON and L. STOCKMEYER, *Some simplified NP-complete graph problems*, Theoretical Computer Science **1** (3) (1976), 237–267.

[5] K. JANSEN *et al.*, *Polynomial time approximation schemes for max-bisection on planar and geometric graphs*, SIAM Journal on Computing **35** (1) (2005), 110–119.

[6] H. RÄCKE, *Optimal hierarchical decompositions for congestion minimization in networks*, In: "Proceedings of the 40th annual ACM symposium on Theory of computing", STOC '08, New York, NY, USA, 2008, ACM, 255–264.

[7] N. ROBERTSON, P. SEYMOUR and R. THOMAS, *Quickly excluding a planar graph*, Journal of Combinatorial Theory, Series B **62** (2) (1994), 323–348.

Colorings

Coloring intersection graphs of arcwise connected sets in the plane

Michał Lasoń[1,2], Piotr Micek[1], Arkadiusz Pawlik[1]
and Bartosz Walczak[3]

Abstract. A family of sets in the plane is simple if the intersection of its any subfamily is arcwise connected. We prove that the intersection graphs of simple families of compact arcwise connected sets in the plane pierced by a common line have chromatic number bounded by a function of their clique number.

1 Introduction

A *proper coloring* of a graph is an assignment of colors to the vertices of the graph such that no two adjacent ones are assigned the same color. The minimum number of colors sufficient to color a graph G properly is called the *chromatic number* of G and denoted by $\chi(G)$. The maximum size of a clique (a set of pairwise adjacent vertices) in a graph G is called the *clique number* of G and denoted by $\omega(G)$. It is clear that $\chi(G) \geqslant \omega(G)$. A class of graphs is χ*-bounded* if there is a function $f : \mathbb{N} \to \mathbb{N}$ such that $\chi(G) \leqslant f(\omega(G))$ holds for any graph G in the class.

In this paper, we focus our attention on the relation between the chromatic number and the clique number for classes of graphs arising from geometry. The *intersection graph* of a family of sets $\mathcal{F}$ is the graph with vertex set $\mathcal{F}$ and edge set consisting of pairs of intersecting elements of $\mathcal{F}$. We consider families $\mathcal{F}$ of arcwise connected compact sets in the plane. For simplicity, we identify the family $\mathcal{F}$ with its intersection graph.

[1] Theoretical Computer Science Department, Faculty of Mathematics and Computer Science, Jagiellonian University. Email: mlason@tcs.uj.edu.pl, micek@tcs.uj.edu.pl, pawlik@tcs.uj.edu.pl.
Supported by the Ministry of Science and Higher Education of Poland under grant no. 884/N-ESF-EuroGIGA/10/2011/0 within the ESF EuroGIGA project GraDR

[2] Institute of Mathematics of the Polish Academy of Sciences, Email: michalason@gmail.com

[3] École Polytechnique Fédérale de Lausanne. Email: bartosz.walczak@epfl.ch. Partially supported by Swiss National Science Foundation Grant no. 200020-144531

In the one-dimensional case of subsets of $\mathbb{R}$, the only arcwise connected compact sets are closed intervals. They define the class of *interval graphs*, for which $\chi(G) = \omega(G)$. The study of the chromatic number of intersection graphs of geometric objects in higher dimensions was initiated in the seminal paper of Asplund and Grünbaum [1], where they proved that the families of axis-aligned rectangles in $\mathbb{R}^2$ are χ-bounded. On the other hand, Burling [2] showed that triangle-free intersection graphs of axis-aligned boxes in $\mathbb{R}^3$ can have arbitrarily large chromatic number.

Gyárfás [3] proved χ-boundedness of the class of graphs defined by intersections of chords of a circle. This was generalized by Kostochka and Kratochvíl [4], who showed that the families of convex polygons inscribed in a circle are χ-bounded. McGuinness [5] proved that the families of L-shapes (shapes consisting of a horizontal and a vertical segments of arbitrary lengths, forming the letter 'L') all of which intersect a fixed vertical line are χ-bounded. Later, McGuinness [6] showed that the simple families $\mathcal{F}$ of compact arcwise connected sets in the plane pierced by a common line with $\omega(\mathcal{F}) \leqslant 2$ have bounded chromatic number. A family is *simple* if the intersection of its any subfamily is arcwise connected, and is *pierced* by a line ℓ if the intersection of its any member with ℓ is a nonempty segment. Suk [8] proved χ-boundedness of the simple families of x-monotone curves intersecting a fixed vertical line.

We generalize the results of McGuinness, allowing any bound on the clique number, and of Suk, getting rid of the x-monotonicity assumption.

Theorem 1.1. *The class of simple families of compact arcwise connected sets in the plane pierced by a common line is χ-bounded.*

This contrasts with a recent result due to Pawlik et al. [7] that there are triangle-free intersection graphs of straight-line segments with arbitrarily large chromatic number. This explains why the assumption of Theorem 1.1 that the sets are pierced by a common line is necessary.

The ultimate goal of this quest is to understand the border line between the classes of graphs (and classes of geometric objects) that are χ-bounded and those that are not. The authors would like to share two open problems in this context.

Problem 1.2. Are the families (not necessarily simple) of x-monotone curves in the plane pierced by a common vertical line χ-bounded?

Problem 1.3. Are the families of curves in the plane pierced by a common line χ-bounded?

2 Preliminaries

First we simplify the setting of Theorem 1.1. Let $\mathcal{F}$ be a simple family of compact arcwise connected sets in the plane pierced by a common line with $\omega(\mathcal{F}) \leqslant k$. We can assume without loss of generality that this piercing line is the horizontal axis $\mathbb{R} \times \{0\}$. Call it the *baseline*. The *base* of a set X, denoted by $\operatorname{base}(X)$, is the intersection of X with the baseline. The intersection graph of the bases of the members of $\mathcal{F}$ is an interval graph and thus can be properly colored with k colors. To find a proper coloring of $\mathcal{F}$, we can restrict our attention to one color class in the coloring of this interval graph. Thus we assume that no two members of $\mathcal{F}$ intersect on the baseline and show that $\mathcal{F}$ can be colored properly with a bounded number of colors.

Let $\mathcal{F}^+ = \{X \cap (\mathbb{R} \times [0, +\infty)) : X \in \mathcal{F}\}$ and $\mathcal{F}^- = \{X \cap (\mathbb{R} \times (-\infty, 0]) : X \in \mathcal{F}\}$. Clearly, $\mathcal{F}^+$ and $\mathcal{F}^-$ are simple families of arcwise connected sets. If we find proper colorings ϕ^+ and ϕ^- of $\mathcal{F}^+$ and $\mathcal{F}^-$, respectively, with bounded numbers of colors, then the coloring of $\mathcal{F}$ by pairs of colors (ϕ^+, ϕ^-) is proper on $\mathcal{F}$. Thus we assume that $\mathcal{F} = \mathcal{F}^+$ (the other case $\mathcal{F} = \mathcal{F}^-$ is symmetric). All geometric objects that we consider from now on are contained in $\mathbb{R} \times [0, +\infty)$. Thus we consider families of compact arcwise connected subsets of $\mathbb{R} \times [0, +\infty)$ all of which are pierced by the baseline. We call such families *attached*.

Theorem 2.1. *For $k \geqslant 1$, there is ξ_k such that $\chi(\mathcal{F}) \leqslant 2^{\xi_k}$ holds for any attached family $\mathcal{F}$ with $\omega(\mathcal{F}) \leqslant k$.*

The case $k = 1$ is trivial, and the case $k = 2$ is the result of McGuinness [6]. Our proof of Theorem 2.1 depends heavily on the techniques developed by McGuinness [6] and Suk [8].

Let $X \prec Y$ denote that $\operatorname{base}(X)$ is entirely to the left of $\operatorname{base}(Y)$. The relation $\prec$ is a total order on any attached family $\mathcal{F}$. For attached sets X_1 and X_2 such that $X_1 \prec X_2$, define $\mathcal{F}(X_1, X_2) = \{Y \in \mathcal{F} : X_1 \prec Y \prec X_2\}$. For an attached family $\mathcal{X}$, we define $\operatorname{ext}(\mathcal{X})$ to be the only unbounded arcwise connected component of $(\mathbb{R} \times [0, +\infty)) \setminus \bigcup \mathcal{X}$.

Lemma 2.2 ([5]). *Let $\mathcal{F}$ be an attached family, let $a, b \geqslant 0$, and suppose $\chi(\mathcal{F}) > 2^{a+b+1}$. Then there exists a subfamily $\mathcal{H}$ of $\mathcal{F}$ such that $\chi(\mathcal{H}) > 2^a$ and for any intersecting $H_1, H_2 \in \mathcal{H}$ we have $\chi(\mathcal{F}(H_1, H_2)) \geqslant 2^b$.*

A subfamily $\mathcal{G}$ of an attached family $\mathcal{F}$ is *externally supported* in $\mathcal{F}$ if for any $X \in \mathcal{G}$ there exists $Y \in \mathcal{F}$ such that $Y \cap X \neq \emptyset$ and $Y \cap \operatorname{ext}(\mathcal{G}) \neq \emptyset$.

Lemma 2.3. *Let $\mathcal{F}$ be an attached family, let $a \geqslant 0$, and suppose $\chi(\mathcal{F}) > 2^{a+1}$. Then there exists a subfamily $\mathcal{G}$ of $\mathcal{F}$ that is externally supported in $\mathcal{F}$ and satisfies $\chi(\mathcal{G}) > 2^a$.*

Let $\mathcal{F}$ be an attached family. A *k-clique* in $\mathcal{F}$ is a family of k pairwise intersecting members of $\mathcal{F}$. For a k-clique $\mathcal{K}$, define $\mathrm{int}(\mathcal{K})$ to be the only arcwise connected component of $(\mathbb{R}\times[0,+\infty))\setminus\bigcup\mathcal{K}$ containing the part of the baseline between the two least members of $\mathcal{K}$. A *k-bracket* in $\mathcal{F}$ is a family $\mathcal{B}\subseteq\mathcal{F}$ consisting of a k-clique $\mathcal{K}$, a set $P\subseteq\mathrm{int}(\mathcal{K})$ called *hook*, and a set S called *support* such that $S\prec\mathcal{K}$ or $\mathcal{K}\prec S$ and $S\cap P\neq\emptyset$. For such a k-bracket $\mathcal{B}$, define $\mathrm{int}(\mathcal{B})$ to be the only arcwise connected component of $(\mathbb{R}\times[0,+\infty))\setminus\bigcup\mathcal{B}$ containing the part of the baseline between S and $\mathcal{K}$. The following lemma exhibits a crucial property of these two constructs.

Lemma 2.4. *If $\mathcal{S}$ is an attached clique or bracket, then any closed curve c such that $\mathcal{S}\cup\{c\}$ is simple, $\mathrm{int}(\mathcal{S})\cap c\neq\emptyset$, and $\mathrm{ext}(\mathcal{S})\cap c\neq\emptyset$ intersects all members of $\mathcal{S}$.*

3 Proof sketch of Theorem 2.1

The proof of Theorem 2.1 proceeds by induction on k. The case $k=1$ is trivial. Thus assume for the remainder of this section that $k\geqslant 2$ and the statement of the theorem holds for $k-1$. A typical application of the induction hypothesis looks as follows: if $\mathcal{F}$ is an attached family with $\omega(\mathcal{F})\leqslant k$, $\mathcal{G}\subseteq\mathcal{F}$, and there is $X\in\mathcal{F}\setminus\mathcal{G}$ intersecting all members of $\mathcal{G}$, then $\omega(\mathcal{G})\leqslant k-1$ and thus $\chi(\mathcal{G})\leqslant 2^{\xi_{k-1}}$.

Define $\beta_k=5\xi_{k-1}+\xi_2+k+7$, $\gamma_k=2\xi_{k-1}+k+5$, $\delta_{k,1}=\xi_{k-1}+\beta_k+\gamma_k+2$, $\delta_{k,i}=\delta_{k,i-1}+\beta_k+\gamma_k+2$ for $i\geqslant 2$, and finally $\xi_k=\delta_{k,k+1}$.

For a k-clique or k-bracket $\mathcal{S}$, define $\mathcal{F}(\mathcal{S})=\{X\in\mathcal{F}:\mathrm{base}(X)\subseteq\mathrm{int}(\mathcal{S})\}$. The following technical fact (considered in the induction context) is a generalization of an analogous statement in [8], with a similar proof.

Lemma 3.1. *Let $\mathcal{F}$ be an attached family with $\omega(\mathcal{F})\leqslant k$ and $\mathcal{S}$ be a k-clique or k-bracket in $\mathcal{F}$. Let $\mathcal{R}=\{R\in\mathcal{F}(\mathcal{S}):R\cap\mathrm{ext}(\mathcal{S})\neq\emptyset\}$ and $\mathcal{D}=\{D\in\mathcal{F}(\mathcal{S}):D\cap\bigcup(\mathcal{R}\cup\mathcal{S})\neq\emptyset\}$. Then $\chi(\mathcal{D})\leqslant 2^{\beta_k}$.*

A *fancy k-clique* in an attached family $\mathcal{F}$ consists of a k-clique $\mathcal{K}$, a set $P\subseteq\mathrm{int}(\mathcal{K})$ called *hook*, two intersecting sets $X_1,X_2\in\mathcal{F}(-\infty,K_1)$ called *left guards*, and two intersecting sets $Y_1,Y_2\in\mathcal{F}(K_k,+\infty)$ called *right guards*, where K_1 and K_k are the least and the greatest elements of $\mathcal{K}$, respectively.

Claim 3.2. Any attached family $\mathcal{F}$ with $\omega(\mathcal{F}) \leqslant k$ and $\chi(\mathcal{F}) > 2^{\gamma_k}$ contains a fancy k-clique.

Proof. Find in $\mathcal{F}$ sets $X_1 \prec X_2 \prec Y_1 \prec Y_2$ so that $X_1 \cap X_2 \neq \emptyset$, $Y_1 \cap Y_2 \neq \emptyset$, and $\chi(\mathcal{F}(X_2, Y_1)) \geqslant \chi(\mathcal{F}) - 4 > 2^{2\xi_{k-1}+k+1}$. Apply Lemma 2.2 to find $\mathcal{H} \subseteq \mathcal{F}(X_2, Y_1)$ such that $\chi(\mathcal{H}) > 2^{\xi_{k-1}}$ and for any intersecting $H_1, H_2 \in \mathcal{H}$ we have $\chi(\mathcal{F}(H_1, H_2)) \geqslant 2^{\xi_{k-1}+k}$. It follows that $\mathcal{H}$ contains a k-clique $\mathcal{K}$ such that $\chi(\mathcal{F}(\mathcal{K})) \geqslant 2^{\xi_{k-1}+k} > 2^{\xi_{k-1}}k$. Since $|\mathcal{K}| = k$, the members of $\mathcal{F}(\mathcal{K})$ that intersect $\bigcup \mathcal{K}$ can be properly colored with $2^{\xi_{k-1}}k$ colors. Hence there exists $P \in \mathcal{F}(\mathcal{K})$ disjoint from $\bigcup \mathcal{K}$, so that $P \subseteq \operatorname{int}(\mathcal{K})$. The clique $\mathcal{K}$ with hook P, left guards X_1, X_2, and right guards Y_1, Y_2 forms a fancy k-clique in $\mathcal{F}$. □

A (k, i)-*bracket system* in an attached family $\mathcal{F}$ consists of k-brackets $\mathcal{B}_1, \ldots, \mathcal{B}_i$ with pairwise intersecting supports, and two intersecting sets $X_1, X_2 \in \mathcal{F}(\mathcal{B}_1) \cap \ldots \cap \mathcal{F}(\mathcal{B}_i)$ called *guards*.

Claim 3.3. Any attached family $\mathcal{F}$ with $\omega(\mathcal{F}) \leqslant k$ and $\chi(\mathcal{F}) > 2^{\delta_{k,i}}$ contains a (k, i)-bracket system.

Proof. The proof goes by induction on i. We start with $i = 1$. First, apply Lemma 2.3 to find $\mathcal{G} \subseteq \mathcal{F}$ that is externally supported in $\mathcal{F}$ and satisfies $\chi(\mathcal{G}) > 2^{\xi_{k-1}+\beta_k+\gamma_k+1}$. Next, apply Lemma 2.2 to find $\mathcal{H} \subseteq \mathcal{G}$ such that $\chi(\mathcal{H}) > 2^{\xi_{k-1}}$ and for any intersecting $H_1, H_2 \in \mathcal{H}$ we have $\chi(\mathcal{G}(H_1, H_2)) \geqslant 2^{\beta_k+\gamma_k}$. It follows that $\mathcal{H}$ contains a k-clique $\mathcal{K}$ such that $\chi(\mathcal{G}(\mathcal{K})) \geqslant 2^{\beta_k+\gamma_k}$. Let $\mathcal{R} = \{R \in \mathcal{F}(\mathcal{K}) : R \cap \operatorname{ext}(\mathcal{K}) \neq \emptyset\}$ and $\mathcal{D} = \{D \in \mathcal{G}(\mathcal{K}) : D \cap \bigcup(\mathcal{K} \cup \mathcal{R}) \neq \emptyset\}$. Lemma 3.1 yields $\chi(\mathcal{D}) \leqslant 2^{\beta_k}$. Let $\mathcal{G}' = \mathcal{G}(\mathcal{K}) \setminus \mathcal{D}$. It follows that $\chi(\mathcal{G}') \geqslant 2^{\beta_k+\gamma_k} - 2^{\beta_k} > 2^{\gamma_k}$. Claim 3.2 guarantees a fancy k-clique $\mathcal{K}'$ with hook P, left guards X_1, X_2, and right guards Y_1, Y_2 in $\mathcal{G}'$. Since $\mathcal{G}$ is externally supported in $\mathcal{F}$, there exists $S \in \mathcal{F}$ that intersects P and $\operatorname{ext}(\mathcal{G})$. Since $P \notin \mathcal{D}$ and $S \cap P \neq \emptyset$, we have $S \notin \mathcal{R}$. This and $S \cap \operatorname{ext}(\mathcal{K}) \supseteq S \cap \operatorname{ext}(\mathcal{G}) \neq \emptyset$ imply $S \notin \mathcal{F}(\mathcal{K})$. Therefore, $\mathcal{K}'$ with hook P, support S, and guards X_1, X_2 or Y_1, Y_2 forms a $(k, 1)$-bracket system in $\mathcal{F}$.

Now, suppose $i \geqslant 2$. As above, find $\mathcal{G} \subseteq \mathcal{F}$ externally supported in $\mathcal{F}$ and $\mathcal{H} \subseteq \mathcal{G}$ such that $\chi(\mathcal{H}) > 2^{\delta_{k,i-1}}$ and for any intersecting $H_1, H_2 \in \mathcal{H}$ we have $\chi(\mathcal{G}(H_1, H_2)) \geqslant 2^{\beta_k+\gamma_k}$. By the induction hypothesis, $\mathcal{H}$ contains a $(k, i-1)$-bracket system with brackets $\mathcal{B}_1, \ldots, \mathcal{B}_{i-1}$ and guards X_1, X_2. Thus $\chi(\mathcal{G}(X_1, X_2)) \geqslant 2^{\beta_k+\gamma_k}$. Again, $\mathcal{G}$ has a fancy k-clique $\mathcal{K}'$ with hook P, left guards X_1, X_2, and right guards Y_1, Y_2, and there exists $S \in \mathcal{F} \setminus \mathcal{F}(X_1, X_2)$ intersecting P and $\operatorname{ext}(\mathcal{G})$. By Lemma 2.4, S intersects all the supports of $\mathcal{B}_1, \ldots, \mathcal{B}_{i-1}$ (it cannot intersect the k-cliques of $\mathcal{B}_1, \ldots, \mathcal{B}_{i-1}$ as $\omega(\mathcal{F}) \leqslant k$). Thus the k-bracket $\mathcal{B}_i$ with k-clique $\mathcal{K}'$, hook P and support S together with $\mathcal{B}_1, \ldots, \mathcal{B}_{i-1}$ and guards X_1, X_2 or Y_1, Y_2 forms a (k,i)-bracket system in $\mathcal{F}$. □

To complete the proof of Theorem 2.1, observe that if $\chi(\mathcal{F}) > 2^{\xi_k} = 2^{\delta_{k,k+1}}$, then by Claim 3.3 $\mathcal{F}$ contains a $(k, k+1)$-bracket system, which contains $k+1$ pairwise intersecting supports, contradicting $\omega(\mathcal{F}) \leqslant k$.

References

[1] E. ASPLUND and B. GRÜNBAUM, *On a colouring problem*, Math. Scand. **8** (1960), 181–188.

[2] J. P. BURLING, "On Coloring Problems of Families of Prototypes", PhD thesis, University of Colorado, 1965.

[3] A. GYÁRFÁS, *On the chromatic number of multiple interval graphs and overlap graphs*, Discrete Math. **55** (2) (1985), 161–166. Corrigendum: Discrete Math. **62** (3) (1986), 333.

[4] A. KOSTOCHKA and J. KRATOCHVÍL, *Covering and coloring polygon-circle graphs*, Discrete Math. **163** (1–3) (1997), 299–305.

[5] S. MCGUINNESS, *On bounding the chromatic number of L-graphs*, Discrete Math. **154** (1–3) (1996), 179–187.

[6] S. MCGUINNESS, *Colouring arcwise connected sets in the plane I*, Graph. Combin. **16** (4) (2000), 429–439.

[7] A. PAWLIK, J. KOZIK, T. KRAWCZYK, M. LASOŃ, P. MICEK, W. T. TROTTER and B. WALCZAK, *Triangle-free intersection graphs of line segments with large chromatic number*, submitted, arXiv:1209.1595.

[8] A. SUK, *Coloring intersection graphs of x-monotone curves in the plane*, Combinatorica, to appear, arXiv:1201.0887.

A characterization of edge-reflection positive partition functions of vertex-coloring models

Guus Regts[1]

Abstract. Szegedy (B. Szegedy, Edge coloring models and reflection positivity, *Journal of the American Mathematical Society* **20**, 2007, 969–988.) showed that the partition function of any vertex-coloring model is equal to the partition function of a complex edge-coloring model. Using some results in geometric invariant theory, we characterize for which vertex-coloring model the edge-coloring model can be taken to be real valued that is, we characterize which partition functions of vertex-coloring models are edge-reflection positive. This answers a question posed by Szegedy.

1 Introduction

Partition functions of vertex- and edge-coloring models are graph invariants introduced by de la Harpe and Jones [4]. In fact, in [4] they are called spin and vertex models respectively. Both models give a rich class of graph invariants. But they do not coincide. For example the number of matchings in a graph is the partition function of a real edge-coloring model but not the partition function of any real vertex-coloring model. This can be deduced from the characterization of partition functions of real vertex-coloring models by Freedman, Lovász and Schrijver [3]. (It is neither the partition function of any complex vertex-coloring model, but we will not prove this here.) Conversely, the number of independent sets is not the partition function of any real edge-coloring model, as follows from Szegedy's characterization of partition functions of real edge-coloring models [7], but it is the partition function of a (real) vertex-coloring model.

However, Szegedy [7] showed that the partition function of any vertex-coloring model can be obtained as the partition function of a complex edge-coloring model. Moreover, he gave examples when the edge-coloring model can be taken to be real valued. This made him ask the question which partition functions of real vertex-coloring models are par-

[1] CWI, Amsterdam. Email: regts@cwi.nl

tition functions of real edge-coloring models (*cf.* [7, Question 3.2]). In fact, he phrased his question in terms of edge-reflection positivity. We will not say anything about this here. See [7] or the full version of the present extended abstract [6].

In this note we completely characterize for which vertex-coloring models there exists a real edge-coloring model such that their partition functions coincide, answering Szegedy's question.

The organization of this paper is as follows. In the next section we give definitions of partition functions of edge and vertex-coloring models and state our main result (*cf.* Theorem 2.4). In Section 3 we give a sketch of the proof of Theorem 2.4.

ACKNOWLEDGEMENTS. I thank Lex Schrijver for his comments on the full version of this paper. In particular, for simplifying some of the proofs.

2 Partition functions of edge and vertex-coloring models

We give the definitions of edge and vertex-coloring models and their partition functions. After that we describe Szegedy's result how to obtain a complex edge-coloring model from a vertex-coloring model such that their partition functions are the same. (The existence also follows from the characterization of partition functions of complex edge-coloring models given in [1], but Szegedy gives a direct way to construct the edge-coloring model from the vertex-coloring model.) And finally we will state our main result saying which partition functions of vertex-coloring models are partition function of real edge-coloring models.

Let $\mathcal{G}$ be the set of all graphs, allowing multiple edges and loops. Let $\mathbb{C}$ denote the set of complex numbers and let $\mathbb{R}$ denote the set of real numbers. If V is a vector space we write V^* for its dual space, but by $\mathbb{C}^*$ we mean $\mathbb{C}\setminus\{0\}$. For a matrix U we denote by U^* its conjugate transpose and by U^T its transpose.

Let $\mathbb{F}$ be a field. An $\mathbb{F}$-valued *graph invariant* is a map $p : \mathcal{G} \to \mathbb{F}$ which takes the same values on isomorphic graphs.

Throughout this paper we set $\mathbb{N} = \{1, 2 \ldots\}$ and for $n \in \mathbb{N}$, $[n]$ denotes the set $\{1, \ldots, n\}$. We will now introduce partition functions of vertex and edge-coloring models.

Let $a \in (\mathbb{C}^*)^n$ and let $B \in \mathbb{C}^{n\times n}$ be a symmetric matrix. We call the pair (a, B) an *n-color vertex-coloring model*[2]. If moreover, a is positive and B is real, then we call (a, B) a *real n-color vertex-coloring model.*

[2] Vertex-coloring models with a equal to the all ones vector were introduced by de la Harpe and Jones in [4] where they are called spin models.

When talking about a vertex-coloring model, we will sometimes omit the number of colors. The *partition function* of an n-color vertex-coloring model (a, B) is the graph invariant $p_{a,B} : \mathcal{G} \to \mathbb{C}$ defined by

$$p_{a,B}(H) := \sum_{\phi: V(H)\to[n]} \prod_{v\in V(H)} a_{\phi(v)} \cdot \prod_{uv\in E(H)} B_{\phi(u),\phi(v)}, \tag{2.1}$$

for $H \in \mathcal{G}$.

We can view $p_{a,B}$ in terms of weighted homomorphisms. Let $G(a, B)$ be the complete graph on n vertices (including loops) with vertex weights given by a and edge weights given by B. Then $p_{a,B}(H)$ can be viewed as counting the number of weighted homomorphisms of H into $G(a, B)$. In this context $p_{a,B}$ is often denoted by $\hom(\cdot, G(a, B))$. So in particular the number of proper k-colorings is the partition function of a vertex-coloring model. The vertex-coloring model can also be seen as a statistical mechanics model where vertices serve as particles, edges as interactions between particles, and colors as states or energy levels.

Let for a field $\mathbb{F}$,

$$R(\mathbb{F}) := \mathbb{F}[x_1, \ldots, x_k] \tag{2.2}$$

denote the polynomial ring in k variables. We will only consider $\mathbb{F} = \mathbb{R}$ and $\mathbb{F} = \mathbb{C}$. Note that there is a one-to-one correspondence between linear functions $h : R(\mathbb{F}) \to \mathbb{F}$ and maps $h : \mathbb{N}^k \to \mathbb{F}$; $\alpha \in \mathbb{N}^k$ corresponds to the monomial $x^\alpha := x_1^{\alpha_1} \cdots x_k^{\alpha_k} \in R(\mathbb{F})$ and the monomials form a basis for $R(\mathbb{F})$. We call any $h \in R(\mathbb{C})^*$ a *k-color edge-coloring model*[3]. Any $h \in R(\mathbb{R})^*$ is called a *real k-color edge-coloring model*. When talking about an edge-coloring model, we will sometimes omit the number of colors. The *partition function* of a k-color edge-coloring model h is the graph invariant $p_h : \mathcal{G} \to \mathbb{C}$ defined by

$$p_h(G) = \sum_{\phi: E(G)\to[k]} \prod_{v\in V(G)} h\Big(\prod_{e\in\delta(v)} x_{\phi(e)}\Big), \tag{2.3}$$

for $G \in \mathcal{G}$. Here $\delta(v)$ is the multiset of edges incident with v. By convention, a loop is counted twice.

Let us give a few examples.

Example 2.1 (Counting perfect matchings). Let $k = 2$. Define the the edge-coloring model $h : \mathbb{R}[x_1, x_2] \to \mathbb{R}$ by $h(x_1^{a_1} x_2^{a_2}) = \delta_{a_1,1}$. Then $p_h(G)$ is equal to the number of perfect matchings of G. To see this,

---

[3] Edge-coloring models were introduced by de la Harpe and Jones in [4] where they are called vertex models.

note that for an assignment of the colors to the edges of G there is a contribution in the sum (2.3) if and only if at each vertex there is a unique edge which is colored with 1, that is, if and only if the edges colored with 1 form a perfect matching.

Example 2.2 (Counting proper k-edge-colorings). Let $k \in \mathbb{N}$. Define the k-color edge-coloring model h by

$$h(x_1^{a_1} \cdots x_k^{a_k}) = \begin{cases} 1 \text{ if } a_i \leq 1 \text{ for all } i \\ 0 \text{ else} \end{cases}.$$

Then $p_h(G)$ is equal to the number of proper k-edge-colorings of G.

For more examples we refer to [4] and [7]. The edge-coloring model can also be considered as a statistical mechanics model, where edges serve as particles, vertices as interactions between particles, and colors as states or energy levels.

We will now describe a result of Szegedy [7] (see also [8]) showing that partition functions of vertex-coloring models are partition functions of edge-coloring models.

Let (a, B) be an n-color vertex-coloring model. As B is symmetric we can write $B = U^T U$ for some $k \times n$ (complex) matrix U, for some k. Let $u_1, \ldots, u_n \in \mathbb{C}^k$ be the columns of U. Define the edge-coloring model h by $h := \sum_{i=1}^n a_i \operatorname{ev}_{u_i}$, where for $u \in \mathbb{C}^k$, $\operatorname{ev}_u \in R(\mathbb{C})^*$ is the linear map defined by $p \mapsto p(u)$ for $p \in R(\mathbb{C})$.

Lemma 2.3 (Szegedy [7]). *Let (a, B) and h be as above. Then $p_{a,B} = p_h$.*

Let (a, B) be an n-color vertex-coloring model. We say that $i, j \in [n]$ are *twins of* (a, B) if $i \neq j$ and the ith row of B is equal to the jth row of B. If (a, B) has no twins we call the model *twin free*. Suppose now $i, j \in [n]$ are twins of (a, B). If $a_i + a_j \neq 0$, let B' be the matrix obtained from B by removing row and column i and let a' be the vector obtained from a by setting $a'_j := a_i + a_j$ and then removing the ith entry from it. In case $a_i + a_j = 0$, we remove the ith and the jth row and column from B to obtain B' and we remove the ith and the jth entry from a to obtain a'. Then $p_{a',B'} = p_{a,B}$. So for every vertex-coloring model with twins, we can construct a vertex-coloring model with fewer colors which is twin free and which has the same partition function.

We need a few more definitions to state our main result. For a $k \times n$ matrix U we denote its columns by $u_1, \ldots, u_n$. Let, for any k, $(\cdot, \cdot)$ denote the standard bilinear form on $\mathbb{C}^k$. We call the matrix U *nondegenerate* if the span of $u_1, \ldots, u_n$ is nondegenerate with respect to $(\cdot, \cdot)$. In other words, if $\mathrm{rk}(U^T U) = \mathrm{rk}(U)$. For $l \in \mathbb{N}$, let $O_l(\mathbb{C})$ be the complex orthogonal group, i.e. $O_l(\mathbb{C}) := \{g \in \mathbb{C}^{l\times l} \mid (gv, gv) = (v, v) \text{ for all } v \in \mathbb{C}^l\}$. We think of vectors in $\mathbb{C}^k$ as vectors in $\mathbb{C}^l$ for any $l \geq k$. We can now state our main result.

Theorem 2.4. *Let (a, B) be a twin-free n-color vertex-coloring model. Let U be a nondegenerate $k \times n$ matrix such that $U^T U = B$. Then the following are equivalent:*

(i) $p_{a,B} = p_y$ *for some real edge-coloring model* y,

(ii) *there exists* $l \geq k$, $g \in O_l(\mathbb{C})$ *such that the set* $\{\begin{pmatrix} gu_i \\ a_i \end{pmatrix} \mid i = 1, \dots, n\}$ *is closed under complex conjugation,*

(iii) *there exists* $l \geq k$, $g \in O_l(\mathbb{C})$ *such that* $\sum_{i=1}^n a_i \,\mathrm{ev}_{gu_i}$ *is real.*

If moreover, $UU^* \in \mathbb{R}^{k\times k}$, *then we can take* g *equal to the identity in* (ii) *and* (iii).

Note that for $h := \sum_{i=1}^n a_i \,\mathrm{ev}_{gu_i}$ in Theorem 2.4, we have, by Lemma 2.3, $p_h = p_{a,B}$. Moreover, observe that if the set of columns of gU are closed under complex conjugation, then $gU(gU)^*$ is real. So the existence of a nondegenerate matrix U such that $U^T U = B$ and UU^* is real, is a necessary condition for $p_{a,B}$ to be the partition function of a real edge-coloring model.

In case B is real, there is an easy way to obtain a $k \times n$ rank k matrix U, where $k = \mathrm{rk}(B)$, such that $UU^* \in \mathbb{R}^{k\times k}$ and $U^T U = B$, using the spectral decomposition of B. So by Theorem 2.4, we get the following characterization of partition functions of real vertex-colorings that are partition functions of real edge-coloring models. We will state it as a corollary.

Corollary 2.5. *Let (a, B) be a twin-free real n-color vertex-coloring model. Then $p_{a,B} = p_h$ for some real edge-coloring model h if and only if for each $i \in [n]$ there exists $j \in [n]$ such that*

(i) $a_i = a_j$,

(ii) *for each eigenvector v of B with eigenvalue* λ : $\begin{cases} \lambda > 0 \Rightarrow v_i = v_j, \\ \lambda < 0 \Rightarrow v_i = -v_j. \end{cases}$

3 A sketch of the proof of Theorem 2.4

Here we give short sketch of the proof of Theorem 2.4. For the full proof see [6]. First of all, Lemma 2.3 shows that (iii) implies (i). The following easy lemma gives the equivalence between (ii) and (iii) for the same g in Theorem 2.4.

Lemma 3.1. *Let $u_1, \ldots, u_n \in \mathbb{C}^k$ be distinct vectors, let $a \in (\mathbb{C}^*)^n$ and let $h := \sum_{i=1}^n a_i \operatorname{ev}_{u_i}$. Then h is a real edge-coloring model if and only if the set $\{\begin{pmatrix} u_i \\ a_i \end{pmatrix} \mid i = 1, \ldots, n\}$ is closed under complex conjugation.*

Proof. Suppose first that the set $\{\begin{pmatrix} u_i \\ a_i \end{pmatrix} \mid i = 1, \ldots, n\}$ is closed under complex conjugation. Then for $p \in R(\mathbb{R})$, $h(p) = \sum_{i=1}^n a_i p(u_i) = \sum_{i=1}^n \overline{a_i p(u_i)} = \overline{h(p)}$. Hence, $h(p) \in \mathbb{R}$. So h is real valued.

Now the 'only if' part. By possibly adding some vectors to $\{u_1, ..., u_n\}$ and extending the vector a with zero's, we may assume that $\{u_1, ..., u_n\}$ is closed under complex conjugation. We must show that $u_i = \overline{u_j}$ implies $a_i = \overline{a_j}$. We may assume that $u_1 = \overline{u_2}$. Using Lagrange interpolating polynomials we find $p \in R(\mathbb{C})$ such that $p(u_j) = 1$ if $j = 1, 2$ and 0 else. Let $p' := 1/2(p + \overline{p})$. Then $p' \in R(\mathbb{R})$ and consequently, $h(p') = \sum_{i=1}^n a_i p(u_i) = a_1 + a_2 \in \mathbb{R}$. Similarly, there exists $q \in R(\mathbb{C})$ such that $q(u_1) = \mathrm{i}$, $q(u_2) = -\mathrm{i}$ and $q(u_j) = 0$ if $j > 2$. Setting $q' := 1/2(q + \overline{q})$ and applying h to it, we find that $\mathrm{i}(a_1 - a_2) \in \mathbb{R}$. So we conclude that $a_1 = \overline{a_2}$. Continuing this way proves the lemma. □

What remains is the proof of (i) implies (iii), which is the hardest part. Our proof of this is based some on fundamental results in geometric invariant theory. We will now sketch our approach.

Let h be a real l-color edge-coloring model such that $p_h = p_{a,B}$. Then, using that U is nondegenerate and that the columns of U are distinct, we can apply a result from [2] and a result from [5] to find $g \in O_l(\mathbb{C})$ such that $h = \sum_{i=1}^n a_i \operatorname{ev}_{gu_i}$.

Suppose now that $UU^* \in \mathbb{R}^{k \times k}$. Define for an $l \times n$ matrix W the function $f_W : O_l(\mathbb{C}) \to \mathbb{R}$ by $g \mapsto \operatorname{tr}(gW(gW)^*)$, where tr denotes the trace. The function f_W was introduced by Kempf and Ness [5] in the context of reductive algebraic groups acting on finite dimensional vector spaces. Denote by $\operatorname{Stab}(W)$ the subgroup of $O_l(\mathbb{C})$ that leaves W invariant. Let $e \in O_l(\mathbb{C})$ denote the identity matrix.

Lemma 3.2. *The function f_W has the following properties:*

(i) $\inf_{g \in O_l(\mathbb{C})} f_W(g) = f_W(e)$ *if and only if* $WW^* \in \mathbb{R}^{l \times l}$,

(ii) *If* $WW^* \in \mathbb{R}^{l \times l}$, *then* $f_W(e) = f_W(g)$ *if and only if* $g \in O_l(\mathbb{R}) \cdot \operatorname{Stab}(W)$.

To prove that we can take g equal to the identity, we first find, as above, $g \in O_l(\mathbb{C})$ such that $h = \sum_{i=1}^n a_i \operatorname{ev}_{gu_i}$. By Lemma 3.1, this implies that $gU(gU)^*$ is real. Hence, by Lemma 3.2 (i) f_{gU} attains its infimum at the identity. Equivalently, f_U attains its infimum at g. So by Lemma 3.2 (ii), $g \in O_l(\mathbb{R}) \cdot \operatorname{Stab}(U)$. From this we can conclude that h is real valued.

References

[1] J. DRAISMA, D. GIJSWIJT, L. LOVÁSZ, G. REGTS and A. SCHRIJVER, *Characterizing partition functions of the vertex model*, Journal of Algebra **350** (2012), 197–206.

[2] J. DRAISMA and G. REGTS, *Tensor invariants for certain subgroups of the orthogonal group*, to appear in *Journal of Algebraic Combinatorics*, doi:10.1007/s10801-012-0408-7.

[3] M. FREEDMAN, L. LOVÁSZ and A. SCHRIJVER, *Reflection positivity, rank connectivity, and homomorphisms of graphs*, Journal of the American Mathematical Society **20** (2007), 37–51.

[4] P. DE LA HARPE and V. F. R. JONES, *Graph invariants related to statistical mechanical models: examples and problems*, Journal of Combinatorial Theory, Series B **57** (1993), 207–227.

[5] G. KEMPF and L. NESS, *The length of vectors in representation spaces*, In: "Algebraic Geometry", Proc. Summer Meeting, Univ. Copenhagen, Copenhagen, 1978. Lecture Notes in Math., Vol. 732, Springer, Berlin 1979, 233–243.

[6] G. REGTS, *A characterization of edge reflection positive partition functions of vertex coloring models*, preprint, http://arxiv.org/pdf/1302.6497.pdf

[7] B. SZEGEDY, *Edge coloring models and reflection positivity*, Journal of the American Mathematical Society **20** (2007), 969–988.

[8] B. SZEGEDY, *Edge coloring models as singular vertex coloring models*, In: "Fete of Combinatorics and Computer Science", G. O. H. Katona, A. Schrijver and T. Szönyi (eds.), Springer, Heidelberg and János Bolyai Mathematical Society, Budapest, 2010, 327–336.

Adjacent vertex-distinguishing edge coloring of graphs

Marthe Bonamy[1], Nicolas Bousquet[1] and Hervé Hocquard[2]

Abstract. An adjacent vertex-distinguishing edge coloring (AVD-coloring) of a graph is a proper edge coloring such that no two neighbors are adjacent to the same set of colors. Zhang *et al.* [17] conjectured that every connected graph on at least 6 vertices is AVD $(\Delta + 2)$-colorable, where Δ is the maximum degree.

In this paper, we prove that $(\Delta + 1)$ colors are enough when Δ is sufficiently larger than the maximum average degree, denoted mad. We also provide more precise lower bounds for two graph classes: planar graphs, and graphs with mad $<$ 3. In the first case, $\Delta \geq 12$ suffices, which generalizes the result of Edwards *et al.* [7] on planar bipartite graphs. No other results are known in the case of planar graphs. In the second case, $\Delta \geq 4$ is enough, which is optimal and completes the results of Wang and Wang [14] and of Hocquard and Montassier [9].

1 Introduction

In the following, a graph is a connected simple graph on at least three vertices. A *(proper) edge k-coloring* of a graph is a coloring of its edges using at most k colors, where any two incident edges receive distinct colors. The *chromatic index* of a graph G, denoted by $\chi'(G)$, is the smallest integer k such that G admits an edge k-coloring. Let $\Delta(G)$ be the maximum degree of G. Since incident edges receive distinct colors in an edge coloring, every graph G satisfies $\chi'(G) \geq \Delta(G)$. The Vizing's theorem ensures that the reverse inequality is nearly true, more precisely:

Theorem 1.1 ([12]). *Every graph G satisfies* $\Delta(G) \leq \chi'(G) \leq \Delta(G)+1$.

An *adjacent vertex-distinguishing edge k-coloring* (*AVD k-coloring* for short) is a proper edge k-coloring such that, for any two adjacent vertices u and v, the set of colors assigned to edges incident to u differs from the set of colors assigned to edges incident to v. The *AVD-chromatic index* of G, denoted by $\chi'_{avd}(G)$, is the smallest integer k such that G admits an AVD k-coloring. It should be noted that, while an isolated edge admits no AVD coloring, the AVD-chromatic index is finite for all connected graphs on at least three vertices. AVD colorings are also known as *adjacent strong edge colorings* [17] and *1-strong edge colorings* [1].

[1] LIRMM (Université Montpellier 2), 161 rue Ada, 34392 Montpellier Cedex, France. Email: marthe.bonamy@lirmm.fr, bousquet@lirmm.fr

[2] LaBRI (Université Bordeaux 1), 351 cours de la Libération, 33405 Talence Cedex, France. Email: hocquard@abri.fr

This research is partially supported by the ANR Grant EGOS (2012-2015) 12 JS02 002 01

Note that AVD colorings are special cases of *vertex-distinguishing proper edge colorings*. Such colorings are proper edge colorings such that no two (not necessarily adjacent) vertices are adjacent to the same set of colors. The corresponding chromatic index is called the *observability* and was studied for different graph classes [3,5,6,8].

Since an AVD coloring is a proper edge coloring, every graph G satisfies $\chi'_{avd}(G) \geq \Delta(G)$. In addition, every graph G with two adjacent vertices of degree $\Delta(G)$ satisfies $\chi'_{avd}(G) \geq \Delta(G)+1$. Zhang *et al.* [17] completely determined the AVD-chromatic index of paths, cycles, trees, complete graphs, and complete bipartite graphs. They noted that a cycle of length five requires five colors, but conjectured that it is the only graph with such a gap between $\chi'_{avd}(G)$ and $\Delta(G)$.

Conjecture 1.2 ([17]). Every graph G on at least 6 vertices satisfies $\chi'_{avd}(G) \leq \Delta(G)+2$.

Balister *et al.* [2] proved Conjecture 1.2 for graphs with $\Delta(G) = 3$ and for bipartite graphs.

For edge coloring, Theorem 1.1 ensures that the chromatic index of a graph can only have two values: $\Delta(G)$ or $\Delta(G)+1$, and the classification of graphs depending on this received considerable interest (for instance [11]). For AVD coloring, if Conjecture 1.2 holds then the AVD chromatic index of a graph can only have three values: $\Delta(G)$, $\Delta(G)+1$ or $\Delta(G)+2$. When considering a given graph class that allows two vertices of maximum degree to be adjacent, there are only two possible upper bounds: $\Delta(G)+1$ or $\Delta(G)+2$. Similarly, the classification of graph classes depending on this received subsequent interest, for instance:

Theorem 1.3 ([7]). *Every (connected) bipartite planar graph G with $\Delta(G) \geq 12$ satisfies $\chi'_{avd}(G) \leq \Delta(G)+1$.*

Let $\mathrm{mad}(G) = \max\left\{\frac{2|E(H)|}{|V(H)|},\ H \subseteq G\right\}$ be the *maximum average degree* of the graph G, where $V(H)$ and $E(H)$ are the sets of vertices and edges of H, respectively. Wang and Wang [14] made the link between maximum average degree and AVD-chromatic index and proved Conjecture 1.2 for graphs with $\Delta(G) \geq 3$ and $\mathrm{mad}(G) < 3$.

Theorem 1.4 ([14]). *Every (connected) graph G with $\Delta(G) \geq 3$ and $\mathrm{mad}(G) < 3$ satisfies $\chi'_{avd}(G) \leq \Delta(G)+2$.*

They also gave sufficient conditions for graphs of bounded maximum average degree to be AVD $(\Delta(G)+1)$-colorable. Combined with results of Hocquard and Montassier [9], we have:

Theorem 1.5 ([9,14]). *Every (connected) graph G with $\Delta(G) \geq 3$ and $\mathrm{mad}(G) < 3 - \frac{2}{\Delta(G)}$ satisfies $\chi'_{avd}(G) \leq \Delta(G)+1$.*

Two main questions arise from these partial results: can this threshold of 3 as an upper-bound on $\mathrm{mad}(G)$ be reached with a sufficiently large lower-bound on $\Delta(G)$ in the case of Theorem 1.5, and broken in the case of Theorem 1.4? We answer positively to these questions with Theorem 1.6: there is no threshold in the case of Theorem 1.5 (and thus in the case of Theorem 1.4).

Theorem 1.6. *Every graph G with* $\Delta(G) > 3 \times (\mathrm{mad}(G))^2$ *satisfies* $\chi'_{avd}(G) \leq \Delta(G) + 1$.

In the case of edge coloring, the best lower bound is due to Woodall [15]: every graph G with $\Delta(G) > \frac{3\times\mathrm{mad}(G)}{2}$ satisfies $\chi'(G) = \Delta(G)$. There is a very large gap between this bound and its AVD counterpart, but this is essentially due to the fact that most methods on edge coloring are not transposable to AVD coloring. On the other hand, the gap between the bound for AVD coloring and its list edge counterpart is a mere constant factor [4] (note that list edge coloring is similarly conjectured to be always possible with $\Delta(G) + 1$ colors [13]).

By Theorem 1.6, planar graphs with sufficiently large maximum degree are AVD $(\Delta(G) + 1)$-colorable. We provide a more refined lower-bound, and prove here that the bipartite hypothesis in Theorem 1.3 is unnecessary.

Theorem 1.7. *Every planar graph G with* $\Delta(G) \geq 12$ *satisfies* $\chi'_{avd}(G) \leq \Delta(G) + 1$.

In the case of graphs with maximum average degree at most 3, we improve Theorem 1.5 by showing that $\Delta(G) \geq 4$ is enough to reach the threshold of 3, as follows.

Theorem 1.8. *Every graph G with* $\Delta(G) \geq 4$ *and* $\mathrm{mad}(G) < 3$ *satisfies* $\chi'_{avd}(G) \leq \Delta(G) + 1$.

Note that Theorem 1.8 is best possible since Figure 1.1 provides a subcubic graph with $\mathrm{mad}(G) = \frac{11}{4} < 3$ that is not AVD $(\Delta + 1)$-colorable.

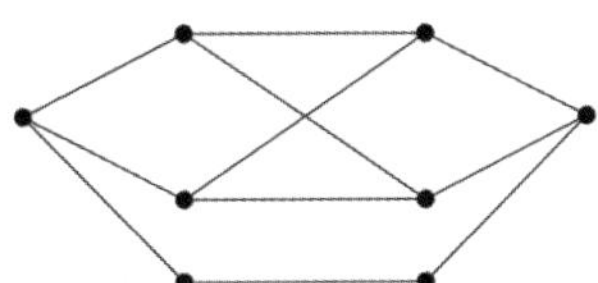

Figure 1.1. A graph G with $\Delta(G) = 3$ and $\mathrm{mad}(G) = \frac{11}{4} < 3$ such that $\chi'_{avd}(G) = 5$.

2 Method

We prove Theorems 1.6, 1.7 and 1.8 using a discharging method. We first choose a partial order on graphs. It depends on the theorem to prove but is basically a customization of the lexicographic order on the number of vertices of each degree. In a first time, we consider by contradiction a "minimal" counter-example, and prove that it has some strong structural properties. We then prove that a graph with those structural properties cannot satisfy the assumptions of the theorem, which provides a contradiction. Due to the limited number of pages, the complete proofs of Theorems 1.6, 1.7 and 1.8 are omitted. We nevertheless sketch the proof of Theorem 1.8.

The four following lemmas provide the most relevant examples of structural properties that we proved on a minimal counter-example of any of Theorems 1.6, 1.7 and 1.8, where the number of colors is $k+1$, for $k \geq 4$. In each case, we assume by contradiction that the graph contains such a configuration, we color by minimality a smaller graph and prove that the coloring can be extended to the whole graph.

Lemma 2.1. *No vertex v_2 is adjacent to two vertices v_1 and v_3, with $d(v_1), d(v_2)$ and $d(v_3) \leq \frac{k}{2}$.*

Lemma 2.1 follows from a recoloring algorithm. The proof is quite involved, when a simple proof exists when $\frac{k}{2}$ is replaced by $\frac{k}{4}$, but this bound is decisive for Theorems 1.6 and 1.7.

Lemma 2.2. *No vertex has at least $\frac{k}{2}$ neighbors of degree 1.*

Lemma 2.3. *No vertex of degree 2 is adjacent to two vertices of degree at most $\frac{k}{2}+1$.*

Lemmas 2.2 and 2.3 follow from a simple combinatorial argument, except in the case of a vertex of degree 2 adjacent to a vertex of degree exactly $\frac{k}{2}+1$, where the result is derived from a recoloring argument and from the 2-connectivity of a minimal counter-example.

Lemma 2.4. *No vertex is adjacent to two vertices u, v with $d(u) = 2$ and $d(v) \leq 2$.*

Lemma 2.4 follows from a simple reduction to a smaller graph, where the choice of the partial order is decisive. Those four lemmas are not enough for Theorems 1.6 and 1.7. Lemma 2.1 is decisive for Theorems 1.6 and 1.7, and Lemmas 2.2 to 2.4 suffice for Theorem 1.8.

We consider a graph G with $\Delta(G) \leq k$ that satisfies Lemmas 2.2, 2.3 and 2.4, and assign to each vertex its degree as weight. We then design discharging rules to rearrange the weight along the graph so as to derive that $\mathrm{mad}(G) \geq 3$. The following observation is instrumental in the proofs of Theorems 1.6 and 1.8.

Observation 2.5. For any vertex partition (V_1, V_2) of a graph G, if every vertex v has an initial weight of $d(v)$, and the weight can be rearranged along the graph so that every vertex v_1 of V_1 has a weight of at least

$2 \times d(v_1)$ and every vertex v_2 of V_2 has a weight of at least m, then $\text{mad}(G) \geq m$.

Thus, to prove Theorem 1.8, we design a single discharging rule stating that a vertex u with $d(u) \geq 3$ that has a neighbor v of degree 1 or 2 gives a weight of 1 to v. We consider V_1 to be the set of vertices of degree 1, and V_2 the set of vertices of degree at least 2. By Lemma 2.2, the vertices incident to a vertex of degree 1 are of degree at least 4, and in particular V_1 is an independent set. Let u be a vertex of G. If $u \in V_1$, since V_1 is an independent set, the vertex u gives nothing and receives $d(u)$, so it has a final weight of $2 \times d(u)$. If $u \in V_2$ with $d(u) = 2$, by Lemma 2.3, the vertex u is adjacent to at least one vertex of degree at least 3, and receives 1 from it. Since it gives nothing, it has a final weight of at least 3. If $u \in V_2$ with $d(u) \geq 3$ and u has a neighbor of degree 2, by Lemmas 2.3 and 2.4, the vertex u is adjacent to no other vertex of degree at most 2, and $d(u) \geq 4$, so u has a final weight of at least 3. If $u \in V_2$ and has no neighbor of degree 2 with $d(u) \geq 3$, by Lemma 2.2, the vertex u has at most $d(u) - 3$ neighbors of degree 1, so u has a final weight of at least 3. By Observation 2.5, $\text{mad}(G) \geq 3$.

For Theorem 1.7, we use a combinatorial argument to prove that a vertex cannot have too many small vertices, with an optimal bound (optimal for a combinatorial argument not involving any recoloring algorithm) depending on the respective degrees and on the number of colors. For Theorem 1.6, we use a method from a beautiful proof by Borodin, Kostochka and Woodall [4] (later simplified in [15]) of a similar result on list edge coloring.

3 Conclusion and perspectives

With Theorem 1.6, we made a significant step towards Conjecture 1.2, by proving that there are many graphs that need one less color. Our methods will however not be sufficient for the conjecture itself, as they require sparsity hypotheses. However, we could aim at proving that all planar graphs are AVD $(\Delta(G) + 2)$-colorable, by further developing the proof of Theorem 1.7. We conclude with two conjectures.

Conjecture 3.1. For any graph G, if the set of vertices of maximum degree in G forms an independent set, then $\chi'_{avd}(G) \leq \Delta(G) + 1$.

Conjecture 3.2. For any graph G, if G admits a subgraph H such that $\chi'_{avd}(H) > \chi'_{avd}(G)$, then either H is not connected or $\Delta(H) = 2$.

References

[1] S. Akbari, H. Bidkhori and N. Nosrati, *r-strong edge colorings of graphs*, Discrete Math. **306** (2006), 3005-3010.

[2] P. N. BALISTER, E. GYŐRI, J. LEHEL and R. H. SCHELP, *Adjacent vertex distinguishing edge-colorings*, SIAM J. Discrete Math. **21** (2007), 237–250.

[3] P. N. BALISTER, O. M. RIORDAN and R. H. SCHELP, *Vertex-distinguishing edge-colorings of graphs*, J. Graph Theory, **42** (2003), 95–109.

[4] O. V. BORODIN, A. V. KOSTOCHKA and D. R. WOODALL, *List Edge and List Total Colourings of Multigraphs*, J. Comb. Theory, Series B **71** (2) (1997), 184–204.

[5] A. C. BURRIS and R. H. SCHELP, *Vertex-distinguishing proper edge-colorings*, J. Graph Theory **26** (1997), 73–83.

[6] J. CERNÝ, M. HORŇÁK and R. SOTÁK, *Observability of a graph*, Mathematica Slovaca **46** (1) (1996), 21–31.

[7] K. EDWARDS, M. HORŇÁK and M. WOŹNIAK, *On the neighbour-distinguishing index of a graph*, Graphs and Comb. **22** (3) (2006), 341–350.

[8] O. FAVARON, H. LI and R. H. SCHELP, *Strong edge coloring of graphs*, Discrete Math. **159** (1996), 103–109.

[9] H. HOCQUARD and M. MONTASSIER, *Adjacent vertex-distinguishing edge coloring of graphs with maximum degree* Δ, DOI: 10.1007/s10878-011-9444-9, 2012.

[10] A. V. KOSTOCHKA and D. R. WOODALL, *Choosability conjectures and multicircuits*, Discrete Math. *240* (2001), 123–143.

[11] D. P. SANDERS and Y. ZHAO, *Planar Graphs of Maximum Degree Seven are Class I*, J. Comb. Theory, Series B **83** (2) (2001), 201–212.

[12] V. G. VIZING, *On an estimate of the chromatic class of a p-graph*, Metody Diskret. Analiz. **3** (1964), 23–30.

[13] V. G. VIZING, *Colouring the vertices of a graph with prescribed colours* (in russian), Diskret. Analiz. **29** (1976), 3–10.

[14] W. WANG and Y. WANG, *Adjacent vertex distinguishing edge-colorings of graphs with smaller maximum average degree*, J. Comb. Optim., **19** (2010), 471–485.

[15] D. R. WOODALL, *The average degree of an edge-chromatic critical graph II*, J. Graph Theory **56** (3) (2007), 194–218.

[16] D. R. WOODALL, *The average degree of a multigraph critical with respect to edge or total choosability*, Discrete Math. **310** (2010), 1167–1171.

[17] Z. ZHANG, L. LIU and J. WANG, *Adjacent strong edge coloring of graphs*, Appl. Math. Lett. **15** (2002), 623–626.

Rainbow path and minimum degree in properly edge colored graphs

Anita Das[1], P. Suresh[1] and S. V. Subrahmanya[1]

Abstract. A rainbow path in a properly edge colored graph is a path in which all the edges are colored with distinct colors. Let G be a properly edge colored graph with minimum degree δ and let t be the maximum length of a rainbow path in G. In this paper, we show that $t \geq \lfloor \frac{3\delta}{5} \rfloor$. It is easy to see that there exist graphs for which $t \leq \delta$; with respect to some proper edge coloring. For example, δ-regular graphs of chromatic index δ are graphs for which $t \leq \delta$, with respect to their optimum proper edge coloring. We leave open the question of getting a lower bound as close to δ as possible.

1 Introduction

Given a graph $G = (V, E)$, a map $c : E \rightarrow \mathbb{N}$ ($\mathbb{N}$ is the set of non-negative integers) is called a *proper edge coloring* of G if for every two adjacent edges e_1 and e_2 of G, we have $c(e_1) \neq c(e_2)$. If G is assigned such a coloring c, then we say that G is a *properly edge colored* graph. We denote the color of an edge $e \in E(G)$ by $color(e)$.

A path in an edge colored graph with no two edges sharing the same color is called a *rainbow path*. Similarly, a cycle in an edge colored graph is called a *rainbow cycle* if no two edges of the cycle share the same color. Given a coloring of the edges of G, a *rainbow matching* is a matching whose edges have distinct colors. Rainbow cycles, rainbow paths and rainbow matching in properly edge colored complete graphs are related to partial transversal of latin squares. Latin squares have been a popular topic in combinatorics at least since the times of Euler, who studied them extensively. A survey on rainbow paths, cycles and other rainbow sub-graphs can be found in [6]. Several theorems and conjectures on rainbow cycles can be found in a paper by Akbari, Etesami, Mahini and Mahmoody in [1].

Maximum length rainbow paths and rainbow cycles are studied extensively in properly edge colored complete graphs. Hahn conjectured that every proper edge coloring of K_n (K_n is a complete graph having n vertices) admits a Hamiltonian rainbow path (a rainbow path visiting every

[1] E-Comm Research Lab, Education & Research, Infosys Limited Bangalore, India. Email: Anita_Das01@infosys.com, Suresh_P01@infosys.com, subrahmanyasv@infosys.com,

vertex of K_n). Later, Maamoun and Meyniel [7] disproved this conjecture by constructing counterexamples for the case where n is a power of two. Still it is widely believed that in every proper edge coloring of K_n, there is a rainbow path on $n-1$ vertices, though this is far from proved. Gyárfás and Mhalla in [5] have shown that the number of vertices in a maximum rainbow path in (a properly edge colored) K_n is at least $(2n+1)/3$. Very recently in [4], H. Gebauer and F. Mousset have improved this bound and have shown that, in every proper edge coloring of K_n, there is a rainbow path of length $(\frac{3}{4} - o(1))n$.

1.1 Our results

In this paper, we consider the problem of finding a maximum length rainbow path in a properly edge colored graph G. Let δ be the minimum degree of G. The main result of this paper is the following:

Theorem 1.1. *If t is the maximum length of a rainbow path in a properly edge colored graph G, then $t \geq \lfloor \frac{3}{5}\delta \rfloor$.*

We are unable to show that the bound given in this paper is tight. On the other hand, it is easy to see that, we cannot expect to get a lower bound greater than δ. For example, consider δ regular graphs of chromatic index δ which are optimally properly edge colored. Clearly for these cases maximum rainbow path length cannot exceed δ. We leave open the question of getting a lower bound of $\delta - c$, for some constant c.

1.2 Preliminaries

All graphs considered in this paper are finite, simple and undirected. A graph is a tuple (V, E), where V is the finite set of vertices and E is the set of edges. For a graph G, we use $V(G)$ and $E(G)$ to denote its vertex set and edge set, respectively. The neighborhood $N(v)$ of a vertex v is the set of vertices adjacent to v but not including v. The degree of a vertex v is $d_v = |N(v)|$. We write $|V(G)|$, δ, Δ for the order, minimum degree and maximum degree of G, respectively. A path is a non-empty graph $P = (V, E)$ of the form $V = \{p_1, p_2, \ldots, p_k\}$ and $E = \{\{p_1, p_2\}, \{p_2, p_3\}, \ldots, \{p_{k-1}, p_k\}\}$, which we usually denote by the sequence $\{p_1, p_2, \ldots, p_k\}$. The *length* of a path is its number of edges. If $P = \{p_1, p_2, \ldots, p_k\}$ is a path, then the graph $C = P \cup \{p_k, p_1\}$ is a cycle, and $|E(C)|$ is the length of C. We represent this cycle by the cyclic sequence of its vertices, for example $C = \{p_1, p_2, \ldots, p_k, p_1\}$.

2 Proof of the main results

Let G be a properly edge colored graph with the length of maximum rainbow path equal to t. In this Section, C stands for the set of colors used in the proper edge coloring of the graph G. The following lemma ensures a rainbow path of length $\lceil \frac{\delta+1}{2} \rceil$ starting from any vertex in a properly edge colored graph G. Due to page restriction we omitted the proof.

Lemma 2.1. *Given any vertex x in G, there exists a rainbow path of length at least $\lceil \frac{\delta+1}{2} \rceil$ starting from x.*

The following lemma ensures that if the maximum length of a rainbow path is small enough, then we can convert the maximum rainbow path into a rainbow cycle by some simple modifications.

Lemma 2.2. *Let G be a properly edge colored graph with a rainbow path of maximum length, say t. If $t < \lfloor \frac{3}{5}\delta \rfloor$, then G contains a rainbow cycle of length $(t+1)$.*

Proof. Assume for contradiction that there is no rainbow cycle of length $t+1$ in G. Let $P = \{u_0(=x), u_1, u_2, \ldots, u_t(=y)\}$ be a rainbow path of length t in G. Let $U = \{color(u_i, u_{i+1}), 0 \le i \le t-1\}$ and $U^c = C \setminus U$, where C is the set of colors used to color the edges of G properly. Clearly $|U| = t$. Let $T_x = \{u_i \ : 0 \le i \le t,\ (x, u_i) \in E(G)$ and $color(x, u_i) \in U^c\}$ and let $T_y = \{u_i \ : \ 0 \le i \le t,\ (y, u_i) \in E(G)$ and $color(y, u_i) \in U^c\}$. First note that $|\{(x,z) \in E(G) : color(x,z) \in U^c\}| \ge \delta - t$. Moreover, if $(x,z) \in E(G)$ with $color(x,z) \in U^c$, then $z \in V(P)$, *i.e.*, $z = u_i$ for some $1 \le i \le t$, since otherwise we would have rainbow path of length $t+1$ in G. It follows that $|T_x| \ge \delta - t$. By a similar argument, we get $|T_y| \ge \delta - t$. Note that $u_0, u_1 \notin T_x$ since $u_0 = x$ and color$(x, u_1) \in U$. Also, $u_t \notin T_x$, since if (x, u_t) is an edge and is colored using a color from U^c, then we already have a $t+1$ length rainbow cycle, contrary to the assumption. So, we can write $T_x = \{u_i \ : 2 \le i \le t-1$ and $color(x, u_i) \in U^c\}$. By similar reasoning, we can write, $T_y = \{u_i \ : 1 \le i \le t-2$ and $color(y, u_i) \in U^c\}$. Define $M_x = \{u_j \ : \ u_{j+1} \in T_x\}$.

Observation 1. $|M_x| = |T_x| \ge \delta - t$.

Claim 1. $M_x \cap T_y \neq \emptyset$.

If possible suppose $M_x \cap T_y = \emptyset$. Now, $|M_x| + |T_y| \le t-1$, as both $M_x \subset V(P)$ and $T_y \subset V(P)$ and number of vertices on P excluding x and y is $t-1$. (Note that $x, y \notin M_x$ and $x, y \notin T_y$.) As $|M_x| \ge \delta - t$ and $|T_y| \ge \delta - t$ and $M_x \cap N_y = \emptyset$ by assumption, we have $\delta - t + \delta - t \le t+1$. That is, $2\delta \le 3t - 1$. So, $t \ge \frac{2\delta+1}{3}$. This is a contradiction to the fact that $t < \lfloor \frac{3}{5}\delta \rfloor$. Hence Claim 1 is true.

Claim 2. If $u_i \in M_x \cap T_y$, then $color(y, u_i) = color(x, u_{i+1})$.

If possible suppose Claim 2 is false. That is, $u_i \in M_x \cap T_y$ and $color(y, u_i) \neq color(x, u_{i+1})$.
Now consider the cycle: $CL = \{x, u_1, \ldots, u_i, y, u_t, u_{t-1}, \ldots, u_{i+1}, x\}$. Clearly CL is a rainbow cycle, as $color(y, u_i) \neq color(x, u_{i+1})$ and $color(y, u_i) \in U^c$ and $color(x, u_{i+1}) \in U^c$. Note that, the length of CL is $t + 1$, as we removed exactly one edge, namely (u_i, u_{i+1}) from P and added two new edges, namely (y, u_i) and (x, u_{i+1}) to CL. So, the length of CL is $t - 1 + 2 = t + 1$, contradiction to the assumption. Hence, we can infer that if $u_i \in M_x \cap T_y$, then $color(y, u_i) = color(x, u_{i+1})$. Let $S_y = \{v \in V(P) - M_x : color(y, v) \in U - \{color(y, u_{t-1})\}\}$.

Observation 2. $|M_x| + |T_y| + |S_y| - |M_x \cap T_y| \leq t - 1$.

Proof: This is because S_y is disjoint from $M_x \cup T_y$ and $S_y \cup M_x \cup T_y \subseteq V(P) - \{y, u_{t-1}\}$. (Note that $y = u_t$ and u_{t-1} do not appear in M_x, T_y or S_y.)

We partition the set $M_x \cap T_y$ as follows. Let $u_i \in M_x \cap T_y$. If $color(u_i, u_{i+1})$ appears in one of the edges incident on y, then $u_i \in A$ otherwise $u_i \in B$.

Observation 3. $|T_y| \geq \delta - t + |B|$.

Proof: To see this first note that there are at least δ edges incident on y and at most $t - |B|$ of them can get the colors from U, since $|B|$ colors in U do not appear on the edges incident on y, by the definition of B. So, at least $\delta - t + |B|$ of the edges incident on y have colors from U^c, and clearly any w, such that (y, w) is an edge, colored by a color in U^c has to be on P, since otherwise we have a longer rainbow path.

Claim 3. If $u_i \in A$, then the edge incident on y with color $color(u_i, u_{i+1})$ has its other end point on the rainbow path P. That is, if w is such that (y, w) is an edge and $color(u_i, u_{i+1}) = color(y, w)$, then $w \in V(P)$.

If possible suppose Claim 3 is false.
Let $(y, w) \in E(G)$ with $color(y, w) = color(u_i, u_{i+1})$ and $w \notin V(P)$. Now, consider the path: $P' = \{w, y, u_{t-1}, u_{t-2}, \ldots, u_{i+1}, x(= u_0), u_1, \ldots, u_i\}$. Clearly P' is a rainbow path as $color(u_i, u_{i+1}) = color(y, w)$, the edge $(u_i, u_{i+1}) \notin E(P')$ and color$(u_{i+1}, x) \in U^c$, since $u_i \in M_x$. Note that, the length of P' is $t + 1$. This is a contradiction to the fact that t is the maximum length rainbow path in G. Hence Claim 3 is true.

Now, partition A as follows: if $u_i \in A$, then by the above claim the edge incident on y with the color $color(u_i, u_{i+1})$ has its other end point say w, on P. If $w \in M_x$, then let $u_i \in A_1$, else $u_i \in A_2$.

Observation 4. $|M_X \cap T_y| = |A| + |B| = |A_1| + |A_2| + |B|$.

Observation 5. $|S_y| \geq |A_2|$. To see this, recall that $S_y = \{v \in V(P) - M_x : color(y, v) \in U - \{color(y, u_{t-1})\}\}$. By definition of A_2, for each $u_i \in A_2$ there exists a unique vertex $w = w(u_i) \in V(P) - M_x$ such that (y, w) is an edge and $color(y, w) = color(u_i, u_{i+1}) \in U$. Since $u_i \in A_2 \subset M_x$, $i < t - 1$ and thus color$(u_i, u_{i+1}) \neq color(y, u_{t-1})$. It follows that $\{w(u_i) : u_i \in A_2\} \subseteq S_y$, and therefore we have $|S_y| \geq |A_2|$.

Claim 4. $|A_1| \leq \frac{|M_x|}{2}$.

Recall that, for each $u_i \in A_1$, there is a unique vertex $w = w(u_i)$ such that (y, w) is an edge with $color(u_i, u_{i+1}) = color(y, w)$. Moreover, $w \in M_x$, by the definition of A_1 and $A_1 \cup \{w(u_i) : u_i \in A_1\} \subseteq M_x$. Note that $w(u_i)$ is uniquely defined for u_i since it is the end point of the edge incident on y colored with the color of the edge (u_i, u_{i+1}). Moreover, $A_1 \cap \{w(u_i) : u_i \in A_1\} = \emptyset$, since A_1 contains vertices which are end points of edges from y, colored by the colors in U^c whereas each $w(u_i)$ is the end point of some edge from y which is colored by a color in U. It follows that $2|A_1| \leq |M_x|$. That is, $|A_1 \leq \frac{|M_x|}{2}$, as required.

Now, substituting $\delta - t + |B|$ for $|T_y|$ (by Observation 3), $|A_2|$ for $|S_y|$ (by Observation 5), and $|A_1| + |A_2| + |B| = |M_x \cap T_y|$ (by Observation 4) in the inequality of Observation 2, and simplifying we get $|M_x| + \delta - t - |A_1| \leq t - 1$. Now using $|A_1| \leq |M_x|/2$ (Claim 4) and simplifying we get $\frac{|M_x|}{2} + \delta - t \leq t - 1$. Recall that $|M_x| \geq \delta - t$ (Observation 1). Substituting and simplifying we get, $t \geq \frac{3\delta+2}{5} \geq \left\lfloor \frac{3}{5}\delta \right\rfloor$, contradicting the initial assumption. Hence the Lemma is true. □

Theorem 2.3. *Let G be a properly edge colored graph with minimum degree δ. If t is the maximum length of a rainbow path in G, then* $t \geq \left\lfloor \frac{3\delta}{5} \right\rfloor$.

Proof. If possible suppose $t < \left\lfloor \frac{3\delta}{5} \right\rfloor$. By Lemma 2.2, G contains a rainbow cycle of length $t + 1$. Let CL be this cycle. Note that, CL contains $(t + 1)$ vertices and $(t + 1)$ edges. Now, $t + 1 \leq \lfloor \frac{3\delta}{5} \rfloor$. Let $CL = \{u_0, u_1, \ldots, u_t, u_0\}$ and $V(CL^c) = V(G) \backslash V(CL)$. Let $U = \{color(e) : e \in E(CL)\}$ and $U^c = C \setminus U$, where C is the set of colors used to color the edges of G properly. Let $F_i = \{z \in V(CL^c) : (u_i, z) \in E(G)\}$.

Claim 1. $|F_i| \geq \left\lceil \frac{2\delta}{5} \right\rceil$. Moreover, for $z \in F_i$, $color(u_i, z) \in U$.

First part follows from the fact that degree$(u_i) \geq \delta$ and there are at most $\left\lfloor \frac{3\delta}{5} \right\rfloor$ vertices in CL. If possible suppose $color(u_i, z) \in U^c$. Now consider the path $P' = \{z, u_i, u_{i+1}, u_{i+2}, \ldots, u_t, u_0, \ldots, u_{i-1}\}$. Clearly, P' is a rainbow path as $color(u_i, z) \in U^c$ and $\{u_i, u_{i+1}, u_{i+2}, \ldots, u_t, u_0, \ldots, u_{i-1}\}$ is already a rainbow path being a part of the rainbow cycle CL. Note that, the length of P' is $t + 1$. This is a contradiction to the assumption that t is the maximum length rainbow path in G. Hence Claim 1 is true.

Let $G' = (V', E')$, where $V'(G') = V(G)$ and $E'(G') = E(G) \setminus \{e \in E(G) : color(e) \in U\}$. Clearly, in G' there is no edge between $V(CL)$ to $V(CL^c)$, since by Claim 1, every such edge is colored by a color in U. Consider the induced subgraph of $V(CL^c)$ in G'. Let $G'' = G'[V(CL^c)]$. Let δ' be the minimum degree of G''.

Observation 1. $\delta' \geq \lceil \frac{2\delta}{5} \rceil$.

Proof: Clearly $\delta' \geq \delta - |U| = \delta - (t+1) \geq \delta - \lfloor \frac{3\delta}{5} \rfloor \geq \lceil \frac{2\delta}{5} \rceil$. Consider the following subset U_0 of U, defined by $U_0 = U_1 \cup U_2$, where $U_1 = \{color(u_i, u_{i+1}) : 0 \leq i \leq \lfloor \frac{\delta}{5} \rfloor - 1\}$ and $U_2 = \{color(u_i, u_{i+1}) : (t+1) - \lfloor \frac{\delta}{5} \rfloor \leq i \leq t-1\} \cup \{color(u_t, u_0)\}$.

Claim 2. $\{color(u_0, z) : z \in F_0\} \cap U_0 = \emptyset$.

Suppose not. Let $z \in F_0$ be such that $color(u_0, z) \in U_0$. Without loss of generality assume that $color(u_0, z) \in U_1$. Then consider the path P^* $= (u_{\lfloor \frac{\delta}{5} \rfloor}, \ldots, u_t, u_0, z)$, which is clearly a rainbow path, since the edge of CL with its color equal to $color(u_0, z)$ is not there in this path. Also the length of P^* is $t + 1 - \lfloor \frac{\delta}{5} \rfloor$. By Observation 1, G'' has minimum degree at least $\lceil \frac{2\delta}{5} \rceil$, and therefore by Lemma 2.1, G'' has a rainbow path of length at least $\lceil \frac{\delta}{5} \rceil$, let us call this path P''. Clearly concatenating the path P'' with P^* we get a rainbow path since colors used in P^* belong to U whereas the colors used in P'' belong to U^c. Moreover, the length of this rainbow path is at least $t+1$, a contradiction, to the assumption that t is the length of the maximum rainbow path in G.

Now we complete the proof as follows: In view of Claim 2, and Claim 1, we know that $|F_0| \leq |U - U_0|$. But $|U - U_0| \leq \lfloor \frac{3\delta}{5} \rfloor - \lceil \frac{2\delta}{5} \rceil \leq \lceil \frac{\delta}{5} \rceil < \lceil \frac{2\delta}{5} \rceil$ (since we can assume $\delta \geq 5$: for smaller values of δ, the Theorem is trivially true). This is a contradiction to the first part of Claim 1. □

References

[1] S. Akbari, O. Etesami, H. Mahini and M. Mahmoody, *On rainbow cycles in edge colored complete graphs*, Australasian Journal of Combinatorics **37** (2007), 33–42.

[2] H. Chen and X. Li, *Long heterochromatic paths in edge-colored graphs*, The electronic journal of combinatorics **12** (2005), # R 33.

[3] J. Diemunsch, M. Ferrara, Allan Lo, C. Moffatt, F. Pfender and P. S. Wenger, *Rainbow matchings of size $\delta(G)$ in properly edge-colored graphs*, The electronic journal of combinatorics **19** (2) (2012), P52.

[4] H. Gebauer and F. Mousset, *Rainbow cycles and paths*, CORR abs/1207.0840 (2012).

[5] A. GYÁRFÁS and M. MHALLA, *Rainbow and orthogonal paths in factorizations of K_n*, Journal of Cominatorial Designs **18** (3) (2010), 167–176.

[6] M. KANO and X. LI, *Monochromatic and heterochromatic subgraphs in edge-colored graphs - a survey*, Graphs Combin. **24** (2008), 237–263.

[7] M. MAAMOUN and H. MEYNIEL, *On a problem of G. Hahn about colored hamiltonian paths in K_{2^t}*, Discrete Mathematics **51** (2) (1984), 213–214.

b-coloring graphs with girth at least 8

Victor Campos[1], Carlos Lima[2] and Ana Silva[1]

Abstract. A b-coloring of a graph is a proper coloring of its vertices such that every color class contains a vertex that has neighbors in all other color classes. The b-chromatic number of a graph is the largest integer $b(G)$ such that the graph has a b-coloring with $b(G)$ colors. This metric is upper bounded by the largest integer $m(G)$ for which G has at least $m(G)$ vertices with degree at least $m(G)-1$. There are a number of results reporting that graphs with high girth have high b-chromatic number when compared to $m(G)$. Here, we prove that every graph with girth at least 8 has b-chromatic number at least $m(G)-1$. This proof is constructive and yields a polynomial time algorithm to find the b-chromatic number of G. Furthermore, we improve known partial results related to reducing the girth requirement of our proof.

1 Introduction

Let G be a simple graph and consider the traditional definitions of proper coloring and chromatic number. Suppose that we have a proper coloring of G and there exists a color c such that every vertex v with color c is not adjacent to at least one other color (which may depend on v); then we can change the color of these vertices and thus obtain a proper coloring with fewer colors. This heuristic can be applied iteratively, but we cannot expect to reach the chromatic number of G, since the coloring problem is $\mathcal{NP}$-hard. On the basis of this idea, Irving and Manlove introduced the notion of b-coloring [6]. Intuitively, a b-coloring is a proper coloring that cannot be improved by the above heuristic, and the b-chromatic number measures the worst possible such coloring. More formally, consider a proper coloring ψ of G. A vertex u is said to be a *b-vertex* in ψ if u has a neighbor in each color class different from its own. A *b-coloring* of G

[1] ParGO, Universidade Federal do Ceará, Brazil. Email: campos@lia.ufc.br, gclima@cos.ufrj.com.br

[2] COPPE, Universidade Federal do Rio de Janeiro, Brazil. Email: anasilva@mat.ufc.br

Partially supported by FUNCAP, CAPES and CNPq - Brasil

is a proper coloring of G such that each color class contains a b-vertex. A *basis* of a b-coloring ψ is a subset of b-vertices of ψ containing one b-vertex of each color class. The *b-chromatic number* of G is the largest integer $b(G)$ for which G has a b-coloring with $b(G)$ colors. Computing the b-chromatic number of a graph G is $\mathcal{NP}$-hard [6], even if G is bipartite [7] or is a chordal graph [5].

Naturally, a proper coloring of G with $\chi(G)$ colors is a b-coloring of G; hence, $\chi(G) \leq b(G)$. For an upper bound, let $m(G)$ be the largest integer such that G has at least $m(G)$ vertices with degree at least $m(G)-1$. Then, $b(G) \leq m(G)$ [6]. This upper bound is called the m-degree of G.

The difference between $b(G)$ and $m(G)$ can be arbitrarily large. For example, the complete bipartite graph $K_{n,n}$ has $b(G) = 2$ and $m(G) = n+1$. Let the *girth* of G, denoted by $g(G)$, be the length of a shortest cycle of G. Even though deciding if $b(G) = m(G)$ is $\mathcal{NP}$-hard [7], there seems to exist a relation between the girth of G and how close $b(G)$ is to $m(G)$. In fact, the following two conjectures have received some attention.

Conjecture 1.1 ([1]). If G is a d-regular graph with girth at least 5 and G is not the Petersen graph, then $b(G) = d+1$.

Conjecture 1.2 ([5]). If $G = (A, B)$ is a C_4-free bipartite graph such that $|A| = m$, $d(u) = m-1$ for all $u \in A$ and $d(v) < m-1$ for all $v \in B$, then $b(G) \geq m-1$.

Note that the d-regular graph in Conjecture 1.1 has girth at least 5 and m-degree $d+1$, while the bipartite graph in Conjecture 1.2 has girth at least 6 and m-degree m.

Many partial results have been given for Conjecture 1.1 ([1, 2, 7, 8, 12]). In particular, Shaebani [12] gives a lower bound of $\left\lfloor \frac{d+3}{2} \right\rfloor$ for the b-chromatic number of a d-regular graph with girth at least 5. Concerning Conjecture 1.2, Lin and Chang [9] proved that the conjecture holds if the famous Erdős-Faber-Lovász Conjecture [4] also holds.

Motivated by these conjectures and by Silva's observation that Irving and Manlove's result on trees actually holds for any graph with girth at least 11 [11], we pose the following question, where g, m, b stands for girth, m-degree and b-chromatic number.

Question g, m, b. *What is the minimum value g^* for which $g(G) \geq g^*$ implies $b(G) \geq m(G) - 1$?*

Observe that the complete bipartite graph implies $g^* \geq 5$. We mention that the previous best bound on g^* is obtained by Campos, Farias and Silva [3] where they prove $g^* \leq 9$. The main result in this paper is to prove that g^* is either 5, 6, 7 or 8. The proof also yields a polynomial time algorithm to find an optimal b-coloring of a graph G with $g(G) \geq 8$.

Theorem 1.3. *If G is a graph with girth at least 8, then $b(G) \geq m(G) - 1$.*

Improving the best known upper bound for g^* has a few consequences for other studied problems. In fact, showing that $g^* \leq 6$ proves Conjecture 1.2. Furthermore, showing that $g^* \leq 5$ improves the lower bound obtained by Shaebani [12] from $\lfloor \frac{d+3}{2} \rfloor$ to d.

Let $C_3 \square C_3$ denote the Cartesian product of two cycles of length 3. In [10], Maffray and Silva make the following conjecture.

Conjecture 1.4. Let G be any graph with no $K_{2,3}$ as subgraph. If $G \neq C_3 \square C_3$, then $b(G) \geq m(G) - 1$.

Although this conjecture is false due to a technical error as the graph $C_3 \square C_3$ together with isolated vertices is a counter-example, we believe Conjecture 1.4 might be true with a few extra constraints. We propose a reformulation of this conjecture as follows.

Conjecture 1.5. Let G be any graph with no $K_{2,3}$ as subgraph. Then $b(G) \geq m(G) - 1$ unless G contains a component isomorphic to $C_3 \square C_3$ and all other components have maximum degree at most 2.

Observe that Conjecture 1.5 implies $g^* \leq 5$. In fact, if $g(G) \geq 5$, then G contains neither a $K_{2,3}$ nor a $C_3 \square C_3$ as subgraphs.

Let $\mathcal{K}_m$ denote the class of graphs obtained from m copies of K_m where any two copies of K_m intersect in at most one vertex. The famous Erdős-Faber-Lovász Conjecture [4] can be stated as follows.

EFL Conjecture. *If $G \in \mathcal{K}_m$, then $\chi(G) = m$.*

Let $H \in \mathcal{K}_m$. Note that each copy of K_m in H contains at least one vertex not contained in any other copy of K_m. Let A be a set of m vertices obtained by picking one vertex from each copy of K_m with this property. Let G be the graph obtained from H by deleting all edges with no endpoint in A and let $B = V(G) \setminus A$. Note that G is bipartite and the bipartition (A, B) satisfies the properties of Conjecture 1.2 with $m(G) = m$ as long as no vertex in B has degree at least $m - 1$. Furthermore, if $b(G) = m$ with A as a basis, then $\chi(H) = m$ and EFL is true for H. Therefore, if $g^* \leq 6$, characterizing the graphs G with $g(G) \geq 6$ and $b(G) = m(G) - 1$ determines the possible counter-examples for the EFL Conjecture. We find this to be a strong partial result which could lead to a proof of EFL.

2 Improving tool lemmas

We say that a vertex $u \in V(G)$ is *dense* if $d(u) \geq m(G) - 1$ and we denote the set of dense vertices of G by $D(G)$. Let W be a subset of $D(G)$ and u be any vertex in $V(G) \setminus W$. If all vertices in W are either adjacent to u or have a common neighbor $v \in W$ with u with $d(v) = m(G) - 1$, then we say that W *encircles* u.A subset W of $D(G)$ of size $m(G)$ is a *good set* if:
(a) W does not encircle any vertex, and
(b) Every vertex $x \in V(G) \setminus W$ with $d(x) \geq m(G)$ is adjacent to W.

Lemma 2.1 ([6]). *Let G be any graph and W be a subset of $D(G)$ with $m(G)$ vertices. If W encircles some vertex $v \in V(G) \setminus W$, then W is not a basis of a b-coloring with $m(G)$ colors.*

Lemma 2.2 ([11]). *If G is a graph with $g(G) \geq 8$, then G does not have a good set if and only if $|D(G)| = m(G)$ and $D(G)$ encircles a vertex. Also, a good set can be found in polynomial time, if one exists.*

Lemma 2.3 ([11]). *Let G be a graph with girth at least* 8. *If G has no good set, then $b(G) = m(G) - 1$.*

First, we note that Lemma 2.1 states that a basis for a b-coloring with $m(G)$ colors cannot encircle any vertex. Therefore, a good set is a possible basis for a b-coloring with $m(G)$ colors. Indeed, Theorem 1.3 easily follows from Lemmas 2.2, 2.3 and the one below.

Lemma 2.4. *If G has girth at least 8 and W is a good set of G, then there exists a b-coloring ψ of G with $m(G)$ colors having W as basis.*

Using the same framework for the proof of Theorem 1.3, we cannot improve this result unless we improve the auxiliary results in Lemmas 2.2 and 2.3. This is done next.

Lemma 2.5. *Let G be a graph with $g(G) \geq 6$. Then G does not have a good set if, and only if, one of the following holds (below $m = m(G)$):*

1. *$|D(G)| = m$ and $D(G)$ encircles a vertex in $V(G) \setminus D(G)$; or*
2. *$g(G) = 6$, $|D(G)| = m+1$, $d(v) = m-1$ for all $v \in D(G)$ and $D(G)$ induces a matching in G. Furthermore, $D(G)$ can be partitioned into stables sets X_0 and X_1 such that $|X_0| = |X_1|$ and there are vertices $u_0, u_1 \in V(G) \setminus D(G)$ such that u_i is adjacent to all vertices in X_i and to no vertices in X_{1-i}, for $i \in \{0, 1\}$.*

Moreover, a good set can be found in polynomial time, if one exists.

Lemma 2.6. *Let G be a graph with girth at least 7. If G has no good set, then* $b(G) = m(G) - 1$.

Lemma 2.7. *If G is a graph satisfyting property (2) of Lemma* 2.5, *then* $b(G) = m(G) - 1$.

3 Comments and open questions

We believe Question g, m, b to be an interesting problem and relate this question to other studied conjectures in the literature in Section 1. Furthermore we show that proving $g^* \leq 6$ proves Conjecture 1.2 and, at the end of Section 1, we point out that this could be an important step in proving the famous Erdős-Faber-Lovász Conjecture. In Section 1 we also restate a conjecture by Silva and Maffray [10], which, if true, implies that $g^* = 5$.

Then, we prove that $g^* \leq 8$ and improve the partial results used in this proof to try to build a stepping stone for proving an upper bound of 6 or 7 for g^* in Section 2. We mention that an equivalent of Lemma 2.4 cannot be obtained for graphs of girth at most 6 [11].

Recall that the EFL Conjecture implies Conjecture 1.2 [9]. We also feel it would be interesting to prove a strengthening of this result. Can we prove $g^* \leq 6$ if we assume that the EFL Conjecture is true?

References

[1] M. Blidia, F. Maffray and Z. Zemir, *On b-colorings in regular graphs Discrete Applied Mathematics* **157** (8) (2009), 1787–1793.

[2] S. Cabello and M. Jakovac, *On the b-chromatic number of regular graphs*, Discrete Applied Mathematics **159** (2011), 1303–1310.

[3] V. Campos, V. Farias and A. Silva, *b-Coloring graphs with large girth*, J. of the Brazilian Computer Society **18** (4) (2012), 375–378.

[4] P. Erdős, *On the combinatorial problems which I would most like to see solved*, Combinatorica **1** (1981), 25–42.

[5] F. Havet, C. Linhares and L. Sampaio, *b-coloring of tight graphs*, Discrete Applied Mathematics **160** (18) (2012), 2709–2715.

[6] R. W. Irving and D. F. Manlove, *The b-chromatic number of a graph*, Discrete Appl. Math. **91** (1999), 127–141.

[7] J. Kratochvíl, Zs. Tuza and M. Voigt, *On the b-chromatic number of graphs*, Lecture Notes In Computer Science **2573** (2002), 310–320.

[8] M. KOUIDER and A. E. SAHILI, *About b-colouring of regular graphs*, Technical Report 1432, Université Paris Sud, 2006.

[9] W-H. LIN and G. J. CHANG, *b-coloring of tight bipartite graphs and the Erdős-Faber-Lovász Conjecture*, Disc. Appl. Math. to appear.

[10] F. MAFFRAY and A. SILVA, *b-colouring the Cartesian product of trees and some other graphs*, Disc. Appl. Math. **161** (2013), 650–669.

[11] A. SILVA, "The b-chromatic Number of Some Tree-like Graphs", PhD Thesis, Université de Grenoble, 2010.

[12] S. SHAEBANI, *On the b-chromatic number of regular graphs without 4-cycle*, Discrete Applied Mathematics **160** (2012), 1610–1614.

The circular chromatic index of k-regular graphs

Barbora Candráková[1] and Edita Máčajová[1]

Abstract. The circular chromatic index of a graph G is the infimum of all rational numbers p/q, such that there exists a circular p/q-edge-coloring of the graph G. A natural problem is to determine the set of possible values of the circular chromatic indices of k-regular graphs. In this work we construct k-regular graphs with certain circular chromatic indices of the form $k + a/p$, in particular graphs for all $a \in \{1, 2, \ldots, \lfloor k/2 \rfloor\}$ and $p \geq 2a^2 + a + 1$.

1 Introduction

Graphs considered in this paper are finite. The notion graph always refers to a simple graph. If parallel edges are allowed, the object in question is called a multigraph.

For a real number $r \geq 1$ we define a *circular r-edge-coloring* of a graph G as a mapping $c\colon E(G) \to [0, r)$ satisfying $1 \leq |c(e) - c(f)| \leq r - 1$ for any pair of adjacent edges e and f of G. The *circular chromatic index* of the graph G, denoted by $\chi'_c(G)$, is the infimum of the set of all real numbers r such that there exists a circular r-edge-coloring of G.

Which real numbers are the circular chromatic indices of k-regular graphs? Since $\chi'(G) = \lceil \chi'_c(G) \rceil$, the Vizing's Theorem implies that $k \leq \chi'_c(G) \leq k + 1$ for any k-regular graph G. Circular edge-colorings are especially interesting for k-regular graphs with the chromatic index $k + 1$, so called k-regular class 2 graphs. While circular edge-colorings of cubic graphs have been extensively studied (e.g. [1–6]), very little is known about circular edge-colorings of k-regular graphs with $k \geq 4$. Re-

[1] Department of Computer Science, Faculty of Mathematics, Physics and Informatics, Comenius University, 842 48 Bratislava, Slovakia.
Email: candrakova@dcs.fmph.uniba.sk, macajova@dcs.fmph.uniba.sk

cently, Lukoťka and Mazák [5] have shown that for each given rational number r, such that $3 < r < 3 + 1/3$, there exists an infinite family of cubic graphs with the circular chromatic index r. On the other hand, by [1], the circular chromatic number of a bridgeless cubic graph is at most $3 + 2/3$. There are known cubic graphs with the circular chromatic index r for infinitely many values r, such that $3+1/3 < r < 3+2/3$ [5], but the global situation in this interval is still open.

It has been conjectured [1] that for any $k \geq 2$ and an $\varepsilon_k > 0$ there exists no graph with $k-\varepsilon_k < \chi'_c < k$. The values of the circular chromatic index we present in this work belong to the interval $(k, k + 1/4)$.

2 k-Regular graphs with $\chi'_c = k + a/p$

In this section we provide a construction of k-regular graphs with the circular chromatic index equal to $k + a/p$ for all integers $k \geq 4, a \in \{1, 2, \ldots, \lfloor k/2 \rfloor\}$ and $p = (2a + 1)m + an$ such that $m, n \geq 1$.

The colors in a circular r-coloring will be taken modulo r. Assume that a and b are two colors in a circular r-coloring. The distance of a and b is denoted by $|a-b|_r$ and defined as $|a-b|_r = \min\{|a-b|, r-|a-b|\}$.

Let us consider a circular r-edge-coloring. For any $a, b \in [0, r)$, the *r-circular interval* $[a, b]_r$ is defined as follows:

$$[a, b]_r = \begin{cases} [a, b] & (a \leq b), \\ [a, r) \cup [0, b] & (a > b). \end{cases}$$

Although our construction aims towards simple graphs, we start with multigraphs. Let $M_{k,a,m}$ be a multigraph with vertex set $\{u_1, u_2, \ldots, u_m, v_1, v_2, \ldots, v_m, w\}$. The vertex w is also denoted as v_0 or u_{m+1}. The edge set of $M_{k,a,m}$ consists of $k - a$ parallel $u_i v_i$-edges denoted by $e_{i,1}, e_{i,2}, \ldots, e_{i,k-a}$ for each $i \in \{1, 2, \ldots, m\}$ and of a parallel $v_i u_{i+1}$-edges denoted by $f_{i,1}, f_{i,2}, \ldots, f_{i,a}$ for each $i \in \{0, 1, \ldots, m\}$, see Figure 2.1.

We create the graph $G_{k,a,m,n}$ from the multigraph $M_{k,a,m}$ by inserting so called ε_k-block between vertices to avoid parallel edges. The ε_k-block contains two dangling edges and is k-regular and k-edge-colorable. To construct this block we take the union of $k - 1$ perfect matchings of the complete graph on $2k-4$ vertices, which is a $(k-1)$-regular graph that is $(k-1)$-edge-colorable. We add one dangling edge to each vertex and we denote the resulting graph by G. Let w_1 and w_2 be two vertices disjoint from G joined by an edge. To create ε_k-block, we adjoin half of the dangling edges in G to the vertex w_1, the other half to the vertex w_2, and add a dangling edge, called the *input edge*, to w_1 and a dangling edge, called the *output edge*, to w_2, see Figure 2.2. Moreover, as we show in

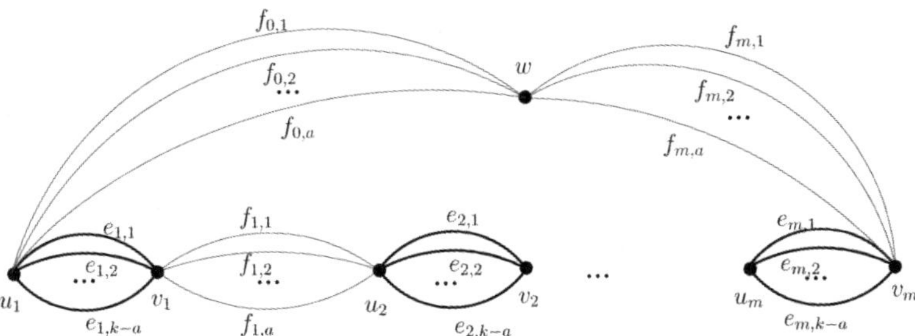

Figure 2.1. The multigraph $M_{k,a,m}$.

the forthcoming paper, it is possible to divide the dangling edges of G into those that will be incident with w_1 or with w_2 in such a way that the resulting block is k-edge-colorable. We will write ε-block instead of ε_k-block when k is fixed.

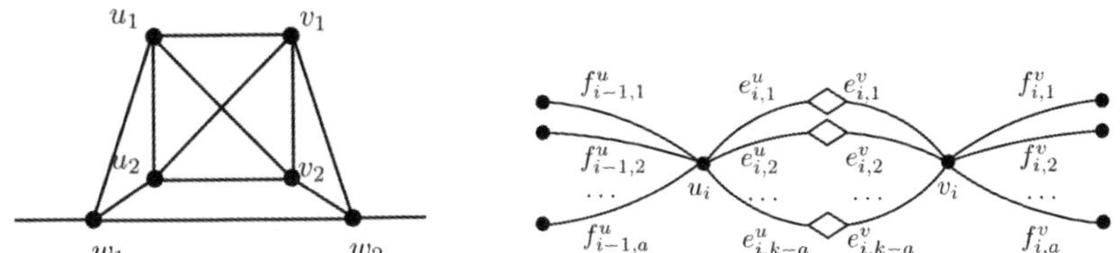

Figure 2.2. The ε_4-block.

Figure 2.3. The subgraph H_i.

The ε-block is inserted into every edge in the multigraph $M_{k,a,m}$ once, except for the edges $f_{m,1}, f_{m,2}, \ldots, f_{m,a}$ where n copies of the ε-block are inserted into each edge. This way we get the graph $G_{k,a,m,n}$ that contains no parallel edges at all. The multigraph $M_{k,a,m}$ contains edges $e_{i,j}$ for all $i \in \{1, 2, \ldots, m\}$ and $j \in \{1, 2, \ldots, k-a\}$ and $f_{b,c}$ for all $b \in \{0, 1, \ldots, m\}$ and $c \in \{1, 2, \ldots, a\}$. In the graph $G_{k,a,m,n}$, we denote the *e-edges* incident with the vertex u_i as $e^u_{i,j}$ and with the vertex v_i as $e^v_{i,j}$. Similarly, the *f-edges* incident with u_i are denoted as $f^u_{i-1,j}$ and incident with v_i are $f^v_{i,j}$. The edges that are incident with ε-blocks inserted into the edge $f_{m,l}$ are denoted by $f^1_{m,l}, f^2_{m,l}, \ldots, f^{n+1}_{m,l}$ in the order as they occur when proceeding form v_m to w.

Let H_i be the subgraph of $G_{k,a,m,n}$ induced by the vertices u_i, v_i, by all ε-blocks lying on $e_{i,j}$ edges for $j \in \{1, 2, \ldots, k-a\}$, as well as by all the vertices incident with the edges u_i and v_i for $i \in \{1, 2, \ldots, m\}$, see Figure 2.3. Let S be the subgraph induced by the vertices v_m, w, and by all ε-blocks lying on the $f_{m,c}$ edges for $c \in \{1, 2, \ldots, a\}$.

Let e_1 be the input edge and e_2 be the output edge of the ε_k-block. Although the edges e_1 and e_2 receive the same color in any k-edge-coloring of ε_k-block, they can slightly differ in a circular $(k+\varepsilon)$-edge-coloring φ. We define the *color change in the ε-block* as $|\varphi(e_1) - \varphi(e_2)|_{k+\varepsilon}$.

Lemma 2.1. *Let $k \geq 4$ and let $\varepsilon < 1/3$. The color change in the ε_k-block is at most ε in any $(k+\varepsilon)$-edge-coloring. Moreover, there exists a $(k+\varepsilon)$-edge-coloring of the ε_k-block with the color change ε.*

Theorem 2.2. *For each integer $k \geq 4$, $a \in \{1, 2, \ldots, \lfloor k/2 \rfloor\}$, and $m, n \geq 1$ there exists a k-regular graph with circular chromatic index $k + \frac{a}{(2a+1)m+an}$.*

Proof. Let k, a, m, n be integers that fulfill conditions of the theorem. Let $p = (2a+1)m + an$, $\varepsilon = a/p$, and $r = k + \varepsilon$. First we show that $\chi'_c(G_{k,a,m,n}) = k + a/p$. Then we can construct infinitely many k-regular graphs, which contain $G_{k,a,m,n}$ as a subgraph, with $\chi'_c = k + a/p$ [5].

It is possible to construct a circular $(k + a/p)$-edge-coloring ψ of $G_{k,a,m,n}$ where $p = (2a+1)m + an$, which gives us the upper bound on the circular chromatic index. The ε-blocks are colored in accordance with Lemma 2.1, hence their maximal color change is used as shown in our forthcoming paper.

We now derive a lower bound on the circular chromatic index of the graph $G_{k,a,m,n}$. Let us consider the subgraph H_i of the graph $G_{k,a,m,n}$. It can be shown that in any circular $(k+\varepsilon)$-edge-coloring φ the f-edges of H_i can be arranged into pairs in such a way that the colors of the edges in every pair differ by at most 2ε. Without loss of generality we will assume that the parallel edges with the same second subscript are the ones that differ in color by at most 2ε. More precisely, we will assume that the edges are denoted in such a way that $|\varphi(f^u_{i-1,g}) - \varphi(f^v_{i,g})| \leq 2\varepsilon$ for every $g \in \{1, 2, \ldots, a\}$.

We say that the edges $f^x_{0,t}, f^x_{1,t}, \ldots, f^x_{m-1,t}, f^1_{m,t}, f^2_{m,t}, \ldots, f^{n+1}_{m,t}$, for $x \in \{u, v\}$ and $t \in \{1, 2, \ldots, a\}$ belong to the same *coloring line* in the graph $G_{k,a,m,n}$.

We say that the edges $f^u_{0,c}, f^u_{0,c+1}, \ldots, f^u_{0,d}$ form a *coloring block* in φ, if $\varphi(f^u_{0,c+q}) \in [\varphi(f^u_{0,c}) + q - \varepsilon, \varphi(f^u_{0,c}) + q + \varepsilon]$, for $c, d \in \{1, 2, \ldots, a\}$, and $q \in \{1, 2, \ldots, d-c\}$ and this set cannot be extended. Roughly speaking, the differences of colors of these edges from $\varphi(f^u_{0,1})$ form a non-extendable sequence of numbers that are near to consecutive integers. We assume that all edges $f^u_{0,l}$ in the coloring φ are divided into r coloring blocks of sizes $n_1, n_2, \ldots, n_r$ such that $\sum n_i = a$. Let $n_{max} = \max_{i \in \{1,2,\ldots,r\}} \{n_i\}$. It can be shown that it is sufficient to consider only the case when $n_{max} = 1$ as the established upper bound is exceeded for greater values of n_{max}. The notion of the coloring block is defined analogously for all subgraphs H_i and the number and sizes of the coloring blocks are the same for all $i \in \{1, 2, \ldots, m\}$.

We define a *local change* in the subgraph H_i as $\sum_{j=1}^{a} |\varphi(f^v_{i,j}) - \varphi(f^u_{i-1,j})|_r$.

We bound the local change by showing that it is at most $(a+1)\varepsilon$.

Let φ be an $(k+\varepsilon)$-edge-coloring of the subgraph H_i. We say that the q-th coloring line is *increasing* in H_i if $\varphi(f^v_{i,q}) \in [\varphi(f^u_{i-1,q}), \varphi(f^u_{i-1,q}) + \varepsilon]$ and *decreasing* otherwise.

If there exist both increasing and decreasing coloring lines in H_i then it can be shown that the local change is at most $a\varepsilon$. More interesting situation happens with the case when all the coloring lines are either increasing or decreasing. We may assume that all coloring lines are increasing and $\varphi(e^u_{i,j}) \le \varphi(e^v_{i,j})$ for every $j \in \{1, 2, \ldots, k-a\}$.

Let $\varphi(e^u_{i,1}) = 0$ and $\varphi(f^u_{i-1,1}) < \varphi(f^u_{i-1,2}) < \cdots < \varphi(f^u_{i-1,a})$. Then there exist integers $P_1, P_2, \ldots, P_a$ from $\{1, 2, \ldots, k-1\}$ and real numbers $p_1, p_2, \ldots, p_a$ such that $\varphi(f^u_{i-1,j}) = P_j + p_j$ for $j \in \{1, 2, \ldots, a\}$ and $0 \le p_j \le \varepsilon$. Since each coloring block is of size 1, there exist integers $Q_1, Q_2, \ldots, Q_{a-1}$ and real numbers $q_1, q_2, \ldots, q_{a-1}$ such that, for $j \in \{1, 2, \ldots, a-1\}$ and $0 \le q_j \le \varepsilon$, $P_j + p_j \le Q_j + q_j \le P_{j+1} + p_{j+1}$, $0 \le p_1 \le q_1 \le p_2 \le q_2 \le \cdots \le p_{a-1} \le q_{a-1} \le p_a$ where $Q_j + q_j$ is the color of $e^u_{i,s}$ for some s. (For the sake of convenience we set $Q_0 = q_0 = 0$ and $\varphi(e^u_{i,1}) = Q_0 + q_0 = 0$). Lemma 2.1 implies that $Q_j + q'_j$ is the color of $e^v_{i,s}$ for some s where $0 \le q'_j - q_j \le \varepsilon$. Finally, we have $\varphi(f^v_{i,j}) = P_j + p'_j$ for real numbers p'_j such that $Q_{j-1} + q'_{j-1} \le P_j + p'_j$ for $j \in \{1, 2, \ldots, a\}$ and $P_j + p'_j \le Q_j + q'_j$ for $j \in \{1, 2, \ldots, a-1\}$ and $0 \le p'_1 \le q'_1 \le p'_2 \le q'_2 \le \cdots \le p'_{a-1} \le q'_{a-1} \le p'_a \le q'_0 + \varepsilon$.

Clearly, the local change is $\sum_{j=1}^{a} p'_j - p_j$ and can be bounded as follows.

$\sum_{j=1}^{a}(p'_j - p_j) = \sum_{j=1}^{a-1}(p'_j - p_j) + p'_a - p_a \le \sum_{j=1}^{a-1}(q'_j - q_{j-1}) + p'_a - p_a \le (a-2)\varepsilon + (q'_{a-1} - q_0) + (p'_a - q_{a-1}) \le (a-1)\varepsilon + (p'_a - q'_0) + (q'_0 - q_0) \le (a-1)\varepsilon + \varepsilon + \varepsilon = (a+1)\varepsilon$.

There are a coloring lines in the graph $G_{k,a,m,n}$ and we need to change the color by at least 1 on each of them. We can obtain the change of at most $(a+1)\varepsilon$ in each H_i. The number of ε-blocks outside H_i subgraphs is $(m+n)a$, each with the change of at most ε. Summing up we have $m(a+1)\varepsilon + an\varepsilon + an\varepsilon \ge a$ and $\varepsilon \ge \frac{a}{m(2a+1)+an}$.

Since the lower and the upper bound coincide, the theorem follows. □

Corollary 2.3. *For each integer $k \ge 3$, $a \in \{1, 2, \ldots, \lfloor k/2 \rfloor\}$, and $p \ge 2a^2 + a + 1$ there exists a k-regular graph with circular chromatic index $k + a/p$.*

References

[1] P. Afshani, M. Ghandehari, M. Ghandehari, H. Hatami, R. Tusserkani and X. Zhu, *Circular chromatic index of graphs of maximum degree 3*, J. Graph Theory **49** (4) (2005), 325–335.

[2] M. Ghebleh, *Circular chromatic index of generalized blanuša snarks*, Electron. J. Combin. **13** (2006).

[3] M. Ghebleh, D. Král, S. Norine and R. Thomas, *The circular chromatic index of flower snarks*, Electron. J. Combin. **13** (2006).

[4] T. Kaiser, D. Král and R. Škrekovski, *A revival of the girth conjecture*, J. Combin. Theory Ser. B **92** (1) (2004), 41–53.

[5] R. Lukoťka and J. Mazák, *Cubic graphs with given circular chromatic index*, SIAM J. Discrete Math. **24** (2010), 1091–1103.

[6] J. Mazák, *Circular chromatic index of type 1 blanusa snarks*, J. Graph Theory **59** (2) (2008), 89–96.

Coloring d-Embeddable k-Uniform Hypergraphs

Carl Georg Heise[1], Konstantinos Panagiotou[2], Oleg Pikhurko[3] and Anusch Taraz[4]

Abstract. We extend the scenario of the Four Color Theorem in the following way. Let $\mathcal{H}_{d,k}$ be the set of all k-uniform hypergraphs that can be linearly embedded into $\mathbb{R}^d$. We investigate lower and upper bounds on the maximum (weak and strong) chromatic number of hypergraphs in $\mathcal{H}_{d,k}$. For example, we can prove that for $d \geq 3$ there are hypergraphs in $\mathcal{H}_{2d-3,d}$ on n vertices whose weak chromatic number is $\Omega(\log n/\log\log n)$, whereas the weak chromatic number for n-vertex hypergraphs in $\mathcal{H}_{d,d}$ is bounded by $\mathcal{O}(n^{(d-2)/(d-1)})$ for $d \geq 3$.

Extended Abstract

The Four Color Theorem [1, 2] has been one of the driving forces in Discrete Mathematics and its theme has inspired many variations. For example, the chromatic number of graphs that are embedabble into a surface of fixed genus has been intensively studied by Heawood [7], Ringel and Youngs [11], and many others.

Here, we consider k-uniform hypergraphs that are embeddable into $\mathbb{R}^d$ in such a way that their edges do not intersect (see Definition 1 below). For $k = d = 2$ the problem specializes to graph planarity. For $k = 2$ and $d \geq 3$ it is not a very interesting question because for any $n \in \mathbb{N}$

[1] Zentrum Mathematik, Technische Universität München, Germany. Email: cgh@ma.tum.de. The author was partially supported by the ENB graduate program TopMath and DFG grant GR 993/10-1. He gratefully acknowledges the support of the TUM Graduate School's Thematic Graduate Center TopMath at the Technische Universität München.

[2] Mathematisches Institut, Ludwig-Maximilians-Universität München, Germany.
Email: kpanagio@math.lmu.de. The author was partially supported by the National Science Foundation, Grant DMS-1100215, and the Alexander von Humboldt Foundation.

[3] University of Warwick, Coventry, UK. Email: o.pikhurko@warwick.ac.uk. The author was partially supported by the National Science Foundation, Grant DMS-1100215, and the Alexander von Humboldt Foundation.

[4] Zentrum Mathematik, Technische Universität München, Germany. Email: taraz@ma.tum.de. The author was partially supported by DFG grant TA 309/2-2.

the vertices of the complete graph K_n can be embedded into $\mathbb{R}^3$ using arbitrary points on the moment curve $t \mapsto (t, t^2, t^3)$.

As a consequence, we focus our attention on hypergraphs, which are in general not embeddable into any specific dimension. Some properties of these hypergraphs (or more generally simplicial complexes) have been investigated (see *e.g.* [4, 8, 9, 15]), but to our surprise, we have not been able to find any bounds on their (vertex-)chromatic number. However, Grünbaum and Sarkaria (see [6, 12]) have differently generalized the concept of graph colorings to simplicial complexes by coloring faces. They also bound this face-chromatic number subject to embeddability constraints.

We now quickly recall and introduce some useful notation. The pair $H = (V, E)$ is a *k-uniform hypergraph* if the vertex set V is a finite set and the edge set E consists of k-element subsets of V.

Let H be a k-uniform hypergraph. A function $\kappa : V(H) \to \{1, \ldots, c\}$ is said to be a *strong c-coloring* if for all $e \in E(H)$ the property $|\kappa(e)| = k$ holds. The function κ is said to be a *weak c-coloring* if $|\kappa(e)| > 1$ for all $e \in E(H)$. The *strong and weak chromatic number* of H is denoted by $\chi^{\mathrm{s}}(H)$ and $\chi^{\mathrm{w}}(H)$, respectively. For graphs, weak and strong colorings are equivalent.

We next define what we mean when we say that a hypergraph is embedabble into $\mathbb{R}^d$. Here, *aff* denotes the affine hull of a set of points and *conv* the convex hull.

Definition 1 (d-embeddings). Let H be a k-uniform hypergraph and $d \in \mathbb{N}$. A (linear) *embedding* of H into $\mathbb{R}^d$ is a function $\varphi : V(H) \to \mathbb{R}^d$, where $\varphi(A)$ for $A \subseteq V(H)$ is to be interpreted pointwise, such that $\dim \mathrm{aff}\varphi(e) = k-1$ for all $e \in E(H)$ and $\mathrm{conv}\varphi\,(e_1 \cap e_2) = \mathrm{conv}\varphi(e_1) \cap \mathrm{conv}\varphi(e_2)$ for all $e_1, e_2 \in E(H)$.

A k-uniform hypergraph H is said to be *d-embeddable* if there exists an embedding of H into $\mathbb{R}^d$. Also, we denote by $\mathcal{H}_{d,k}$ the set of all d-embeddable k-uniform hypergraphs. By Fáry's theorem (see [5]), we have that the $k = d = 2$ case of this notion of embeddability coincides with the classical concept of planarity.

Our main results are summarized in the Tables 1 and 2, which contain upper or lower bounds for the maximum weak chromatic number of a d-embeddable k-uniform hypergraph on n vertices. All results which only follow non-trivially from prior knowledge are indexed with a number of the theorem in this extended abstract. On the other hand, the trivial entries in the tables are direct consequences of the Menger-Nöbeling Theorem (see [9, page 295] and [10]) which characterizes for which d all k-uniform hypergraphs are d-embeddable. The main results for the maximum strong chromatic number are Proposition 1 and Theorem 2.

$d \diagdown k$	3	4	5	6	7
1	1	1	1	1	1
2	2	1	1	1	1
3	$\Omega\left(\frac{\log n}{\log\log n}\right)_{\langle 5\rangle}$	1	1	1	1
4	$\Omega\left(\frac{\log n}{\log\log n}\right)_{\langle 5\rangle}$	1	1	1	1
5	$\lceil n/2\rceil$	$\Omega\left(\frac{\log n}{\log\log n}\right)_{\langle 6\rangle}$	1	1	1
6	$\lceil n/2\rceil$	$\Omega\left(\frac{\log n}{\log\log n}\right)_{\langle 6\rangle}$	1	1	1
7	$\lceil n/2\rceil$	$\lceil n/3\rceil$	$\Omega\left(\frac{\log n}{\log\log n}\right)_{\langle 6\rangle}$	1	1
8	$\lceil n/2\rceil$	$\lceil n/3\rceil$	$\Omega\left(\frac{\log n}{\log\log n}\right)_{\langle 6\rangle}$	1	1

Table 1. Currently known *lower bounds* for the maximum weak chromatic number of a d-embeddable k-uniform hypergraph on n vertices as $n \to \infty$. The number in chevrons indicates the theorem number where we prove this bound.

$d \diagdown k$	3	4	5	6	7
1	1	1	1	1	1
2	2	1	1	1	1
3	$\mathcal{O}(n^{\frac{1}{2}})_{\langle 3\rangle}$	$\mathcal{O}(n^{\frac{1}{2}})_{\langle 3\rangle}$	1	1	1
4	$\lceil n/2\rceil$	$\mathcal{O}(n^{\frac{2}{3}})_{\langle 3\rangle}$	$\mathcal{O}(n^{\frac{2}{3}})_{\langle 3\rangle}$	1	1
5	$\lceil n/2\rceil$	$\mathcal{O}(n^{\frac{26}{27}})_{\langle 4\rangle}$	$\mathcal{O}(n^{\frac{3}{4}})_{\langle 3\rangle}$	$\mathcal{O}(n^{\frac{3}{4}})_{\langle 3\rangle}$	1
6	$\lceil n/2\rceil$	$\lceil n/3\rceil$	$\mathcal{O}(n^{\frac{35}{36}})_{\langle 4\rangle}$	$\mathcal{O}(n^{\frac{4}{5}})_{\langle 3\rangle}$	$\mathcal{O}(n^{\frac{4}{5}})_{\langle 3\rangle}$
7	$\lceil n/2\rceil$	$\lceil n/3\rceil$	$\mathcal{O}(n^{\frac{107}{108}})_{\langle 4\rangle}$	$\mathcal{O}(n^{\frac{44}{45}})_{\langle 4\rangle}$	$\mathcal{O}(n^{\frac{5}{6}})_{\langle 3\rangle}$
8	$\lceil n/2\rceil$	$\lceil n/3\rceil$	$\lceil n/4\rceil$	$\mathcal{O}(n^{\frac{134}{135}})_{\langle 4\rangle}$	$\mathcal{O}(n^{\frac{53}{54}})_{\langle 4\rangle}$

Table 2. Currently known *upper bounds* for the maximum weak chromatic number of a d-embeddable k-uniform hypergraph on n vertices as $n \to \infty$. The number in chevrons indicates the theorem number where we prove this bound.

For $d, k, n \in \mathbb{N}$ we define $\chi^{\mathsf{s}}_{d,k}(n) = \max\{\chi^{\mathsf{s}}(H) : H \in \mathcal{H}_{d,k}, |V(H)| = n\}$ to be the maximum strong chromatic number of a d-embeddable k-uniform hypergraph on n vertices. The maximum weak chromatic number $\chi^{\mathsf{w}}_{d,k}(n)$ is defined analogously.

Proposition 1. *For large n, $d \geq 3$, and $d+1 \geq k$ we have that* $\chi^{\mathsf{s}}_{d,k}(n) \geq \lfloor\sqrt{n-d+3}\rfloor + d - 3$.

Theorem 2. *For large n, $d \geq 3$, and $d \geq k$ we have that* $\chi^{\mathsf{s}}_{d,k}(n) = n$.

This bound is shown by constructing a sequence of hypergraphs in $\mathcal{H}_{d,k}$, which have strong chromatic number equal to the number of their vertices. Thus, except for the cases where $k = d + 1$, the maximum strong coloring problem was solved. In particular, we have shown that an unbounded number of colors can be necessary for any strong coloring of a d-embeddable hypergraph if $d > 2$.

Theorem 3. *Let $d \geq 3$. Then one has*

$$\chi^{\mathrm{w}}_{d,d}(n) \leq \left\lceil \left(\frac{6ed}{(d-1)!}\right)^{\frac{1}{d-1}} n^{\frac{d-2}{d-1}} \right\rceil = \mathcal{O}\left(\left(\frac{n}{d}\right)^{\frac{d-2}{d-1}}\right).$$

Theorem 4. *Let $d \geq l \geq 3$. Then one has*

$$\chi^{\mathrm{w}}_{2d-l,d}(n) \leq \left\lceil (ed)^{\frac{1}{d-1}} n^{1-\frac{3^{l}-1-d}{d-1}} \right\rceil = \mathcal{O}\left(n^{1-\frac{3^{l}-1-d}{d-1}}\right).$$

These two results also holds for piecewise linear embeddings (for a definition *e.g.* see [8]). To prove them, we first limit the number of edges (relative to the number of vertices) that a hypergraph in $\mathcal{H}_{d,d}$ and $\mathcal{H}_{2d-l,d}$ can have. Then, an easy application of the Lovász Local Lemma [3, 14] yields the existence of a weak c-coloring if c is as high as requested.

Theorem 5. *As $n \to \infty$ one has $\chi^{\mathrm{w}}_{3,3}(n) = \Omega\left(\frac{\log n}{\log\log n}\right)$.*

Sketch of proof. We inductively construct a sequence of 3-uniform, 3-embeddable hypergraphs H_m which are weakly m-chromatic. Each new H_m consists of several copies of H_{m-1} and a few additional vertices (see Figure 1). The vertices are then arranged on the moment curve and the embedabbility is proven using a theorem by Shephard [13].

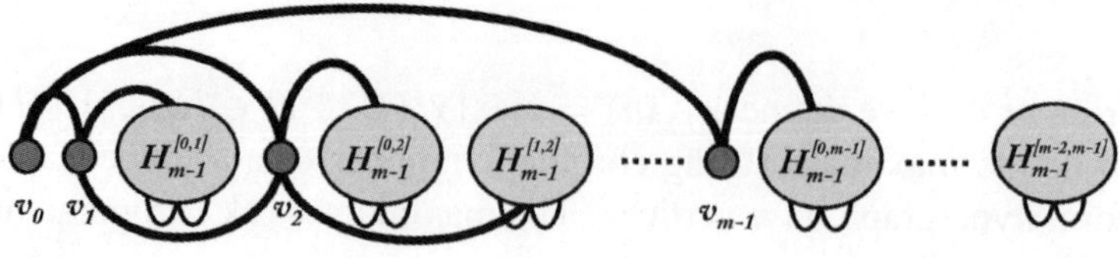

Figure 1. Construction of H_m.

Note that by monotonicity also $\chi^{\mathrm{w}}_{4,3}(n) = \Omega\left(\frac{\log n}{\log\log n}\right)$. Furthermore, it is possible to generalize this result for higher dimensions as follows.

Theorem 6. *Let $d \geq 3$. Then, as $n \to \infty$ one has $\chi^{\mathsf{W}}_{2d-3,d}(n) = \Omega\left(\frac{\log n}{\log\log n}\right)$.*

Note that in general there are several other notions of embeddability, the most popular being piecewise linear embeddings and general topological embeddings. A short and comprehensive introduction is given in Section 1 in [8]. Since piecewise linear and topological embeddings are more general than linear embeddings, all lower bounds for chromatic numbers can easily be transferred. Furthermore, we prove all our results on upper bounds for piecewise linear embeddings.

ACKNOWLEDGEMENTS. The authors wish to thank Penny Haxell for helpful discussions and an anonymous referee for valuable remarks concerning the presentation of this work.

References

[1] K. APPEL and W. HAKEN, *Every planar map is four colorable. Part I: Discharging*, Illinois Journal of Mathematics **21** (3) (1977), 429–490.

[2] K. APPEL, W. HAKEN and J. KOCH, *Every planar map is four colorable. Part II: Reducibility*, Illinois Journal of Mathematics **21** (3) (1977), 491–567.

[3] P. ERDŐS and L. LOVÁSZ, *Problems and results on 3-chromatic hypergraphs and some related questions*, Infinite and Finite Sets (to Paul Erdős on his 60th birthday), Vol. II, North-Holland (1975), 609–627.

[4] A. FLORES, *Über n-dimensionale Komplexe, die im $\mathbb{R}_{2n+1}$ absolut selbstverschlungen sind*, Ergebnisse Eines Mathematischen Kolloquiums **6** (1934), 4–7.

[5] I. FÁRY, *On straight-line representation of planar graphs*, Acta Scientiarum Mathematicarum Szeged **11** (1948), 229–233.

[6] B. GRÜNBAUM, *Higher-dimensional analogs of the four-color problem and some inequalities for simplicial complexes*, Journal of Combinatorial Theory **8** (2) (1970), 147–153.

[7] J. C. HEAWOOD, *Map-colour theorem*, The Quarterly Journal of Pure and Applied Mathematics **24** (1890), 332–338.

[8] J. MATOUŠEK, M. TANCER and U. WAGNER, *Hardness of Embedding Simplicial Complexes in $\mathbb{R}^d$*, Journal of the European Mathematical Society **13** (2011), 259–295.

[9] K. MENGER, *Dimensionstheorie*, Teubner, Leipzig, 1928.

[10] G. NÖBELING, *Über eine n-dimensionale Universalmenge im* $\mathbb{R}_{2n+1}$, Mathematische Annalen **104** (1) (1931), 71–80.

[11] G. RINGEL and J. W. T. YOUNGS, *Solution of the Heawood map-coloring problem*, Proceedings of the National Academy of Sciences **60** (2) (1968), 438–445.

[12] K. S. SARKARIA, *Heawood inequalities*, Journal of Combinatorial Theory, Series A **46** (1) (1987), 50–78.

[13] G. C. SHEPHARD, *A Theorem on Cyclic Polytopes*, Israel Journal of Mathematics **6** (4) (1968), 368–372.

[14] J. SPENCER, *Asymptotic lower bounds for Ramsey functions*, Discrete Mathematics **20** (1977), 69–76.

[15] E. R. VAN KAMPEN, *Komplexe in euklidischen Räumen*, Abhandlungen aus dem Mathematischen Seminar der Universität Hamburg **9** (1) (1933), 72–78, corrections ibidem, 152–153.

Homomorphisms of signed bipartite graphs

Reza Naserasr[1], Edita Rollová[2] and Éric Sopena[3]

Abstract. We study the homomorphism relation between signed graphs where the underlying graph G is bipartite. We show that this notion captures the notions of chromatic number and graph homomorphisms. In particular we will study Hadwiger's conjecture in this setting. We show that for small values of the chromatic number there are natural strengthening of this conjecture but such extensions will not work for larger chromatic numbers.

1 Homomorphisms

A signature on a graph G is an assignment of negative or positive sign to the edges. Resigning at a vertex v is to change the sign of all edges incident to v. Two signatures are equivalent if one can be obtained from the other by a sequence of resigning. The set of negative edges is normally denoted by Σ. A signed graph, denoted (G, Σ) is a graph G together with the set of signatures all equivalent to Σ. A signed minor of (G, Σ) is a signed graph obtained from (G, Σ) by a sequence of deleting vertices or edges, contracting positive edges and resigning. Given two signed graphs (G, Σ) and (H, Σ_1) we say there is a homomorphism of (G, Σ) to (H, Σ_1) if there is a signature Σ' equivalent to Σ and a mapping $\phi : V(G) \to V(H)$ such that ϕ preserves both adjacency and the sign of an edge with respect to Σ'. If there is a homomorphism of (G, Σ) to (H, Σ_1) we write $(G, \Sigma) \to (H, \Sigma_1)$. This relation is a quasi order on the class of all signed graphs. Thus we may use terms such as *bound* and *maximum*.

A cycle with only one, equivalently odd number of, negative edges is called an *unbalanced cycle* and is denoted by UC_k if the length of the

[1] CNRS, LRI, UMR8623, Univ. Paris-Sud 11, F-91405 Orsay Cedex, France. Email: reza@lri.fr

[2] Department of Mathematics, Faculty of Applied Sciences, University of West Bohemia, Univerzitní 22, 306 14 Plzeň, Czech Republic. Email: rollova@kma.zcu.cz

[3] Univ. Bordeaux, LaBRI, UMR5800, F-33400 Talence, France, CNRS, LaBRI, UMR5800, F-33400 Talence, France. E-mail: sopena@labri.fr

cycle is k. The following is one of the first theorems in the theory of signed graph homomorphisms.

Lemma 1.1. *There is a homomorphism of UC_k to UC_ℓ if and only if $k \geq \ell$ and $k = \ell \pmod 2$.*

Let G be a graph; the signed graph $S(G) = (G^*, \Sigma)$ is obtained by replacing each edge uv of G by an unbalanced 4-cycle on four vertices $ux_{uv}vy_{uv}$, where x_{uv} and y_{uv} are new and distinct vertices. Let $(K_{k,k}, M)$ be the signed graph where edges of a perfect matching M are negative. The following theorem shows how to define $\chi(G)$ using only the notion of homomorphism between signed bipartite graphs.

Theorem 1.2. *For every $k \geq 3$ and every graph G, $\chi(G) \leq k$ if and only if $S(G) \to (K_{k,k}, M)$.*

Proof. The main idea is that if $S(G) \to (K_{k,k}, M)$, then adjacent vertices of G are mapped to a same side of $K_{k,k}$ and to distinct vertices. □

In a similar way we show below that the problem of the existence of a homomorphism of a graph G into a graph H is captured by the notion of homomorphism between signed bipartite graphs. For a comprehensive study of graph homomorphisms we refer to [3].

Theorem 1.3. *For every two graphs G and H, $G \to H$ if and only if $S(G) \to S(H)$.*

Proof. The main idea again is to show that a mapping $S(G) \to S(H)$ will map $V(G)$ to $V(H)$ while preserving adjacency of G in H. □

2 Minors

We prove the following minor relation between graphs and their corresponding signed bipartite graphs:

Theorem 2.1. *For every integer n and every graph G, G has a K_n-minor if and only if $S(G)$ has a (K_n, Σ)-minor for some Σ.*

Proof. First assume (K_n, Σ) is a signed minor of $S(G)$ for some Σ. We would like to prove that K_n is a minor of G. This is clear for $n = 1, 2$. So we assume $n \geq 3$. Thus, in producing (K_n, Σ) as a signed minor of $S(G)$ each vertex of degree 2 in $S(G)$ is either deleted or identified with one of its neighbours as a result of contracting an incident edge. We define a minor of G as follows: For each edge uv of G, if the corresponding unbalanced 4-cycle is deleted in the process of producing (K_n, Σ) as a signed minor of $S(G)$, then delete uv. If u and v are identified through contraction of edges in producing (K_n, Σ) as a signed minor of $S(G)$, then contract the edge uv. Otherwise uv remains an edge. The resulting minor then must be K_n.

For the opposite direction, suppose K_n is a minor of G. Let uv be an edge of G. If the edge uv is deleted in producing K_n-minor from G, then delete all the four edges of corresponding unbalanced 4-cycle. If uv is contracted, then contract two positive edges of the corresponding unbalanced 4-cycle in $S(G)$ in such a way that u and v are identified after these contractions and delete the other two edges of the unbalanced 4-cycle. Otherwise contract two positive edges of the corresponding unbalanced 4-cycle in such a way that there are two new parallel edges between u and v, one positive and one negative. Finally delete all isolated vertices. By allowing multiple edges at the end of this process we get a signed minor of $S(G)$ which has n vertices and for each pair x and y of vertices two xy edges, one positive and one negative. For each such pair we delete the negative edge unless $xy \in \Sigma$ in which case we delete the positive edge. The result is (K_n, Σ) obtained as a signed minor of G. □

3 Hadwiger's conjecture for signed bipartite graphs

Conjecture 3.1 (Hadwiger, [2]). If a graph G has no K_n-minor, then it is $(n-1)$-colorable.

By Theorem 1.2, Hadwiger's conjecture can be restated as follows:

Conjecture 3.2. Given $n \geq 4$, the class $\mathcal{C} = \{S(G) \mid G \text{ is } K_n\text{-minor-free}\}$ of signed bipartite graphs is bounded by $(K_{n-1,n-1}, M)$ in the signed graph homomorphism order.

If the conjecture holds, then the next question would be: what is a natural superclass of $\mathcal{C}$ which is still bounded by $(K_{n-1,n-1}, M)$?
Hadwiger's conjecture is known to be true for $n \leq 6$, thus Conjecture 3.2 is also true for $n \leq 6$. For $n = 4$ we have the following generalization.

Theorem 3.3. *If G is a bipartite graph with no K_4-minor and Σ is any signature on G, then $(G, \Sigma) \to (K_{3,3}, M)$.*

Proof. By adding more edges, if needed, we may assume that G is edge maximal with respect to being bipartite and having no K_4-minor. Obviously it is enough to prove the theorem for such edge maximal graphs.

As mentioned before, a classical decomposition theorem for edge-maximal K_4-minor-free graphs states that every such graph is built from a sequence of triangles starting by one triangle and pasting each new triangle to the graph previously built along an edge. To use the decomposition theorem we add new edges to G, of green color, until we reach a maximal K_4-minor-free graph G', which obviously is not bipartite anymore. Let G'' be the edge-colored graph obtained from G' by coloring original positive edges of (G, Σ) in blue, original negative edges of (G, Σ) in red and keeping the green color for edges not in G.

We claim that there is no triangle in G'' with exactly two green edges. To see this, suppose that v_1v_2 and v_1v_3 are both green and that v_2v_3 is an edge of G. Since G is bipartite v_2 and v_3 are in two different parts and thus v_1 is in a different part with respect to one of them. Without loss of generality assume v_1 and v_2 are in different parts. Consider the graph $G + \{v_1v_2\}$. By the choice of v_2 this graph is bipartite and since it is a subgraph of G', it has also no K_4-minor but this contradicts the edge maximality of G.

We now build a new edge-colored graph F from $(K_{3,3}, M)$. The blue and red edges of F are defined as before and we add green edges between every pair of vertices non adjacent in $(K_{3,3}, M)$. The edge-colored graph F has three types of triangles: (i) triangles with three green edges, (ii) triangles with one green edge and two blue edges, and (iii) triangles with no two edges of the same color. Furthermore it is not hard to verify that each red edge only belongs to triangles of type (iii), each blue edge belongs to triangles of type (ii) or (iii) and each green edge is contained in triangles of each of the three types.

To prove the theorem we now prove the following stronger statement: there exists a suitable "resigning" G^* of G'' such that G^* admits a color-preserving homomorphism to F. By resigning here we mean exchanging the colors red and blue on edges of an edge cut, this can be regarded as a sequence of vertex resigning.

To prove this stronger statement, let $T_1, \dots, T_k$ be the sequence of triangles obtained from the decomposition of G'' mentioned above. Note that since G was bipartite, each such triangle contains a green edge. Consider the triangle T_1. Either it is one of the three types (i), (ii) or (iii), in which case we simply map it to F, or it has one green and two red edges. Let u be the common vertex of these two red edges. After resigning at u we have a triangle of type (ii) and thus we can map it to F.

By induction, assume now that the graph G_i'', obtained by pasting the triangles $T_1, \dots T_i$, $i < k$, is mapped to F and assume that T_{i+1} is pasted to G_i'' along the edge e. Let v be the vertex of T_{i+1} not incident to e. If T_{i+1} is a triangle of one the three types, because of the above mentioned property of F, we can extend the mapping of G_i'' to G_{i+1}'', where the colors of the two edges of T_{i+1} incident with v are preserved. Otherwise T_{i+1} has exactly two red edges and one green edge. By resigning at v we get a triangle that has either one or no red edge, thus obtaining a triangle of type (ii) or (iii). We now extend the homomorphism thanks to the properties of F. In this process, resigning a vertex would be done at most once, when it is added to the already built part of the graph, so our process is well-defined and the stronger claim is proved. □

We note that our proof has an algorithmic feature. Given a signed bipartite graph (G, Σ), where G is a K_4-minor-free graph, we can find, in polynomial time, a homomorphism of (G, Σ) to $(K_{3,3}, M)$.

Furthermore, we believe that the following stronger statement should also be true:

Conjecture 3.4. If G is bipartite and (G, Σ) has no $(K_4, E(K_4))$ as a signed minor, then $(G, \Sigma) \to (K_{3,3}, M)$.

For $n = 5$ it is shown in [4], using the four-color theorem and a result of [1], that the following holds.

Theorem 3.5. *If G is a bipartite planar graph and* Σ *is any signature on G, then* $(G, \Sigma) \to (K_{4,4}, M)$.

For large values of n ($n \geq 7$) Conjecture 3.2 does not extend so nicely. To show this we use the following signed bipartite graph, $Fano$. That is signed graph on $K_{7,7}$ where vertices on one side are labeled with points of Fano plane and on the other side with lines of the Fano plane. An edge is negative if it connects a line to a point of a line.

Theorem 3.6. *There exists no value of n for which* $Fano$ *admits a homomorphism to* $(K_{n,n}, M)$.

A proof can be obtained mainly by counting number of distinct copies of UC_4 containing a given edge. For more details we refer to [4]. The following then is an immediate corollary.

Corollary 3.7. *The class* $\mathcal{C}=\{(G,\Sigma)|G$ *is bipartite and has no H-minor*$\}$ *is not bounded by* $(K_{n,n}, M)$ *(for no values of n) if H is a graph on at least* 15 *vertices.*

This shows that for $n \geq 15$ the reformulation of Hadwiger's conjecture given in Conjecture 3.2 cannot be extended to a general minor closed class of signed bipartite graphs. Even though such an extension was possible for small values of n.

References

[1] B. Guenin, *Packing T-joins and edge-colouring in planar graphs*, Mathematics of Operations Research, to appear.

[2] H. HADWIGER, *Über eine Klassifikation der Streckenkomplexe*, Vierteljschr. Naturforsch. Ges. Zürich **88** (1943), 133–143.

[3] P. HELL and J. NEŠETŘIL, "Graphs and Homomorphisms", Oxford Lecture Series in Mathematics and its Applications, Vol. 28, Oxford University Press, Oxford, 2004.

[4] R. NASERASR, E. ROLLOVÁ and É. SOPENA, *Homomorphisms of signed graphs*, submitted.

[5] R. NASERASR, E. ROLLOVÁ and É. SOPENA, *Homomorphisms of planar signed graphs to signed projective cubes*, manuscript.

Games

A threshold for the Maker-Breaker clique game

Tobias Müller[1] and Miloš Stojaković[2]

Abstract. We study Maker-Breaker k-clique game played on the edge set of the random graph $G(n, p)$. In this game, two players alternately claim unclaimed edges of $G(n, p)$, until all the edges are claimed. Maker wins if he claims all the edges of a k-clique; Breaker wins otherwise. We determine that the threshold for the graph property that Maker can win is at $n^{-\frac{2}{k+1}}$, for all $k > 3$, thus proving a conjecture from [5]. More precisely, we conclude that there exist constants $c, C > 0$ such that when $p > Cn^{-\frac{2}{k+1}}$ the game is Maker's win a.a.s., and when $p < cn^{-\frac{2}{k+1}}$ it is Breaker's win a.a.s.
For the triangle game, when $k = 3$, we give a more precise result, describing the hitting time of Maker's win in the random graph process. We show that, with high probability, Maker can win the triangle game exactly at the time when a copy of K_5 with one edge removed appears in the random graph process. As a consequence, we are able to give an expression for the limiting probability of Maker's win in the triangle game played on the edge set of $G(n, p)$.

1 Introduction

Let X be a finite set and let $\mathcal{F} \subseteq 2^X$ be a family of subsets of X. In the positional game $(X, \mathcal{F})$, two players take turns in claiming one previously unclaimed element of X. The set X is called the "board", and the members of $\mathcal{F}$ are referred to as the "winning sets". In a *Maker-Breaker* positional game, the two players are called Maker and Breaker. Maker wins the game if he occupies all elements of some winning set; Breaker wins otherwise. A game $(X, \mathcal{F})$ is said to be a *Maker's win* if Maker has a strategy that ensures his win against any strategy of Breaker; otherwise it is a *Breaker's win*. Note that $\mathcal{F}$ alone determines whether the game is Maker's win or Breaker's win.

[1] Mathematical Institute, Utrecht University, the Netherlands. Email: tobias@cwi.nl

[2] Department of Mathematics and Informatics, University of Novi Sad, Serbia. Email: milos.stojakovic@dmi.uns.ac.rs.

The first author was supported in part by a VENI grant from Netherlands Organization for Scientific Research. The second author was partly supported by Ministry of Education and Science, Republic of Serbia, and Provincial Secretariat for Science, Province of Vojvodina.

A well-studied class of positional games are the *games on graphs*, where the board is the set of edges of a graph. The winning sets are usually representatives of some graph theoretic structure. The first game studied in this area was the connectivity game, a generalization of the well-known Shannon switching game, where Maker's goal is to claim a spanning connected graph by the end of the game. We denote the game by $(E(K_n), \mathcal{T})$. Another important game is the Hamilton cycle game $(E(K_n), \mathcal{H})$, where $\mathcal{H} = \mathcal{H}_n$ consists of the edge sets of all Hamilton cycles of K_n.

In the clique game the winning sets are the edge sets of all k-cliques, for a fixed integer $k \geq 3$. We denote this game with $(E(K_n), \mathcal{K}_k)$. Note that the size of the winning sets is fixed and does not depend on n, which distinguishes it from the connectivity game and the Hamilton cycle game. A simple Ramsey argument coupled with the strategy stealing argument ensures Maker's win if n is large.

All three games that we introduced are straightforward Maker's wins when n is large enough. This is however not the end of the story, as there are two general approaches to even out the odds, giving Breaker more power – *biased games* and *random games*. Here, we will stick with the latter.

2 Our results

A way to give Breaker more power in a positional game, introduced by the second author and Szabó in [5], is to randomly thin out the board before the game starts, thus eliminating some of the winning sets.

For games on graphs, given a game $\mathcal{F}$ that is Maker's win when played on $E(K_n)$, we want to find the *threshold probability* $p_{\mathcal{F}}$ so that, if the game is played on $E(G(n, p))$, an almost sure Maker's win turns into an almost sure Breaker's win. Such a threshold $p_{\mathcal{F}}$ exists, as "being Maker's win" is clearly a monotone increasing graph property.

The threshold probability for the connectivity game was determined to be $\frac{\log n}{n}$ in [5], and shown to be sharp. As for the Hamilton cycle game, the order of magnitude of the threshold was given in [4]. Using a different approach, it was proven in [3] that the threshold is $\frac{\log n}{n}$ and it is sharp. Finally, as a consequence of a hitting time result, Ben-Shimon *et al.* [2] closed this question by giving a very precise description of the low order terms of the limiting probability.

Moving to the clique game, it was shown in [5] that for every $k \geq 4$ and every $\varepsilon > 0$ we have $n^{-\frac{2}{k+1}-\varepsilon} \leq p_{\mathcal{K}_k} \leq n^{-\frac{2}{k+1}}$. Moreover, it was proved that there exist a constant $C > 0$ such that for $p \geq Cn^{-\frac{2}{k+1}}$ Maker wins the k-clique game on $G(n, p)$ a.a.s. The threshold for the triangle game was determined to be $p_{\mathcal{K}_3} = n^{-\frac{5}{9}}$, showing that the behavior of the triangle game is different from the k-clique game for $k \geq 4$.

Our main result is the following theorem. It gives a lower bound on the threshold for the k-clique game, when $k \geq 4$, which matches the upper bound from [5] up to the leading constant.

Theorem 2.1. *Let $k \geq 4$. There exists a constant $c > 0$ such that for $p \leq cn^{-\frac{2}{k+1}}$ Breaker wins the Maker-Breaker k-clique game played on the edge set of $G(n, p)$ a.a.s.*

The threshold probability for the k-clique game for $k \geq 4$ was conjectured to be $p_{\mathcal{K}_k} = n^{-\frac{2}{k+1}}$ in [5]. The previous theorem resolves this conjecture in the affirmative. Summing up the results of Theorem 2.1 and Theorem 19 from [5], we now have the following.

Corollary 2.2. *Let $k \geq 4$ and consider the Maker-Breaker k-clique game on the edge set of $G(n, p)$. There exist constants $c, C > 0$ such that the following hold:*

1. *If $p \geq Cn^{-\frac{2}{k+1}}$, then Maker wins a.a.s.;*
2. *If $p \leq cn^{-\frac{2}{k+1}}$, then Breaker wins a.a.s.*

A result of this type is sometimes called a "semi-sharp threshold" in the random graphs literature.

Hitting time of Maker's win. Let V be a set of cardinality n, and let π be a permutation of the set $\binom{V}{2}$. If by G_i we denote the graph on the vertex set V whose edges are the first i edges in the permutation π, $G_i = (V, \pi^{-1}([i]))$, then we say that $\tilde{G} = \{G_i\}_{i=0}^{\binom{n}{2}}$ is a *graph process*. Given a monotone increasing graph property $\mathcal{P}$ and a graph process $\tilde{G}$, we define the hitting time of $\mathcal{P}$ with $\tau(\tilde{G}; \mathcal{P}) = \min\{t : G_t \in \mathcal{P}\}$. If π is chosen uniformly at random from the set of all permutations of the set $\binom{V}{2}$, we say that $\tilde{G}$ is a *random graph process*. Such processes are closely related to the model of random graph we described above.

Given a positional game, our general goal is to describe the hitting time of the graph property "Maker's win" in a typical graph process. For a game $\mathcal{G}$, by $\mathcal{M}_{\mathcal{F}}$ we denote the graph property "Maker wins $\mathcal{G}$". It was shown in [5] that in the connectivity game (with the technical assumption that Breaker is the first to play), for a random graph process $\tilde{G}$, we have $\tau(\tilde{G}; \mathcal{M}_{\mathcal{T}}) = \tau(\tilde{G}; \delta_2)$, where δ_ℓ is the graph property "minimum degree at least ℓ". Recently, Ben-Shimon *et al.* [2] resolved the same question for the Hamilton cycle game, obtaining $\tau(\tilde{G}; \mathcal{M}_{\mathcal{H}}) = \tau(\tilde{G}; \delta_4)$. Note that inequality in one direction for both of these equalities holds trivially.

Moving on to the clique game, we denote the property "the graph contains $K_5 - e$ as a subgraph" with $\mathcal{G}_{\mathcal{K}_5^-}$. We are able to show the following hitting time result for Maker's win in the triangle game.

Theorem 2.3. *For a random graph process $\tilde{G}$, the hitting time for Maker's win in the triangle game is asymptotiaclly almost surely the same as the hitting time for appearance of $K_5 - e$, i.e., $\tau(\tilde{G}; \mathcal{M}_{\mathcal{K}_3}) = \tau(\tilde{G}; \mathcal{G}_{\mathcal{K}_5^-})$ a.a.s.*

Using this, we are able to give a precise expression for the probability for Maker's win in the triangle game on $G(n, p)$.

Corollary 2.4. *Let $p = p(n)$ be an arbitrary sequence of numbers $\in [0, 1]$ and let us write $x = x(n) = p \cdot n^{\frac{2}{k+1}}$. Then*

$$\lim_{n\to\infty} Pr[\textit{Maker makes triangle on } G(n,p)] = \begin{cases} 0 & \textit{if } x \to 0, \\ 1 - e^{-\frac{c^5}{3}} & \textit{if } x \to c \in \mathbb{R}, \\ 1 & \textit{if } x \to \infty. \end{cases}$$

3 Conclusion and open problems

Random graph intuition. In the 70s, Erdős observed the following paradigm which is referred to as the *random graph intuition* in positional game theory. As it turns out for many games on graphs, the inverse of the threshold bias $b_{\mathcal{G}}$ in the game played on the complete graph is "closely related" to the probability threshold for the appearance of a member of $\mathcal{G}$ in $G(n, p)$. Another parameter that is often "around" is the threshold probability $p_{\mathcal{G}}$ for Maker's win when played on $G(n, p)$. As we saw, for the two games mentioned in the introduction, the connectivity game and the Hamilton cycle game, all three parameters are equal to $\frac{\log n}{n}$.

In the k-clique game, for $k \geq 4$, the threshold bias is $b_{\mathcal{K}_k} = \Theta(n^{\frac{2}{k+1}})$ and the threshold probability for Maker's win is the inverse (up to the leading constant), $p_{\mathcal{K}_k} = n^{-\frac{2}{k+1}}$, supporting the random graph intuition. But, the threshold probability for appearance of a k-clique in $G(n, p)$ is not at the same place, it is $n^{-\frac{2}{k-1}}$. And in the triangle game there is even more disagreement, as all three parameters are different – they are, respectively, $n^{\frac{1}{2}}, n^{-\frac{5}{9}}$ and n^{-1}. Now, more than thirty years after Chvátal and Erdős formulated the paradigm, there is still no general result that would make it more formal. We are curious to the reasons behind the total agreement between the three thresholds in the connectivity game and the Hamilton cycle game, partial disagreement in k-clique game for $k \geq 4$, and the total disagreement in the triangle game.

Random clique game vs. biased clique game. Our Corollary 2.2 gives two constants $c > 0$ and $C > 0$, stating that the probability threshold for Maker's win in the k-clique game on $G(n, p)$ for $k \geq 4$ is between $cn^{-\frac{2}{k+1}}$ and $Cn^{-\frac{2}{k+1}}$. In a way, with this result, the game played on the random graph catches up with the biased k-clique game played on the complete graph, as a result of Bednarska and Łuczak [1] guarantees the existence of constants $c' > 0$ and $C' > 0$, such that the bias threshold for this game is between $c'n^{\frac{2}{k+1}}$ and $C'n^{\frac{2}{k+1}}$, for all $k \geq 3$. Both pairs of constants, c, C and c', C', are quite far apart. Also, in both games, the best known strategy for Maker's exploits the same derandomized random strategy approach, proposed in [1].

We know much more for the triangle game on the random graph, as Corollary 2.4 gives the threshold probability quite accurately, and it turns out to be a coarse threshold. The reason for such different behavior (compared to $k > 3$) may lie behind the fact that $K_3 = C_3$.

A more precise result for the k-clique game when $k \geq 4$? As we saw, we can say a lot about the threshold probability for the triangle game, the connectivity game and the Hamilton cycle game when the game is played on the random graph. We do not know that much about the k-clique game, when $k \geq 4$, and it would be interesting to see what happens between the bounds given in Corollary 2.2. Also, a graph-theoretic description of the hitting time of Maker's win on the random graph process would be of great importance, as we know very little about Maker's winning strategy at the threshold. What we know is that we cannot hope for a result analogous to Theorem 2.3 – the reason for Maker's win cannot be the appearance of a fixed graph, as we know that Breaker wins on every typical (fixed) subgraph of the random graph on the probability threshold. Hence, Maker's optimal strategy must be of "global nature", taking into account a non-constant part of the random graph to win the game. Having that in mind we propose the following conjecture.

Conjecture 3.1. For every $k \geq 4$ there exists a $c = c(k)$ such that for any fixed $\varepsilon > 0$, if $p \leq (c - \varepsilon)n^{-\frac{2}{k+1}}$, then Breaker wins the k-clique game on $G(n, p)$ a.a.s, and if $p \geq (c + \varepsilon)n^{-\frac{2}{k+1}}$, then Maker wins a.a.s.

References

[1] M. BEDNARSKA and T. ŁUCZAK, *Biased positional games for which random strategies are nearly optimal*, Combinatorica **20** (2000), 477–488.

[2] S. BEN-SHIMON, A. FERBER, D. HEFETZ and M. KRIVELEVICH, *Hitting time results for Maker-Breaker games*, Random Structures and Algorithms **41** (2012), 23–46.

[3] D. HEFETZ, M. KRIVELEVICH, M. STOJAKOVIĆ and T. SZABÓ, *A sharp threshold for the Hamilton cycle Maker-Breaker game*, Random Structures and Algorithms **34** (2009), 112–122.

[4] M. STOJAKOVIĆ, "Games on Graphs", PhD Thesis, ETH Zürich, 2005.

[5] M. STOJAKOVIĆ and T. SZABÓ, *Positional games on random graphs*, Random Structures and Algorithms **26** (2005), 204–223.

On the threshold bias in the oriented cycle game

Dennis Clemens[1] and Anita Liebenau[2]

Abstract. In the Oriented cycle game, the two players, called OMaker and OBreaker, alternately direct edges of K_n. OMaker directs exactly one edge, whereas OBreaker is allowed to claim between one and b edges. OMaker wins if the final tournament contains a directed cycle, otherwise OBreaker wins. It was shown recently [1] that OMaker has a winning strategy for this game whenever $b \leq n/2 - 2$. We show that OBreaker has a strategy whenever $b > 5n/6$, and give a non-trivial upper bound when OBreaker is asked to direct exactly b edges in each of his moves.

1 Introduction and Results

We study the oriented cycle game, which is a particular orientation game. Orientation games were studied by Ben-Eliezer, Krivelevich and Sudakov in [1], and we follow their notation. In orientation games, the board consists of the edges of the complete graph K_n. In the $(p : q)$ orientation game, the two players called OMaker and OBreaker, orient previously undirected edges alternately. OMaker starts, and in each round, OMaker directs between one and p edges, and then OBreaker directs between one and q edges. At the end of the game, the final graph is a tournament on n vertices. OMaker wins the game if this tournament has some predefined property $\mathcal{P}$. Otherwise, OBreaker wins. We study the $(1 : b)$-game and refer to it as the *b-biased orientation game*. Increasing b can only help OBreaker, so the game is *bias monotone*. Therefore, any such game has a threshold $t(n, \mathcal{P})$ such that OMaker wins the b-biased game when $b \leq t(n, \mathcal{P})$ and OBreaker wins the game when $b > t(n, \mathcal{P})$.

[1] Department of Mathematics and Computer Science, Freie Universität Berlin, Germany. Email: d.clemens@fu-berlin.de. Research supported by DFG, project SZ 261/1-1.

[2] Department of Mathematics and Computer Science, Freie Universität Berlin, Germany. Email: liebenau@math.fu-berlin.de. Research supported by the Berlin Mathematical School (BMS).

In a variant, OBreaker is required to direct exactly b edges. We refer to this variant as the *strict* b-biased orientation game. Playing the exact bias in every round may be disadvantageous for OBreaker, so the existence of a threshold as for the monotone rules is not guaranteed in general. We therefore define $t^+(n, \mathcal{P})$ to be the largest value b such that OMaker has a strategy to win the strict b-biased orientation game, and $t^-(n, \mathcal{P})$ to be the largest integer such that for every $b \leq t^-(n, \mathcal{P})$, OMaker has a strategy to win the strict b-biased orientation game. Trivially, $t(n, \mathcal{P}) \leq t^-(n, \mathcal{P}) \leq t^+(n, \mathcal{P})$. The threshold bias $t(n, \mathcal{P})$ was investigated in [1] for several orientation games. However, the relation between all three parameters in question is still widely open. It is not even clear whether $t^-(n, \mathcal{P})$ and $t^+(n, \mathcal{P})$ need to be distinct values.

We focus on the oriented cycle game, in which OMaker wins if the final tournament contains a directed cycle. Let $\mathcal{P}$ be the property of containing a directed cycle. The strict version of this game was studied by Bollobás and Szabó in [2]. They show that $t^+(n, \mathcal{P}) \geq \lfloor (2 - \sqrt{3})n \rfloor$. Moreover, they remark that the proof also works for the monotone rules, which implies that $t(n, \mathcal{P}) \geq \lfloor (2 - \sqrt{3})n \rfloor$. For an upper bound, it is rather simple to see that OBreaker wins the b-biased oriented cycle game for $b \geq n - 2$, even when the strict rules apply. Therefore, $t^+(n, \mathcal{P}) \leq n - 3$. Bollobás and Szabó conjecture that this upper bound is tight. In [1], Ben-Eliezer, Krivelevich and Sudakov show that for $b \leq n/2 - 2$, OMaker has a strategy guaranteeing a cycle in the b-biased orientation game, i.e. $t(n, \mathcal{P}) \geq n/2 - 2$. We give a strategy for OBreaker in the b-biased oriented cycle game when $b \geq 5n/6 + 1$.

Theorem 1.1. *For $b \geq 5n/6 + 1$, OBreaker has a strategy to prevent OMaker from closing a directed cycle in the b-biased orientation game. In particular, $t(n, \mathcal{P}) \leq 5n/6$.*

Furthermore, we adjust our strategy to the strict rules and show the following.

Theorem 1.2. *For $b \geq n - c\sqrt{n}$, where $0 < c < 1$ is a constant, OBreaker has a strategy to prevent OMaker from closing a directed cycle in the strict b-biased orientation game. In particular, $t^+(n, \mathcal{P}) \leq n - c\sqrt{n} - 1$.*

Theorem 1.2 refutes the above conjecture of Bollobás and Szabó.

2 Outline of the proofs

For both proofs, we need to provide OBreaker with a strategy to prevent OMaker from closing a directed cycle, no matter how she plays. This is equivalent to constructing the transitive tournament.

There are two essential concepts to our proofs, so called UDB's and α-structures. Suppose the game is in play, and let G denote the subgraph of already directed edges (by either player). For a directed edge $e \in G$, we write e^+ for its tail and e^- for its head, i.e. $e = (e^+, e^-)$. For two disjoint subsets $A, B \subseteq V$, we call the pair (A, B) a *uniformly directed biclique* (or short UDB), if for all $a \in A, b \in B$ the edge $(a, b) \in E(G)$ is present in G already. Our goal is to create a UDB (A, B) such that both parts fulfil $|A|, |B| \leq b$ and $A \cup B = V$. Suppose both sets A and B would be independent. OBreaker could then follow the "trivial strategy" inside A and B respectively (as OBreaker wins on K_{b+2}). However, while building such a UDB, OMaker will direct edges inside these sets, and OBreaker needs to control those. To handle this obstacle, we introduce α-structures. Let $V' \subseteq V$ and $E = E(G[V'])$. Then the set E is called an *α-structure in V' of size k* if there exist edges $e_1, \dots, e_k \in E$ such that

(α_1) for every directed path $P = (e_{i_1}, \dots, e_{i_k})$: $i_1 > \dots > i_k$;
(α_2) for every $1 \leq i < j \leq k$: $(e_j^+, e_i^-) \in E$;
(α_3) and no other edges are present in $G[V']$.

In our strategy, the edges $e_1, \dots, e_k$ will be the edges directed by OMaker (though not necessarily in that order), and the edges of "type" (α_2) are the ones directed by OBreaker. It is easy to verify that (α_1)-(α_3) imply that E does not contain a directed cycle, nor can OMaker close one in her next move inside V'. Suppose $E = E(G[V'])$ is an α-structure inside some subset $V' \subseteq V$, and let $e = (v, w)$ be the edge OMaker directed in her previous move. If $v, w \in V'$, we provide OBreaker with a *procedure α* to add e to the α-structure. This includes providing e with an appropriate index $\ell \in [k + 1]$ (and an index shift of the existing edges $e_1, \dots, e_k$) such that

(i) OBreaker needs to direct at most k new edges $\{f_1, \dots, f_\ell\}$ and
(ii) $E \cup \{e, f_1, \dots, f_\ell\}$ forms an α-structure in V' again.

The details are straight-forward though technical, so we omit them here. We are now ready to describe the *global* strategies of OBreaker.

Strategy for Theorem 1.1
The strategy is divided into three stages. In Stage I, OBreaker maintains a UDB (A, B) such that after each of his moves

- $E(G[A])$ is an α-structure of size k in $V \setminus B$,
- $E(G[B])$ is an α-structure of size ℓ in $V \setminus A$,
- $|A| - k$ and $|B| - \ell$ increase by at least 1 in every round,
- $k + \ell \leq |A| - k = |B| - \ell \leq \frac{n}{6}$.

Note that property (α_3) implies that $E(G) \subseteq (A \cup B) \times (A \cup B)$. Let $e = (v, w)$ be the edge OMaker directed in her previous move. Since (A, B) is a UDB, either $\{v, w\} \subseteq V \setminus A$ or $\{v, w\} \subseteq V \setminus B$. Let $\{v, w\} \subseteq V \setminus A$. Then OBreaker adds e to the α-structure $E(G[A])$ by procedure α. As noted above, this takes him at most k edges to direct. He then directs all edges (v, b), (w, b) for $b \in B$ and thus adds v and w to A. If v (or w respectively) was already an element of A, OBreaker picks an arbitrary new vertex $v' \in V \setminus (A \cup B)$ (or w' respectively), and directs all edges (v, b) for $b \in B$. Furthermore, he picks an arbitrary element $b' \in V \setminus (A \cup B)$ and directs all edges (a, b') for $a \in A$. This way, $|A| - k$ and $|B| - \ell$ increase by 1. Furthermore, since $|A| - k$, $|B| - \ell$, and $k + \ell$ are bounded by $\frac{n}{6}$ and since $b \geq \frac{5n}{6} + 1$, OBreaker can follow the strategy in Stage I. The analysis for $\{v, w\} \subseteq V \setminus B$ is similar. As soon as $|A| - k, |B| - \ell \geq n/6$, OBreaker proceeds to Stage II.

In Stage II, OBreaker stops increasing the values $|A| - k$ and $|B| - \ell$. He now maintains a UDB (A, B) such that after each of his moves

- $E_A := E(G[V \setminus B])$ is an α-structure of size k in $V \setminus B$,
- for all $e^+ \in e \in E_A$: $e^+ \in A$,
- $E(G[B])$ is an α-structure of size ℓ in $V \setminus A$,
- $|A| - k$ and $|B| - \ell$ do not decrease,
- i.e. $|A| - k = |B| - \ell \geq \frac{n}{6} = n - b$.

Again, let $e = (v, w)$ be the edge OMaker directed in her previous move and assume w.l.o.g. that $\{v, w\} \subseteq V \setminus B$. Then OBreaker adds e to the α-structure $E(G[A])$ by procedure α. Furthermore, he adds v to A by directing all edges (v, b) for $b \in B$. If $v \in A$ already, OBreaker picks a new vertex $v' \in V \setminus (A \cup B)$ and directs all edges (v', b) for $b \in B$. This way, $|A| - k$ does not decrease. Stage II ends when $A \cup B = V$. Since the strategy asks to direct at most $|B| + k \leq V - (|A| - k) \leq b$ edges, OBreaker can follow that strategy.

In the final Stage, the situation is as follows: The vertex set V can be partitioned into two sets A and B such that

- all edges $(a, b) \in E(G)$ are already directed,
- $E(G[A])$ is an α-structure in $V \setminus B = A$,
- $E(G[B])$ is an α-structure in $V \setminus A = B$,
- $|A|, |B| \leq b$ (since $|A|, |B| \geq n - b$ and $V = A \dot{\cup} B$).

OBreaker now follows procedure α inside A or B respectively, depending on the part OMaker plays in. It is evident that OMaker cannot close a cycle: Since she only plays one edge in every round, since $G[V \setminus (A \cup B)] = \emptyset$ throughout the whole game, and since (A, B) is a UDB, any cycle she could close lies completely inside $V \setminus A$ or $V \setminus B$. But the edge sets $E_A := E(G[V \setminus B])$ and $E_B := E(G[V \setminus A])$ form α-structures inside $V \setminus B$ and $V \setminus A$ respectively, so OMaker cannot close a cycle inside these sets.

Strategy for Theorem 1.2

In our strategy for Theorem 1.1, we heavily use that OBreaker may direct fewer than b edges in each round. This way, we have complete control over the structures that evolve. For the strict game, OBreaker needs to be a lot more careful where to put the remaining edges, as OMaker could make use of them to create a directed cycle. Here, we split the strategy into two stages. Stage I consists of exactly one move in which OBreaker claims a UDB (A, B) such that $|A|, |B| \geq n - b \sim \sqrt{n}$. Similar to the proof for the monotone rules, he now plays either inside $V \setminus A$, or $V \setminus B$, depending on the placement of OMaker's directed edge. For the exact details, we refer the reader to our paper [3].

3 Concluding remarks

The upper bound of $5n/6$ in Theorem 1.1 is not tight. As one might expect, playing almost the full bias from the beginning is advantageous for OBreaker. However, the upper bound improves only to roughly $0.82n$ when optimizing Stage I. Our strategy for OBreaker utterly fails when $b \leq 2n/3$. We therefore conjecture that $t(n, \mathcal{P}) \geq 2n/3 + o(n)$.

Concerning the strict rules, OBreaker has to be a lot more careful where to put remaining edges, since any additional edge can be used by OMaker to her advantage. We conjecture that there is a constant $\varepsilon > 0$ such that $t^+(n, \mathcal{P}) \leq (1 - \varepsilon)n$.

References

[1] I. BEN-ELIEZER, M. KRIVELEVICH and B. SUDAKOV, *Biased orientation games*, Discrete Mathematics 312.10 (2012), 1732–1742.

[2] B. BOLLOBÁS and T. SZABÓ, *The oriented cycle game*, Discrete Mathematics 186.1 (1998), 55–67.

[3] D. CLEMENS and A. LIEBENAU, *On the threshold bias in the oriented cycle game*, in preparation

Building spanning trees quickly in Maker-Breaker games

Dennis Clemens[1], Asaf Ferber[2], Roman Glebov[3], Dan Hefetz[4] and Anita Liebenau[5]

Abstract. For a tree T on n vertices, we study the Maker-Breaker game, played on the edge set of the complete graph on n vertices, which Maker wins as soon as the graph she builds contains a copy of T. We prove that if T has bounded maximum degree, then Maker can win this game within $n+1$ moves. Moreover, we prove that Maker can build almost every tree on n vertices in $n-1$ moves and provide non-trivial examples of families of trees which Maker can build in $n-1$ moves.

1 Introduction

Let X be a finite set and let $\mathcal{F} \subseteq 2^X$ be a family of subsets. In the Maker-Breaker game $(X, \mathcal{F})$, two players, called Maker and Breaker, take turns in claiming a previously unclaimed element of X, with Breaker going first. The set X is called the board of the game and the members of $\mathcal{F}$ are referred to as the winning sets. Maker wins this game as soon as she claims all elements of some winning set. If Maker does not fully

[1] Department of Mathematics and Computer Science, Freie Universität Berlin, Germany. Email: d.clemens@fu-berlin.de. Research supported by DFG, project SZ 261/1-1.

[2] School of Mathematical Sciences, Raymond and Beverly Sackler Faculty of Exact Sciences, Tel Aviv University, Tel Aviv, 69978, Israel. Email: ferberas@post.tau.ac.il.

[3] Mathematics Institute and DIMAP, University of Warwick, Coventry CV4 7AL, UK. Previous affiliation: Department of Mathematics and Computer Science, Freie Universität Berlin, Germany. Email: glebov@zedat.fu-berlin.de. Research supported by DFG within the research training group "Methods for Discrete Structures".

[4] School of Mathematics, University of Birmingham, Edgbaston, Birmingham B15 2TT, United Kingdom. Email: d.hefetz@bham.ac.uk. Research supported by an EPSRC Institutional Sponsorship Fund.

[5] Department of Mathematics and Computer Science, Freie Universität Berlin, Germany. Email: liebenau@math.fu-berlin.de. Research supported by DFG within the graduate school Berlin Mathematical School.

claim any winning set by the time every board element is claimed by some player, then Breaker wins the game. We say that the game $(X, \mathcal{F})$ is Maker's win if Maker has a strategy that ensures her win in this game (in some number of moves) against any strategy of Breaker, otherwise the game is Breaker's win. One can also consider a *biased* version in which Maker claims p board elements per move (instead of just 1) and Breaker claims q board elements per move. We refer to this version as a $(p : q)$ game. For a more detailed discussion, we refer the reader to [3].

The following game was studied in [8]. Let T be a tree on n vertices. The board of the *tree embedding game* $(E(K_n), \mathcal{T}_n)$ is the edge set of the complete graph on n vertices and the minimal (with respect to inclusion) winning sets are the labeled copies of T in K_n. Several variants of this game were studied by various researchers (see *e.g.* [2,4]).

It was proved in [8] that for any real numbers $0 < \alpha < 0.005$ and $0 < \varepsilon < 0.05$ and a sufficiently large integer n, Maker has a strategy to win the $(1 : q)$ game $(E(K_n), \mathcal{T}_n)$ within $n + o(n)$ moves, for every $q \leq n^{\alpha}$ and every tree T with n vertices and maximum degree at most n^{ε}. The bounds on the duration of the game, on Breaker's bias and on the maximum degree of the tree to be emdedded, do not seem to be best possible. Indeed, it was noted in [8] that it would be interesting to improve each of these bounds, even at the expense of the other two. In this paper we focus on the duration of the game, while we restrict our attention to the case of bounded degree trees and to unbiased games (that is, the case $q = 1$). The smallest number of moves Maker needs in order to win some Maker-Breaker game is an important game invariant which has received a lot of attention in recent years (see *e.g.* [5–10]).

It is obvious that Maker cannot build any tree on n vertices in less than $n - 1$ moves. This trivial lower bound can be attained for some trees. For example, it was proved in [9] that Maker can build a Hamilton path of K_n in $n - 1$ moves. On the other hand it is not hard to see that there are trees on n vertices which Maker cannot build in less than n moves, *e.g.* the complete binary tree on n vertices. In this paper we prove the following general upper bound which is only one move away from the aforementioned lower bound.

Theorem 1.1. *Let Δ be a positive integer. Then there exists an integer $n_0 = n_0(\Delta)$ such that for every $n \geq n_0$ and for every tree $T = (V, E)$ with $|V| = n$ and $\Delta(T) \leq \Delta$, Maker has a strategy to win the game $(E(K_n), \mathcal{T}_n)$ within $n + 1$ moves.*

As mentioned before, it can be shown that there exist trees on n vertices which Maker cannot claim in less than n moves. Nevertheless, the following theorem suggests that such examples are quite rare.

Theorem 1.2. *Let T be a tree, chosen uniformly at random from the class of all labeled trees on n vertices. Then asymptotically almost surely, T is such that Maker has a strategy to win the game $(E(K_n), \mathcal{T}_n)$ in $n - 1$ moves.*

Moreover, we construct a non-trivial family of trees for which Maker wins the tree embedding game within $n - 1$ moves. We call a path P inside a tree T a *bare path* if all its inner vertices have degree 2 in T. Our result then is the following generalization of Theorem 1.4 from [9].

Theorem 1.3. *Let Δ be a positive integer. Then there exists an integer $m_1 = m_1(\Delta)$ and an integer $n_1 = n_1(\Delta, m_1)$ such that the following holds for every $n \geq n_1$ and for every tree $T = (V, E)$ such that $|V| = n$ and $\Delta(T) \leq \Delta$. If T admits a bare path of length m_1, such that one of its endpoints is a leaf of T, then Maker has a strategy to win the game $(E(K_n), \mathcal{T}_n)$ in $n - 1$ moves.*

2 Outline of the proofs

A fast winning strategy. The proof of Theorem 1.1 highly depends on the existence of a long bare path.

Assume first that there is a bare path $P \subseteq T$ whose length is at least $C_1 = C_1(\Delta)$, where C_1 is a large constant depending only on Δ. Then we split T into this bare path P and a forest of two subtrees $T_1 = (V_1, E_1)$ and $T_2 = (V_2, E_2)$. In a first step, we show that Maker can claim a copy of $T_1 \cup T_2$ within $|V_1| + |V_2| - 2$ moves, while ensuring that Breaker does not claim too many edges that might become dangerous for Maker with respect to the still necessary embedding of P. To do so, we use a nice trick and consider the *method of potential functions,* studied intensively in [3]. Throughout the first step, Maker consistently increases a set $S \subseteq V_1 \cup V_2$ of embedded vertices, which means that Maker claims a copy of the induced subgraph $(T_1 \cup T_2)[S]$, and she also considers a set U consisting of all *available* vertices (*i.e.* vertices in K_n that are not part of the embedding) plus two vertices x_1, x_2 corresponding to the endpoints of P. For every vertex u she defines its potential $\phi(u)$ that measures those Breaker edges incident to u which might become dangerous in the proceeding game. Further, she considers a cumulative potential $\psi = e_B(U) + \sum_{u \text{ open}} \phi(u)$, where $e_B(U)$ denotes the number of Breaker edges inside U and where *open* means that u is already part of the embedding, while Maker still needs to claim edges incident to u (in order to complete the embedding of T). Now Maker's strategy essentially is based on the idea to embed $T_1 \cup T_2$ step by step and to ensure in parallel that ψ never exceeds a given constant $C_2 = C_2(\Delta)$. In a second step, using $e_B(U) \leq$

C_2 after $T_1 \cup T_2$ is fully embedded, we prove that Maker can claim a path from x_1 to x_2 through all remaining available vertices, wasting at most one move. On the one hand this gives us a strengthening of the fast winning result in the Hamilton cycle game from [10], on the other hand this finishes our proof in the first case.

Assume then that the tree T does not admit a bare path of length C_1. We conclude that T contains $\Theta(n)$ leaves and, since its maximum degree is bounded, T also contains a matching M of size $\Theta(n)$ where each edge is incident with some leaf of T. In a first step, we introduce a *danger function*, calling a vertex dangerous if its Breaker degree exceeds some large constant $C_3 = C_3(\Delta)$. With some technical argumentation we prove that Maker can claim a copy of some subtree $T'' \subseteq T$ within $|V(T'')| - 1$ moves such that $E(T) \setminus E(T'')$ is a subset of M of size $\Theta(n)$. Moreover, she can do it in such a way that, at the time when T'' is fully embedded, there is neither an open nor an available vertex that is dangerous. In a second step, giving a stronger inductive statement, we prove that Maker can claim a perfect matching between the available and the open vertices, wasting at most two moves. This way, we complete our proof in the second case, but also give a fast winning result for the perfect matching game on nearly complete bipartite graphs which strengthens the results from [9].

Building trees in optimal time. Recall that in the case where Maker wants to claim a tree with a long bare path, she only wastes one move when she tries to create a path between two designated vertices. A central ingredient in the proofs of Theorem 1.2 and Theorem 1.3 is Maker's ability to build a Hamilton path with exactly one designated vertex as an endpoint in optimal time. The latter is proven by taking a closer look at the Hamilton path game, initiated in [9]. Here, we generalize the known result and give a Maker strategy consisting of five stages that looks more carefully at the actions of Breaker throughout the Hamilton path game.

Assume now that T is a tree, chosen uniformly at random from the class of all labeled trees on n vertices. In order to show that Maker can win in optimal time, we use a nice mixture of results from general graph theory and random graph theory, coupled with our methods introduced in the previous proof. By [1] and [11] we at first observe that asymptotically almost surely the maximum degree is given by $\Delta(T) = (1 + o(1)) \log(n)/\log\log(n)$, while T also contains a large family $\mathcal{P}$ of edge disjoint bare paths, with each path being of size at least $C_4 = C_4(\Delta)$ and having a leaf of T as one of its endpoints. Then, in the first two steps of the game Maker claims a copy of some subtree $T' \subseteq T$ within $|V(T')| - 1$ moves such that $T \setminus T'$ is a union of $\Theta(n)$ bare paths from $\mathcal{P}$.

Moreover, again using a danger function, we guarantee that, when T' is embedded, the Breaker degree at every open/available vertex is bounded by some sufficiently small function $f(n)$. In a third step, using general graph theory, we partition the set of open vertices and available vertices into sets of size $C_4 + 1$, each having the property that so far Breaker did not claim any edge inside and each containing exactly one open vertex. Finally, Maker plays on each of these sets seperately, always claiming a Hamilton path with one designated vertex in optimal time. This finalizes the embedding of T.

3 Concluding remarks and open problems

Building trees in the shortest possible time. As noted in the introduction, there are trees T on n vertices with bounded maximum degree which Maker cannot build in $n - 1$ moves. In this paper we proved that Maker can build such a tree T in at most $n + 1$ moves. We do not believe that there are bounded degree trees that require Maker to waste more than one move. This leads us to make the following conjecture.

Conjecture 3.1. Let Δ be a positive integer. Then there exists an integer $n_0 = n_0(\Delta)$ such that for every $n \geq n_0$ and for every tree $T = (V, E)$ with $|V| = n$ and $\Delta(T) \leq \Delta$, Maker has a strategy to win the game $(E(K_n), \mathcal{T}_n)$ within n moves.

Strong tree embedding games. In the *strong game* $(X, \mathcal{F})$, two players, called Red and Blue, take turns in claiming one previously unclaimed element of X, with Red going first. The winner of the game is the first player to fully claim some $F \in \mathcal{F}$. If neither player is able to fully claim some $F \in \mathcal{F}$ by the time every element of X has been claimed by some player, the game ends in a *draw*.

Strong games are notoriously hard to analyze. However, the use of explicit very fast winning strategies for Maker in a weak game for devising an explicit winning strategy for Red in the corresponding strong game was initiated in [6]. This idea was used to devise such strategies for the strong perfect matching and Hamilton cycle games [6] and for the k-vertex-connectivity game [7]. Since it was proved in [8] that Maker has a strategy to win the weak tree embedding game $(E(K_n), \mathcal{T}_n)$ within $n + o(n)$ moves, it was noted in [7] that one could be hopeful about the possibility of devising an explicit winning strategy for Red in the corresponding strong game. The first step towards this goal is to find a much faster strategy for Maker in the weak game $(E(K_n), \mathcal{T}_n)$. This was accomplished in the current paper.

References

[1] N. ALON, S. HABER and M. KRIVELEVICH, *The number of F-matchings in almost every tree is a zero residue*, The Electronic Journal of Combinatorics **18** (1) (2011), P30.

[2] J. BECK, *Deterministic graph games and a probabilistic intuition*, Combinatorics, Probability and Computing **3** (1994), 13–26.

[3] J. BECK, "Combinatorial Games: Tic-Tac-Toe Theory", Cambridge University Press, 2008.

[4] M. BEDNARSKA, *On biased positional games*, Combinatorics, Probability and Computing **7** (1998), 339–351.

[5] D. CLEMENS, A. FERBER, M. KRIVELEVICH and A. LIEBENAU, *Fast strategies in Maker-Breaker games played on random boards*, Combinatorics, Probability and Computing **21** (2012), 897–915.

[6] A. FERBER and D. HEFETZ, *Winning strong games through fast strategies for weak games*, The Electronic Journal of Combinatorics **18** (1) (2011), P144.

[7] A. FERBER and D. HEFETZ, *Weak and strong k-connectivity games*, European Journal of Combinatorics, to appear.

[8] A. FERBER, D. HEFETZ and M. KRIVELEVICH, *Fast embedding of spanning trees in biased Maker-Breaker games*, European Journal of Combinatorics **33** (2012), 1086–1099.

[9] D. HEFETZ, M. KRIVELEVICH, M. STOJAKOVIĆ and T. SZABÓ, *Fast winning strategies in Maker-Breaker games*, Journal of Combinatorial Theory, Ser. B. **99** (2009), 39–47.

[10] D. HEFETZ and S. STICH, *On two problems regarding the Hamilton cycle game*, The Electronic Journal of Combinatorics **16** (1) (2009), R28.

[11] J. W. MOON, *On the maximum degree in a random tree*, Michigan Math. J. **15** (1968), 429–432.

Dicots, and a taxonomic ranking for misère games

Paul Dorbec[1], Gabriel Renault[1], Aaron Siegel[2] and Éric Sopena[1]

Abstract. We study combinatorial games in misère version. In a general context, little can be said about misère games. For this reason, several universes were earlier considered for their study, which can be ranked according to their inclusion ordering. We study in particular a special universe of games called dicots, which turns out to be the known universe of lowest rank modulo which equivalence in misère version implies equivalence in normal version. We also prove that modulo the dicot universe, we can define a canonical form as the reduced form of a game that can be obtained by getting rid of dominated options and most reversible options. We finally count the number of dicot equivalence classes of dicot games born by day 3.

We study combinatorial games in misère version, and in particular a special universe (*i.e.* family) of games called dicots. We first recall basic definitions, following [1,3,4].

A combinatorial game is a finite two-player game with no chance and perfect information. The players, called Left and Right, alternate moves until one player has no available move. Under the normal convention, the last player to move wins the game while under the misère convention, that player loses the game.

A game can be defined recursively by its sets of options $G = \{G^L | G^R\}$, where G^L is the set of games reachable in one move by Left (called Left options), and G^R the set of games reachable in one move by Right (called Right options). The zero game $0 = \{\cdot|\cdot\}$, is the game with no options. The birthday of a game is defined recursively as $birthday(G) = 1 + \max_{G' \in G^L \cup G^R} birthday(G')$, with 0 being the only game with birthday 0. We say a game G is born on day n if $birthday(G) = n$, and that it is born by day n if $birthday(G) \leq n$. The games born on day 1 are

[1] Univ. Bordeaux, LaBRI, UMR 5800, F-33400 Talence, France CNRS, LaBRI, UMR 5800, F-33400 Talence, France. Email: paul.dorbec@u-bordeaux1.fr, gabriel.renault@labri.fr, Eric.Sopena@labri.fr

[2] Institute for Advanced Study, 1 Einstein Drive, Princeton, NJ 08540.
Email: aaron.n.siegel@gmail.com

$\{0|\cdot\} = 1$, $\{\cdot|0\} = \bar{1}$ and $\{0|0\} = *$.

Given two games $G = \{G^L|G^R\}$ and $H = \{^L H|^R H\}$, we recursively define the (disjunctive) sum of G and H as $G + H = \{G^L + H, G +^L H|G^R + H, G +^R H\}$ (where $G^L + H$ is the set of sums of H and an element of G^L), *i.e.* the game where each player chooses on his turn which one of G and H to play on. One of the main objectives of combinatorial game theory is to determine for a game G the outcome of its sum with any other game.

For both conventions, there are four possible outcomes for a game. Games for which Left player has a winning strategy whatever Right does have outcome $\mathcal{L}$ (for *left*). Similarly, $\mathcal{N}$, $\mathcal{P}$ and $\mathcal{R}$ (for *next*, *previous* and *right*) denote respectively the outcomes of games for which the first player, the second player, and Right has a winning strategy. We note $o^+(G)$ the normal outcome of a game G *i.e.* its outcome under the normal convention and $o^-(G)$ the misère outcome of G. Outcomes are partially ordered according to Figure 1, with greater games being more advantageous for Left. Note that there is no general relationship between the normal outcome and the misère outcome of a game.

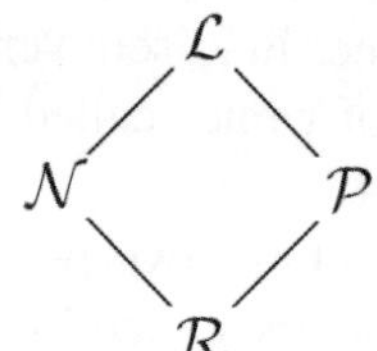

Figure 1. Partial ordering of outcomes.

Given two games G and H, we say that G is greater than or equal to H in misère play whenever Left prefers the game G rather than the game H, that is $G \geq^- H$ if for every game X, $o^-(G + X) \geq o^-(H + X)$. We say that G and H are equivalent in misère play, denoted $G \equiv^- H$, when for every game X, $o^-(G + X) = o^-(H + X)$ (*i.e.* $G \geq^- H$ and $H \geq^- G$). Inequality and equivalence are defined similarly in normal convention, using superscript $+$ instead of $-$.

General equivalence and comparison are very limited in misère play (see [5, 10]), this is why Plambeck and Siegel defined in [8, 9] an equivalence relationship under restricted universes, leading to a breakthrough in the study of misère play games.

Definition 1 ([8,9]). Let $\mathcal{U}$ be a universe of games, G and H two games. We say G is greater than or equal to H modulo $\mathcal{U}$ in misère play and write $G \geq^- H \pmod{\mathcal{U}}$ if $o^-(G + X) \geq o^-(H + X)$ for every $X \in \mathcal{U}$. We say G is equivalent to H modulo $\mathcal{U}$ in misère play and write $G \equiv^- H \pmod{\mathcal{U}}$ if $G \geq^- H \pmod{\mathcal{U}}$ and $H \geq^- G \pmod{\mathcal{U}}$.

For instance, Plambeck and Siegel [8,9] considered the universe of all positions of given games, especially octal games. Other universes have been considered, including the universes of impartial games $\mathcal{I}$ [3,4], dicot games $\mathcal{D}$ [2,6], dead-ending games $\mathcal{E}$ [7], and all games $\mathcal{G}$ [10]. These classes are ordered (ranked) by inclusion as follows:

$$\mathcal{I} \subset \mathcal{D} \subset \mathcal{E} \subset \mathcal{G}.$$

The canonical form of a game is the simplest game of its equivalence class. It is therefore natural to consider canonical forms modulo a given universe. In normal play, impartial games have the same canonical form when considered modulo the universe of impartial games or modulo the universe of all games. In misère play, the corresponding canonical forms are different.

In the following, we focus on the universe of dicots. A game is said to be dicot either if it is $\{\cdot|\cdot\}$ or if it has both Left and Right options and all these options are dicot. Note that the universe of dicots, denoted $\mathcal{D}$ is closed under sum of games and taking option.

Theorem 2. *Let G and H be any games. If $G \geq^-_{\mathcal{D}} H$, then $G \geq^+ H$.*

The dicot universe is the universe of lowest rank known to have this property.

Modulo the dicot universe, we propose a reduced form of a game that can be obtained by getting rid of dominated options and most reversible options.

Theorem 3. *Consider two dicot games G and H. If $G \equiv^-_{\mathcal{D}} H$ and both are in reduced form, then either G and H are the games $0 = \{\cdot|\cdot\}$ and $\{*|*\}$, or there exists a bijection between the Left (resp. Right) options of G and of H such that an option and its image are equivalent modulo $\mathcal{D}$.*

As a consequence, we can define the canonical form of a game as its reduced form, except when the game reduces to $\{*|*\}$, in which case the canonical form is 0.

Thanks to that result, we are able to count the number of dicot equivalence classes (modulo $\mathcal{D}$) of games born by day 3, improving the bound of 5041 proposed by Milley in [6].

Theorem 4. *The* 1046530 *dicot games born by day* 3 *are distributed among* 1214 *equivalence classes modulo $\mathcal{D}$.*

By comparison, Milley proved in [6] that the number of misère dicot equivalence classes of dicot games born by day 2 is 9. In normal play, there are 50 non-equivalent dicot games born by day 3 (both modulo the universe of all games or the universe of dicots).

References

[1] H. ALBERT MICHAEL, R. J. NOWAKOWSKI and D. WOLFE, "Lessons in Play", 2007, A K Peters Ltd.

[2] A. R. MEGHAN, "An Investigation of Misère Partizan Games", PhD thesis, Dalhousie University, 2009.

[3] E. R. BERLEKAMP, J. H. CONWAY and R. K. GUY, "Winning Ways for your Mathematical Plays" (2nd edition), 2001, A K Peters Ltd.

[4] J. H. CONWAY, "On Numbers and Games" (2nd edition), 2001, A K Peters Ltd.

[5] G. A. MESDAL and P. OTTAWAY, *Simplification of partizan games in misère play*, INTEGERS, 7:#G06, 2007. G.A. Mesdal is comprised of M. Allen, J. P. Grossman, A. Hill, N. A. McKay, R.J. Nowakowski, T. Plambeck, A. A. Siegel, D. Wolfe.

[6] R. MILLEY, "Restricted Universes of Partizan Misère Games", PhD thesis, Dalhousie University, 2013.

[7] R. MILLEY and GABRIEL RENAULT, Dead ends in misère play: the misère monoid of canonical numbers, preprint, 2013; available at arxiv 1212.6435.

[8] T. E. PLAMBECK, *Taming the wild in impartial combinatorial games, INTEGERS*, 5:#G5, 36pp., Comb. Games Sect., 2005.

[9] T. E. PLAMBECK and A. N. SIEGEL, *Misère quotients for impartial games*, J. Combin. Theory, Ser. A **115** (4) (2008), 593–622.

[10] A. N. SIEGEL, *Misère canonical forms of partizan games*, arxiv preprint math/0703565.

Avoider-Enforcer star games

Andrzej Grzesik[1], Mirjana Mikalački[2], Zoltán Lóránt Nagy[3], Alon Naor[4], Balázs Patkós[5] and Fiona Skerman[6]

Abstract. We study $(1 : b)$ Avoider-Enforcer games played on the edge set of the complete graph K_n, on n vertices, where Avoider's goal is to avoid claiming a copy of some small fixed graph G. In particular, we give explicit winning strategies for both players in the k-star game, where G is a $K_{1,k}$, for constant $k \geq 2$ under both strict and monotone rules. We also give the winning strategies for both players in another two related monotone games.

1 Introduction

Let V be a finite set and let $\mathcal{F} \subseteq 2^V$. Consider a hypergraph $\mathcal{H} = (V, \mathcal{F})$ with vertex set V and edge set $\mathcal{F}$. The set V is called the *board* and $\mathcal{F}$ the family of *losing sets*. Two players, Avoider and Enforcer, take turns in claiming unoccupied vertices of V until all vertices are claimed and Avoider starts the game. Avoider's goal in the game is to avoid claiming all the elements of any losing set in $\mathcal{F}$, while Enforcer's goal is to force him to do so before the end of the game. Avoider-Enforcer games can be played by two different sets of rules [9]. Let a and b be positive integers called the *biases* of Avoider, respectively Enforcer. By the first set of rules, in the $(a : b)$ Avoider-Enforcer game, Avoider claims exactly a

[1] Theoretical Computer Science Department, Faculty of Mathematics and Computer Science, Jagiellonian University, ul. Prof. St. Lojasiewicza 6, 30-348 Krakow, Poland.
Email: andrzej.grzesik@uj.edu.pl

[2] Department of Mathematics and Informatics, University of Novi Sad, Serbia.
Email: mirjana.mikalacki@dmi.uns.ac.rs. Research partly supported by Ministry of Education and Science, Republic of Serbia, and Provincial Secretariat for Science, Province of Vojvodina.

[3] Alfréd Rényi Institute of Mathematics, P.O.B. 127, Budapest H-1364, Hungary.
Email: nagy.zoltan.lorant@renyi.mta.hu. Supported by Hungarian National Scientific Research Funds (OTKA) grant 81310.

[4] School of Mathematical Sciences, Raymond and Beverly Sackler Faculty of Exact Sciences, Tel Aviv University, Tel Aviv, 69978, Israel. Email: alonnaor@post.tau.ac.il

[5] Alfréd Rényi Institute of Mathematics, P.O.B. 127, Budapest H-1364, Hungary.
Email: patkos.balazs@renyi.mta.hu. Research is supported by OTKA Grant PD-83586 and the János Bolyai Research Scholarship of the Hungarian Academy of Sciences.

[6] University of Oxford, Department of Statistics, 1 South Parks Road, Oxford OX1 3TG, United Kingdom. Email: skerman@stats.ox.ac.uk

vertices and Enforcer claims exactly b vertices per move. This set of rules is called the *strict rules*, and games played by these rules are referred to as *strict games*. By the second set of rules, the *monotone rules*, Avoider and Enforcer claim at least a, respectively at least b vertices, per move, and games played by monotone rules are called *monotone games*. In any move of either player, if there are less unclaimed vertices than what a player should claim in his turn, he must claim all the remaining vertices and the game is then over. The game is an *Enforcer's win* if, at any point of the game Avoider has claimed all the vertices of at least one losing set $F \in \mathcal{F}$. Otherwise, the game is an *Avoider's win*.

One of the main advantages of the monotone rules is that they are bias monotone. In monotone Avoider-Enforcer games, if the $(a : b)$ game is an Enforcer's win, then the $(a + 1 : b)$ and $(a : b - 1)$ games are also won by Enforcer. Similarly, if the $(a : b)$ game is an Avoider's win, then the $(a : b + 1)$ and $(a - 1 : b)$ games are also won by Avoider.

When $a = b = 1$, we call such games *unbiased*. Unbiased Avoider-Enforcer games were studied *e.g.* in [3] and [11]. In an unbiased game, it is also interesting to see how fast Enforcer can force Avoider to lose, or to see how long can Avoider defend himself before losing. These type of problems were considered in [1,7].

In [11], Lu proved that Beck's [2] generalization of the Erdős-Selfridge criterion [6] gave sufficient conditions for an Avoider's win in the $(1 : 1)$ Avoider-Enforcer game. In [10], Hefetz, Krivelevich and Szabó gave a general winning criterion for Avoider in $(a : b)$ Avoider-Enforcer games played by both sets of rules. This criterion takes only Avoider's bias into account. In [4] a new criterion for Avoider's win in both strict and monotone $(a : b)$ games on $\mathcal{H}$ is introduced, which depends on both biases a and b. Note, however, that these criteria are non-constructive and do not give the strategy of the winning player.

The focus of our research is on $(1 : b)$ Avoider-Enforcer games, played by both monotone and strict rules on a graph where the board is the set of edges of the complete graph on n vertices, *i.e.* $V = E(K_n)$. This type of game appears frequently in the literature (see, for example [8] and [9]).

We follow the terminology for strict Avoider-Enforcer games introduced by Hefetz, Krivelevich and Szabó in [10]. The *upper threshold bias* $f^+_{\mathcal{H}}$ is the smallest integer such that for every integer b, $b > f^+_{\mathcal{H}}$, the $(1 : b)$ game on $\mathcal{H}$ is an Avoider's win. The *lower threshold bias* $f^-_{\mathcal{H}}$ is the largest integer such that for every integer b, $b \leq f^-_{\mathcal{H}}$, the $(1 : b)$ game on $\mathcal{H}$ is an Enforcer's win. The inequality $f^-_{\mathcal{H}} \leq f^+_{\mathcal{H}}$ always holds, but the upper and lower threshold biases can be close to each other in the case of some games, like for example in Connectivity game [10] or be far apart. When $f^-_{\mathcal{H}} = f^+_{\mathcal{H}}$ we call this number $f_{\mathcal{H}}$ and refer to it as the *threshold bias* of the game $\mathcal{H}$. This threshold bias may not exist for some games.

For monotone $(1 : b)$ Avoider-Enforcer game on $\mathcal{H}$, there is a unique *monotone threshold bias* $f_{\mathcal{H}}^{mon}$ defined as the largest integer value such that for every integer b, $b \leq f_{\mathcal{H}}^{mon}$, the game is an Enforcer's win. The question that arises is whether $f_{\mathcal{H}}^{-} \leq f_{\mathcal{H}}^{mon} \leq f_{\mathcal{H}}^{+}$ hold for all $(1 : b)$ Avoider-Enforcer games. The results from [9] and [10] show that it does not hold in general.

We are interested in $(1 : b)$ Avoider-Enforcer games where Avoider wants to avoid claiming a copy of some small fixed graph G. This problem was studied in [9] for some graphs G. Let $\mathcal{K}_G$ denote the hypergraph whose edges are the edge sets of all the copies of G in K_n. In [9] the authors analysed games where G is K_3 and P_3 respectively and gave the thresholds for both the monotone and strict game $\mathcal{K}_{P_3}$ and for the monotone game $\mathcal{K}_{K_3}$. They showed that
$f_{\mathcal{K}_{P_3}}^{mon} = \binom{n}{2} - \lfloor \frac{n}{2} \rfloor - 1$, $f_{\mathcal{K}_{P_3}}^{+} = \binom{n}{2} - 2$, $f_{\mathcal{K}_{P_3}}^{-} = \Theta(n^{\frac{3}{2}})$ and $f_{\mathcal{K}_{K_3}}^{mon} = \Theta(n^{\frac{3}{2}})$.
Bednarska-Bzdęga in [4] showed that $f_{\mathcal{K}_{K_3}}^{-} = \Omega(n^{\frac{1}{2}})$. Clemens et al. in [5] showed that in a monotone game where losing sets are the edges of all P_4, the threshold bias is $f_{\mathcal{K}_{P_4}}^{mon} = \frac{1}{2}\binom{n}{2} - \frac{n}{2}\left(\frac{1}{\sqrt{2}} - o(1)\right)$. Moreover, in [9], the authors conjectured that in general $f_{\mathcal{K}_G}^{+}$ and $f_{\mathcal{K}_G}^{-}$ are not of the same order and asked about the strategies in games where G is some fixed graph on more than 4 vertices. Bednarska-Bzdęga established in [4] general upper and lower bounds on $f_{\mathcal{K}_G}^{+}$, $f_{\mathcal{K}_G}^{-}$ and $f_{\mathcal{K}_G}^{mon}$ for every fixed graph G, but these bounds are not tight.

2 Results

In the present paper, we study the game $\mathcal{K}_G$ where G is a k-star $K_{1,k}$, for some fixed $k \geq 2$ and denote it by $\mathcal{S}_k$. We call the game *k-star game*.

In order to state our main result, we have to introduce some functions: let us define $r = r(n, b)$ by $1 \leq r \leq b+1$ and $\binom{n}{2} \equiv r \bmod (b+1)$. Note that r is the number of edges that Avoider is allowed to choose from in his last move when playing the strict $(1 : b)$ game. Let
$e_{n,k}^{+} = \max\left\{b : r(b+1) < \frac{1}{8}\frac{n^k}{(2b)^{k-2}}\right\}$, and
$e_{n,k}^{-} = \max\left\{b' : \forall b, \frac{1}{4}n^{\frac{k+1}{k}} \leq b \leq b' : \ r(b+1) < \frac{1}{8}\frac{n^k}{(2b)^{k-2}}\right\}$.
The main result in our paper is the following theorem.

Theorem 2.1. *In $(1 : b)$ k-star game $\mathcal{S}_k$, $k \geq 2$, we have*

(i) $f_{\mathcal{K}_{\mathcal{S}_k}}^{mon} = \Theta(n^{\frac{k}{k-1}})$,

(ii) $e^+_{n,k} \leq f^+_{\mathcal{K}_{\mathcal{S}_k}} = O(n^{\frac{k}{k-1}})$ *holds for all values of* n, *and*

$f^+_{\mathcal{K}_{\mathcal{S}_k}} = \Theta(n^{\frac{k}{k-1}})$ *holds for infinitely many values of* n,

(iii) $\max\{\frac{1}{2}n^{\frac{k+1}{k}}, e^-_{n,k}\} \leq f^-_{\mathcal{K}_{\mathcal{S}_k}} = O(n^{\frac{k+1}{k}} \log n)$ *holds for all values of* n, *and* $f^-_{\mathcal{K}_{\mathcal{S}_k}} = \Theta(n^{\frac{k+1}{k}})$ *holds for infinitely many values of* n.

We also consider two more monotone games similar to the k-star game. Let the *double star* $\mathcal{S}_{k,k}$ be a graph on $2k$ vertices $u, u_1, \ldots, u_{k-1}, v, v_1, \ldots, v_{k-1}$ such that the edge set of $\mathcal{S}_{k,k}$ is $\{(uv)\} \cup \{(uu_i) : 1 \leq i \leq k-1\} \cup \{(vv_i) : 1 \leq i \leq k-1\}$ (see Figure 2.1) and let $\mathcal{K}_{\mathcal{S}_{k,k}}$ be the hypergraph of the game.

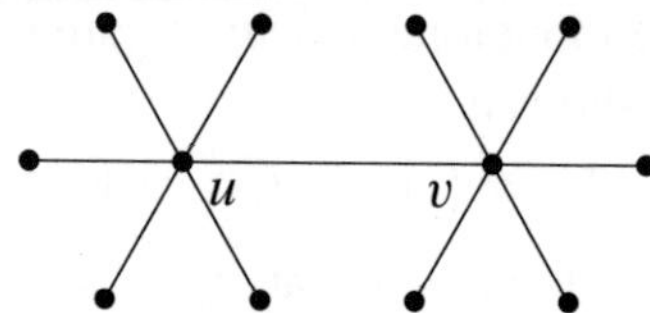

Figure 2.1. $\mathcal{S}_{6,6}$ on vertices (u, v).

Let the *path double star* $\mathcal{PS}_{k,k}$ be a graph on $2k+1$ vertices $w, u, u_1, \ldots, u_{k-1}, v, v_1, \ldots, v_{k-1}$ such that $E(\mathcal{PS}_{k,k}) = \{\{(u, u_i), 1 \leq i \leq k-1\} \cup \{(v, v_i), 1 \leq i \leq k-1\} \cup (v, w) \cup (u, w)\}$, as shown in Figure 2.2, and let $\mathcal{K}_{\mathcal{PS}_{k,k}}$ be the hypergraph of the game.

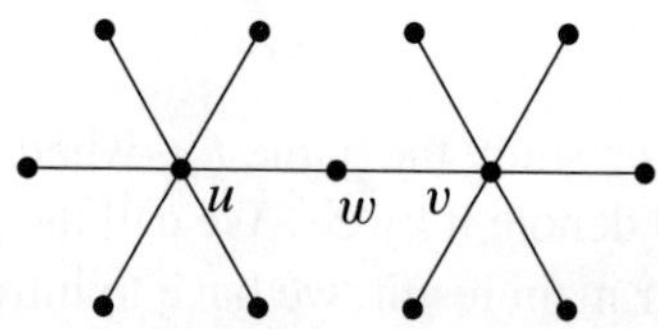

Figure 2.2. $\mathcal{PS}_{6,6}$ on vertices (u, v, w).

Theorem 2.2. *Let* $k \geq 2$. *In* $(1 : b)$ *double star* $\mathcal{S}_{k,k}$ *and path double star* $\mathcal{PS}_{k,k}$ *games, we have*

(i) $f^{mon}_{\mathcal{K}_{\mathcal{S}_{k,k}}} = \Theta(n^{\frac{k}{k-1}})$,

(ii) $f^{mon}_{\mathcal{K}_{\mathcal{PS}_{k,k}}} = \Theta(n^{\frac{k+1}{k}})$.

References

[1] J. BARÁT and M. STOJAKOVIĆ, *On winning fast in Avoider-Enforcer games*, The Electronic Journal of Combinatorics **17** (2010), R56

[2] J. BECK, *Remarks on positional games. I*, Acta Math. Acad. Sci. Hungar. **40** (1982), 65–71.

[3] J. BECK, "Combinatorial Games: Tic-Tac-Toe Theory", 1st ed., Encyclopedia of Mathematics and its Applications, vol. 114, Cambridge University Press, 2008.

[4] M. BEDNARSKA-BZDĘGA, *Degree and Small-graph Avoider-Forcer games*, manuscript.

[5] D. CLEMENS, R. HOD, A. LIEBENAU, D. VU and K. WELLER, personal communication.

[6] P. ERDŐS and J. SELFRIDGE, *On a combinatorial game*, J. of Combinatorial Theory, Ser. A **14** (1973) 298–301.

[7] D. HEFETZ, M. KRIVELEVICH, M. STOJAKOVIĆ and T. SZABÓ, *Fast winning strategies in Avoider-Enforcer games*, Graphs and Combinatorics **25** (2009), 533–544.

[8] D. HEFETZ, M. KRIVELEVICH, M. STOJAKOVIĆ and T. SZABÓ, *Planarity, colorability and minor games*, SIAM Journal on Discrete Mathematics **22** (2008), 194–212.

[9] D. HEFETZ, M. KRIVELEVICH, M. STOJAKOVIĆ and T. SZABÓ, *Avoider-Enforcer: the rules of the game*, Journal of Combinatorial Theory Series A **117** (2010), 152–163.

[10] D. HEFETZ, M. KRIVELEVICH and T. SZABÓ, *Avoider-Enforcer games*, Journal of Combinatorial Theory Series A **114** (2007), 840–853.

[11] X. LU, *A Matching Game*, Discrete Mathematics **94** (1991), 199–207.

Algebra and Polynomials

Fooling-sets and rank in nonzero characteristic

Mirjam Friesen[1] and Dirk Oliver Theis[2]

Abstract. An $n \times n$ matrix M is called a *fooling-set matrix of size n*, if its diagonal entries are nonzero, whereas for every $k \neq \ell$ we have $M_{k,\ell} M_{\ell,k} = 0$. Dietzfelbinger, Hromkovič, and Schnitger (1996) showed that $n \le (\mathrm{rk} M)^2$, regardless of over which field the rank is computed, and asked whether the exponent on $\mathrm{rk} M$ can be improved.
We settle this question for nonzero characteristic by constructing a family of matrices for which the bound is asymptotically tight. The construction uses linear recurring sequences.

1 Introduction

An $n \times n$ matrix M over some field $\mathbb{K}$ is called a *fooling-set matrix of size n* if

$$M_{kk} \neq 0 \quad \text{for all } k \text{ (its diagonal entries are all nonzero), and} \tag{1.1a}$$

$$M_{k,\ell}\, M_{\ell,k} = 0 \quad \text{for all } k \neq \ell. \tag{1.1b}$$

Note that the definition depends only on the zero-nonzero pattern of M. The word "fooling set" originates from Communication Complexity, but the concept is used under different names in other contexts.

In Communication Complexity and Combinatorial Optimization, one is interested in finding a large fooling-set (sub-)matrix contained in a given matrix A (permutation of rows and columns is allowed), as its size provides a lower bound to other numerical properties of the matrix. Since large fooling-set submatrices are typically difficult to identify (the problem is equivalent to finding a large clique in a graph of a certain type), one would like to upper-bound the size of a fooling-set matrix one may possibly hope for in terms of easily computable properties of A.

Dietzfelbinger, Hromkovič, and Schnitger ([4, Thm. 1.4], or see [10, Lemma 4.15]; *cf.* [5, 8]) proved that the rank of a fooling-set matrix of size n is at least $\sqrt{n}$, *i.e.*,

$$n \le (\mathrm{rk}_{\mathbb{K}} M)^2. \tag{1.2}$$

This inequality gives such an upper bound on the largest fooling-set submatrix in terms of the easily computable rank of A.

[1] Faculty of Mathematics, Otto von Guericke University Magdeburg, Germany

[2] Faculty of Mathematics and Computer Science, University of Tartu, Estonia.
Email: dirk.oliver.theis@ut.ee

However, it is an open question whether the exponent on the rank in the right-hand side of (1.2) can be improved or not. Dietzfelbinger *et al.* [4, Open Problem 2] were particularly interested in 0/1-matrices and $\mathbb{K} = \mathbb{F}_2$, which corresponds to the Communication Complexity situation they dealt with.

Klauck and de Wolf [8] have pointed out the importance for Communication Complexity of the question regarding general (*i.e.*, not 0/1) matrices.

Currently, the examples (attributed to M. Hühne in [4]) of 0/1 fooling-set matrices M with smallest rank are such that $n \approx (\mathrm{rk}_{\mathbb{F}_2} M)^{\log_4 6}$ ($\log_4 6 = 1.292\dots$); for general matrices, Klauck and de Wolf [8] have given examples with $n \approx (\mathrm{rk}_{\mathbb{Q}} M)^{\log_3 6}$ ($\log_3 6 = 1.63\dots$).

In our paper, we settle the question for fields $\mathbb{K}$ of nonzero characteristic. We prove that inequality (1.2) is asymptotically tight if the characteristic of $\mathbb{K}$ is nonzero. Notably, not only is the exponent on the rank in inequality (1.2) best possible, but so is the constant (one) in front of the rank.

Organization of this extended abstract. In the next section we will explain some of the connections of the fooling-set vs. rank problem with Combinatorial Optimization and Graph Theory concepts. In Section 3, we will sketch the proof of our result. In the final section, we point to some questions which remain open.

2 Some remarks on the importance of fooling-set matrices

While the fooling-set size vs. rank problem is of interest in its own right as a minimum-rank type problem in Combinatorial Matrix Theory, fooling-set matrices are connected to other areas of Mathematics and Computer Science.

In Polytope Theory, given a polytope P, sizes of fooling-set submatrices of appropriately defined matrices provide lower bounds to the number of facets of any polytope Q which can be mapped onto P by a projective mapping ([14], *cf.* [5]). Similarly, in **Combinatorial Optimization,** sizes of fooling-set matrices are lower bounds to the minimum sizes of Linear Programs for combinatorial optimization problems ([14]). For example, it is an open question whether Edmond's matching polytope for a complete graph on n vertices admits a fooling-set matrix whose size grows quicker in n than the dimension of the polytope. Such a fooling-set matrix would yield a fairly spectacular improvement on the currently known lower bounds of sizes of Linear Programming formulations for the matching problem. See [5] for bounds based on fooling sets for a number of combinatorial optimization problems, including bipartite matching.

In the Polytope Theory / Combinatorial Optimization applications, we typically have $\mathbb{K} = \mathbb{Q}$, and the rank of the large matrix A is known. However, since the definition of a fooling-set matrix depends only on the zero-nonzero pattern, changing the field from $\mathbb{Q}$ to $\mathbb{K}'$ and replacing the nonzero rational entries of A by nonzero numbers in $\mathbb{K}'$ may yield a lower rank and hence a better upper bound on the size of a fooling-set matrix.

In Computational Complexity, fooling-set matrices provide lower bounds for the communication complexity of Boolean functions (see, *e.g.*, [1, 4, 8, 10, 12]), and for the number of states of an automaton accepting a given language (*e.g.*, [6]).

As an example from Communication Complexity where the "fooling-set method" can be seen to yield a poor lower bound is the inner product function[3]

$$f(x, y) = \sum_{j=1}^{n} x_j y_j, \qquad \text{for } x, y \in \mathbb{Z}_2^n.$$

The rank of the associated $2^n \times 2^n$-matrix is n, hence, by (1.2), there is no fooling-set sub-matrix larger than n^2.

In Graph Theory, a fooling-set matrix (up to permutation of rows and columns) can be understood as the incidence matrix of a bipartite graph containing a perfect cross-free matching. Recall that a matching in a bipartite graph H is called *cross-free* if no two matching edges induce a C_4-subgraph of H.

Cross-free matchings are best known as a lower bound on the size of biclique coverings of graphs (*e.g.* [3,7]). A *biclique covering* of a graph G is a collection of complete bipartite subgraphs of G such that each edge of G is contained in at least one of these bipartite subgraphs. If a cross-free matching of size n is contained as a subgraph in G, then at least n bicliques are needed to cover all edges of G. (For some classes of graphs, this is a sharp lower bound on the biclique covering number [3, 13]).

In Matrix Theory, the maximum size of a fooling-set sub-matrix is known under a couple of different names, *e.g.* as independence number [2, Lemma 2.4]), or as the intersection number. For some semirings, this number provides a lower bound for the so-called factorization rank of the matrix over the semiring.

In each of these areas, fooling-set matrices are used as lower bounds. Upon embarking on a search for a big fooling-set matrix in a large, complicated matrix A, one is interested in an *a priori* upper bound on their sizes and thus the potential usefulness of the lower bound method.

[3] Thanks to one of the referees for pointing us to this example.

3 Fooling-Set Matrices from Linear Recurring Sequences

For a prime number p, we denote by $\mathbb{F}_p$ the finite field with p elements. The following is an accurate statement of our result.

Theorem 3.1. *For every prime number p, there is a family of fooling-set matrices $M^{(t)}$ over $\mathbb{F}_p$ of size $n^{(t)}$, $t = 1, 2, 3, \dots$, such that $n^{(t)} \to \infty$, and*

$$\frac{n^{(t)}}{(\mathrm{rk}_{\mathbb{F}_p} M^{(t)})^2} \longrightarrow 1.$$

The method used in all the *earlier* examples (mentioned in the introduction) of fooling-set matrices with small rank was the following: One conjures up a single, small fooling-set matrix M^0 (of size, say, 6), determines its rank (say, 3), and then uses the tensor-powers of M^0 (which are fooling-set matrices, too). With these numerical values, from M^0, one obtains $\log_3 6$ as a lower bound on the exponent on the rank in (1.2).

Our technique is a departure from that approach. As noted above, we use linear recurring sequences. For every t, we construct an $n^{(t)}$-periodic function, which gives us a fooling-set matrix of size $n^{(t)}$.

We now describe that construction. Let p be a prime number and $r \geq 2$ an integer. Define the function $f : \mathbb{Z} \to \mathbb{F}_p$ by the recurrence relation

$$f(k+r) = -f(k) - f(k+1) \quad \text{for all } k \in \mathbb{Z} \tag{3.1a}$$

and the initial conditions

$$f(0) = 1, \text{ and } f(1) = \ldots = f(r-1) = 0. \tag{3.1b}$$

Fix an integer $n > r$. From the sequence, we define an $n \times n$ matrix as follows. For ease of notation, the matrix indices are taken to be in $\{0, \dots, n-1\} \times \{0, \dots, n-1\}$. We let

$$M_{k,\ell} := f(k-\ell). \tag{3.2}$$

It is fairly easy to see that $\mathrm{rk} M \leq r$.

Lemma 3.2. *The rank of M is at most r.*

Proof. From (3.1a), for $k \geq r$, we deduce the equation $M_{k,\star} = -M_{k-r,\star} - M_{k-r+1,\star}$. Hence, each of the rows $M_{k,\star}$, $k \geq r$, is a linear combination of the first r rows of M. □

It can be seen that the rank is, in fact, equal to r: The top-left $r \times r$ sub-matrix is non-singular because it is upper-triangular with nonzeros along the diagonal.

Next, we reduce the fooling-set property (1.1) to a property of the function f.

Lemma 3.3. *The matrix M defined in* (3.2) *is a fooling-set matrix, if and only if,*

$$f(k)f(-k) = 0 \quad \text{for all } k \in \{1, \ldots, n-1\}. \tag{3.3}$$

Proof. It is clear from (3.1b) and (3.2) that $M_{j,j} = f(0) = 1$ for all $j = 0, \ldots, n-1$, so it remains to verify (1.1b). Since

$$M_{i,j}M_{j,i} = f(i-j)f(j-i) = f(i-j)f(-(i-j)),$$

if $f(k)f(-k) = 0$ for all $k = 1, \ldots, n-1$, then $M_{i,j}M_{j,i}$ is zero whenever $i \neq j$. This proves (1.1b). □

Given appropriate conditions on r and n (depending on p), this condition on f can indeed be verified:

Lemma 3.4. *For all integers $t \geq 1$, if we let $r := p^t + 1$ and $n := r(r-1)+1$, then $f(k)f(-k) = 0$ for all $k \in \mathbb{Z} \setminus n\mathbb{Z}$.*

Combining the above three lemmas, we can complete the proof of Theorem 3.1.

Proof of Theorem 3.1. Let p be a prime number. For every integer $t \geq 1$, let $r := p^t + 1$ and $n^{(t)} := r(r-1)+1$, and define the matrix $M^{(t)} := M$ over $\mathbb{F}_p$ as in (3.2). By Lemma 3.2, the rank of $M^{(t)}$ is at most r, and from Lemmas 3.3 and 3.4 we conclude that $M^{(t)}$ is a fooling-set matrix. Hence, we have

$$1 \geq \frac{n^{(t)}}{\mathrm{rk}_{\mathbb{F}_p}(M^{(t)})^2} \geq \frac{r^2 - r + 1}{r^2} \geq 1 - p^{-t}/4 \xrightarrow{t \to \infty} 1,$$

where the left-most inequality is from (1.2). □

To prove Lemma 3.4, we need two more lemmas. The first one states that in every section $\{jr, \ldots, (j+1)r - 1\}$, $j = 0, 1, \ldots,$ there is a block of zeros whose length decreases with j.

Lemma 3.5. *For $j = 0, \ldots, r-2$, we have*

$$f(jr + i) = 0 \quad \text{for } i = 1, \ldots, r-1-j. \tag{3.4}$$

Proof. Equation (3.4) is true for $j = 0$ by (3.1b). Suppose (3.4) holds for some $j < r-2$. Then $f((j+1)r+i) = 0$ for $i = 1, \ldots, r-1-(j+1)$, because, by (3.1a),

$$f((j+1)r+i) = f(jr+i+r) = -f(jr+i) - f(jr+(i+1)) = -0-0$$

holds. □

Every function on $\mathbb{Z}$ with values in a finite field which is defined by a (reversible) linear recurrence relation is periodic (*cf. e.g.* [11]). The second lemma establishes that a specific number n is a period of f as defined in (3.1).

Lemma 3.6. *If* $r = p^t + 1$ *for some integer* $t \geq 1$, *then* $n := r(r-1)+1$ *is a period of the function* f.

This lemma is the difficult part of the proof of Theorem 3.1. Due to the space limitations, for its proof, we have to refer to the full paper. At this point, suffice it to say that the argument proceeds by identifying binomial coefficients among the values of f, and then uses the known periodicity of the binomial coefficients modulo p.

Lemmas 3.5 and 3.6 allow us to prove Lemma 3.4.

Proof of Lemma 3.4. We need to show $f(k)f(-k) = 0$ whenever $n \nmid k$. By Lemma 3.6, this is equivalent to showing $f(k)f(n-k) = 0$ for $k = 1, \ldots, n-1$. Given such a k, let j, i be such that $k = jr + i$ and $0 \leq i \leq r-1$.

If $i \leq r-1-j$, then $f(k) = 0$ by Lemma 3.5, and we are done. If, on the other hand, $i > r-1-j$, then

$$n - k = r^2 - r + 1 - jr - i = (r - 1 - (j+1))r + (r - i + 1),$$

and $r - i + 1 \leq j + 1$, so, by Lemma 3.5, we have $f(n-k) = 0$. □

4 Conclusion

Dietzfelbinger *et al.*'s original question regarding the tightness of inequality (1.2) for 0/1-matrices remains open in characteristic $p > 2$. For these matrices, it may still be possible that the exponent on the rank in the inequality (1.2) can be improved.

For characteristic zero, Klauck and de Wolf [8] have given an example of a fooling-set matrix of size 6 with entries in $\{0, \pm 1\}$ which has rank 3. Thus, using the method sketched above (following Theorem 3.1), the exponent on the rank in inequality (1.2) with $\mathbb{K} := \mathbb{Q}$ for general (*i.e.*, not 0/1) matrices is at least $\log_3 6 = 1.63\ldots$, while the best known bound for 0/1-matrices is $\log_4 6 = 1.292\ldots$.

We would like to point out the possibility that, in characteristic zero, the minimum achievable rank on the right hand side of inequality (1.2) may depend not only on the characteristic, but on the field $\mathbb{K}$ itself. Indeed, there are examples of zero-nonzero patterns for which the minimum rank of a matrix with that zero-nonzero pattern differs between $\mathbb{K} = \mathbb{Q}$ and $\mathbb{K} = \mathbb{R}$, see *e.g.* [9]. Hence, for characteristic zero, we ask the following weaker version of Dietzfelbinger *et al.*'s question.

Question 4.1. Is there a field $\mathbb{K}$ (of characteristic zero) over which the fooling-set matrix size vs. rank inequality in (1.2) can be improved?

As mentioned in Section 2, another question in characteristic zero comes from polytope theory. Let P be a polytope. Let A be a matrix whose rows are indexed by the facets of P and whose columns are indexed by the vertices of P, and which satisfies $A_{F,v} = 0$, if $v \in F$, and $A_{F,v} \neq 0$, if $v \notin F$. For any fooling-set submatrix of size n of A, the following inequality follows from (1.2) (*cf.* [5]):

$$n \leq (\dim P + 1)^2. \tag{4.1}$$

The following variant of Dietzfelbinger *et al.*'s question is of pertinence in Polytope Theory and Combinatorial Optimization (see Section 2).

Question 4.2. Can the fooling-set size vs. dimension inequality (4.1) be improved (for polytopes)?

To our knowledge, the best known lower bound for the best possible exponent on the dimension in inequality (4.1) is 1.

Finally, the complexity of the Fooling-Set-Submatrix problem is still open:

Conjecture 4.3. The *Fooling-Set-Submatrix problem*

Input: Integers n, m and $m \times m$ 0/1-matrix A

Output: "Yes", if a fooling-set submatrix of size n of A exists, "No" otherwise.

is NP-hard.

References

[1] S. ARORA and B. BARAK, "Computational Complexity", Cambridge University Press, Cambridge, 2009, A modern approach. MR 2500087 (2010i:68001) 2

[2] J. E. COHEN and U. G. ROTHBLUM, *Nonnegative ranks, decompositions, and factorizations of nonnegative matrices*, Linear Algebra Appl. **190** (1993), 149–168. MR 1230356 (94i:15015) 3

[3] M. DAWANDE, *A notion of cross-perfect bipartite graphs*, Inform. Process. Lett. **88** (4) (2003), 143–147. MR 2009283 (2004g:05118) 3

[4] M. DIETZFELBINGER, J. HROMKOVIČ and G. SCHNITGER, *A comparison of two lower-bound methods for communication complexity*, Theoret. Comput. Sci. **168** (1) (1996), 39–51, 19th International Symposium on Mathematical Foundations of Computer Science (Košice, 1994). MR 1424992 (98a:68068) 1, 2

[5] S. Fiorini, V. Kaibel, K. Pashkovich and D. O. Theis, *Combinatorial bounds on nonnegative rank and extended formulations*, http://arxiv.org/abs/1111.0444, arXiv:1111.0444) Discrete Math. (2013), to appear.

[6] H. Gruber and M. Holzer, *Finding lower bounds for nondeterministic state complexity is hard (extended abstract)*, Developments in language theory, Lecture Notes in Comput. Sci., Vol. 4036, Springer, Berlin, 2006, pp. 363–374. MR 2334484 2

[7] S. Jukna and A. S. Kulikov, *On covering graphs by complete bipartite subgraphs*, Discrete Math. **309** (10) (2009), 3399–3403. MR 2526759 (2010h:05231) 3

[8] H. Klauck and R. de Wolf, *Fooling one-sided quantum protocols*, http://arxiv.org/abs/1204.4619, arXiv:1204.4619, 2012. 1, 2, 5

[9] S. Kopparty and K. P. S. Bhaskara Rao, *The minimum rank problem: a counterexample*, Linear Algebra Appl. **428** (7) (2008), 1761–1765. MR 2388655 (2009a:15002) 6

[10] E. Kushilevitz and N. Nisan, "Communication Complexity", Cambridge University Press, Cambridge, 1997. MR 1426129 (98c:68074) 1, 2

[11] R. Lidl and H. Niederreiter, "Introduction to Finite Fields and their Applications", first ed., Cambridge University Press, Cambridge, 1994. MR 1294139 (95f:11098) 5

[12] L. Lovász and M. Saks, *Möbius functions and communication complexity*, Proc. 29th IEEE FOCS, IEEE, 1988, 81–90.

[13] J. A. Soto and C. Telha, *Jump number of two-directional orthogonal ray graphs*, Integer programming and combinatorial optimization, Lecture Notes in Comput. Sci., vol. 6655, Springer, Heidelberg, 2011, 389–403. MR 2820923 (2012j:05305) 3

[14] M. Yannakakis, *Expressing combinatorial optimization problems by linear programs*, J. Comput. System Sci. **43** (3) (1991), 441–466. MR 1135472 (93a:90054) 2

Krasner near-factorizations and 1-overlapped factorizations

Tadashi Sakuma[1] and Hidehiro Shinohara[2]

Abstract. Near and/or 1-overlapped factorizations on cyclic groups play important roles both in perfect graph theory and ideal clutter theory. Such a factorization is *Krasner* if its construction does not need any modulo operation (i.e. every addition can be thought as the addition of integers). In this paper, we characterize Krasner near-factorizations and 1-overlapped factorizations, which solves a problem posed by S. Szabó and A.D. Sands [9].

1 Introduction

In this paper, let G denote an abelian group. For two subsets S_1 and S_2 of G, let $S_1 + S_2$ denote the multiset $\{a + b \mid a \in S_1, b \in S_2\}$. If S_2 has only one element g, we use $S_1 + g$ instead of $S_1 + \{g\}$. A sum of two sets $S_1 + S_2$ is *direct* if all elements are distinct. A subset $S \subset G$ is *symmetric* respect to an element $x \in G$ if $S + x = -S - x$, where this x is called the *center* of A. Especially, a subset $S \subset G$ is *symmetric* if it is symmetric respect to 0. A subset $S \subset G$ is *shift-symmetric* if there exists an element $g \in G$ such that $S = -S + g$. For a set $S \subset G$ and an element $g \in G$, let gS or Sg denote the set $\{gs \mid s \in S\}$.

A pair (A, B) of a finite cyclic group $\mathbb{Z}_n$ with $\min\{|A|, |B|\} \geq 2$ is a *factorization* if $A + B$ equals $\mathbb{Z}_n$, and a near (resp. 1-overlapped) factorization if $A + B$ equals $G \setminus \{g\}$ (resp. $G \cup \{g\}$) for some element $g \in \mathbb{Z}_n$ which is called the *uncovered element* (resp. the *doubly covered element*) of (A, B). For two integers a and b, let $[a, b]$ be the set $\{i \in \mathbb{Z} \mid a \leq i \leq b\}$. A pair (A, B) of subsets of $\mathbb{Z}$ is a *factorization of*

[1] Faculty of Education, Art and Science, Yamagata University, Yamagata, Japan.
Email: sakuma@e.yamagata-u.ac.jp

[2] Graduate School of Information Science, Tohoku University, Sendai, Japan.
Email: shinohara@math.is.tohoku.ac.jp

an interval $[a, b]$ if $A + B = [a, b]$. Let ϕ_n be the natural bijection from $\mathbb{Z}_n$ to $[0, n-1]$, and ψ_n be the inverse map of ϕ_n. Let us use the abbreviations $\bar{0}$ and $\bar{1}$ for $\psi_n(0)$ and $\psi_n(1)$, respectively. A factorization (resp. a near factorization and 1-overlapped factorization) (A, B) is called *Krasner* if $0 \le \phi_n(a) + \phi_n(b) \le n$ for all $a \in A$ and $b \in B$. Let J be the n by n all one matrix, and I be the identity matrix. A simple graph is a *normalized partitionable* if its clique incidence matrix is a 0-1 solution of $XY = J - I$. A square 0-1 matrix is a *thin Lehman matrix* if it is a solution of $XY = J + I$. Normalized partitionable graphs and thin Lehman matrices are rare known methods for constructing minimally imperfect graphs and minimally non-ideal clutters. (See [1], [6].)

Let $\rho(\ge 1)$ and $m_1, m_2, \dots, m_{2\rho}(\ge 2)$ $r, s(\ge 2)$ be integers such that $r = \prod_{i=1}^{\rho} m_{2i-1}$, $s = \prod_{i=1}^{\rho} m_{2i}$ and $n = \prod_{i=1}^{2\rho} m_i$. Let $\mu_j = \prod_{i=1}^{j-1} m_i$ for $1 \le j \le 2\rho$. Define a subset M_i of $\mathbb{N}$ by $\{0, 1, 2, \dots, m_i - 1\}\mu_i$, $A' := M_1 + M_3 + \cdots + M_{2\rho-1}$ and $B' := M_2 + M_4 + \cdots + M_{2\rho}$. It is clear that $(\psi_n(A'), \psi_n(B'))$, $(\psi_{n+1}(A'), \psi_{n+1}(B'))$, $(\psi_{n-1}(A'), \psi_{n-1}(B'))$ are a factorization, a near factorization, and 1-overlapped factorization of cyclic groups with corresponding orders. Let (A, B) be one of a factorization, a near factorization, or a 1-overlapped factorization of $\mathbb{Z}_n$. Then the following three operations carry each factorization to another factorization of the same kind.

- *Shifting*: Consider $(A + a, B + b)$ for some $a, b \in \mathbb{Z}_n$.
- *Scaling*: Consider $(\lambda A, \lambda B)$ for some $\lambda \in \mathbb{Z}_n^{\times}$.
- *Swapping*: Consider $(-A, B)$.

We call a factorization (near factorization, 1-overlapped factorization) constructed by the above method (with shifting, scaling and swapping) a *DBNS factorization* (*DBNS near factorization*, *DBNS 1-overlapped factorization*) of $\mathbb{Z}_n$.

N. G. De Bruijn proved the following theorem.

Theorem 1.1. *Let A and B be two subsets of $\mathbb{Z}$. If $A + B$ is direct, and equals* $[\min A + \min B, \max A + \max B]$, *and is direct, there exist parameters $m_1, m_2, \dots, m_r$ such that*

$$A = \sum_{i:\text{odd}} [0, m_i - 1] \prod_{k=1}^{i-1} m_k + \min A, \quad B = \sum_{i:\text{even}} [0, m_i - 1] \prod_{k=1}^{i-1} m_k + \min B,$$

$$\max A + \max B - \min A - \min B + 1 = \prod_{k=1}^{r} m_k.$$

Especially, each Krasner factorization is a DBNS factorization.

In comparison to the above, S. Szabó and A.D. Sands [9] recently posed the following problem.

Problem 1.2. Characterize Krasner near factorization.

On the other hand, in 1984, C. Grinstead raised the following conjecture.

Conjecture 1.3 (C. Grinstead (1984)). Every near-factorization of a finite cyclic group is a DBNS near-factorization.

Certainly, the problem of S. Szabó et al. is corresponding to the most basic case (sub-conjecture) of the above.

2 Preliminaries

Here, we introduce some theorems which we use in this article.

Theorem 2.1 (D. De Caen and el. [3], H. Shinohara [8]). *Let (A, B) be a near (resp. 1-overlapped) factorization of G. There exist two elements $a, b \in G$ such that the uncovered element (resp. the doubly covered element) of $(A + a, B + b)$ is 0, and that both of $A + a$ and $B + b$ are symmetric.*

A subset S of G is an (a, i, k) *arithmetic progression* if $S = \{a + il \mid 0 \le l \le k - 1\}$. A set $S \subset G$ is *partially* (i, k) *arithmetic* if there exists an (a, i, k) arithmetic progression S_1 and a subset $S_2 \subset G$ such that $S = S_1 + S_2$.

Theorem 2.2 (K. Kashiwabara *et al.* [5], T. Sakuma *et al.* [7]). *Let (A, B) be a near (resp. 1-overlapped) factorization of $\mathbb{Z}_n$. If A is partially (i, k) arithmetic for some i, k, then (A, B) is a DBNS near (resp. 1-overlapped) factorization.*

3 Results

In this section, we prove the following theorem, which not only solves the problem of S. Szabó et al. but also settles positively the most basic cases of Grinstead's conjecture and its ideal clutter theoretical analog [7], at the same time.

Theorem 3.1. *Every Krasner near (resp. 1-overlapped) factorization is a DBNS near (resp. 1-overlapped) factorization.*

Lemma 3.2. *Let A, B be two sets of $\mathbb{Z}$ such that A is shift-symmetric, $\min A = \min B = 0$ and $\max B < \max A$. If $A + B$ is direct and $[0, \max A] \subseteq A + B$, then $A + B$ is an interval.*

Proof of Theorem 3.1. Let (A, B) be a Krasner near factorization of $\mathbb{Z}_n$ with uncovered element u. If u is 0 then $(\phi_n(A), \phi_n(B))$ is a factorization of either $[1, n]$ or $[0, n-1]$ and hence this (A, B) is a DBNS near-factorization. Therefore without loss of generality, we can assume that u does not equal 0. Furthermore, we can assume that $\bar{0} \in A \cap B$, for otherwise we can use either the Krasner factorization $(A - \bar{1}, B)$ (if $\bar{0} \notin A$) or $(A, B - \bar{1})$ (if $\bar{0} \notin B$) as a substitute for the (A, B). From the assumption $\bar{0} \in A \cap B$, we have $\phi_n(A) + \phi_n(B) \subset [0, n-1]$. Thus, either $\max \phi_n(A)$ or $\max \phi_n(B)$ is strictly less than $n/2$. Without loss of generality, we assume $\max \phi_n(A) < n/2$.

Let a' and b' be two elements of $\mathbb{Z}_n$ such that $(A - a', B - b')$ is a near factorization of $\mathbb{Z}_n$ whose uncovered element is the identity, and that both of $A - a'$ and $B - b'$ are symmetric. Then the uncovered element of (A, B) is $a' + b'$. Since a' and b' are the centers of symmetry of A and B respectively, we have $2a' \in A$ and $2b' \in B$. Combining this and the assumption $\max \phi_n(A) < n/2$, we also have $\phi_n(2a') = \max(\phi_n(A))$. Without loss of generality, we can assume that $\phi_n(a'+b') \leq n/2$, for otherwise we can use the other Krasner near factorization $(\psi_n(\max \phi_n(A)) - A, \psi_n(\max \phi_n(B)) - B))$ as a substitute for the (A, B). (Note that shifting carries (A, B) to $(\psi_n(\max \phi_n(A)) - A, \psi_n(\max \phi_n(B)) - B))$ by Theorem 2.1).

Suppose that there exists an element b_1 such that $\phi_n(b_1) \leq 2\phi_n(a' + b') < \phi_n(b_1) + \phi_n(2a')$. Then, since b' is the center of B, we have $2b' - b_1 \in B$ and hence $(2b' - b_1) + 2a' \in B + A$. Furthermore, since $(a' + b') - (2b' - b_1) = (b_1 + 2a') - (a' + b')$, we have that $\phi_n(2b' - b_1) < n < \phi_n((2b' - b_1)) + \phi_n(2a')$, which contradicts the assumption that (A, B) is Krasner. Thus, there exists a subset $B' \subset B$ such that $\phi_n(A) + \phi_n(B') = [2\phi_n(a' + b') + 1, n - 1]$. In other words, (A, B) is a DBNS factorization. If $2\phi_n(a' + b') + 1 < n - 1$, then, from Theorem 1.1, $\phi_n(A)$ turns to be partially arithmetic in $\mathbb{Z}$. If $2\phi_n(a' + b') + 1 = n - 1$, then $A \subset [0, a' + b' - 1]$ holds. Hence, combining Lemma 3.2 and Theorem 1.1, we have that A is partially arithmetic in $\mathbb{Z}$ again. Thus, in any case, A is also partially arithmetic in $\mathbb{Z}_n$. Therefore, (A, B) is a DBNS near-factorization from Lemma 2.2. We can prove similarly to the above for the case of 1-overlapped factorizations. □

We also have the following complete descriptions for the DBNS parameters of the Krasner near factorizations; let (A, B) be a Krasner near factorization (resp. 1-overlapped factorization) of $\mathbb{Z}_n$ such that $0 \in A$, and that $\max \phi_n(A) \leq \max \phi_n(B) - \min \phi_n(B)$. Then one of the following holds;

- there are $2\rho + 1$ integers such that $m_1(\geq 1)$, and that $m_i \geq 2$ for $i \in [2, 2\rho - 1]$, and that $m_i \geq 0$ for $i = 2\rho, 2\rho + 1$ which satisfies

$$A=\psi_n\left(\sum_{j=1}^{\rho}[0, m_{2j-1}-1]\prod_{k=1}^{2j-2} m_k\right)$$
$$B=\psi_n\left(\sum_{j=1}^{\rho}[0, m_{2j}-1]\prod_{k=1}^{2j-1} m_k\right)\cup\psi_n\left(\sum_{j=1}^{\rho}[0, m_{2j}-1]\prod_{k=1}^{2j-1} m_k+1\right),$$

- there is an integer m_1 at least 2, and nonnegative integers m_2, m_3 such that

$$A = \psi_n(\{0, 2m_1 - 1\})$$
$$B = \psi_n(([0, 2m_1 - 2] + ([0, m_2]2(2m_1 - 1)$$
$$\cup ([m_2 + 1, m_3 - 1]2(2m_1-1)-1))) \setminus \{(2m_1 + 1)(2m_1-1)-1\}).$$

Although we also specified the parameters for Krasner 1-overlapped factorizations, its details are omitted here due to space limitations.

References

[1] G. Cornuéjols, *Combinatorial optimization. Packing and covering*, In: "CBMS-NSF Regional Conference Series in Applied Mathematics", 74, Society for Industrial and Applied Mathematics (SIAM), Philadelphia, PA, 2001.

[2] N. G. De Bruijn, *On number systems*, Nieuw Archief voor Wiskunde **3** (1956), 15–17.

[3] D. De Caen, D. A. Gregory, I. G. Hughes and D. L. Kreher, *Near-factors of finite groups*, Ars Combin. **29** (1990), 53–63.

[4] C. M. Grinstead, *On circular critical graphs*, Discrete Math. **51** (1984), 11–24.

[5] K. Kashiwabara and T. Sakuma, *Grinstead's conjecture is true for graphs with a small clique number*, Discrete Math. **306** (2006), 2572–2581.

[6] A. Lehman, *Width-length inequality*, Math. Programming **17** (1979), 403–417.

[7] T. Sakuma and H. Shinohara, *On circulant thin Lehman matrices*, Electronic Notes on Discrete Mathematics **38** (2011), 783–788.

[8] H. Shinohara, *Thin Lehman matrices arising from finite groups*, Linear Algebra and Appl. **436** (2012), 850–857.

[9] S. Szabó and A. D. Sands, "Factoring Groups into Subsets", CRC Press, Boca Raton, 2009.

Correlation inequality for formal series

Vladimir Blinovsky[1]

Abstract. We extend the considerations of the paper [1] and prove two correlation inequalities (statement of lemma below and inequality (8)) for totally ordered set.

First we introduce class of correlation inequalities.

Assume that $f_1, \ldots, f_n$ are nonnegative nondecreasing functions $2^X \to R$. The expectation of a random variable $f : 2^X \to R$ with respect to μ we denote by $\langle f \rangle_\mu$. For a subset $\delta \in [n]$ define

$$E_\delta = \left\langle \prod_{i \in \delta} f_i \right\rangle_\mu .$$

Let

$$\sigma = \{\sigma_1, \ldots, \sigma_\ell\}$$

be a partition of $[n]$ into disjoint subsets. Define

$$E_\sigma = \prod_{i=1}^{\ell} E_{\sigma_i}.$$

Let $\lambda_1 = |\sigma_i|$. We have $\sum_{i=1}^{\ell} \lambda_i = n$. Let $\lambda(\sigma) = (\lambda_1, \ldots, \lambda_\ell)$ and $\lambda_1 \geq \ldots \geq \lambda_\ell$. For a partition λ of number n define

$$E_\lambda = \sum_{\sigma : \lambda(\sigma) = \lambda} E_\sigma .$$

We need the following

Lemma 1. *Consider the totally odered set* $1, \ldots, N$ *with probability measure* μ *on it and let's functions* f_i, $i = 1, \ldots, n$ *are nonnegative and monotone nondecreasing. Functional*

$$E_n(f_1, \ldots, f_n) = \sum_{\lambda \vdash n} c_\lambda E_\lambda \tag{1}$$

[1] Institute for Information Transmission Problems, B. Karetnyi 19, Moscow, Russia,
and
Instituto de Matematica e Statistica, USP, Rua do Matao 1010, 05508- 090, Sao Paulo, Brazil
Email: vblinovs@yandex.ru

where

$$c_\lambda = (-1)^{\ell+1} \prod_{i=1}^{\ell} (\lambda_i - 1)!$$

is nonnegative.

In [2] was conjectured that statement of lemma (along with (8)) is true when probability measure μ on 2^X satisfies FKG conditions

$$\mu(A \cap B)\mu(A \cup B) \geq \mu(A)\mu(B) \tag{2}$$

and functions f_i are nonnegative and monotone.

Note that under conditions from the lemma in particular case $n = 2$ lemma gives Chebyshev inequality

$$\langle f_1 f_2 \rangle_\mu \geq \langle f_1 \rangle_\mu \langle f_2 \rangle_\mu \,.$$

Hence our proof can be considered as extention of Chebyshev inequality to multiple variables. For monotone functions $f_i(j),\ i = 1, \dots, n;\ j = 1, \dots, N$ we put

$$f_i(1) = a_{i,1},\ f_i(j) = f_i(j-1) + a_{i,j},\ j = 2, \dots, N,\ a_{i,j} \geq 0. \tag{3}$$

Then substituting in the formula

$$E_n(f_1, \dots, f_n) = \sum_{\lambda \vdash n} c_\lambda \sum_{\sigma:\ \lambda(\sigma)=\lambda} \prod_{i=1}^{\ell} \langle \prod_{j \in \sigma_i} f_j \rangle_\mu \tag{4}$$

coefficients c_λ one can show that to prove lemma it is sufficient to prove the inequality

$$\sum_{\{m_s\}} F_{m_1,\dots,m_N}(\mu) \prod_{j=0}^{N-1} \prod_{i=\sum_{s=1}^{j} m_s+1}^{\sum_{s=1}^{j+1} m_s} \left(\sum_{t=1}^{j+1} a_{i,t} \right) \geq 0, \tag{5}$$

where

$$F_{m_1,\dots,m_N}(\mu) = -\prod_{j=1}^{N} \prod_{i=1}^{m_j} (i - 1 - \mu(j)) \,.$$

One can check that coefficient before the monomial

$$\prod_{j=0}^{N-1} \prod_{i=\sum_{s=1}^{j} m_s+1}^{\sum_{s=1}^{j+1} m_s} a_{i,j+1}$$

in the lhs of (5) is

$$B(m_1, \ldots, m_{N-1})$$

$$\triangleq -\sum_{\{i_j\}} \prod_{j=1}^{N-1} \binom{\sum_{s=1}^{j} m_s - \sum_{s=1}^{j-1} i_s}{i_j} \prod_{i=1}^{i_j} (i-1-\mu(j)) \tag{6}$$

$$\times \prod_{i=1}^{n-\sum_{s=1}^{N-1} i_s} (i-1-\mu(N)).$$

Thus to prove (5) and complete the proof of lemma it is sufficient to prove the inequality

$$B(m_1, \ldots, m_{N-1}) \geq 0. \tag{7}$$

We prove this inequality by induction on m_j. This proves lemma.

Next we consider the set of formal series $P[[t]]$, whose coefficients are monotone nondecreasing nonnegative functions on 2^X. Then $p(A) = p_1(A)t + p_2(A)t^2 + \ldots \in P[[t]]$. In [2] was formulated the following

Conjecture 1. *For FKG probability measure μ the following inequality is true*

$$1 - \prod_{A \in 2^X} (1-p(A))^{\mu(A)} \geq 0. \tag{8}$$

The inequality (8) is understood as non negativeness of coefficients of formal series obtained by series expansion of the product on the left-hand side of this inequality.

We will prove, that inequality (8) follows from inequalities

$$E_n(f_1, \ldots, f_n) \geq 0 \tag{9}$$

for all n and hence it is sufficient to prove last inequalities and then inequality (8) follows under the same conditions on μ.

We make some transformations of the expression in the lhs of (8). We have

$$1 - \prod_{A \in 2^X} (1-p(A))^{\mu(A)} = 1 - \exp\left\{\langle \ln(1-p) \rangle_\mu\right\}$$

$$= 1 - \exp\left\{-\sum_{i=1}^{\infty} \frac{1}{i} \langle p^i \rangle_\mu\right\} = \sum_{j=1}^{\infty} \frac{(-1)^{j+1}}{j!} \left(\sum_{i=1}^{\infty} \frac{1}{i} \langle p^i \rangle_\mu\right)^j$$

$$= \sum_{j=1}^{\infty} \frac{(-1)^{j+1}}{j!} \sum_{\{q_s\}: \sum q_s = j} \binom{j}{q_1, \ldots, q_j} \sum_{\{i_s\}} \frac{\prod_{s=1}^{j} \langle p^{i_s} \rangle_\mu^{q_s}}{(i_1)^{q_1} (i_2)^{q_2} \ldots (i_j)^{q_j}}.$$

Next remind that the number of partitions of n with given set $\{q_i\}$ of occurrence of i is equal to

$$\frac{n!}{\prod_i (i!)^{q_i} q_i!}.$$

Continuing the last chain of identities and using last formula we obtain

$$\begin{aligned}
&1 - \prod_{A\in 2^X} (1 - p(A))^{\mu(A)} \\
&= \sum_{n=1}^{\infty} \frac{1}{n!} \sum_{\lambda\vdash n} \sum_{\sigma:\,\lambda(\sigma)=\lambda} (-1)^{\sum_i q_i+1} \frac{n! \prod((i-1)!)^{q_i}}{\prod_i (i!)^{q_i} q_i!} \prod_i \langle p^i\rangle_\mu^{q_i} \\
&= \sum_{n=1}^{\infty} \frac{1}{n!} \sum_{\lambda\vdash n} (-1)^{\ell(\lambda)+1} \prod_i (\lambda_i - 1)! \sum_{\sigma:\,\lambda(\sigma)=\lambda} E_\sigma(p, \dots, p) \qquad (10) \\
&= \sum_{n=1}^{\infty} \frac{1}{n!} \sum_{\lambda\vdash n} c_\lambda E_\lambda(p, \dots, p) \\
&= \sum_{n=1}^{\infty} \frac{1}{n!} E_n(p, \dots, p).
\end{aligned}$$

Hence now to prove the conjecture 3 we need to show that

$$E_n(p, \dots, p) \geq 0. \qquad (11)$$

But the coefficients of the formal series $E_n(p, \dots p)$ are the sums of $E_n(p_{i_1}, \dots, p_{i_n})$ for multisets $\{i_1, \dots, i_n\}$. This completes the proof that inequality (8) follows from inequalities (9) under the same conditions on μ.

Thus because we prove lemma, we prove inequality (8) for totally ordered lattice and this is our main result.

Remark. To extend lemma to the conditions (2) one can try to find proper expansion for the monotone functions f_i which extend expansion (3) to the case of poset 2^X.

References

[1] V. BLINOVSKY, *A proof of one correlation inequality*, Problems of Inform. Transm. **45** (2009), 264–269.

[2] S. SAHI, *Higher correlation inequalities*, Combinatorica **28** (2008), 209–227.

[3] J. RIORDAN, "Combinatorial Identities", Wiley, Ney York, 1968.

Covariants of spherical Θ-orbits for types E_6, E_7, E_8

Witold Kraśkiewicz[1] and Jerzy Weyman[2]

Abstract. We calculate the rings of covariants for spherical orbits in the class of representations of reductive algebraic groups associated to various gradings on simple Lie algebras of type E_6, E_7 and E_8.

1 Introduction

Let X_n be a Dynkin diagram with a distinguished node $x \in X_n$ and let $\mathfrak{g}$ be a complex semisimple Lie algebra corresponding to X_n. Then $\mathfrak{g} = \mathfrak{h} \oplus \bigoplus_{\beta \in \Delta} \mathfrak{g}_\beta$ where $\mathfrak{h}$ is the Cartan subalgebra and Δ is a root system of type X_n. Choose a basis for Δ and let α be a simple root corresponding to x. For a root $\beta \in \Delta$, let the rank of β be the multiplicity of α in decomposition of β in the basis of simple roots. For $i \neq 0$, let $\mathfrak{g}_i$ be the sum of $\mathfrak{g}_\beta$ with β of rank i and let $\mathfrak{g}_0$ be the sum of $\mathfrak{h}$ and the sum of $\mathfrak{g}_\beta$ with β of rank 0. In this way, x defines grading of $\mathfrak{g}$ with $\mathfrak{g}_0$ being a Lie subalgebra of $\mathfrak{g}$. Let G_0 be the adjoint group corresponding to $\mathfrak{g}_0$. The group $G = G_0 \times \mathbb{C}^*$ acts on $\mathfrak{g}_1$ with finitely many orbits. Moreover, V. Kac proved in [1] that almost all actions of reductive algebraic groups with finitely many orbits can be obtained in such a way. The orbits of G in $\mathfrak{g}_1$ where parameterized by E.B. Vinberg ([5,6]) using combinatorics of root subsystems of Δ. We call them Θ–orbits.

Systematic study of algebraic–geometric properties of orbits connected with exceptional roots systems was undertaken by authors in a series of papers [2–4]. In the following, we use previous results to describe the action of G on the ring of regular functions on spherical Θ-orbits.

2 Methodology

Let $\mathcal{O}$ be an orbit of the action of G on $\mathfrak{g}_1$. Its closure $\overline{\mathcal{O}}$ is an affine algebraic variety and the group G acts on the ring $\mathbb{C}[\overline{\mathcal{O}}]$ of regular functions

[1] Faculty of Mathematics and Computer Science, Nicholas Copernicus University, Toruń, Poland. Email: wkras@mat.umk.pl

[2] Department of Mathematics, Northeastern University, Boston, USA. Email: j.weyman@neu.edu

in a standard way. Let $B \subset G$ be a Borel subgroup. B-invariant regular functions on $\overline{\mathcal{O}}$ are called covariants of $\mathcal{O}$. The orbit is spherical if and only if it contains dense B-orbit, or equivalently if the action of G on $\mathbb{C}[\overline{\mathcal{O}}]$ is multiplicity free.

Let $\mathcal{M} = \mathcal{M}_{\mathcal{O}}$ be the set of dominant weights λ of G such that the irreducible representation $V(\lambda, G)$ of highest weight λ occurs in $\mathbb{C}[\overline{\mathcal{O}}]$. The set $\mathcal{M}$ is in fact a monoid with respect to usual addition of weights.

Assume that the orbit $\mathcal{O}$ is spherical. For $\lambda \in \mathcal{M}$, we write $\lambda[d]$ to indicate the degree d of the homogeneous component of $\mathbb{C}[\overline{\mathcal{O}}]$ that contains the only copy of $V(\lambda, G)$. In that way we obtain a grading $\mathcal{M} = \oplus_{d=0}^{\infty} \mathcal{M}_d$ and

$$\mathbb{C}_d[\overline{\mathcal{O}}] = \oplus_{\lambda \in \mathcal{M}_d} V(\lambda, G), \quad d = 0, 1, 2, \ldots. \tag{2.1}$$

The ring of covariants of $\mathcal{O}$ is isomorphic to the monoid ring $\mathbb{C}[\mathcal{M}]$. We use standard notation e^{λ} for a weight $\lambda \in \mathcal{M}$ regarded as an element of $\mathbb{C}[\mathcal{M}]$.

Let X_n be a Dynkin diagram of type E_6, E_7 or E_8 and let $\mathcal{O}$ be a Θ-orbit of type X_n. The key result of [2–4] is an explicit construction of a desingularisation $\mathcal{S}$ of $\overline{\mathcal{O}}$ which is a vector bundle over a homogeneous space X of G. Using that desingularisation and a geometric technique described in [7] algebraic properties of the variety $\overline{\mathcal{O}}$ were derived. In particular Hilbert–Poincaré series $\mathcal{H}_{\mathcal{O}}(t) = \sum_{n=0}^{\infty} p(n)t^n = \sum_{n=0}^{\infty} dim\, \mathbb{C}_n[\overline{\mathcal{O}}]t^n$ was calculated. It is of the form

$$\mathcal{H}_{\mathcal{O}}(t) = \frac{N(t)}{(1-t)^c} = N(t) \cdot \sum_{n=0}^{\infty} \binom{n+c-1}{c-1} t^n,$$

where $N(t)$ is a polynomial given in [2–4] and $c = \dim\mathcal{O}$. It follows that $p(n)$ is a polynomial function of n of degree $\leq c-1$.

On the other hand $\mathbb{C}_n[\overline{\mathcal{O}}] = H^0(X, Sym_n(\mathcal{S}^*))$ and it can be computed by Bott algorithm, at least for small n. Assume that we know s covariants of $\mathcal{O}$ with lineary independent weights $\lambda_i[d_i]$, $i = 1, 2, \ldots, s$, and we want to show that in fact the ring of covariants is generated by them. For $\mathbf{k} = (k_1, k_2, \ldots, k_s) \in \mathbb{N}^s$, let

$$W(k_1, k_2, \ldots, k_s) = \dim V(k_1\lambda_1 + k_2\lambda_2 + \cdots + k_s\lambda_s, G) \tag{2.2}$$

and let

$$q(n) = \sum_{\substack{(k_1,k_2,\ldots,k_s) \\ k_1d_1+k_2d_2+\cdots+k_sd_s=n}} W(k_1, k_2, \ldots, k_s), \quad n = 0, 1, 2, \ldots. \tag{2.3}$$

In order to prove the statement, it is enough to show that

$$N(t) = (1-t)^c \sum_{n=0}^{\infty} q(n)t^n. \tag{2.4}$$

It follows from Weyl character formula that $W(\mathbf{k})$ is polynomial in $\mathbf{k}$ of degree less or equal to

$$\delta(\lambda_1, \lambda_2, \ldots, \lambda_s) = \mid \Delta^+ \setminus \{\alpha \in \Delta^+ : (\lambda_1, \alpha) = (\lambda_2, \alpha) = \ldots (\lambda_s, \alpha) = 0\} \mid .$$

Let $\mathcal{P} \subset \mathbb{R}^s$ be a polytope with vertices

$$A_i = \left(0, \ldots, 0, \frac{1}{d_i}, 0, \ldots, 0\right), \; i = 1, 2, \ldots, s.$$

Then (2.3) is equivalent to

$$q(n) = \sum_{\mathbf{k} \in n\mathcal{P} \cap \mathbb{N}^s} W(k_1, k_2, \ldots, k_s). \tag{2.5}$$

Recall that a function $g : \mathbb{N} \to \mathbb{R}$ is quasi–polynomial of period d if there exist d polynomials $G_0(t), G_1(t), \ldots, G_{d-1}(t) \in \mathbb{R}[t]$ such that $g(n) = G_{n \bmod d}(n)$ for every natural n. It is known that if $\mathcal{P} \subset \mathbb{R}^s$ is a rational polytope and $f = f(x_1, x_2, \ldots, x_s)$ is a polynomial function on $\mathbb{R}^s$ then $\overline{f}(n) = \sum_{x \in n\mathcal{P} \cap \mathbb{N}^s} f(x_1, x_2, \ldots, x_s)$ is quasi–polynomial in n of degree $\leq \deg(f) + \dim(\mathcal{P})$. Moreover, if d is common denominator of all coordinates of vertices of $\mathcal{P}$ then d is a period of $\overline{f}$.

Let $d = \operatorname{lcm}(d_1, d_2, \ldots, d_s)$. Then the equation (2.4) is equivalent to

$$N(t) \equiv (1-t)^c \sum_{n=0}^{\infty} q(n)t^n \pmod{t^L}, \tag{2.6}$$

where $L = d \cdot (\delta(\lambda_1, \lambda_2, \ldots, \lambda_s) + s)$. The condition (2.6) can be checked by computer calculation (in Maple).

The approach can be modified to cover cases of generators with relation since they are of very special form. It turns out that we can divide a set of generators of $\mathcal{M}$ into two groups

$$\lambda_1[d_1], \; \lambda_2[d_2], \ldots, \; \lambda_s[d_s]$$

and

$$\mu_1[e_1], \; \mu_2[e_2], \; \ldots, \; \mu_t[e_t]$$

in such a way that

1. λ_i's are linearly independent; let $\mathcal{M}^0$ be a monoid spanned by them;
2. the sets $\mathcal{M}^0$ and $\mathcal{M}^i = \mu_i + \mathcal{M}^0, i = 1, 2 \dots, t$, are disjoint and

$$\mathcal{M} = \mathcal{M}^0 \cup \mathcal{M}^1 \cup \dots \cup \mathcal{M}^t$$

(*i.e.* minimal relations express sums of two not necessarily different μ_i's as a positive combination of λ_i's possibly plus another μ_j).

For $i = 1, 2, \dots, t$, let

$$W_i(k_1, k_2, \dots, k_s) = \dim V\left(\mu_i + k_1\lambda_1 + k_2\lambda_2 + \dots + k_s\lambda_s, G\right),$$

and let

$$q(n) = \sum_{\mathbf{k}\in n\mathcal{P}\cap\mathbb{N}^s} W(k_1, k_2, \dots, k_s) + \sum_{i=1}^{t} \sum_{\mathbf{k}\in(n-e_i)\mathcal{P}\cap\mathbb{N}^s} W_i(k_1, k_2, \dots, k_s).$$

Denote $e = \max(e_1, e_2, \dots, e_t)$. For $n \geq e$, $q(n)$ is quasi-polynomial and in order to prove that $\lambda_1, \dots, \lambda_s; \mu_1, \dots \mu_t$ generate the ring of covariants, it is enough to check (2.6) with L substituted with $L' = e + d \cdot (\delta(\lambda_1, \lambda_2, \dots, \lambda_s) + s)$.

3 Results

There are 15 spherical Θ-orbits of type E_6, 33 of type E_7 and 42 of type E_8. For each of them we have found explicit generators and relations of the monoid $\mathcal{M}$.

As a corollary, we obtained the following result.

Theorem 3.1. *Let $\mathcal{O}$ be a spherical Θ-orbit of type E_6, E_7 or E_8. Then the ring of covariants of $\mathcal{O}$ is either a polynomial ring or a quotient of a polynomial ring by the ideal generated by 2×2 minors of a symmetric matrix of dimension 2×2 or 3×3.*

Example 3.2. Let $\alpha = \alpha_2$ in a root system of type E_7 (in Bourbaki enumeration of simple roots). Then $G = GL(7)$ and $\mathfrak{g}_1 = \bigwedge^3 \mathbb{C}^7$. There are 5 spherical orbits of that type denoted by $\mathcal{O}_1, \mathcal{O}_2, \mathcal{O}_3, \mathcal{O}_4$ and $\mathcal{O}_6$ in [3] of dimension 13, 20, 21, 25 and 28, respectively. Let

$$\begin{aligned}
&\lambda_1 = (1, 1, 1, 0, 0, 0, 0)[1],\ \lambda_2 = (2, 1, 1, 1, 1, 0, 0)[2],\\
&\lambda_3 = (3, 1, 1, 1, 1, 1, 1)[3],\ \lambda_4 = (2, 2, 2, 1, 1, 1, 0)[3],\\
&\lambda_5 = (3, 3, 2, 2, 1, 1, 0)[4],\ \lambda_6 = (3, 3, 3, 3, 1, 1, 1)[5],\\
&\lambda_7 = (5, 5, 3, 3, 3, 1, 1)[7];\\
&\mu_1 = (3, 2, 2, 2, 1, 1, 1)[4],\ \mu_2 = (4, 3, 2, 2, 2, 1, 1)[5],\\
&\mu_3 = (4, 4, 3, 3, 2, 1, 1)[6];
\end{aligned}$$

The monoids $\mathcal{M}$ for the first four orbits are freely generated and their generators are given in the following table.

$\mathcal{O}_1$	λ_1;
$\mathcal{O}_2$	λ_1, λ_2;
$\mathcal{O}_3$	$\lambda_1, \lambda_2, \lambda_3$;
$\mathcal{O}_4$	$\lambda_1, \lambda_2, \lambda_4, \lambda_5$;

The monoid $\mathcal{M}$ for $\mathcal{O}_6$ is generated by $\lambda_1, ..\lambda_7$ and $\mu_1, .., \mu_3$ with relations

$$\begin{array}{lll} 2\mu_1=\lambda_3+\lambda_6\,[8], & \mu_1+\mu_2=\lambda_3+\mu_3\,[9], & 2\mu_2=\lambda_3+\lambda_7\,[10], \\ \mu_1+\mu_3=\mu_2+\lambda_6\,[10], & \mu_2+\mu_3=\mu_1+\lambda_7\,[11], & 2\mu_3=\lambda_6+\lambda_7\,[12]. \end{array}$$

The ring of covariants of $\mathcal{O}_6$ is isomorphic to a quotient of the polynomial ring $\mathbb{C}[e^{\lambda_1}, ..e^{\lambda_7}, e^{\mu_1}, .., e^{\mu_3}]$ by the ideal generated by 2×2 minors of the matrix

$$\begin{bmatrix} e^{\lambda_3} & e^{\mu_1} & e^{\mu_2} \\ e^{\mu_1} & e^{\lambda_6} & e^{\mu_3} \\ e^{\mu_2} & e^{\mu_3} & e^{\lambda_7} \end{bmatrix}.$$

The case of (E_8, α_2) is analogous: $G = GL(8)$ and $\mathfrak{g}_1 = \bigwedge^3 \mathbb{C}^8$. There are also 5 spherical orbits and their rings of covariants are isomorphic to the rings of corresponding orbits of type (E_7, α_2) (the orbits are degenerate in a sense of [2, Section 4]).

Example 3.3. For (E_7, α_5) we have $G = SL(3) \times SL(5) \times \mathbb{C}^*$, $\mathfrak{g}_1 = \mathbb{C}^3 \otimes \bigwedge^2 \mathbb{C}^5$. The monoid $\mathcal{M}$ for the orbit $\mathcal{O}_8$ of dimension 19 is generated by

$$\begin{array}{ll} \lambda_1 = (1,0,0 \mid 1,1,0,0,0)[1], & \lambda_2 = (2,0,0 \mid 1,1,1,1,0)[2], \\ \lambda_3 = (1,1,0 \mid 2,1,1,0,0)[2], & \lambda_4 = (1,1,1 \mid 2,2,2,0,0)[3], \\ \lambda_5 = (1,1,1 \mid 3,1,1,1,0)[3], & \lambda_6 = (2,2,0 \mid 3,3,1,1,0)[4]; \\ \mu_1 = (2,1,0 \mid 2,2,1,1,0)[3], & \mu_2 = (2,1,1 \mid 3,2,2,1,0)[4], \\ \mu_3 = (2,2,1 \mid 4,3,2,1,0)[5] & \end{array}$$

with relations

$$\begin{array}{ll} 2\mu_1 = \lambda_2+\lambda_6\,[6], & \mu_1+\mu_2 = \mu_3+\lambda_2\,[7], \\ 2\mu_2 = \lambda_2+\lambda_4+\lambda_5\,[8], & \mu_1+\mu_3 = \mu_2+\lambda_6\,[8], \\ \mu_2+\mu_3 = \mu_1+\lambda_4+\lambda_5\,[9], & 2\mu_3 = \lambda_4+\lambda_5+\lambda_6\,[10]. \end{array}$$

The ring of covariants is a quotient of the ring of polynomials in $e^{\lambda_1}, ..., e^{\lambda_6}$ and $e^{\mu_1}, ..., e^{\lambda_3}$ by the ideal generated by 2×2 minors of the matrix

$$\begin{bmatrix} e^{\lambda_2} & e^{\mu_2} & e^{\mu_1} \\ e^{\mu_2} & e^{\lambda_4}e^{\lambda_5} & e^{\mu_3} \\ e^{\mu_1} & e^{\mu_3} & e^{\lambda_7} \end{bmatrix}.$$

References

[1] V. KAC, *Some remarks on nilpotent orbits*, J. of Algebra, **64** (1982), 190–213.

[2] W. KRAŚKIEWICZ and J. WEYMAN, *Geometry of orbit closures for the representations associated to gradings of Lie algebras of types E_6, F_4 and G_2*, arXiv:1201.1102.

[3] W. KRAŚKIEWICZ and J. WEYMAN, *Geometry of orbit closures for the representations associated to gradings of Lie algebras of type E_7*, arXiv:1301.0720.

[4] W. KRAŚKIEWICZ and J. WEYMAN, *Geometry of orbit closures for the representations associated to gradings of Lie algebras of type E_8*, in preparation.

[5] E. B. VINBERG, *Weyl group of a graded Lie algebra,* Izv. Akad. Nauk SSSR **40** (1975), 488–526.

[6] E. B. VINBERG, *Classification of homogeneous nilpotent elements of a semisimple graded Lie algebra,* Selecta Mathematica Sovietica **6** no.1 (1987).

[7] J. WEYMAN, "Cohomology of Vector Bundles and Syzygies", Cambridge Tracts in Mathematics **149**, Cambridge University Press, 2003.

Partition regularity of nonlinear polynomials: a nonstandard approach

Lorenzo Luperi Baglini[1]

Abstract. While the linear partition regular polynomials have been characterized by Richard Rado in the 1930's, very few results are known in the nonlinear case. Our aim is to prove that, on $\mathbb{N}$, there are at least two interesting classes of partition regular nonlinear polynomials.
Our approach is based on the study of ultrafilters on $\mathbb{N}$ from the point of view of Nonstandard Analysis. A particularity of this technique is that the proofs of the main results can be carried out by almost elementary algebraic considerations.

Our aim is to expose a few results regarding the partition regularity of nonlinear polynomials, which have been presented in [1]. We recall the following definitions:

Definition. *A polynomial $P(x_1, ..., x_n)$ is*

- ***partition regular** (on $\mathbb{N}$) if for every natural number r, for every partition $\mathbb{N} = \bigcup_{i=1}^{r} A_i$, there is an index $j \leq r$ and nonzero natural numbers $a_1, ..., a_n \in A_j$ such that $P(a_1, ..., a_n) = 0$;*
- ***injectively partition regular** (on $\mathbb{N}$) if for every natural number r, for every partition $\mathbb{N} = \bigcup_{i=1}^{r} A_i$, there is an index $j \leq r$ and mutually distinct nonzero natural numbers $a_1, ..., a_n \in A_j$ such that $P(a_1, ..., a_n) = 0$.*

The problem of determining which polynomials are partition regular has been studied since Issai Schur's work [2], and the linear case was settled by Richard Rado in [3]:

Theorem (Rado). *Let $P(x_1, ..., x_n) = \sum_{i=1}^{n} a_i x_i$ be a linear polynomial with nonzero coefficients. The following conditions are equivalent:*
1. *$P(x_1, ..., x_n)$ is partition regular on $\mathbb{N}$;*
2. *there is a nonempty subset J of $\{1, ..., n\}$ such that $\sum_{j \in J} a_j = 0$.*

In his work, Rado also characterized the partition regular finite systems of linear equations and, since then, one of the main streams of the research

[1] University of Vienna, Universitätsring 1, 1010 Vienna, Austria. Email: lorelupe@gmail.com

in this field has been the study of infinite systems of linear equations (for a general background on many notions related to this subject see, *e.g.*, [4]). One other possible direction is the study of the partition regularity for nonlinear polynomials, which is the problem that we want to face.
As far as we know, perhaps the most interesting known result in the context of partition regularity of nonlinear polynomials is the following (see [5]):

Theorem (Hindman). *For every natural numbers $n, m \geq 1$, with $n + m \geq 3$, the nonlinear polynomial*

$$\sum_{i=1}^{n} x_i - \prod_{j=1}^{m} y_j$$

is injectively partition regular.

Our main results generalize the previous Theorem.

Our first result states that, if we start with a partition regular linear polynomial $P(x_1, ..., x_n)$, a finite set of variables $Y = \{y_1, ..., y_m\}$ and we multiply each variable x_i for some of the variables in Y then we obtain a partition regular polynomial. To precisely state this result we need the following definition:

Definition. *Let m be a positive natural number, and $\{y_1, ..., y_m\}$ a set of mutually distinct variables. For every finite set $F \subseteq \{1, .., m\}$, we denote by $Q_F(y_1, ..., y_m)$ the monomial*

$$Q_F(y_1, ..., y_m) = \begin{cases} \prod\limits_{j \in F} y_j, & \text{if } F \neq \emptyset; \\ 1, & \text{if } F = \emptyset. \end{cases}$$

Theorem (1). *Let $n \geq 2$ be a natural number, $R(x_1, ..., x_n) = \sum\limits_{i=1}^{n} a_i x_i$ a partition regular polynomial, and m a positive natural number. Then, for every $F_1, ..., F_n \subseteq \{1, .., m\}$ (with the request that, when $n = 2$, $F_1 \cup F_2 \neq \emptyset$), the polynomial*

$$P(x_1, ..., x_n, y_1, ..., y_m) = \sum_{i=1}^{n} a_i x_i Q_{F_i}(y_1, ..., y_m)$$

is injectively partition regular.

E.g., as a consequence of Theorem (1) we have that the polynomial

$$P(x_1, x_2, x_3, x_4, y_1, y_2, y_3) = 2x_1 + 3x_2 y_1 y_2 - 5x_3 y_1 + x_4 y_2 y_3$$

is injectively partition regular. We observe that the exposed Theorem of Hindman is a particular case of our result. A particularity of the polynomials considered in Theorem (1) is that the degree of each of their variables is one; to state a result that ensures the partition regularity for many polynomials having variables with degree greater than one we need to introduce a few definitions:

Definition 1) *Let*

$$P(x_1, ..., x_n) : \sum_{i=1}^{k} a_i M_i(x_1, ..., x_n)$$

be a polynomial, and let $M_1(x_1, ..., x_n), ..., M_k(x_1, ..., x_n)$ *be the distinct monic monomials of* $P(x_1, ..., x_n)$. *We say that* $\{v_1, ..., v_k\} \subseteq V(P)$ *is a* ***set of exclusive variables*** *for* $P(x_1, ..., x_n)$ *if, for every* $i, j \leq k$, $d_{M_i}(v_j) \geq 1 \Leftrightarrow i = j$, *where* $d_{M_i}(v_j)$ *denotes the degree of the variable* v_j *in the monomial* M_i.
In this case we say that the variable v_i *is* ***exclusive*** *for the monomial* $M_i(x_1, ..., x_n)$ *in* $P(x_1, ..., x_n)$.

2) *We say that* $P(x_1, ..., x_n)$ *satisfies* ***Rado's Condition*** *if there is a nonempty subset* F *of the set of coefficients of* $P(x_1, ..., x_n)$ *such that* $\sum_{c \in F} c = 0$.

Definition. *Let* $P(x_1, ..., x_n) = \sum_{i=1}^{k} a_i M_i(x_1, ..., x_n)$ *be a polynomial, and let* $M_1(x_1, ..., x_n), ..., M_k(x_1, ..., x_n)$ *be the monic monomials of* $P(x_1, ..., x_n)$. *Then*

- ***NL(P)***$=\{x \in V(P) \mid d(x) \geq 2\}$ *is the set of nonlinear variables of* $P(x_1, ..., x_n)$;
- *for every* $i \leq k$, $\mathbf{l_i} = \max\{d(x) - d_i(x) \mid x \in NL(P)\}$.

Theorem (2). *Let*

$$P(x_1, ..., x_n) = \sum_{i=1}^{k} a_i M_i(x_1, ..., x_n)$$

be a polynomial, and let $M_1(x_1, ..., x_n)$,...,$M_k(x_1, ..., x_n)$ *be the monic monomials of* $P(x_1, ..., x_n)$. *Suppose that* $k \geq 3$, *that* $P(x_1, ..., x_n)$ *satisfies Rado's Condition and that, for every index* $i \leq k$, *in the monomial* $M_i(x_1, ..., x_n)$ *there are at least* $m_i = \max\{1, l_i\}$ *exclusive variables with degree equal to* 1.
Then $P(x_1, ..., x_n)$ *is injectively partition regular.*

E.g., in conseque of Theorem (2) we get that the polynomial

$$P(x, y, z, t_1, t_2, t_3, t_4, t_5, t_6) = t_1t_2x^2 + t_3t_4y^2 - t_5t_6z^2$$

is injectively partition regular.

The technique we use to prove our main results is based on an approach to combinatorics by means of nonstandard analysis (with something in common with, *e.g.*, [6–8]): the idea behind this approach is that, as it is well-known, problems talking about partition regularity of properties on $\mathbb{N}$ can be reformulated in terms of ultrafilters and it can be showed that some properties of ultrafilters can be translated, and studied, in terms of sets of hyperintegers. This can be obtained by associating, in particular hyperextensions ${}^*\mathbb{N}$ of $\mathbb{N}$, to every ultrafilter $\mathcal{U}$ its monad $\mu(\mathcal{U})$:

$$\mu(\mathcal{U}) = \{\alpha \in^* \mathbb{N} \mid \alpha \in^* A \text{ for every } A \in \mathcal{U}\},$$

and then proving that some of the properties of $\mathcal{U}$ can be deduced by properties of $\mu(\mathcal{U})$.

From the point of view of the partition regularity of polyomials, this approach works as follows: first of all, we recall that, given a polynomial $P(x_1, ..., x_n)$, we have that $P(x_1, ..., x_n)$ is partition regular if and only if there exists an ultrafilter $\mathcal{U}$ such that for every set A in $\mathcal{U}$ there are elements $a_1, ..., a_n \in A$ such that $P(a_1, ..., a_n) = 0$. Then we reformulate this property in terms of monads of ultrafilters, giving one of the basic results that we use to prove Theorems (1) and (2) (as usual, we denote by $\beta\mathbb{N}$ the space of ultrafilters on $\mathbb{N}$):

Theorem (3). *Let $P(x_1, ..., x_n)$ be a polynomial, and $\mathcal{U} \in \beta\mathbb{N}$ an ultrafilter. Then the following two conditions are equivalent:*

1. *for every set A in $\mathcal{U}$ there are elements $a_1, ..., a_n \in A$ such that $P(a_1, ..., a_n) = 0$;*
2. *there are elements $\alpha_1, ..., \alpha_n$ in $\mu(\mathcal{U})$ such that $P(\alpha_1, ..., \alpha_n) = 0$.*

The interesting fact is that, whenever it is given an ultrafilter $\mathcal{U}$ and elements $\alpha_1, ..., \alpha_n$ in $\mu(\mathcal{U})$ for which there is a polynomial $P(x_1, ..., x_n)$ such that $P(\alpha_1, ..., \alpha_n) = 0$, we automatically know that the polynomial is partition regular.

Usually, the ultrafilters that we use to prove our results are (or are related to) idempotent ultrafilters which, we recall, are defined as follows:

Definition. *An ultrafilter $\mathcal{U} \in \beta\mathbb{N}$ is called idempotent if $\mathcal{U} \oplus \mathcal{U} = \mathcal{U}$, where $\oplus$ denotes the usual sum of ultrafilters, defined as follows:*

$\forall \mathcal{U}, \mathcal{V} \in \beta\mathbb{N}$,

$\forall A \in \wp(\mathbb{N}), A \in \mathcal{U} \oplus \mathcal{V} \Leftrightarrow \{n \in \mathbb{N} \mid \{m \in \mathbb{N} \mid n + m \in A\} \in \mathcal{V}\} \in \mathcal{U}$;

similarly, an ultrafilter $\mathcal{U}$ is called a multiplicative idempotent if $\mathcal{U} \odot \mathcal{U} = \mathcal{U}$, where $\odot$ denotes the usual product of ultrafilters, defined as follows:

$$\forall \mathcal{U}, \mathcal{V} \in \beta\mathbb{N},$$
$$\forall A \in \wp(\mathbb{N}),\ A \in \mathcal{U} \oplus \mathcal{V} \Leftrightarrow \{n \in \mathbb{N} \mid \{m \in \mathbb{N} \mid n \cdot m \in A\} \in \mathcal{V}\} \in \mathcal{U}.$$

The last general result about ultrafilters that we need to apply the technique to the problem of the partition regularity of nonlinear polynomials regards the existence of particular ultrafilters on $\beta\mathbb{N}$:

Theorem (4). *There is a multiplicative idempotent ultrafilter $\mathcal{U}$ such that, for every linear partition regular polynomial $P(x_1, ..., x_n)$, for every set A in $\mathcal{U}$ there are elements $a_1, ..., a_n \in A$ such that $P(a_1, ..., a_n) = 0$.*

Theorems (1) and (2) can be proved by easy algebraical considerations on the elements in the monads of the ultrafilter given by Theorem (4) in particular hyperextensions of $\mathbb{N}$, called ω-hyperextensions (or iterated-hyperexensions). These hyperextensions are constructed by considering an universe $\mathbb{U}$ containing $\mathbb{N}$ and a star map $* : \mathbb{U} \to \mathbb{U}$ satisfying the transfer property. Since $* : \mathbb{U} \to \mathbb{U}$, this map can be iterated: the ω-hyperextension of $\mathbb{N}$ (denoted by ${}^\bullet\mathbb{N}$) is

$${}^\bullet\mathbb{N} = \bigcup_{n \in \mathbb{N}} S_n(\mathbb{N}),$$

where $S_0(\mathbb{N}) = \mathbb{N}$ and $S_{n+1}(\mathbb{N}) =^*(S_n(\mathbb{N}))$. As a consequence of the elementary chain condition, it is easy to prove that ${}^\bullet\mathbb{N}$ is elementarily equivalent to $\mathbb{N}$.

The particularity of these hyperextensions is that the possibility of iterating the map $*$ translates in a few interesting algebraical properties of the monads, in particular of the monads of idempotent ultrafilters. In fact we have the following useful facts regarding the monads in ${}^\bullet\mathbb{N}$:

Theorem (5). *If $\mathcal{U} \in \beta\mathbb{N}$ is an ultrafilter and $\mu(\mathcal{U})$ is the monad of $\mathcal{U}$ in ${}^\bullet\mathbb{N}$, then we have the following three properties:*

1. *if $\alpha \in \mu(\mathcal{U})$ then ${}^*\alpha \in \mu(\mathcal{U})$;*
2. *$\mathcal{U}$ is idempotent if and only if, for every α, β in $\mu(\mathcal{U}) \cap {}^*\mathbb{N}$, $\alpha + {}^*\beta \in \mu(\mathcal{U})$;*
3. *$\mathcal{U}$ is a multiplicative idempotent if and only if, for every α, β in $\mu(\mathcal{U}) \cap {}^*\mathbb{N}$, $\alpha \cdot {}^*\beta \in \mu(\mathcal{U})$.*

The nonstandard technique to prove results in combinatorics that we just described has been applied in [9], [10] to obtain a few results in combinatorics.

As an example of application of this nonstandard approach, we conclude by proving that the nonlinear polynomial

$$P(x, y, z, w) = x + y - zw$$

is injectively partition regular: by Theorems (3) and (4) it follows that there is a multiplicative idempotent ultrafilter $\mathcal{U}$ and elements $\alpha, \beta, \gamma \in \mu(\mathcal{U})$ with $\alpha + \beta - \gamma = 0$. Then

$$P(\alpha \cdot^* \alpha, \beta \cdot^* \alpha, \gamma, ^*\gamma) = \alpha \cdot^* \alpha + \beta \cdot^* \alpha - \gamma \cdot^* \alpha = 0$$

and, by Theorem (5), $\alpha{\cdot}^*\alpha$, $\beta{\cdot}^*\alpha$, γ and $^*\alpha$ are in $\mu(\mathcal{U})$. So we can conclude by applying Theorem (3).

The proofs of Theorems (1) and (2) can be carried out by similar considerations.

References

[1] L. Luperi Baglini, *Partition regularity of nonlinear polynomials: a nonstandard approach*, avaible online at http://arxiv.org/abs/1301.1467.

[2] I. Schur, *Uber die Kongruenz $x^m + y^m = z^m \mod (p)$*, Jahresber. Deutsch. Math.-Verein. **25** (1916), 114–117.

[3] R. Rado, *Studien zur kombinatorik*, Math. Z. **36** (1933), 242–280.

[4] N. Hindman, *Partition regularity of matrices*, Integers **7(2)** (2007), A–18.

[5] N. Hindman, *Monochromatic sums equal to products in* $\mathbb{N}$, Integers **11A** (2011), Article 10, 1–10.

[6] J. Hirschfeld, *Nonstandard combinatorics*, Studia Logica **47**, n. 3 (1988), 221–232.

[7] C. Puritz, *Ultrafilters and standard functions in nonstandard analysis*, Proc. London Math. Soc. **22** (1971), 706–733.

[8] C. Puritz, *Skies, constellations and monads*, In: "Contributions to Non-Standard Analysis", WAJ Luxemburg and A. Robinson (eds.), North Holland 1972, 215–243.

[9] M. Di Nasso, *Iterated hyper-extensions and an idempotent ultrafilter proof of Rado's theorem*, in preparation.

[10] L. Luperi Baglini, "Hyperintegers and Nonstandard Techniques in Combinatorics of Numbers", PhD Dissertation (2012), University of Siena, avaible online at http://arxiv.org/abs/1212.2049.

Randomness

Random subgraphs make identification affordable

Florent Foucaud[1], Guillem Perarnau[1] and Oriol Serra[1]

Abstract. An identifying code of a graph is a dominating set which uniquely determines all the vertices by their neighborhood within the code. Whereas graphs with large minimum degree have small domination number, this is not true for the identifying code number.

We show that every graph G with n vertices, maximum degree $\Delta = \omega(1)$ and minimum degree $\delta \geq c \log \Delta$, for some constant $c > 0$, contains a large spanning subgraph which admits an identifying code of size $O(n \log \Delta / \delta)$. The result is best possible both in terms of code size and in number of edges deleted. The proof is based on the study of random subgraphs of G using standard concentration tools and the local lemma.

1 Introduction

Consider any graph parameter that is not monotone with respect to graph inclusion. Given a graph G, a natural problem in this context is to study the minimum value of this parameter over all spanning subgraphs of G. In particular, how many edge deletions are sufficient in order to obtain from G a graph with optimal value of the parameter? Herein, we study this question with respect to the identifying code number of a graph, a well-studied non-monotone parameter.

An *identifying code* of a graph is a subset of vertices which is a dominating set C such that each vertex is uniquely determined by its neighborhood within C. More formally, each vertex x of $V(G) \setminus C$ has at least one neighbor in C (x is *dominated*) and for each pair u, v of vertices of G, $N[u] \cap C \neq N[v] \cap C$; u, v are *separated*. The minimum size of an identifying code in a graph G, denoted by $\gamma^{\text{ID}}(G)$, is the *identifying code number* of G. Identifying codes were introduced in [5], motivated by

[1] Universitat Politècnica de Catalunya, BarcelonaTech, C/ Jordi Girona 1-3, 08034 Barcelona, Spain. Email: florent.foucaud@ma4.upc.edu, guillem.perarnau@ma4.upc.edu, oserra@ma4.upc.edu

several applications. Generally speaking, if the graph models a facility or a computer network, identifying codes can be used to detect dangers in facilities [9] or failures in networks [5]. In this context, deleting edges to the underlying graph is particularly meaningful since it may represent the sealing of a door in a facility network, and the removal of a wire in a computer network.

Note that some graphs may not admit an identifying code, in particular when they have pairs of twin vertices (*i.e.* which have the same closed neighborhood). However any twin-free graph is easily seen to admit an identifying code (*e.g.* its vertex set). The identifying code number of a graph G on n vertices satisfies $\log_2(n+1) \leq \gamma^{\text{ID}}(G) \leq n$.

There are very dense graphs that have a huge identifying code number; sparse graphs, such as trees and planar graphs, also have a linear identifying code number [10]. On the other hand, one can also find sparse and dense graphs with identifying code number $O(\log n)$ [4,8].

It shall be observed from the previous facts that the identifying code number is not a monotone function with respect to the addition (or deletion) of edges. This motivates the following question:

Given any sufficiently dense graph, can we delete a small number of edges to get a spanning subgraph with a small identifying code? If the answer is positive, how many edges are sufficient (and necessary)?

In other words, we would like to study the minimum size of an identifying code among all spanning subgraphs of a given graph, and to determine the largest spanning subgraph with an asymptotically optimal identifying code.

Despite being dense, the random graph $G(n, p)$ (for $0 < p < 1$) has a logarithmic size identifying code, as with high probability,

$$\gamma^{\text{ID}}(G(n,p)) = (1+o(1))\frac{2\log n}{\log(1/q)},$$

where $q = p^2 + (1-p)^2$ [4]. This suggests that in a dense graph, the lack of structure implies the existence of a small identifying code number. Hence, introducing some randomness to the structure of a dense graph having large identifying code number might decrease this number. Indeed, this intuition is used in this work.

By studying the behavior of a *random subgraph of a graph with large minimum degree* (see for example [1,6]), we prove the following:

Theorem 1.1. *For any graph G on n vertices (n large enough) with maximum degree $\Delta = \omega(1)$ and minimum degree $\delta \geq 66\log\Delta$, there exists* *a subset of edges $F \subset E(G)$ of size*

$$|F| = O(n \log \Delta) ,$$

such that

$$\gamma^{ID}(G \setminus F) = O\left(\frac{n \log \Delta}{\delta}\right) .$$

In particular, when the minimum degree is linear, $\delta = \Theta(n)$, this shows that it is enough to delete $O(n \log n)$ edges to get a logarithmic size identifying code. The next theorem shows that Theorem 1.1 cannot be improved much.

Theorem 1.2. *For any $d \geq 2$, there exists a d–regular graph G_n^d on n vertices with the following properties.*

1. *For any $M \geq 0$, there exists a constant $c > 0$ such that for any set of edges $F \subset E(G_n^d)$ satisfying $\gamma^{ID}(G_n \setminus F) \leq M\frac{n \log d}{d}$, $|F| \geq cn \log d$.*
2. *For any spanning subgraph H of G_n^d, $\gamma^{ID}(H) = \Omega\left(\frac{n \log d}{d}\right)$.*

When $\delta = \text{Poly}(\Delta)$, Theorem 1.2 shows that Theorem 1.1 is tight, that is, we cannot hope for having a smaller identifying code by deleting any set of edges. Moreover, if Δ is bounded or $\delta \leq c' \log \Delta$, for some small constant $c' > 0$, there is no way to improve the size of the identifying code of G by deleting edges.

2 Methods and proofs

The complete proofs can be found in [3].

The proof of Theorem 1.1 focuses on the study of the random spanning subgraph $G(B, f)$ of G, where $B \subseteq V$ and $f : V(G) \to \mathbb{R}^+ \cup \{0\}$ is a function. Edges non incident to B are always present in $G(B, f)$, while each incident edge uv appears in $G(B, f)$ independently with probability $1 - p_{uv}$, where p_{uv} depends on $f(u)$, $f(v)$ and the degree of u and v in B, $d_B(u)$ and $d_B(v)$ respectively. The next lemma gives an exponential upper-bound on the probability that two vertices of $G(B, f)$ are not separated by B.

Lemma 2.1. *Given a graph G and a subset $B \subseteq V(G)$, consider the random subgraph $G(B, f)$. For every pair u, v of distinct vertices with $d_B(u) \geq d_B(v)$,*

$$\Pr\left(N_{G(B,f)}[u] \cap B = N_{G(B,f)}[v] \cap B\right) < e^{-3f(u)/16} .$$

The proof of Theorem 1.1 is structured in the following steps:

1. Select a set $C \subseteq V$ at random, where each vertex is selected independently with probability p. Using the Chernoff inequality, estimate the probability of the event A_C that C is small enough for our purposes. From C, construct the spanning subgraph $G(C, f)$ of G, with $f(u) = \min(66 \log \Delta, d_B(u))$.
2. Use Lovász Local Lemma and Lemma 2.1 to lower-bound the probability that the following events (whose disjunction we call A_{LL}) hold jointly: 1. in $G(C, f)$, each pair of vertices that are at distance at most 2 from each other are separated by C; 2. for each such pair and each neighbor of this pair in G, its degree within C in G is close to its expected value $d(v)p$. Show that with nonzero probability, A_C and A_{LL} hold jointly.
3. Find a dominating set D with $|D| = O(|C|)$; if A_{LL} holds, $C \cup D$ is an identifying code.
4. Show that, if A_C and A_{LL} hold, the expected number of deleted edges is small.

To prove Theorem 1.2, we first study the complete graph on n vertices. We combine the following two lemmata to get as a direct corollary Proposition 2.4.

Lemma 2.2. *For any $M \geq 0$, there exists a constant $c_0 > 0$ such that any graph G with $\gamma^{ID}(G) \leq M \log n$ contains at least $c_0 n \log n$ many edges.*

Lemma 2.3. *Let G be a graph and $\overline{G}$ its complement. If G and $\overline{G}$ are twin-free,*

$$\frac{1}{2} \leq \frac{\gamma^{ID}(\overline{G})}{\gamma^{ID}(G)} \leq 2 .$$

Proposition 2.4. *Let K_n be the complete graph on n vertices. For any $M \geq 0$, there exists a constant $c > 0$ such that for any set of edges $F \subset E(K_n)$ satisfying $\gamma^{ID}(K_n \setminus F) \leq M \log n$, $|F| \geq cn \log n$.*

Now, consider the graph G_n^d to be the disjoint union of cliques of order $d+1$. Since each clique is a connected component, an asymptotically optimal identifying code for G_n^d must be also asymptotically optimal for each component. By Proposition 2.4, we must delete at least $\Omega(d \log d)$ edges from each clique to get an identifying code of size $O(\log d)$ in each component. Thus one must delete at least $\Omega(n \log d)$ edges from G_n^d to get an identifying code of size $O\left(\frac{n \log d}{d}\right)$, thus, proving Theorem 1.2.

3 Concluding remarks

1. In [2], the notion of a *watching system* has been introduced as a relaxation of identifying codes: in a watching system, code vertices ("watchers") are allowed to identify any subset of their closed neighborhood, and several watchers can be placed in one vertex. Hence, for any spanning subgraph G' of G and denoting by $w(G)$ the minimum size of a watching system of G, we have $w(G) \leq w(G') \leq \gamma^{\text{ID}}(G')$. In particular, the watching number is a monotone parameter with respect to graph inclusion. From Theorem 1.1 we have:

Corollary 3.1. *Under the hypothesis of Theorem* 1.1,

$$w(G) = O\left(\frac{n \log \Delta}{\delta}\right).$$

2. Using Lemma 2.3, Theorem 1.1 can be adapted to the case where we want to *add* edges rather than deleting them.

3. Given a graph property $\mathcal{P}$, the *resilience* of G with respect to $\mathcal{P}$ is the minimum number of edges one has to delete to obtain a graph not satisfying $\mathcal{P}$. The resilience of monotone properties is well studied, in particular, in the context of random graphs [11]. Our result can be understood in terms of the resilience of the property $\mathcal{P}$, G does not admit an small identifying code. For any graph G satisfying the hypothesis of Theorem 1.1 and Theorem 1.2, the resilience with respect to $\mathcal{P}$ is at most $O(n \log \Delta)$, and this upper bound is attained.

4. The proof of Theorem 1.1 just provides an exponentially small lower bound on the probability that we can find the desired object. However, if we assume that $\Delta = n$, this probability can be shown to be $1 - o(1)$. In such a case our proof provides a randomized algorithm which constructs a good spanning subgraph and a small identifying code meeting the bounds on Theorem 1.1.

References

[1] N. ALON, *A note on network reliability*, In: "Discrete probability and algorithms", IMA Vol. Math. Appl. **72**, Springer, New York, 1995, 11–14.

[2] D. AUGER, I. CHARON, O. HUDRY and A. LOBSTEIN, *Watching systems in graphs: an extension of identifying codes*, Discrete Applied Mathematics, to appear.

[3] F. FOUCAUD, G. PERARNAU and O. SERRA, *Random subgraphs make identification affordable*, available in ArXiv e-prints, 2013.

[4] A. FRIEZE, R. MARTIN, J. MONCEL, M. RUSZINKÓ and C. SMYTH, *Codes identifying sets of vertices in random networks*, Discrete Mathematics **307** (9-10) (2007), 1094–1107.

[5] M. G. KARPOVSKY, K. CHAKRABARTY and L. B. LEVITIN, *On a new class of codes for identifying vertices in graphs*, IEEE Transactions on Information Theory **44** (1998), 599-611.

[6] M. KRIVELEVICH, C. LEE and B. SUDAKOV, *Long paths and cycles in random subgraphs of graphs with large minimum degree*, ArXiv e-prints, 2012.

[7] A. LOBSTEIN, *Watching systems, identifying, locating-dominating and discriminating codes in graphs: a bibliography*, http://www.infres.enst.fr/ lobstein/debutBIBidetlocdom.pdf

[8] J. MONCEL, *On graphs on n vertices having an identifying code of cardinality* $\log_2(n+1)$, Discrete Applied Mathematics **154** (14) (2006), 2032–2039.

[9] S. RAY, R. UNGRANGSI, F. DE PELLEGRINI, A. TRACHTENBERG and D. STAROBINSKI, *Robust location detection in emergency sensor networks*, IEEE Journal on Selected Areas in Communications **22** (6) (2004), 1016–1025.

[10] P. J. SLATER and D. F. RALL, *On location-domination numbers for certain classes of graphs*, Congressus Numerantium **45** (1984), 97–106.

[11] B. SUDAKOV and V. H. VU, *Local resilience of graphs*, Random Structures Algorithms **33** (4) (2008), 409–433.

On two-point configurations in subsets of pseudo-random sets

Elad Aigner-Horev[1] and Hiệp Hàn[2]

Abstract. We prove a transference type result for pseudo-random subsets of $\mathbb{Z}_N$ that is analogous to the well-known Fürstenberg-Sárközy theorem. More precisely, let $k \geq 2$ be an integer and let β and γ be real numbers satisfying

$$\gamma + (\gamma - \beta)/(2^{k+1} - 3) > 1.$$

Let $\Gamma \subseteq \mathbb{Z}_N$ be a set with size at least N^γ and linear bias at most N^β. Then, every $A \subseteq \Gamma$ with relative density $|A|/|\Gamma| \geq (\log\log N)^{-\frac{1}{2}\log\log\log\log\log N}$ contains a pair of the form $\{x, x + d^k\}$ for some nonzero integer d.

For instance, for squares, *i.e.*, $k = 2$, and assuming the best possible pseudo-randomness $\beta = \gamma/2$ our result applies as soon as $\gamma > 10/11$.

Our approach uses techniques of Green as seen in [6] relying on a Fourier restriction type result also due to Green.

1 Introduction

A classical result in additive combinatorics, proved independently by Sárközy [10] and Fürstenberg [4], states that subsets of the first N integers with positive density contains a pair which differ by a perfect k-th power, *i.e.*, a pair $\{x, x + d^k\}$ for some $d > 0$. As a quantitative version it is known due to [1,9] that this conclusion already holds for sets A of density

$$|A|/N \geq (\log N)^{-c\log\log\log\log N} \quad \text{for some } c > 0 \tag{1.1}$$

In this note we consider the problem of extending (1.1) to hold for subsets of vanishing relative density of sparse pseudo-random subsets of

[1] Department of Mathematics, University of Hamburg, Bundesstrasse 55 D-20146 Hamburg, Germany. Email: elad.horev@math.unihamburg.de

[2] Instituto de Matemática e Estatística, Universidade de São Paulo, Rua do Matão 1010, 05508–090 São Paulo, Brazil. Email: hh@ime.usp.br. Supported by FAPESP (Proc. 2010/16526-3) and by CNPq (Proc. 477203/2012-4) and by NUMEC/USP.

$\mathbb{Z}_N$. This type of extensions are commonly called transference results in which an extremal problem known for dense objects is transferred or carried over to sparse objects taken from a well-behaved universe like a random or a pseudo-random set. We refer to [2, 3, 6–8, 11] for further information.

Qualitative transference of (1.1) to sparse random sets were first considered in [7]. Later on, Nguyen [8] proved that with high probability every relatively dense subset of random sets $R \subset [N]$ of size $\Omega(N^{1-1/k})$ contains the configuration $\{x, x + d^k\}$, for some nonzero d. Up to a multiplicative constant this density attained for the random host is best possible.

It is also interesting to note that Tao and Ziegler proved that the polynomial Szemerédi theorem also holds in the primes. Their proof relies heavily on pseudo-random properties of the primes, thus, can be seen in the scheme mentioned above. For a classical notion of pseudo-randomness defined by small non-trivial Fourier coefficients, however, nothing is known concerning extensions of (1.1) and our Theorem 1.1 shall give the first nontrivial bound for this setting.

Our main result. Before stating the result we require some notation. For the purposes of Fourier analysis we endow $\mathbb{Z}_N$ with the counting measure and, consequently, endow its dual group $\widehat{\mathbb{Z}_N}$ with the uniform measure. As a result, given a function $f{:}\mathbb{Z}_N \to \mathbb{C}$ the Fourier transform of f is defined to be the function $\widehat{f} : \widehat{\mathbb{Z}_N} \to \mathbb{C}$ given by $\widehat{f}(\xi) = \sum_{x \in \mathbb{Z}} f(x)e(-\xi x)$ where $e(x) = e^{2\pi i x/N}$. We write $\|f\|_u = \sup_{\widehat{0} \neq \xi \in \widehat{\mathbb{Z}}_N} |\widehat{f}(\xi)|$ to denote magnitude of the second largest Fourier coefficient of f, and call $\|f\|_u$ the *linear bias* of f. Given $f, g : \mathbb{Z}_N \to \mathbb{C}$, the convolution of f and g is given by $f * g(x) = \sum_{y \in \mathbb{Z}_N} f(y)g(x - y)$.

We identify a set with its characteristic function, *i.e.*, if $A \subseteq \mathbb{Z}_N$ then A also denotes a 0, 1-function with $A(x) = 1$ if and only if $x \in A$. Finally, let $Q_k = \{x^k : x \leq N^{1/k}, x \text{ is integer}\}$ denote the set of kth powers. Our main result reads as follows.

Theorem 1.1. *Let $k \geq 2$ be an integer and let $\gamma > \beta > 0$ be reals with $\gamma + (\gamma - \beta)/(2^{k+1} - 3) > 1$. Then there is an n_0 such that for all $N > n_0$ the following holds. Let $\Gamma \subseteq \mathbb{Z}_N$ satisfy*

$$|\Gamma| \geq N^\gamma \text{ and } \|\Gamma\|_u \leq N^\beta,$$

and let $\alpha = \alpha(N) \geq (\log\log N)^{-\frac{1}{2}\log\log\log\log\log N}$. Then, every subset $A \subseteq \Gamma$ satisfying $|A| \geq \alpha|\Gamma|$ contains a pair $\{x, x + d\} \subseteq A$ where $d \in Q_k$.

It is worth to note that the proof of Theorem 1.1 does not merely guarantee one desired configuration but many. Indeed, the number of such configurations found in the subset A is at least $\eta|\Gamma|^2|Q_k|N^{-1}$, where $\eta = \eta(\alpha) > 0$ if $\alpha \neq 0$. Up to $\eta(\alpha)$ this bound is clearly best possible. Moreover, let us emphasise that in Theorem 1.1 we can handle subsets of Γ whose relative density is vanishing as N goes to infinity. This is due to the fact that we are transferring the *quantitative* version of the dense case of the Fürstenberg-Sárközy theorem, namely (1.1).

Due to Parseval's equality the the parameter β in Theorem 1.1 controlling the pseudo-randomness of Γ satisfies $\beta \geq \gamma/2$. That is, one may think of $\beta = \gamma/2$ as though Γ is "as pseudo-random as possible". In this case, *i.e.*, $\beta = \gamma/2$, we have that Theorem 1.1 is applicable as long as $\gamma > 1 - \frac{1}{2^{k+2}-5}$.

2 Sketch of the proof of Theorem 1.1

Our approach follows that of Green [6]. Given $\alpha = \alpha(N)$ we introduce functions η, δ, and ε depending on N so that $\alpha \gg \eta \gg \delta \gg \varepsilon > 0$. Given the set $A \subseteq \Gamma$ and a set of frequencies $\emptyset \neq S \subset \mathbb{Z}_N$ we know that there is a $\varrho = \varrho(N) \in [\varepsilon/2, \varepsilon]$ such that the Bohr set $B = B(S, \varrho)$ is a *regular* (see, *e.g.* [12], chapter 4.4). Define the function $a : \mathbb{Z}_N \to \mathbb{R}$ given by $a(x) = \frac{N}{|\Gamma||B|}(A * B)(x)$ which can be shown to have certain attributes seen in characteristic functions of *d*ense sets; that is $\|a\|_{\ell^1} \geq \alpha N$ and $\|a\|_\infty \leq 3$, provided $|\Gamma|\|\Gamma\|_u^{-1} > 2(\varepsilon/(20|S|^{1/2}))^{-|S|}$.

Applying(1.1) combined with a Varnavides type argumentthen yields a lower bound on the of number of desired configurations "in a"

$$\Lambda(a, Q_k) = \sum_{x,d \in \mathbb{Z}_N} a(x)a(x+d)Q_k(d) \geq \eta N|Q_k|. \tag{2.1}$$

Due to the convolution property and assuming that $A \subseteq \Gamma$ contains less than $\eta N^{1+1/k}$ pairs of the form $\{x, x^k\}$, we have

$$\Lambda(a, Q_k) < \frac{N}{|\Gamma|^2} \sum_{\xi \in \widehat{\mathbb{Z}}_N} |\widehat{A}(\xi)|^2 |\widehat{Q_k}(\xi)| \left| \frac{|\widehat{B}(\xi)|^2}{|B|^2} - 1 \right| + \frac{\eta}{2} N^{1+1/k} \tag{2.2}$$

from which we then derive a contradiction to (2.1).

To this end, we split the sum into two sums; one ranging over $\xi \in S = Spec_\delta(A) = \{\xi \in \widehat{\mathbb{Z}}_N : |\widehat{A}(\xi)| \geq \delta|\Gamma|\}$ and another ranging over $\xi \notin S$. It can be shown that $\left|\frac{|\widehat{B}(\xi)|^2}{|B|^2} - 1\right| \leq 2\varepsilon$ for all $\xi \in S$ and $|S| \leq (2/\delta)^{(2-2\beta)/(\gamma-\beta)}$ from which we conclude that for the first sum ranging over S is at most $2\varrho|S||\widehat{A}(0)|^2|\widehat{Q_k}(0)| \leq \eta|\Gamma|^2|Q_k|/4$.

Handling the sum over $\widehat{\mathbb{Z}}_N \setminus S$, however, is more complicated. In fact, its proof is a central part of the proof of Theorem 1.1 but due to space limitation we omit it here. Nevertheless, we mention that the main tools in the proof are Waring's theorem and a restriction type result due to Green [5] which is similar to that used in [6]. It is a close adaption of the restriction theorem due to Stein-Tomas [13]. We omit the details.

References

[1] A. BALOG, J. PELIKÁN, J. PINTZ and E. SZEMERÉDI, *Difference sets without κth powers*, Acta Math. Hungar. **65** (2) (1994), 165–187.

[2] D. CONLON and W. T. GOWERS, *Combinatorial theorems in sparse random sets*, submitted.

[3] D. CONLON, J. FOX and Y. ZHAO, *Extremal results in sparse pseudorandom graphs*, submitted.

[4] H. FURSTENBERG, *Ergodic behavior of diagonal measures and a theorem of Szemerédi on arithmetic progressions*, J. Analyse Math. **31** (1977), 204–256.

[5] B. GREEN, *Roth's theorem in pseudorandom sets*, manuscript.

[6] B. GREEN, *Roth's theorem in the primes*, Ann. of Math. (2) **161** (3) (2005), 1609–1636.

[7] I. ŁABA and M. HAMEL, *Arithmetic structures in random sets*, Integers **8** (2008).

[8] H. H. NGUYEN, *On two-point configurations in random set*, Integers **9** (2009), 41–45.

[9] J. PINTZ, W. L. STEIGER and E. SZEMERÉDI, *On sets of natural numbers whose difference set contains no squares*, J. London Math. Soc. (2) **37** (2) (1988), 219–231.

[10] A. SÁRKŐZY, *On difference sets of sequences of integers. I*, Acta Math. Acad. Sci. Hungar. **31** (1-2) (1978), 125–149.

[11] M. SCHACHT, *Extremal results for random discrete structures*, submitted.

[12] T. TAO and V. H. VU, "Additive Combinatorics", Cambridge Studies in Advanced Mathematics, Vol. 105, Cambridge University Press, Cambridge, 2010.

[13] P. A. TOMAS, *A restriction theorem for the Fourier transform*, Bull. Amer. Math. Soc. **81** (1975), 477–478.

On the giant component of random hyperbolic graphs

Michel Bode[1], Nikolaos Fountoulakis[1] and Tobias Müller[2]

1 Introduction

The theory of geometric random graphs was initiated by Gilbert [2] already in 1961 in the context of what is called *continuum percolation*. In 1972, Hafner [4] focused on the typical properties of large but finite random geometric graphs. Here N points are sampled within a certain region of $\mathbb{R}^d$ following a certain distribution and any two of them are joined when their Euclidean distance is smaller than some threshold which, in general, is a function of N. In the last two decades, this class of random graphs has been studied extensively – see the monograph of Penrose [6].

However, what structural characteristics emerge when one considers these points distributed on a curved space where distances are measured through some (non-Euclidean) metric? Such a model was introduced by Krioukov et al. [5] and some typical properties of these random graphs were studied with the use of non-rigorous methods.

1.1 Random geometric graphs on a hyperbolic space

The most common representations of the hyperbolic space is the upper-half plane representation $\{z : \Im z > 0\}$ as well as the Poincaré unit disc which is simply the open disc of radius one, that is, $\{(u, v) \in \mathbb{R}^2 : 1 - u^2 - v^2 > 0\}$. Both spaces are equipped with the hyperbolic metric; in the former case this is $\frac{1}{(\zeta y)^2} dy^2$ whereas in the latter this is $\frac{4}{\zeta^2} \frac{du^2+dv^2}{(1-u^2-v^2)^2}$, where ζ is some positive real number. It can be shown that the (Gaussian) curvature in both cases is equal to $-\zeta^2$.We will denote by $\mathbb{H}^2_\zeta$ the class of these spaces.

[1] School of Mathematics, University of Birmingham, United Kingdom.
Email: michel.bode@gmx.de, n.fountoulakis@bham.ac.uk. This research has been supported by a Marie Curie Career Integration Grant PCIG09-GA2011-293619.

[2] Mathematical Institute, Utrecht University, The Netherlands. Email: t.muller@uu.nl

In this paper, following the definitions in [5], we shall be using the native representation of $\mathbb{H}^2_\zeta$. Under this representation, the ground space of $\mathbb{H}^2_\zeta$ is $\mathbb{R}^2$ and every point $x \in \mathbb{R}^2$ whose polar coordinates are (r, θ) has hyperbolic distance from the origin O equal to r. Also, a circle of radius r around the origin has length equal to $\frac{2\pi}{\zeta}\sinh(\zeta r)$ and area equal to $\frac{2\pi}{\zeta^2}(\cosh(\zeta r) - 1)$.

Let $N = \nu e^{\zeta R/2}$, where ν is a positive real number that controls the average degree of the random graph. We create a random graph by selecting randomly N points from the disc of radius R centred at the origin O, which we denote by $\mathcal{D}_R$. If such a random point u has polar coordinates (r, θ), then θ is uniformly distributed in $(0, 2\pi]$, whereas the probability density function of r, which we denote by $\rho(r)$, is determined by a parameter $\alpha > 0$ and is equal to

$$\rho(r) = \alpha \frac{\sinh \alpha r}{\cosh \alpha R - 1}. \tag{1.1}$$

When $\alpha = \zeta$, then this is the uniform distribution. This set of points will be the vertex set of the random graph and we denote it by V_N. The random graph $\mathcal{G}(N; \zeta, \alpha)$ is formed when we join two vertices, if they are within (hyperbolic) distance R.

Krioukov et al. [5] focus on the degree distribution of $\mathcal{G}(N; \zeta, \alpha)$, showing that when $0 < \zeta/\alpha < 2$ this follows a power law with exponent $2\alpha/\zeta + 1$. They also discuss clustering on a smooth version of the above model. Their results have been verified rigorously by Gugelmann et al. [3]. When $1 < \zeta/\alpha < 2$, the exponent is between 2 and 3, as is the case in a number of networks that emerge in applications such as computer networks, social networks and biological networks (see for example [1]). Krioukov et al. [5] introduce this model as a geometric framework for the study of complex networks. In fact, they view the degree distribution as well as the existence of clustering at a local level as "natural reflections of the underlying hyperbolic geometry".

1.2 Component structure of $\mathcal{G}(N; \zeta, \alpha)$

This paper focuses on the component structure of $\mathcal{G}(N; \zeta, \alpha)$ and, in particular, on the size of its largest component. We also denote by $|L_1|$ the size of a largest connected component of $\mathcal{G}(N; \zeta, \alpha)$.

In this contribution, we show that when ζ/α crosses 1 a "phase transition" occurs. More specifically, if $\zeta/\alpha < 1$, then *asymptotically almost surely (a.a.s.)*, that is, with probability $1 - o(1)$ as $N \to \infty$, $|L_1|$ is bounded by a sublinear function, whereas if $\zeta/\alpha > 1$, then $|L_1|$ is linear.

Theorem 1.1. *Let ζ, α be positive real numbers. The following hold:*

- *If $\zeta/\alpha > 1$, then there exists $c = c(\zeta, \alpha, \nu) > 0$ such that a.a.s. $|L_1| > cN$.*
- *If $\zeta/\alpha < 1$, then a.a.s. $|L_1| < CR^2N^{\zeta/\alpha}$, where $C = C(\zeta, \alpha) > 0$.*

Furthermore, one can show that when $\zeta/\alpha > 2$, then $\mathcal{G}(N; \zeta, \alpha)$ is a.a.s. connected. We now proceed with a brief sketch of the proof of each part of the above theorem.

2 The supercritical regime

For any given point $v \in \mathcal{D}_R$, we let $t_v = R - r_v$, where r_v denotes the radius of v – we call this the *type* of point v. We define a partition of $\mathcal{D}_R$ into homocentric bands $\mathcal{B}_i$, for $i = 1, \ldots, T$, where $T = T(N)$ is a suitably defined function of N. More specifically, the central band $\mathcal{B}_0$ consists of all points in $\mathcal{D}_R$ whose type is larger than $R/2$, that is, their radius is less than $R/2$. Note that any two vertices in $\mathcal{B}_0$ are connected by an edge as their hyperbolic distance is less than R. In other words, the subgraph of $\mathcal{G}(N; \zeta, \alpha)$ induced by the vertices in $\mathcal{B}_0$ is the complete graph. To define the remaining bands, we define a decreasing sequence of positive real numbers $t_0 > t_1 > \cdots$, where $t_0 := R/2$ and for $i \geq 1$ we have

$$t_i - \frac{2}{\zeta} \ln\left(\frac{8\pi}{\nu} \ln t_i\right) = \lambda t_{i-1}, \tag{2.1}$$

where $\lambda := \frac{2}{\zeta}\left(\alpha - \frac{\zeta}{2}\right) < 1$ (as $\zeta/\alpha > 1$) and

$$\mathcal{B}_i = \{v \in \mathcal{D}_R \ : \ t_i \leq t_v < t_{i-1}\}.$$

We define T as the largest i such that $e^{-\alpha(t_{i-1}-t_i)} < 1/2$ and $t_i > e$. Let $\mathcal{N}_0$ denote the set of vertices that belong to the set $\mathcal{B}_0$. In turn, for $i > 0$ we let $\mathcal{N}_i$ denote the set of vertices in $\mathcal{B}_i$ that have at least one neighbour in $\mathcal{N}_{i-1}$. Since $\mathcal{B}_0$ is a clique, the subgraph of $\mathcal{G}(N; \zeta, \alpha)$ that is induced by $\cup_{i=0}^{T} \mathcal{N}_i$ is connected and has size $\sum_{i=0}^{T} |\mathcal{N}_i|$.

We establish bounds on the sizes of the sets $\mathcal{N}_i$, for $i = 1, \ldots, T$. In particular, we show that the number of vertices in $\mathcal{N}_i$ stochastically dominates the number of vertices that are contained in a subset of $\mathcal{B}_i$ that has arc Θ_i, where the sequence of Θ_i satisfies for $i \geq 1$

$$\Theta_i \geq \Theta_{i-1}\left(1 - \exp\left(-\frac{\nu}{4\pi}\left(e^{-\alpha t_{i-1}} - e^{-\alpha t_{i-2}}\right)\theta^{(i)}\right)\right), \quad \Theta_0 := 2\pi$$

and $\theta^{(i)} := e^{\frac{\zeta}{2}(t_{i-1}+t_i)}$ (here $t_{-1} := R$). We denote this number by N_i'. A concentration argument shows that a.a.s. $N_i' \geq \frac{1}{2} N \frac{\Theta_i}{2\pi}\left(e^{-\alpha t_i} - e^{-\alpha t_{i-1}}\right)$.

As $e^{-\alpha(t_{i-1}-t_i)} < 1/2$, it follows that

$$\left(e^{-\alpha t_{i-1}} - e^{-\alpha t_{i-2}}\right)\theta^{(i)} = e^{-\alpha t_{i-1}}\left(1 - e^{-\alpha(t_{i-2}-t_{i-1})}\right)\theta^{(i)}$$
$$> \frac{1}{2} e^{-\alpha t_{i-1} + \frac{\zeta}{2}(t_{i-1}+t_i)} = \frac{4\pi}{\nu}\ln t_i,$$

whereby

$$\Theta_i \geq \Theta_{i-1}\left(1 - \frac{1}{t_i}\right) \overset{(2.1)}{>} \Theta_{i-1}\left(1 - \frac{1}{\lambda^i t_0}\right).$$

It then follows that for some $c = c(\zeta, \alpha, \nu) > 0$ a.a.s.

$$\sum_{i=0}^{T} N_i' \geq N\left(e^{-\alpha t_T} - e^{-\alpha t_0}\right)\prod_{i=0}^{T}\left(1 - \frac{1}{\lambda^i t_0}\right) > cN.$$

The stochastic domination implies this part of the theorem.

3 The subcritical regime

A first moment argument shows that all vertices have type at most $\frac{\zeta}{2\alpha}R + \omega(N)$, where $\omega(N)$ is a function such that $\omega(N) \to \infty$ as $N \to \infty$. Hence, since $\zeta/\alpha < 1$, it follows that all vertices have types which are smaller than and bounded away from $R/2$. We consider a vertex v which has this type and we analyse a *breadth exploration process* through which we bound the *total angle of the component* which v belongs to. We define the *total angle* of the component of v to be the largest relative angle between any two of its vertices – if the component has only one vertex, then this is equal to zero. We denote this angle by $\Theta(v)$. We show that the assumption that the type of v is $\frac{\zeta}{2\alpha}R + \omega(N)$ gives a stochastic upper bound on the total angle of the component of any vertex in V_N. Our bound is as follows.

Lemma 3.1. *Let $v \in \mathrm{V}_N$ be a vertex having $t_v = \frac{\zeta}{2\alpha}R + \omega(N)$. There exists a constant $C' = C'(\zeta, \alpha, \nu) > 0$ such that with probability $1 - o\left(\frac{1}{N^{1-\zeta/\alpha}}\right)$ we have*

$$\Theta(v) \leq C'\frac{R^2 N^{\zeta/\alpha}}{N}.$$

The *breadth exploration process* is a process that is somewhat similar to the breadth-first search algorithm. More specifically, starting from vertex v with type as in the above lemma, we expose the vertex of the largest type among those vertices that are within distance R from v in clockwise direction. Subsequently, we continue this procedure until a vertex of type K is reached, where K is a large constant. We repeat this in anticlockwise direction. This completes the first phase of the process. Thereafter, we bound the contribution that comes from vertices of type smaller than K. If these have also vertices within distance R that have not been covered previously, then we start the first phase again. We show that the number of repetitions of this phase is bounded in probability. A first moment argument shows that the probability that there is a vertex v with $\Theta(v)$ larger than the bound of the above lemma is $o(1)$. The result provides a bound on the total angle of each component. This needs to be complemented by a result which associates the total angle of a component with the number of vertices. We show that a.a.s. there is no component with total angle at most $C' R^2 N^{\zeta/\alpha}/N$ that has more than $C R^2 N^{\zeta/\alpha}$ vertices, where $C = C(\zeta, \alpha, \nu) > 0$ is another constant. This completes the proof of Theorem 1.1.

4 Conclusions - Further directions

This contribution focuses on the size of the largest component of random geometric graphs on the hyperbolic plane. We show that when the ratio ζ/α crosses 1 a giant component emerges. But is this component unique? What is the size of the largest component in each case? Moreover, our results do not cover the critical case $\zeta/\alpha = 1$, that is, when the points are uniformly distributed on $\mathcal{D}_R$. This is a natural direction that goes along the lines of the theory of random geometric graphs on Euclidean spaces.

References

[1] R. Albert and A.-L. Barabási, *Statistical mechanics of complex networks*, *Rev. Mod. Phys.* **74** (2002), 47–97.

[2] E. N. Gilbert, *Random plane networks*, J. Soc. Indust. Appl. Math. **9** (1961), 533–543.

[3] L. Gugelmann, K. Panagiotou and U. Peter, *Random hyperbolic graphs: degree sequence and clustering*, In: "Proceedings of the 39th International Colloquium on Automata, Languages and Programming", A. Czumaj *et al.* (eds.), Lecture Notes in Computer Science 7392, 2012, 573–585.

[4] R. Hafner, *The asymptotic distribution of random clumps*, Computing **10** (1972), 335–351.

[5] D. Krioukov, F. Papadopoulos, M. Kitsak, A. Vahdat and M. Boguñá, *Hyperbolic geometry of complex networks*, Phys. Rev. E **82** (2010), 036106.

[6] M. Penrose, "Random Geometric Graphs", Oxford University Press, 2003.

Discontinuous bootstrap percolation in power-law random graphs

Hamed Amini[1], Nikolaos Fountoulakis[2]
and Konstantinos Panagiotou[3]

1 Introduction

Bootstrap percolation was introduced by Chalupa, Leath and Reich [6] during the 1970's in the context of magnetic disordered systems and has been re-discovered since then by several authors mainly due to its connections with various physical models. A *bootstrap percolation process* with *activation threshold* an integer $r \geq 2$ on a graph $G = G(V, E)$ is a deterministic process which evolves in rounds. Every vertex has two states: it is either *infected* or *uninfected*. Initially, there is a subset $\mathcal{A}_0 \subseteq V$ which consists of infected vertices, whereas every other vertex is uninfected. Subsequently, in each round, if an uninfected vertex has at least r of its neighbours infected, then it also becomes infected and remains so forever. This is repeated until no more vertices become infected. We denote the final infected set by $\mathcal{A}_f$. Our general assumption will be that the initial set of infected vertices $\mathcal{A}_0$ is chosen randomly among all subsets of vertices of a certain size.

These processes have been studied on a variety of graphs, such as trees, grids, hypercubes, as well as on several distributions of random graphs. A short survey regarding applications of bootstrap percolation processes can be found in [1].

During the last decade, there has been significant experimental evidence on the structural characteristics of networks that arise in applications such as the Internet, the World Wide Web as well as social networks or even biological networks. One of the fundamental features is their

[1] Swiss Finance Institute, EPFL, Lausanne, Switzerland. Email: hamed.amini@epfl.ch

[2] School of Mathematics, University of Birmingham, United Kingdom.
Email: n.fountoulakis@bham.ac.uk. This research has been supported by the EPSRC Grant EP/K019740/1.

[3] Mathematishes Institut, LMU, Munich, Germany. Email: kpanagio@math.lmu.de. The author was partially supported by DFG grant PA 2080/1.

degree distribution, which is most cases appears to follow a power law with exponent between 2 and 3 (see for example the article of Albert and Barabási [2]). The theme of this contribution is the study of the evolution of bootstrap percolation processes on random graphs which exhibit this characteristic. We show that this boosts the evolution of the process, resulting in large infected sets starting from a small set of infected vertices.

2 Models and results

The random graph model that we consider is asymptotically equivalent to a model considered by Chung and Lu [7], and is a special case of the so-called *inhomogeneous random graph*, which was introduced by Söderberg [9] and was studied in detail by Bollobás, Janson and Riordan in [5].

2.1 Inhomogeneous random graphs – The Chung-Lu model

In order to define the model we consider for any $n \in \mathbb{N}$ the vertex set $[n] := \{1, \ldots, n\}$. Each vertex i is assigned a positive weight $w_i(n)$, and we will write $\mathbf{w} = \mathbf{w}(n) = (w_1(n), \ldots, w_n(n))$. We assume in the remainder that the weights are deterministic, and we will suppress the dependence on n, whenever this is obvious from the context. However, note that the weights could be random variables; we will not consider this case here, although it is very likely that under suitable technical assumptions our results generalize to this case as well. For any $S \subseteq [n]$, set

$$W_S(\mathbf{w}) := \sum_{i \in S} w_i.$$

In our random graph model, the event of including the edge $\{i, j\}$ in the resulting graph is independent of the events of including all other edges, and its probability equals

$$p_{ij}(\mathbf{w}) = \min\left\{\frac{w_i w_j}{W_{[n]}(\mathbf{w})}, 1\right\}. \tag{2.1}$$

We will refer to this model as the *Chung-Lu* model, and we shall write $CL(\mathbf{w})$ for a random graph in which each possible edge $\{i, j\}$ is included independently with probability as in (2.1).

2.2 Power-law degree distributions

Following van der Hofstad [10], we write for any $n \in \mathbb{N}$ and any sequence of weights $\mathbf{w} = (w_1(n), \ldots, w_n(n))$

$$F_n(x) = n^{-1} \sum_{i=1}^{n} \mathbf{1}[w_i(n) < x], \quad \forall x \in [0, \infty)$$

for the empirical distribution function of the weight of a vertex chosen uniformly at random. We will assume that F_n satisfies the following two conditions.

Definition 2.1. We say that $(F_n)_{n\geq 1}$ is *regular*, if it has the following two properties.

- **[Weak convergence of weight]** There is a distribution function $F : [0, \infty) \to [0, 1]$ such that for all x at which F is continuous $\lim_{n\to\infty} F_n(x) = F(x)$;
- **[Convergence of average weight]** Let W_n be a random variable with distribution function F_n, and let W_F be a random variable with distribution function F. Then we have $\lim_{n\to\infty} \mathbb{E}[W_n] = \mathbb{E}[W_F]$.

The regularity of $(F_n)_{n\geq 1}$ guarantees two important properties. Firstly, the weight of a random vertex is approximately distributed as a random variable that follows a certain distribution. Secondly, this variable has finite mean and therefore the resulting graph has bounded average degree. Apart from regularity, our focus will be on weight sequences that give rise to power-law degree distributions.

Definition 2.2. We say that a regular sequence $(F_n)_{n\geq 1}$ is *of power law with exponent* β, if there are $0 < \gamma_1 < \gamma_2, x_0 > 0$ and $0 < \zeta \leq 1/(\beta - 1)$ such that for all $x_0 \leq x \leq n^\zeta$

$$\gamma_1 x^{-\beta+1} \leq 1 - F_n(x) \leq \gamma_2 x^{-\beta+1},$$

and $F_n(x) = 0$ for $x < x_0$, but $F_n(x) = 1$ for $x > n^\zeta$.

We consider the random graph $CL(\mathbf{w})$ where the weight sequence $\mathbf{w} = \mathbf{w}(n)$ gives rise to a regular sequence of empirical distribution functions that are of power law with exponent β. We assume that a random set of $a(n)$ vertices is initially infected. We say that an event occurs *asymptotically almost surely (a.a.s.)*, if it occurs with probability$\to$ 1 as $n \to \infty$, in the product space of the random graph and the choice of the initially infected vertices.

2.3 Results

We determine explicitly a critical function which we denote by $a_c(n)$ such that when we infect randomly $a(n)$ vertices in $[n]$, then the following threshold phenomenon occurs. If $a(n) \ll a_c(n)$, then a.a.s. the infection spreads no further than $\mathcal{A}_0$, but when $a(n) \gg a_c(n)$, then a linear number of vertices become eventually infected. We remark that $a_c(n) = o(n)$. We define the function $\psi_r(x)$ for $x \geq 0$ to be equal to the probability that a Poisson-distributed random variable with parameter x is at least r.

Also, for a random variable X with finite expected value and distribution function F, we (informally) say that X^* follows the F-*size-biased* distribution function, if the distribution of X^* is weighted by the value of X.

Theorem 2.3. *For any $\beta \in (2,3)$ and any integer $r \geq 2$, we let $a_c(n) = n^{\frac{r(1-\zeta)+\zeta(\beta-1)-1}{r}}$ for all $n \in \mathbb{N}$. Let $a : \mathbb{N} \to \mathbb{N}$ be a function such that $a(n) \to \infty$, as $n \to \infty$, but $a(n) = o(n)$. Let also $\zeta \leq \frac{1}{\beta-1}$. If we initially infect uniformly at random $a(n)$ vertices in $[n]$, then the following holds:*

- *if $a(n) \ll a_c(n)$, then a.a.s. $\mathcal{A}_f = \mathcal{A}_0$;*
- *if $a(n) \gg a_c(n)$ and also $\frac{r-1}{2r-\beta+1} < \zeta \leq \frac{1}{\beta-1}$, then*

$$\frac{|\mathcal{A}_f|}{n} \xrightarrow{p} \mathbb{E}\left[\psi_r(U\hat{y})\right], \text{ as } n \to \infty,$$

where U is a random variable with F as its distribution function and $\hat{y}$ is the smallest positive solution of

$$y = \mathbb{E}\left[\psi_r(Wy)\right],$$

with W being a random variable whose law follows the F-size-biased distribution function.

When $0 < \zeta \leq \frac{r-1}{2r-\beta+1}$ the second part of the above statement holds with $a_c^+(n) = n^{1-\zeta\,\frac{r-\beta+2}{r-1}}$ instead of $a_c(n)$.

Note that the above theorem implies that when the maximum weight of the sequence is $n^{1/(\beta-1)}$, then the threshold function becomes equal to $n^{\frac{\beta-2}{\beta-1}}$ and does not depend on r.

This result is in sharp contrast with the behaviour of the bootstrap percolation process in $G(n,p)$ random graphs, where every edge on a set of n vertices is included independently with probability p. Recently, Janson, Łuczak, Turova and Vallier [8] (see Theorem 5.2 there) showed that when $p = d/n$, with $d > 0$ fixed, if $|\mathcal{A}_0| = o(n)$, then typically no evolution occurs. In other words, the density of the initially infected vertices must be positive in order for the density of infected vertices to grow. We note that similar behavior to the case of $G(n,p)$ has been observed in the case of random regular graphs [4], and in random graphs with given vertex degrees constructed through the configuration model, studied by the first author in [3], when the sum of the square of degrees scales linearly with n, the size of the graph. The later case includes random graphs with power-law degree sequence with exponent $\beta > 3$. Our results imply that the two regimes $2 < \beta < 3$ and $\beta > 3$ have completely different behaviors.

The next theorem complements the above theorem, as it gives a law of large numbers for the size $\mathcal{A}_f$ when a positive fraction of vertices are initially infected.

Theorem 2.4. *Let $2 < \beta < 3$ and $r \geq 2$. If $a(n) = pn$, where $p \in (0, 1)$ is fixed, then*

$$\frac{|\mathcal{A}_f|}{n} \xrightarrow{p} (1-p)\mathbb{E}\left[\psi_r(U\hat{y})\right] + p, \text{ as } n \to \infty,$$

where U is a random variable having F as its distribution function with $\hat{y}$ being the smallest positive solution of

$$y = (1-p)\mathbb{E}\left[\psi_r(Wy)\right] + p$$

and W is a random variable whose law follows the F-size-biased distribution function.

References

[1] J. ADLER and U. LEV, *Bootstrap percolation: visualizations and applications* Brazilian Journal of Physics, **33** (3) (2003), 641–644.

[2] R. ALBERT and A. BARABÁSI, *Statistical mechanics of complex networks*, Reviews of Modern Physics **74** (1) (2002), 47–97.

[3] H. AMINI, *Bootstrap percolation and diffusion in random graphs with given vertex degrees*, Electronic Journal of Combinatorics **17** (2010), R25.

[4] J. BALOGH and B. G. PITTEL, *Bootstrap percolation on the random regular graph*, Random Structures & Algorithms **30** (1-2) (2007), 257–286.

[5] B. BOLLOBÁS, S. JANSON and O. RIORDAN, *The phase transition in inhomogeneous random graphs*, Random Structures & Algorithms **31** (1) (2007), 3–122.

[6] J. CHALUPA, P. L. LEATH and G. R. REICH, *Bootstrap percolation on a bethe lattice*, Journal of Physics C: Solid State Physics **12** (1979), L31–L35.

[7] F. CHUNG and L. LU, *The average distance in a random graph with given expected degrees*, Internet Mathematics **1** (1) (2003), 91–113.

[8] S. JANSON, T. ŁUCZAK, T. TUROVA and T. VALLIER, *Bootstrap percolation on the random graph $G_{n,p}$*, Annals of Applied Probability **22** (2012), 1989–2047.

[9] B. SÖDERBERG, *General formalism for inhomogeneous random graphs*, Physical Review E **66** (2002), 066121.
[10] R. VAN DER HOFSTAD, "Random Graphs and Complex Networks" 2011, Book in preparation, www.win.tue.nl/rhofstad/NotesRGCN2011.pdf.

On a conjecture of Graham and Häggkvist for random trees

Michael Drmota[1] and Anna Lladó[2]

Abstract. A conjecture of Graham and Häggkvist says that every tree with m edges decomposes the complete bipartite graph $K_{m,m}$. By establishing some properties of random trees with the use of singularity analysis of generating functions, we prove that asymptotically almost surely a tree with m edges decomposes the complete bipartite graph $K_{2m,2m}$.

1 Introduction

Given two graphs H and G we say that H decomposes G if G is the edge–disjoint union of isomorphic copies of H. The following is a well–known conjecture of Ringel.

Conjecture 1.1 (Ringel [12]). Every tree with m edges decomposes the complete graph K_{2m+1}.

The conjecture has been verified by a number of particular classes of trees, see the dynamic survey of Gallian [5]. By using the polynomial method, the conjecture was verified by Kézdy [7] for the more general class of so–called *stunted* trees. As mentioned by the author, this class is still small among the set of all trees.

The following bipartite version of the conjecture was formulated by Graham and Häggkvist.

Conjecture 1.2 (Graham and Häggkvist [6]). Every tree with m edges decomposes the complete bipartite graph $K_{m,m}$.

Again the conjecture has been verified by a number of cases; see *e.g.* [9]. Approximate versions of the two conjectures have been also proved [6, 8–10]. However, to our knowledge, there are no results stating that

[1] Institute of Discrete Mathematics and Geometry, Vienna University of Technology, Vienna, Austria. Email: drmota@tuwien.ac.at

[2] Dept. Matemàtica Aplicada 4, Universitat Politècnica de Catalunya-BarcelonaTech, Barcelona, Spain. Email: allado@ma4.upc.edu

every tree decomposes K_{cm+1} or $K_{cm,cm}$ for some absolute constant c of reasonable size. The purpose of this paper is to show such a result for almost all trees.

Let $\mathcal{T}$ denote the class of (unlabelled) trees and let $\mathcal{T}_m$ be the class of trees with m edges. By a random tree with m edges we mean a tree chosen from $\mathcal{T}_m$ with the uniform distribution. We say that a random tree satisfies a property $\mathbb{P}$ asymptotically almost surely (a.a.s) if the probability that a random tree with m edges satisfies $\mathbb{P}$ tends to one with $m \to \infty$.

Theorem 1.3. *Asymptotically almost surely a tree with m edges decomposes $K_{2m,2m}$.*

The proof of Theorem 1.3 combines a structural analysis of random trees with combinatorial techniques for graph decompositions. In the following two sections we discuss the results and tools used in this proof.

2 Stable sets of random trees

The first property we use in the proof of Theorem 1.3 concerns the number of leaves in a random tree. Robinson and Schwenk [13] proved that the average number of leaves in an (unlabelled) random tree with m edges is asymptotically cm with $c \approx 0.438$. Drmota and Gittenberger [2] showed that the distribution of the number of leaves in a random tree with m edges is asymptotically normal with variance $c_2 m$ for some positive constant c_2. Thus, asymptotically almost surely a random tree with m edges has more than $2m/5$ leaves.

The second property we use deals with the size of a stable set in the base tree of T (the tree obtained from T by deleting its leaves.) Unfortunately this is not a parameter whose analysis can be explicitly found in the literature. We prove the following result.

Theorem 2.1. *The stable sets A, B of the base tree of a random tree with m edges satisfy a.a.s.*

$$||A| - |B|| \leq \epsilon m,$$

for every fixed $\epsilon > 0$.

The proof of Theorem 2.1 is based on the use of generating functions. We first consider the case of rooted trees. Let

$$t(x, w_0, w_1) = \sum_{m,k_0,k_1} t_{m,k_0,k_1} x^{m+1} w_0^{k_0} w_1^{k_1}, \tag{2.1}$$

where t_{m,k_0,k_1} denotes the number of rooted trees with m edges and k_0 inner vertices (including the root if the tree has at least one edge even if the root has degree one) with even distance to the root and k_1 inner vertices with odd distance to the root. Then, by using the recursive description of a tree as a collection of trees hanging from a root (iterated twice to get the proper alignment of stable sets), one can obtain an explicit expression for (2.1).

Recall that we are interested in the difference $|A| - |B|$ which we can do by setting $w_0 = w$ and $w_1 = w^{-1}$. Hence, if $T(x, w) = \sum_{m,\ell} T_{m,\ell} x^{m+1} w^{\ell}$ denotes the generating function, where $T_{m,\ell}$ denotes the number of rooted trees with m edges and $|A| - |B| = \ell$ (where ℓ is some – possibly negative – integer and the root is contained in A even if the root has degree one) then an epxlicit expression for $T(x, w) = t(x, w, w^{-1})$ is also obtained.

As usual we denote by $a_n = [x^n]a(x)$ the n-th coefficient of a power series $a(x) = \sum_{n \geq 0} a_n x^n$. With the help of this notation it follows that

$$\mathbb{E}\, w^{|A|-|B|} = \frac{[x^{m+1}]\,T(x, w)}{[x^{m+1}]\,T(x, 1)}.$$

This magnitude can be determined asymptotically if w is close to 1 with the help of standard singularity analysis tools.

From a version of Hwang's Quasi-Power-Theorem (see [1]), one can deduce that the variable $Z = |A| - |B|$ follows a normal distribution which in our case has zero mean and variance linear in m from which

$$\Pr\left(||A| - |B|| \geq \varepsilon m\right) \leq C e^{-c\varepsilon^2 m}$$

for some positive constants c and C and for sufficiently small $\varepsilon > 0$. Of course this is precisely the statement that we want to prove for unlabelled trees. We translate the above analysis to unrooted unlabeled trees via Otter's bijection (see [11] or [1]). Unfortunately we cannot prove something like a central limit theorem for $|A| - |B|$ in this case, but it is still possible to keep track of the second moment which, by using Chebyshev inequality, provides a proof of Theorem 2.1.

3 The embedding

The general approach to show that a tree T decomposes a complete graph or a complete bipartite graph consists in showing that T cyclically decomposes the corresponding graphs. We next recall the basic principle behind this approach in slightly different terminology.

A rainbow embedding of a graph H into an oriented arc–colored graph X is an injective homomorphism f of some orientation $\vec{H}$ of H in X such that no two arcs of $f(\vec{H})$ have the same color.

Let $X = Cay(G, S)$ be a Cayley digraph of an abelian group G with respect to an antisymmetric subset $S \subset G$ (that is, $S \cap -S = \emptyset$). We consider X as an arc–colored oriented graph, by giving to each arc $(x, x+s)$, $x \in G, s \in S$, the color s. Suppose that H admits a rainbow embedding f in X. For each $a \in G$ the translation $x \to x + a$, $x \in G$, is an automorphism of X which preserves the colors and has no fixed points. Therefore, each translation sends $f(\vec{H})$ to an isomorphic copy which is edge disjoint from it. Thus the sets of translations for all $a \in G$ give rise to $n := |G|$ edge–disjoint copies of $\vec{H}$ in X. By ignoring orientations and colors, we thus have n edge disjoint copies of H in the underlying graph of X.

We will use the above approach with the Cayley graph $X = Cay(\mathbb{Z}_m \times \mathbb{Z}_4, \mathbb{Z}_m \times \{1\})$. We note that the underlying graph of X is isomorphic to $K_{2m,2m}$. The strategy of the proof is to show first that the base tree T_0 of a random tree with m edges admits a rainbow embedding f into X in such a way that $f(T_0) \subset \mathbb{Z}_m \times \{1, 2\}$. This can actually be achieved greedily as shown in the proof of next Lemma.

Lemma 3.1. *Let m be a positive integer. Let T be a tree with $n < 3m/5$ edges and stable sets A, B. If $||A| - |B|| \leq m/10$ then there is a rainbow embedding f of T into $X = Cay(\mathbb{Z}_m \times \mathbb{Z}_4, \mathbb{Z}_m \times \{1\})$ such that $f(V(T)) \subset (\mathbb{Z}_m \times \{1\}) \cup (\mathbb{Z}_m \times \{2\})$.*

The second step involves a proper embedding of the leaves of T. For this we use Häggkvist [6, Corolary 2.8] to get:

Lemma 3.2. *Let T be a tree with m edges. If the base tree T_0 of T admits a rainbow embedding f in $X = Cay(\mathbb{Z}_m \times \mathbb{Z}_4, \mathbb{Z}_m \times \{1\})$ such that $f(V(T_0)) \subset (\mathbb{Z}_m \times \{1\}) \cup (\mathbb{Z}_m \times \{2\})$ then T decomposes $K_{2m,2m}$.*

The proof of Theorem 1.3 follows now directly from Lemma 3.1 and Lemma 3.2 and the results on random trees from Section 2.

Proof. As it has been mentioned in Section 2, a random tree T with m edges has a.a.s. more than $2m/5$ leaves. Furthermore, by Theorem 2.1, the cardinalities of the stable sets of the base tree of T differ less than $m/10$ in absolute value a.a.s. By Lemma 3.1, the base tree of T admits a.a.s. a rainbow embedding in $Cay(\mathbb{Z}_m \times \mathbb{Z}_4, \mathbb{Z}_m \times \{1\})$ in such a way that the image of the embedding sits in $\mathbb{Z}_m \times \{1, 2\}$. In that case, Lemma 3.2 ensures that the tree T decomposes $K_{2m,2m}$. □

References

[1] M. DRMOTA, "Random Trees", Springer-Verlag, 2009.

[2] M. DRMOTA and B. GITTENBERGER, *The distribution of nodes of given degree in random trees*, Journal of Graph Theory **31** (1999), 227–253.

[3] M. DRMOTA and B. GITTENBERGER, *The shape of unlabeled rooted random trees*, European J. Combin. **31** (2010), 2028–2063

[4] P. FLAJOLET and A. ODLYZKO, *Singularity analysis of generating functions*, SIAM J. Discrete Math. **3** (1990), no. 2, 216–240.

[5] J. A. GALLIAN, *A Dynamic Survey of Graph Labeling*, The Electronic Journal of Combinatorics **5** (2007), # DS6.

[6] R. L. HÄGGKVIST, *Decompositions of complete bipartite graphs*, In: "Surveys in Combinatorics", Johannes Siemons (ed.), Cambridge University Press (1989), 115–146.

[7] A. E. KÉZDY, *ρ–valuations for some stunted trees*, Discrete Math. **306** (21) (2006), 2786–2789.

[8] A. E. KÉZDY and H. S. SNEVILY, "Distinct Sums Modulo n and Tree Embeddings", Combinatorics, Probability and Computing, 1, Issue 1, (2002).

[9] A. LLADÓ and S. C. LÓPEZ, *Edge-decompositions of $K_{n,n}$ into isomorphic copies of a given tree*, J. Graph Theory **48** (2005), no. 1, 1–18.

[10] A. LLADÓ, S.C. LÓPEZ and J. MORAGAS, *Every tree is a large subtree of a tree that decomposes K_n or $K_{n,n}$*, Discrete Math. **310** (2010), 838–842

[11] R. OTTER, *The number of trees*, Ann. of Math. (2) **49** (1948), 583–599.

[12] G. RINGEL, "Problem 25, Theory of Graphs and its Applications", Nakl. CSAV, Praha, 1964, 162.

[13] R. W. ROBINSON and A. J. SCHWENK, *The distribution of degrees in a large random tree*, Discrete Math. **12** (4) (1975), 359–372.

[14] H. SNEVILY, *New families of graphs that have α–labelings*, Discrete Math. **170** (1997), 185–194.

[15] R. YUSTER, *Packing and decomposition of graphs with trees*, J. Combin. Theory Ser. B **78** (2000), 123–140.

Sharp threshold functions via a coupling method

Katarzyna Rybarczyk[1]

Abstract. We will present a new method used to establish threshold functions in the random intersection graph model. The method relies on a coupling of a random intersection graph with a random graph similar to an Erdős and Rényi random graph. Formerly a simple version of the technique was used in the case of homogeneous random intersection graphs. Now it is considerably modified and extended in order to be applied in the general case. By means of the method we are able to establish threshold functions for the general random intersection graph model for monotone properties. Moreover the new approach allows to sharpen considerably the best known results concerning threshold functions for homogeneous random intersection graph. We outline the main results obtained in [10].

1 Introduction

Various comparison methods such as equivalence or contiguity have shown to be very helpful in the analysis of the structure of random graph models. For example one may consider a random graph with independent edges instead of the model considered in the seminal papers of Erdős and Rényi [3,4]. In some cases, when the models are neither equivalent nor contiguous the comparison by coupling still may simplify the arguments. This is so in the case of finding threshold functions for random intersection graphs.

The first random intersection graph model was introduced by Karoński, Scheinerman, and Singer–Cohen [7]. Several generalisations of the model have been proposed, mainly in order to adapt it to use in some particular purpose. We consider the $\mathcal{G}(n, m, \overline{p})$ model studied for example in [1, 2, 8]. In a random intersection graph $\mathcal{G}(n, m, \overline{p})$ there is a set of n vertices $\mathcal{V} = \{v_1, \ldots, v_n\}$, an auxiliary set of $m = m(n)$ features $\mathcal{W} = \{w_1, \ldots, w_{m(n)}\}$, and a vector $\overline{p}(n) = (p_1, \ldots, p_{m(n)})$ such that $p_i \in (0, 1)$, for each $1 \leq i \leq m$. Each vertex $v \in \mathcal{V}$ adds a feature $w_i \in \mathcal{W}$ to its feature set $W(v)$ with probability p_i independently of all other properties and features. Any two vertices $v, v' \in \mathcal{V}$

[1] Faculty of Mathematics and Computer Science, Adam Mickiewicz University, 60–769 Poznań, Poland. Email: kryba@amu.edu.pl

are connected by an edge in $\mathcal{G}(n, m, \overline{p})$ if $W(v)$ and $W(v')$ intersect. If $\overline{p}(n) = (p, \ldots, p)$ for some $p \in (0; 1)$ then $\mathcal{G}(n, m, \overline{p})$ is a random intersection graph defined in [7]. We denote it by $\mathcal{G}(n, m, p)$.

For $m = n^{\alpha}$ with $\alpha > 6$, $\mathcal{G}(n, m, p)$ is equivalent to a random graph with independent edges (see [5]) but in some cases the models differ (see [7]). However, even in the case when there is a large correlation between appearance of edges in $\mathcal{G}(n, m, p)$ (as well as in $\mathcal{G}(n, m, \overline{p})$), interesting results may be obtained by coupling with an auxiliary graph with almost independent edges. It is shown in [10] that the coupling may be used to obtain sharp results on threshold functions for $\mathcal{G}(n, m, \overline{p})$ for monotone properties. This extends results obtained in [9].

Let $G_2(n, \hat{p}_2)$ be a random graph with the vertex set $\mathcal{V}$ in which each edge appears independently with probability $\hat{p}_2$. Similarly let $H_3(n, \hat{p}_3)$ be a random hypergraph in which each 3–element subset of the set of vertices $\mathcal{V}$ is added to the hyperedge set independently with probability $\hat{p}_3$. Define $G_3(n, \hat{p}_3)$ to be a graph with the vertex set $\mathcal{V}$ and an edge set consisting of those two element subsets of $\mathcal{V}$ which are subsets of at least one hyperedge of $H_3(n, \hat{p}_3)$. For the family $\mathcal{G}$ of all graphs with the vertex set $\mathcal{V}$, we call $\mathcal{A} \subseteq \mathcal{G}$ a property if it is closed under isomorphism. Moreover $\mathcal{A}$ is increasing if $G \in \mathcal{A}$ implies $G' \in \mathcal{A}$ for all $G' \in \mathcal{G}$ such that $E(G) \subseteq E(G')$ and decreasing if $\mathcal{G} \setminus \mathcal{A}$ is increasing.

Let $\overline{p} = (p_1, \ldots, p_m)$ be such that $p_i \in (0, 1)$, for all $1 \leq i \leq m$. Moreover for all $1 \leq i \leq m$, let X_i be the number of vertices which contain w_i in the feature set and $\mathbb{I}_{i1}$, $\mathbb{I}_{i2}$ and $\mathbb{I}_{i\,\mathrm{odd}}$ be indicator random variables of the events $\{X_i = 1\}$, $\{X_i = 2\}$ and $\{X_i \text{ is odd}\}$, respectively. Define

$$S_1 = \mathbb{E}\left(\sum_{i=1}^{m} (X_i - \mathbb{I}_{i1})\right) = \sum_{i=1}^{m} np_i \left(1 - (1-p_i)^{n-1}\right);$$

$$S_2 = \mathbb{E}\left(\sum_{i=1}^{m} (X_i - \mathbb{I}_{i\mathrm{odd}})\right) = \sum_{i=1}^{m} np_i \left(1 - \frac{1-(1-2p_i)^n}{2np_i}\right);$$

$$\begin{aligned} S_3 &= \mathbb{E}\left(\sum_{i=1}^{m} (\mathbb{I}_{i\mathrm{odd}} - \mathbb{I}_{i1})\right) \\ &= \sum_{i=1}^{m} np_i \left(\frac{1-(1-2p_i)^n}{2np_i} - (1-p_i)^{n-1}\right); \end{aligned} \tag{1.1}$$

$$S_{1,2} = \mathbb{E}\left(\sum_{i=1}^{m} 2\mathbb{I}_{i2}\right) = \sum_{i=1}^{m} n(n-1)p_i^2(1-p_i)^{n-2},$$

The following theorem is the main tool of the presented coupling technique. It is an extension of the result obtained in [9].

Theorem 1.1. *Let S_1, S_2 and S_3 be given by* (1.1). *For some function ω tending to infinity let*

$$\hat{p} = \frac{S_2 - \omega\sqrt{S_2} - 2S_2^2 n^{-2}}{2\binom{n}{2}};$$

$$\hat{p}_2 = \begin{cases} \frac{S_1 - 3S_3 - \omega\sqrt{S_1} - 2S_1^2 n^{-2}}{2\binom{n}{2}}, & \textit{for } S_3 \gg \sqrt{S_1} \textit{ and } \omega^2 \ll S_3/\sqrt{S_1}; \\ \frac{S_1 - \omega\sqrt{S_1} - 2S_1^2 n^{-2}}{2\binom{n}{2}}, & \textit{for } S_3 = O(\sqrt{S_1}); \end{cases}$$

$$\hat{p}_3 = \begin{cases} \frac{S_3 - \omega\sqrt{S_1} - 6S_3^2 n^{-3}}{\binom{n}{3}}, & \textit{for } S_3 \gg \sqrt{S_1} \textit{ and } \omega^2 \ll S_3/\sqrt{S_1}; \\ 0, & \textit{for } S_3 = O(\sqrt{S_1}). \end{cases}$$

If $S_1 \to \infty$ and $S_1 = o\left(n^2\right)$ then for any increasing property $\mathcal{A}$.

$$\liminf_{n\to\infty} \Pr\left\{G_2\left(n, \hat{p}\right) \in \mathcal{A}\right\} \leq \limsup_{n\to\infty} \Pr\left\{\mathcal{G}\left(n, m, \overline{p}\right) \in \mathcal{A}\right\},$$

$$\liminf_{n\to\infty} \Pr\left\{G_2\left(n, \hat{p}_2\right) \cup G_3\left(n, \hat{p}_3\right) \in \mathcal{A}\right\} \leq \limsup_{n\to\infty} \Pr\left\{\mathcal{G}\left(n, m, \overline{p}\right) \in \mathcal{A}\right\}.$$

We use Theorem 1.1 to establish threshold functions for $\mathcal{G}\left(n, m, \overline{p}\right)$ for k-connectivity (denoted $\mathcal{C}_k$), a Hamilton cycle containment (denoted $\mathcal{HC}$), and a perfect matching containment (denoted $\mathcal{PM}$).

Theorem 1.2. *Let* $\max_{1\leq i\leq m} p_i = o((\ln n)^{-1})$ *and S_1 and $S_{1,2}$ be given by* (1.1).

(i) *If $S_1 = n(\ln n + c_n)$, then*

$$\lim_{n\to\infty} \Pr\left\{\mathcal{G}\left(n, m, \overline{p}\right) \in \mathcal{C}_1\right\} = \begin{cases} 0 & \textit{for } c_n \to -\infty; \\ e^{-e^{-c}} & \textit{for } c_n \to c \in (-\infty; \infty); \\ 1 & \textit{for } c_n \to \infty \end{cases}$$

(ii) *Let k be a positive integer and $a_n = \frac{S_{1,2}}{S_1}$. If*

$$S_1 = n(\ln n + (k-1)\ln\ln n + c_n),$$

then

$$\lim_{n\to\infty} \Pr\left\{\mathcal{G}\left(n, m, \overline{p}\right) \in \mathcal{C}_k\right\} = \begin{cases} 0 & \textit{for } c_n \to -\infty \textit{ and } a_n \to a \in (0; 1]; \\ 1 & \textit{for } c_n \to \infty. \end{cases}$$

Theorem 1.3. *Let* $\max_{1\le i\le m} p_i = o((\ln n)^{-1})$ *and* S_1 *be given by* (1.1). *If* $S_1 = n(\ln n + c_n)$ *then*

$$\lim_{n\to\infty} \Pr\{\mathcal{G}(2n, m, \overline{p}(2n)) \in \mathcal{PM}\} = \begin{cases} 0 & \text{for } c_{2n} \to -\infty; \\ e^{-e^{-c}} & \text{for } c_{2n} \to c \in (-\infty; \infty); \\ 1 & \text{for } c_{2n} \to \infty. \end{cases}$$

Theorem 1.4. *Let* $\max_{1\le i\le m} p_i = o((\ln n)^{-1})$, S_1 *and* $S_{1,2}$ *be given by* (1.1) *and* $a_n = \frac{S_{1,2}}{S_1}$. *If* $S_1 = n(\ln n + \ln\ln n + c_n)$, *then*

$$\lim_{n\to\infty} \Pr\{\mathcal{G}(n, m, \overline{p}) \in \mathcal{HC}\} = \begin{cases} 0 & \text{for } c_n \to -\infty \text{ and } a_n \to a \in (0; 1]; \\ 1 & \text{for } c_n \to \infty. \end{cases}$$

Even simple corollaries of Theorems 1.2–1.4 give sharp threshold functions for $\mathcal{G}(n, m, p)$. However the method of the proof is strong enough to enable to improve the best known results concerning $\mathcal{G}(n, m, p)$ even more.

Theorem 1.5. *Let* $m \gg \ln^2 n$ *and*

$$p(1-(1-p)^{n-1}) = \frac{\ln n + \ln\left(\max\left\{1, \ln\left(\frac{npe^{-np}\ln n}{1-e^{-np}}\right)\right\}\right) + c_n}{m}.$$

Then

$$\lim_{n\to\infty} \Pr\{\mathcal{G}(n, m, p) \in \mathcal{HC}\} = \begin{cases} 0 & \text{for } c_n \to -\infty; \\ 1 & \text{for } c_n \to \infty. \end{cases}$$

Theorem 1.6. *Let* $m \gg \ln^2 n$, k *be a positive integer, and*

$$a_n = (np)^{k-1}\left(\left(\frac{e^{-np}\ln n}{1-e^{-np}}\right)^{k-1} + \frac{e^{-np}\ln n}{1-e^{-np}}\right)$$

If

$$p(1-(1-p)^{n-1}) = \frac{\ln n + \ln(\max\{1, a_n\}) + c_n}{m},$$

then

$$\lim_{n\to\infty} \Pr\{\mathcal{G}(n, m, p) \in \mathcal{C}_k\} = \begin{cases} 0 & \text{for } c_n \to -\infty; \\ 1 & \text{for } c_n \to \infty. \end{cases}$$

In the proof we modify and extend the techniques used in [9]. First of all, to get the general result, we couple $\mathcal{G}(n, m, \overline{p})$ with a sum of independent random graphs $G_2(n, \hat{p}_2)$ and $G_3(n, \hat{p}_3)$. Edges in $G_2(n, \hat{p}_2) \cup G_3(n, \hat{p}_3)$ are not fully independent, therefore we prove some additional facts about it. Moreover we need sharp bounds on the minimum degree threshold function for $\mathcal{G}(n, m, \overline{p})$. Due to edge correlation, estimation of moments of the random variable counting vertices with a given degree in $\mathcal{G}(n, m, \overline{p})$ is complicated. Therefore we suggest a different approach to resolve the problem. We divide $\mathcal{G}(n, m, \overline{p})$ into subgraphs and use coupling to relate construction of those subgraphs with a coupon collector process. Finally the solution of a coupon collector problem combined with the method of moments provides the answer.

Concluding, we provide a general method to establish bounds on threshold functions for many properties for $\mathcal{G}(n, m, \overline{p})$. By means of the method we are able to obtain sharp thresholds for k–connectivity, perfect matching containment and hamiltonicity for the general model. Last but not least we considerably improve known results concerning $\mathcal{G}(n, m, p)$. In general, the coupling technique provides a very elegant method to get bounds on threshold functions for random intersection graphs for a large class of properties. The full proofs are presented in [10].

References

[1] M. BRADONJIĆ, A. HAGBERGY, N. W. HENGARTNERZ, N. LEMONS and A. G. PERCUS, *The phase transition in inhomogeneous random intersection graphs*, submitted.

[2] M. BLOZNELIS and J. DAMARACKAS, *Degree distribution of an inhomogeneous random intersection graph*, submitted.

[3] P. ERDŐS and A. RÉNYI, *On random graphs I*, Publicationes Mathematicae Debrecen **6** (1959), 290–297.

[4] P. ERDŐS and A. RÉNYI, *On the evolution of random graphs*, Publ. Math. Inst. Hungar. Acad. Sci. **5** (1960), 17–61.

[5] J. A. FILL, E. R. SCHEINERMAN and K. B. SINGER–COHEN, *Random Intersection Graphs when $m = \omega(n)$: An Equivalence Theorem Relating the Evolution of the G(n, m, p) and G(n, p) Models*, Random Structures and Algorithms **16** (2000), 156–176.

[6] S. JANSON, *Random regular graphs: asymptotic distributions and contiguity*, Combinatorics, Probability and Computing **4** (1995), 369–405.

[7] M. KAROŃSKI, E. R. SCHEINERMAN and K. B. SINGER-COHEN, *On random intersection graphs: The subgraph problem*, Combinatorics, Probability and Computing **8** (1999), 131–159.

[8] S. NIKOLETSEAS, C. RAPTOPOULOS andP. SPIRAKIS, *Large independent sets in general random intersection graphs*, Theoretical Computer Science **406** (2008), 215–224.

[9] K. RYBARCZYK, *Sharp Threshold Functions for Random Intersection Graphs via a Coupling Method*, The Electronic Journal of Combinatorics **18** (2011), R36.

[10] K. RYBARCZYK, *The coupling method for inhomogeneous random intersection graphs*, submitted (2013+).

Analytic description of the phase transition of inhomogeneous multigraphs

Élie de Panafieu[1] and Vlady Ravelomanana[1]

A random graph from the $G(n, p)$ model has n vertices and each pair of vertices is linked with probability p. In [6], Erdös and Rényi located the density of edges at which the first connected component with more than one cycle - called a *complex* component - appears. Using analytic tools, Janson, Knuth, Łuczak and Pittel derived in [11] more precise informations on the structure of a random graph or multigraph near the birth of the first complex component. Söderberg introduced in [13] a model of inhomogeneous random graphs, extended by Bollobás, Janson and Riordan [3]. This model generalizes $G(n, p)$ in the following way: each vertex receives a type among a set of q types, and the probability that a vertex of type i and one of type j are linked is the coefficient (i, j) of a symmetric matrix R of dimension $q \times q$. Among other results, they located the birth of the complex component. We combine here the accuracy of the approach of [11] with the generality of the inhomogeneous random graph model.

Phase transitions for Boolean Satisfiability (SAT) and for Constraint Satisfaction Problems (CSP) are fundamental problems arising in different communities [1,5,9]. Several polynomial SAT and CSP problems can be encoded into the inhomogeneous (multi)graph model, and their probability of satisfiability linked to the phase transition corresponding to the birth of the complex component. In this paper, we derive a complete picture of the finite size scaling and the critical exponents associated to the birth of the complex component of inhomogeneous multigraphs, using analytic methods. Equivalent results can be obtained for graphs. As applications, we present a new proof of an already known result from [12] on the probability of 2-coloriability, and new results on the probability of satisfiability of quantified 2-XOR-formulas [4].

[1] Univ. Paris Diderot, Sorbonne Paris Cité, LIAFA, UMR 7089, 75013, Paris, France.
Email: depanafieuelie@gmail.com, vlad@liafa.jussien.fr

1 Model

We consider a family of labelled multigraphs - loops and multiple edges are allowed - with colored vertices and weighted edges, close to the inhomogeneous random graph model [13]. Let R be a symmetric square matrix of dimension $q \times q$ with non-negative coefficients and σ a fixed positive constant. Let $\{c_1, \dots, c_q\}$ be a set of q distinct colors. A multigraph M is a (R, σ)-*multigraph* if each vertex x of M is colored with color $c(x) \in \{c_1, \dots, c_q\}$ and each edge $e = (x, y)$ of M is weighted with $R_{c(x), c(y)}$.

Following [11], the *compensation factor* $\kappa(M)$ of a multigraph M with n vertices is defined as $\prod_{x=1}^{n} (2^{m_{xx}} \prod_{y=x}^{n} m_{xy}!)^{-1}$ where m_{xy} is the number of edges binding x to y in M and $m = \sum_{x=1}^{n} \sum_{y=x}^{n} m_{xy}$ is the total number of edges. For a multigraph M with m edges, the number of sequences of couples of vertices $(x_1, y_1), \dots, (x_m, y_m)$ that lead to M is exactly $2^m \, m! \, \kappa(M)$.

Given a (R, σ)-multigraph M, we define its weight $\omega(M)$ as

$$\omega(M) = \kappa(M) \times \sigma^{\sharp\mathrm{cc}(M)} \times \prod_{(x,\, y) \in \{\text{edges of } M\}} R_{c(x),\, c(y)},$$

where $\sharp\mathrm{cc}(M)$ denotes the number of connected components of M. The parameter σ is not present in [13], but classic in statistic physics, for instance in the random cluster model. Finally, $g_{R,\sigma}(n, m)$ denotes the sum of the weights of the (R, σ)-multigraphs built with n vertices and m edges

$$g_{R,\sigma}(n, m) = \sum_{|G|=n,\ \|G\|=m} \omega(G).$$

An edge-weighted multigraph is *vertex-transitive* if its automorphism group is transitive and also preserve the weights – see for instance Godsil and Gordon [10]. Intuitively, the vertex-transitive property means that, using only the topology of the multigraph, no vertex can be distinguished from another.

Let G be a multigraph with q vertices and weighted edges (loops are also allowed). The *weighted adjacency matrix* $R = R(G)$ of G is built as follows: R is a $q \times q$ matrix with entry $R_{i,j}$ equal to the total weight of the edges between vertex i and vertex j. For simplicity, we say that a matrix R is *transitive* if the weighted multigraph associated is vertex-transitive. Due to size constraints, we only detail the results corresponding to transitive matrices R, although general theorems have also been derived and will be part of a longer version of the paper.

2 Main theorem

The asymptotic growth of $g_{R,\sigma}(n,m)$ with respect to n is different whether m/n stays at a fixed distance below $1/2$ or is close to it. We call this rupture the *phase transition* of inhomogeneous multigraphs. The evolution of the probability of satisfiability of several SAT and CSP problems can be explained by this phase transition: two examples are presented in the next section.

Theorem 2.1. *Let R be a transitive matrix with greatest eigenvalue δ and σ a positive fixed constant. Let c be the number of connected components in the multigraph associated to R. Let $\chi(X)$ be the polynomial $\prod_{\lambda \in Spect(R)\setminus\delta}\left(1-\frac{\lambda}{\delta}X\right)$. For any small fixed $\epsilon > 0$, as n is large and m is such that $\frac{2m}{n} < 1-\epsilon$, then*

$$g_{R,\sigma}(n,m) \sim \frac{\delta^m(\sigma q)^{n-m}}{\chi(\frac{2m}{n})^{\sigma/2}}\left(1-\frac{2m}{n}\right)^{\frac{1-c\sigma}{2}}\sqrt{1-\frac{m}{n}}\frac{n^{2m}}{2^m m!}.$$

As n is large and $m=\frac{n}{2}(1+\mu n^{-1/3})$ with μ bounded,

$$g_{R,\sigma}(n,m) \sim \frac{\delta^m(\sigma q)^{n-m}}{\chi(1)^{\sigma/2}}\, n^{(c\sigma-1)/6}\,\frac{n^{2m}}{2^m m!}\,\phi_{c\sigma}(\mu)$$

where $\phi_\sigma(\mu)$ is equal to $\sqrt{2\pi}\sum_k e_k^{(\sigma)}\sigma^k A(3k+\frac{\sigma}{2},\mu)$, $e_k^{(\sigma)}$ is the $(2k)$-th coefficient of $\left(\sum_n \frac{(6n)!z^{2n}}{(2n)!(3n)!2^n(3!)^n}\right)^\sigma$ and $A(y,\mu) = \frac{e^{-\mu^3/6}}{3^{(y+1)/3}}\sum_k \frac{(3^{2/3}\mu/2)^k}{k!\Gamma((y+1-2k)/3)}$.

Proof. Let δ denote the greatest eigenvalue of R. We can assume it to be equal to 1 without loss of generality, replacing R by $\frac{1}{\delta}R$ and $g(n,m)$ by $\delta^m g(n,m)$. We can also assume that the number c of connected components of the multigraph encoded by R is 1: the (R,σ)-multigraphs are in a one-to-one mapping with the $(S,c\sigma)$-multigraphs where S is the adjacency matrix of one of the connected components.

Since R is symmetric, there exist an orthogonal matrix Q and a diagonal matrix Δ such that $R = Q\Delta Q^T$. Furthermore, the vector $\vec{1}$ is an eigenvector for the greatest eigenvalue 1. We choose $\Delta_{1,1} = 1$.

A R-Cayley tree is a R-rooted non-planar labelled tree. Let $T_i(z)$ denote its generating function when the color of the root is i and $\vec{T}(z) = (\,T_1(z) \cdots T_q(z)\,)^T$. The generating function of the R-unrooted trees and R-unicyclic graphs are denoted by $U(z)$ and $V(z)$. A R-path of trees is a colored path that links two vertices (that may not be distinct) of color i and j. Each internal vertex of the path is the root of a colored R-Cayley tree. Its generating function is $P_{i,j}(z)$. Using the analytic combinatorics tools (a good reference is [8]), the combinatorial specification of R-Cayley trees translates into the following equations:

$$\vec{T}(z) = T(z)\vec{1} \qquad V(z) = -\tfrac{1}{2}\log(1 - T(z)) - \tfrac{1}{2}\log(\chi(T(z)))$$
$$U(z) = q(T(z) - \tfrac{1}{2}T(z)^2) \quad P_{i,j}(z) = \tfrac{1}{q(1-T(z))} + \textstyle\sum_{l=2}^{q} Q_{i,l}Q_{j,l}\tfrac{\Delta_{l,l}}{1-\Delta_{l,l}T(z)}$$

Remark that at the first order, $U(z)$, $V(z)$ and $P_{i,j}(z)$ behave as their non-colored counterparts $T(z) - \frac{1}{2}T(z)^2$, $-\frac{1}{2}\log(1 - T(z))$ and $\frac{1}{1-T(z)}$. Furthermore, the first order of $P_{i,j}(z)$ is independent of i and j.

When $\frac{2m}{n} < 1 - \epsilon$, with high probability a graph with n vertices and m edges contains only trees and unicyclic components (see [7]). Therefore, $g_{R,\sigma}(n, m) \sim n![z^n]\frac{(\sigma U(z))^{n-m}}{(n-m)!}e^{\sigma V(z)}$. The *Large Powers* Theorem of [8] ends the proof.

Let us remind that the *excess* of a graph is the difference between the number of edges and of vertices. The *complex* part of a graph is the set of its connected components that have positive excess. Deleting the vertices of degree one and fusioning the vertices of degree two, each graph can be reduced to a simpler graph with same excess, called its *kernel*, with minimum degree at least three. Reciprocally, any such graph can be developed by replacing edges by paths and adding trees to the vertices.

When $m = \frac{n}{2}(1 + \mu n^{-1/3})$, [11] proved that the proportion of multigraphs with non-cubic kernel is negligible. This holds true for inhomogeneous graphs. The set $\mathbb{K}_k$ of the cubic multigraphs of excess k is finite. The sum of their compensation factors with a weight σ for each connected component is $(2k)!e_k^{(\sigma)}$ as defined in the theorem. The generating function of the R-developed cubic graphs of fixed excess k is

$$K_{k,\sigma}(z) = \sum_{K\in\mathbb{K}_k}\sum_{\vec{c}\in[1,q]^{2k}} \kappa(K)\sigma^{\sharp cc(K)}\frac{1}{(2k)!}\Big(\prod_{i\in K} T_{c_i}(z)\Big)\prod_{(i,j)\in K} P_{c_i,c_j}(z).$$

Replacing $P_{i,j}(z)$ by its first order term, we obtain that near its dominant singularity, $K_{k,\sigma}(z) \sim \frac{e_k^{(\sigma)}T(z)^{2k}}{q^k(1-T(z))^{3k}}$. The number of (R, σ)-multigraphs with n vertices, m edges and kernel of excess k is

$$g(n, m, k) \sim n![z^n]\frac{(\sigma U(z))^{n-m+k}}{(n - m + k)!}e^{\sigma V(z)}K_{k,\sigma}(z).$$

The asymptotic is derived using the same tools as in Theorem 11 of [2] or Lemma 3 of [11]. Finally, $g(n, m) = \sum_{k\geq 1} g(n, m, k)$. When R is not transitive, but the multigraph encoded by R is connected, a similar result holds. □

3 Examples of application of the main theorem

The asymptotic probability for a multigraph with a large number n of vertices and m edges to be bipartite evolves from 1 to 0 when the quotient m/n increases from 0 to $1/2$. This phase transition can be precisely described using our main theorem. The bipartite multigraphs are in a one-to-one mapping with the $\left(\left(\begin{smallmatrix} 0 & 1 \\ 1 & 0 \end{smallmatrix}\right), \frac{1}{2}\right)$-multigraphs. Since the sum of the compensator factors of the multigraphs with n vertices and m edges is $\frac{n^{2m}}{m!2^m}$, the probability for a multigraph to be bipartite is $\frac{2^m m!}{n^{2m}} g_{\left(\begin{smallmatrix} 0 & 1 \\ 1 & 0 \end{smallmatrix}\right), \frac{1}{2}}(n,m)$, given by the Theorem 2.1. This result has already been derived by Pittel and Yeum [12] when m is around $n/2$.

In [4], the authors analyse 2-QXOR-SAT formulas. Those are quantified conjunctions of m XOR-clauses with β universal and n existential variables

$$\forall x_1 \ldots x_\beta \, \exists y_1 \ldots y_n \bigwedge_{i=1}^{m} \left(y_{f_{i,1}} \oplus y_{f_{i,2}} = \left(e_i(1) \cdots e_i(\beta) \right) \cdot \left(x_1 \cdots x_\beta \right)^T \right)$$

where each e_i is a β-tuple of bits and $\left(a_1 \cdots a_k \right) \cdot \left(b_1 \cdots b_k \right)^T = \bigoplus_{i=1}^{k} a_i \wedge b_i$. A clause φ is characterized by a triplet $(y_{\varphi,1}, y_{\varphi,2}, e_\varphi)$. The authors study how the probability of satisfiability $P_{\mathrm{SAT}}(n, m)$ evolves with the number m of clauses when the number n of existential variables is large, and locate the value of m at which the phase transition occurs.

We consider clauses φ such that e_φ is in a certain fixed multiset E of $beta$-tuples of bits. We call the formulas that contain only those clauses the E-formulas. Let x be an integer in $[1, 2^\beta]$ and $[x]_2$ denote the binary decomposition of $x - 1$, interpreted as a β-tuple of bits. To a multiset E of β-tuples of bits, we associate a square symmetric transitive matrix $R^{(E)}$ of dimension $2^\beta \times 2^\beta$ such that $R^{(E)}_{x,y}$ is the number of occurences of $[x]_2 \oplus [y]_2$ in E: $R^{(E)}_{x,y} = \#\{e \in E \mid [x]_2 \oplus [y]_2 = e\}$. Using similar idees as in [4], a one-to-one mapping is built between the satisfiable E-formulas with n existential variables and m clauses and the $(R^{(E)}, 2^{-\beta})$-multigraphs with n vertices and m edges. Therefore, $P_{\mathrm{SAT}}(n, m)$ is equal to $\frac{2^m m!}{n^{2m}|E|^m} g_{R^{(E)}, 2^{-\beta}}(n, m)$ and its asymptotic is derived by the theorem 2.1, using the parameters q, $\sigma = q^{-1}$, δ, c and $\chi(X)$ of the matrix $R^{(E)}$. In [4], the authors located the phase transition and described it qualitatively - coarse or sharp - while our result is more precise and quantifies the evolution of $P_{\mathrm{SAT}}(n, m)$ from the subcritical to the critical range of m.

References

[1] D. Achlioptas and C. Moore, *Random k-SAT: Two moments suffice to cross a sharp threshold*, SIAM Journal of Computing **36** (2006), 740–762.

[2] C. BANDERIER, P. FLAJOLET, G. SCHAEFFER and M. SORIA, *Random maps, coalescing saddles, singularity analysis, and airy phenomena*, Random Struct. Algorithms **19** (3-4) (2001), 194–246.

[3] B. BOLLOBÁS, S. JANSON and O. RIORDAN, *The phase transition in inhomogeneous random graphs*, Random Struct. Algorithms **31** (1) (2007), 3–122.

[4] N. CREIGNOU, H. DAUDÉ and U. EGLY, *Phase transition for random quantified XOR-formulas*, Journal . Artif. Intell. Res. **29** (2007), 1–18.

[5] O. DUBOIS, R. MONASSON, B. SELMAN and R. ZECCHINA, "Editorial", volume 265(1-2). Elsevier, 2001.

[6] P. ERDŐS and A RÉNYI, *On the evolution of random graphs*, In: "Publication of the Mathematical Institute of the Hungarian Academy of Sciences", 1960, 17–61.

[7] P. FLAJOLET, D. E. KNUTH and B. PITTEL, *The first cycles in an evolving graph*, Discrete Mathematics **75** (1-3) (1989), 167–215.

[8] P. FLAJOLET and R. SEDGEWICK, "Analytic Combinatorics", Cambridge University Press, 2009.

[9] E. FRIEDGUT, *Sharp thresholds of graph properties, and the k-sat problem*, Journal of the A.M.S. **12** (4) (1999), 1017–1054.

[10] C. GODSIL and R. GORDON, "Algebraic Graph Theory, Graduate Texts in Mathematics", New-York : Springer-Verlag, 2001.

[11] S. JANSON, D. E. KNUTH, T. ŁUCZAK and B. PITTEL, *The birth of the giant component*, Random Structures and Algorithms **4** (3) (1993), 233–358.

[12] B. PITTEL and J-A YEUM, *How frequently is a system of 2-linear boolean equations solvable?* The Electronic Journal of Combinatorics, Volume 17, Research paper R92, 2010.

[13] B. SÖDERBERG, *General formalism for inhomogeneous random graphs*, Phys. Rev. E **66** (Dec. 2002), 066121.

Fixed-point

On the Bruhat-Chevalley order on fixed-point-free involutions

Mahir Bilen Can[1], Yonah Cherniavsky[2] and Tim Twelbeck[3]

1 Introduction

The purpose of this paper is twofold. First is to prove that the Bruhat-Chevalley ordering restricted to fixed-point-free involutions is a lexicographically shellable poset. Second is to prove that the Deodhar-Srinivasan poset is a graded subposet of the Bruhat-Chevalley poset structure on fixed-point-free involutions.

In this work we are concerned with the interaction between two well known subgroups of the special linear group SL_{2n}, namely a Borel subgroup and a symplectic subgroup. Without loss of generality, we choose the Borel subgroup B to be the group of invertible upper triangular matrices, and define the *symplectic group*, Sp_{2n} as the subgroup of fixed elements of the involutory automorphism $\theta : SL_{2n} \to SL_{2n}$, $\theta(g) = J(g^{-1})^{\top} J^{-1}$, where J denotes the skew form $J = \begin{pmatrix} 0 & \omega_0 \\ -\omega_0 & 0 \end{pmatrix}$, and ω_0 is the $n \times n$, 0/1 matrix with 1's on its main anti-diaonal.

It is clear that B acts by left-multiplication on SL_{2n}/Sp_{2n}. We investigate the covering relations of the poset F_{2n} of inclusion relations among the B-orbit closures. To further motivate our discussion and help the reader to place our work appropriately we look at a related situation. It is well known that the symmetric group of permutation matrices, S_m parameterizes the orbits of the Borel group of upper triangular matrices $B \subset SL_m$ in the flag variety SL_m/B. For $u \in S_m$, let $\dot{u}$ denote the right coset in SL_m/B represented by u. The classical *Bruhat-Chevalley ordering* is defined by $u \leq_{S_m} v \iff B \cdot \dot{u} \subseteq \overline{B \cdot \dot{v}}$ for $u, v \in S_m$.

[1] Yale University, New Haven, USA. Email: mahir.can@yale.edu

[2] Ariel University, Israel. Email: yonahch@ariel.ac.il

[3] Tulane University, New Orleans, USA. Email: ttwelbec@tulane.edu

The full-text paper can be found in [6].

A permutation $u \in S_m$ is said to be an *involution*, if $u^2 = id$, or equivalently, its permutation matrix is a symmetric matrix. We denote by I_m the set of all involutions in S_m, and consider it as a subposet of the Bruhat-Chevalley poset $(S_m, \leq_{S_m})$. Let m be an even number, $m = 2n$. An element $x \in I_{2n}$ is called *fixed-point-free*, if the matrix of x has no non-zero diagonal entries. In [15], Example 10.4, Richardson and Springer show that there exists a poset isomorphism between F_{2n} and a subposet of fixed-point-free involutions in I_{2n}. Unfortunately, F_{2n} does not form an interval in I_{2n}, hence it does not immediately inherit nice properties therein. Let $\leq$ denote the restriction of the Bruhat-Chevalley ordering on F_{2n}. Our first main result is that $(F_{2n}, \leq)$ is "EL-shellable," which is a property that well known to be true for many other related posets. See [3–5, 8, 11, 12]. See [10], also. Notice that there are several versions of lexicographic shellability and EL-shellability is the strongest one. In [13] it is shown that F_{2n} is CL-shellable. So, our result is a strengthening of this result of [13] in the special case of the poset F_{2n}. Recall that a finite graded poset P with a maximum and a minimum element is called *EL-shellable*, if there exists a map $f = f_\Gamma : C(P) \to \Gamma$ between the set of covering relations $C(P)$ of P into a totally ordered set Γ satisfying: 1) In every interval $[x, y] \subseteq P$ of length $k > 0$ there exists a unique saturated chain $\mathfrak{c} : x_0 = x < x_1 < \cdots < x_{k-1} < x_k = y$ such that the entries of the sequence $f(\mathfrak{c}) = (f(x_0, x_1), f(x_1, x_2), \ldots, f(x_{k-1}, x_k))$ is weakly increasing. 2) The sequence $f(\mathfrak{c})$ of the unique chain $\mathfrak{c}$ from (1) is the smallest among all sequences of the form $(f(x_0, x_1'), f(x_1', x_2'), \ldots, f(x_{k-1}', x_k))$, where $x_0 \leq x_1' \leq \cdots \leq x_{k-1}' \leq x_k$. Recall that the *order complex* of a poset P is the abstract simplicial complex $\Delta(P)$ whose simplexes are the chains in P. For a lexicographically shellable poset the order complex is shellable, in particular it implies that $\Delta(P)$ is Cohen-Macaulay [2]. These, of course, are among the most desirable properties of a topological space. One of the reasons the EL-shellability of F_{2n} is not considered before is that there is a closely related EL-shellable partial order studied by Deodhar and Srinivasan in [7] which we denote here as $\leq_{DS}$. By some authors the Deodhar-Srinivasan's ordering is thought to be the same as Bruhat-Chevalley ordering on F_{2n}. A careful inspection of the Hasse diagrams of $(F_{2n}, \leq)$ and $(\tilde{F}_{2n}, \leq_{DS})$ reveals that these two posets are "almost" the same but different. Our second main result is that the rank functions of these posets are the same, and furthermore, the latter is a graded subposet of the former.

2 Preliminaries

We denote the set $\{1, \ldots, m\}$ by $[m]$. In this work, all posets are assumed to be finite and assumed to have a minimal and a maximal element, denoted by $\hat{0}$ and $\hat{1}$, respectively. Recall that in a poset P, an element y is said to *cover* another element x, if $x < y$ and if $x \leq z \leq y$ for some $z \in P$, then either $z = x$ or $z = y$. In this case, we write $y \to x$. Given P, we denote by $C(P)$ the set of all covering relations of P. An (increasing) *chain* in P is a sequence of distinct elements such that $x = x_1 < x_2 < \cdots < x_{n-1} < x_n = y$. A chain in a poset P is called *saturated* (or, *maximal*), if it is of the form $x = x_1 \leftarrow x_2 \leftarrow \cdots \leftarrow x_{n-1} \leftarrow x_n = y$. Recall also that a poset is called *graded* if all maximal chains between any two comparable elements $x \leq y$ have the same length. This amounts to the existence of an integer valued function $\ell_P : P \to \mathbb{N}$ satisfying 1) $\ell_P(\hat{0}) = 0$, 2) $\ell_P(y) = \ell_P(x) + 1$ whenever y covers x in P. ℓ_P is called the *length function* of P.

Let Sym_n denote the affine space of symmetric matrices and let Sym_n^0 denote its closed subset consisting of symmetric matrices with determinant 1. Similarly, let $Skew_{2n}$ denote the affine space of skew-symmetric matrices, and let $Skew_{2n}^0$ denote its closed subset consisting of elements with determinant 1. Let X denote any of the spaces Sym_n, Sym_n^0, $Skew_{2n}$, or $Skew_{2n}^0$. Then the special linear group of appropriate rank acts on X via $g \cdot A = (g^{-1})^\top A g^{-1}$. Define $SO_n := \{g \in SL_n : gg^\top = id_n\}$. The symmetric spaces $\mathrm{SL}_{2n}/\mathrm{Sp}_{2n}$ and $\mathrm{SL}_n/\mathrm{SO}_n$ can be canonically identified with the spaces $Skew_{2n}^0$ and Sym_n^0, respectively (for details see [9]). Recall that an $n \times n$ *partial permutation matrix* (or, a *rook matrix*) is a 0/1 matrix with at most one 1 in each row and each column. The set of all $n \times n$ rook matrices is denoted by R_n. In [14], Renner shows that R_n parameterizes the $B \times B$-orbits on the monoid of $n \times n$ matrices. It is known that the partial ordering on R_n induced from the containment relations among the $B \times B$-orbit closures is a lexicographically shellable poset (see [4]). On the other hand, for the purposes of this paper, it is more natural for us to look at the inclusion poset of $B^\top \times B$-orbit closures in R_n, which we denote by $(R_n, \leq_{Rook})$. A symmetric rook matrix is called a *partial involution*. The set of all partial involutions in R_n is denoted by PI_n. It is known that each Borel orbit in Sym_n contains a unique element of PI_n. A rook matrix is called a *partial fixed-point-free involution*, if it is symmetric and does not have any non-zero entry on its main diagonal. We denote by PF_{2n} the set of all partial fixed-point-free involutions. It is known that PF_{2n} parameterizes the Borel orbits in $Skew_{2n}$.

Containment relations among the closures of Borel orbits in $Skew_{2n}$ define a partial ordering on PF_{2n}. We denote its opposite by $\leq_{Skew}$. Similarly, on PI_n we have the opposite of the partial ordering induced from the containment relations among the Borel orbit closures in Sym_n. We denote this opposite partial ordering by $\leq_{Sym}$. In [11], Incitti, studying the restriction of the partial order $\leq_{Sym}$ on I_n, finds an EL-labeling for I_n. Let us mention that in a recent preprint, Can and Twelbeck, using an extension of Incitti's edge-labeling show that PI_n is EL-shellable. See [5].

There is a combinatorial method for deciding when two elements x and y from $(R_n, \leq_{Rook})$ (respectively, from $(PI_n, \leq_{Sym})$, or from $(PF_{2n}, \leq_{Skew})$) are comparable with respect to $\leq_{Rook}$ (respectively, with respect to $\leq_{Sym}$, or $\leq_{Skew}$). Denote by $Rk(x)$ the matrix whose i, j-th entry is the rank of the upper left $i \times j$ submatrix of x. We call $Rk(x)$, the *rank-control matrix* of x. Let $A = (a_{i,j})$ and $B = (b_{i,j})$ be two matrices of the same size with real number entries. We write $A \leq B$ if $a_{i,j} \leq b_{i,j}$ for all i and j. Then $x \leq_{Rook} y \iff Rk(y) \leq Rk(x)$. The same criterion holds for the posets $\leq_{Sym}$ and $\leq_{Skew}$. Now, suppose x is an $m \times m$ matrix with the rank-control matrix $Rk(x) = (r_{i,j})_{i,j=1}^{m}$. Set $r_{0,i} = 0$ for $i = 0, \dots, m$, and define $\rho_{\leq}(x) = \#\{(i, j) : 1 \leq i \leq j \leq 2n \text{ and } r_{i,j} = r_{i-1,j-1}\}$, $\rho_{<}(x) = \#\{(i, j) : 1 \leq i < j \leq 2n \text{ and } r_{i,j} = r_{i-1,j-1}\}$. Then the length function $\ell_{PF_{2n}}$ of the poset PF_{2n} is equal to the restriction of $\rho_{<}$ to PF_{2n}. Furthermore, y covers x if and only if $Rk(y) \leqslant Rk(x)$ and $\ell_{PF_{2n}}(y) - \ell_{PF_{2n}}(x) = 1$. Similarly, $\ell_{PI_{2n}}$ is the restriction of $\rho_{\leq}$ to PI_{2n}, and that y covers x if and only if $Rk(y) \leqslant Rk(x)$ and $\ell_{I_{2n}}(y) - \ell_{I_{2n}}(x) = 1$. For details, see [1].

3 Results

It turns out that the intersection $PF_{2n} \cap I_{2n}$ is equal to F_{2n}, and furthermore, $(F_{2n}, \leq_{Sym})$ and $(F_{2n}, \leq_{Skew})$ are isomorphic. The relationships between the posets PI_{2n}, PF_{2n},I_{2n} and F_{2n} are as follows.

Let $w_0 \in PI_{2n}$ denote the "longest permutation," namely, the $2n \times 2n$ anti-diagonal permutation matrix, and let $j_{2n} \in F_{2n}$ denote the $2n \times 2n$ fixed-point-free involution having non-zero entries at the positions

$$(1, 2), (2, 1), (3, 4), (4, 3), \dots, (2n-1, 2n), (2n, 2n-1), \text{ only}.$$

In other words, j_{2n} is the fixed-point-free involution with the only non-zero entries along its super-diagonals. Then I_{2n} is an interval in PI_{2n} with the smallest element id_{2n} and the largest element w_0. Similarly, F_{2n} is an interval in PF_{2n} with the smallest element j_{2n} and the largest element w_0.

Consider F_{2n} as a subposet of I_{2n} and let $x, y \in F_{2n}$ be two elements such that $x \leq y$. It turns out there exits a saturated chain in I_{2n} from x to y consisting of fixed-point-free involutions only. With the help of this observation, we prove

Theorem 3.1. *F_{2n} is an EL-shellable poset.*

Sketch of the proof. Recall that F_{2n} is a connected graded subposet of I_{2n}. Therefore, its covering relations are among the covering relations of I_{2n} described in [11]. Let x and y be two fixed-point-free involutions. We know the existence of a saturated chain between x and y that is entirely contained in F_{2n}. Since lexicographic ordering is a total order on maximal chains, there exists a unique largest such chain, say $\mathfrak{c}$. The idea of the proof is showing that $\mathfrak{c}$ is the unique decreasing chain. Once this is done, by switching the order of our totally ordered set $\mathbb{Z}^2$, we obtain the lexicographically smallest chain, which is the unique increasing chain.

As an important consequence of Theorem 3.1, we further show that

Theorem 3.2. *The order complex $\Delta(F_{2n})$ triangulates a ball of dimension $n(n-1)-2$.*

As it is mentioned in the introduction, the posets $(\tilde{F}_{2n}, \leq_{DS})$ and $(F_{2n}, \leq)$ are different. Indeed, for $2n = 6$ the Hasse diagrams of these two posets differ by an edge. Contrary to this observation, we have the following

Theorem 3.3. *The length functions of $(F_{2n}, \leq)$ and $(\tilde{F}_{2n}, \leq_{DS})$ are the same. Covering relations of the poset $\tilde{F}_{2n}$ are among the covering relations of F_{2n}.*

References

[1] E. Bagno and Y. Cherniavsky, *Congruence B-orbits and the Bruhat poset of involutions of the symmetric group*, Discrete Math. **312** (1) (2012), 1289–1299.

[2] A. Björner, *Shellable and Cohen-Macaulay partially ordered sets*, Trans. Amer. Math. Soc. **260** (1) (1980), 159–183.

[3] A. Björner and M. Wachs, *Bruhat order of Coxeter groups and shellability*, Adv. in Math. **43** (1) (1982), 87–100.

[4] M. Bilen Can, *The rook monoid is lexicographically shellable*, Preprint available at http://arxiv.org/abs/1001.5104. accepted for publication, 2012.

[5] M. Bilen Can and T. Twelbeck, *Lexicographic shellability of partial involutions*, preprint available at http://arxiv.org/abs/1205.0062.

[6] M. BILEN CAN, Y. CHERNIAVSKY and T. TWELBECK, *Bruhat-Chevalley order on fixed-point-free involutions*, preprint available at http://arxiv.org/abs/1211.4147.

[7] R. S. DEODHAR and M. K. SRINIVASAN, *A statistic on involutions*, J. Algebraic Combin. **13** (2) (2001), 187–198.

[8] P. H. EDELMAN, *The Bruhat order of the symmetric group is lexicographically shellable*, Proc. Amer. Math. Soc. **82** (3) (1981), 355–358.

[9] R. GOODMAN and N. R. WALLACH, "Symmetry, Representations, and Invariants", volume 255 of Graduate Texts in Mathematics. Springer, Dordrecht, 2009.

[10] A. HULTMAN, *Fixed points of involutive automorphisms of the Bruhat order*, Adv. Math. **195** (1) (2005), 283–296.

[11] F. INCITTI, *The Bruhat order on the involutions of the symmetric group*, J. Algebraic Combin. **20** (3) (2004), 243–261.

[12] R. A. PROCTOR, *Classical Bruhat orders and lexicographic shellability*, J. Algebra **77** (1) (1982), 104–126.

[13] E. M. RAINS and M. VAZIRANI, *Deformations of permutations of Coxeter groups*, J. Algebraic Combin., 2012.

[14] L. E. RENNER, *Analogue of the Bruhat decomposition for algebraic monoids*, J. Algebra, **101** (2) (1986), 303–338.

[15] R. W. RICHARDSON and T. A. SPRINGER, *The Bruhat order on symmetric varieties*, Geom. Dedi- cata **35** (1-3) (1990), 389–436.

A geometric approach to combinatorial fixed-point theorems: extended abstract

Elyot Grant[1] and Will Ma[2]

Abstract. We develop a geometric framework that unifies several different combinatorial fixed-point theorems related to Tucker's lemma and Sperner's lemma, showing them to be different geometric manifestations of the same topological phenomena. In doing so, we obtain (1) new Tucker-like and Sperner-like fixed-point theorems involving an exponential-sized label set; (2) a generalization of Fan's parity proof of Tucker's Lemma to a much broader class of label sets; and (3) direct proofs of several Sperner-like lemmas from Tucker's lemma via explicit geometric embeddings.

1 Introduction

Combinatorial fixed-point theorems such as the Sperner and Tucker lemmas have generated a wealth of interest in recent decades, in part due to the discovery of important new applications in economics and theoretical computer science (see [2, 12, 15]). Extensive research has examined the construction of *direct proofs* of the implications among these and other similar theorems (and generalizations), yielding many different proofs via a variety of methods (see [4, 6, 8]). Some of this work has succeeded in connecting fixed-point theorems in the (Brouwer, Sperner)-family to the seemingly unrelated antipodality theorems in the (Borsuk-Ulam, Tucker)-family; for example, Su has shown that it is possible to prove the Brouwer fixed-point theorem directly from the Borsuk-Ulam theorem via an explicit topological construction [13], and Živalcević [16] has shown how Ky Fan's [3] generalization of Tucker's lemma implies Sperner's lemma. However, the construction of a direct proof that Tucker's lemma implies Sperner's lemma appears to remain an open question [11].

To shed some light on this question, we investigate the Tucker and Sperner lemmas from a *geometric* viewpoint. Cast in this light, it be-

[1] Computer Science and Artificial Intelligence Laboratory, Massachusetts Institute of Technology. Email: elyot@mit.edu

[2] Operations Research Center, Massachusetts Institute of Technology. Email: willma@mit.edu

The full version of this paper is available on arXiv.org as [7].

comes apparent that the Tucker and Sperner lemmas are actually members of a much larger family of combinatorial fixed-point theorems sharing a common topological structure, but having different geometric manifestations. Our approach hence unifies many known combinatorial fixed-point theorems, and yields a new Tucker-type lemma and a new Sperner-type lemma, both with an exponential number of labels. In doing so, we generalize the technique in [3] to obtain a framework that proves our new Tucker-like theorem without any topological fixed-point theorems. As a bonus, our framework also permits us to prove some of the Sperner-like theorems directly from Tucker's lemma via explicit geometric embeddings. Moreover, we derive some insight into why Sperner's lemma may be difficult to prove directly from Tucker's lemma—the analogy between Borsuk-Ulam and Tucker results is geometrically different from the analogy between Brouwer and Sperner results, and alternate Sperner-like theorems provide a more direct analogy.

ACKNOWLEDGEMENTS. We thank Rob Freund for helpful discussions, and referee 2 for several useful comments. Both authors were partially supported by NSERC PGS-D awards. The second author was supported in part by NSF grant CCF-1115849 and ONR grants N00014-11-1-0053 and N00014-11-1-0056.

Combinatorial fixed-point theorems

We omit definitions of standard terminology and notation; details can be found in the full version of the paper, or a standard text such as that of Matoušek [10]. We shall use the following conventions for embeddings: B^n is defined as $\{x \in R^n, ||x||_2 \leq 1\}$; Δ^{n-1} is defined as $\text{conv}\{e_i : 1 \leq i \leq n\}$; $\lozenge^n$ is defined as $\text{conv}\{\pm e_i\}$; $\square^n$ is defined as $\text{conv}\{(x_1, \ldots, x_n) : x_i \in \{-1, 1\}\}$; here, $\text{conv}(S)$ is the convex hull of the set S.

We define the *vertices* $V(T)$ of a triangulation T to be the set of all 0-simplices in T. A *label function* λ is a mapping from $V(T)$ to a finite *label set* L. In the case of the Tucker and Sperner lemmas, the sets $\{1, \ldots, n+1\}$ or $\{1, -1, \ldots, n, -n\}$ are typically used for L. However, in our paper, we will instead represent these labels as the sets of *extreme points* $\text{ext}(\Delta^n) = \{e_1, \ldots, e_{n+1}\}$ and $\text{ext}(\lozenge^n) = \{e_1, -e_1 \ldots, e_n, -e_n\}$. Cast in this framework, we shall state Sperner's lemma as follows:

Theorem 1.1 (Sperner's lemma). *Let T be a triangulation of Δ^n. Let $\lambda : V(T) \to \text{ext}(\Delta^n)$ be a label function with the property that for all $x = (x_1, \ldots, x_{n+1}) \in V(T)$, for all $1 \leq i \leq n+1$, if $x_i = 0$, then $\lambda(x) \neq e_i$ (such a λ is sometimes called a* proper colouring*). Then T*

contains a panchromatic simplex*—that is, a simplex* σ *such that* $\{\lambda(v) : v \in V(\sigma)\} = \{e_1, \ldots, e_{n+1}\}$.

We can establish results similar to Sperner's lemma in other geometric spaces as long as we have suitable analogies of the notions of *proper colouring* and *panchromatic simplex.* The following are immediate:

Proposition 1.2. *Let* σ *be a simplex and let* $\lambda : V(\sigma) \to \operatorname{ext}(\Diamond^n)$ *be a label function. Define a* complementary edge *to be two vertices* $v_1, v_2 \in V(\sigma)$ *with* $\lambda(v_1) = -\lambda(v_2)$. *Then* $\operatorname{conv}\{\lambda(v) : v \in V(\sigma)\}$ *intersects the interior of* $\Diamond^n$ *if and only if* σ *contains a complementary edge.*

Proposition 1.3. *Let* σ *be a simplex and let* $\lambda : V(\sigma) \to \operatorname{ext}(\Box^n)$ *be a label function. We say* σ *is a* neutral simplex *if for all* $1 \leq i \leq n$, *there exist vertices* $v_1, v_2 \in V(\sigma)$ *such that* $\lambda_i(v_1) = -1$, $\lambda_i(v_2) = +1$, *where* $\lambda_i(v)$ *is the* i^{th} *coordinate of* $\lambda(v)$. *Then* $\operatorname{conv}\{\lambda(v) : v \in V(\sigma)\}$ *intersects the interior of* $\Box^n$ *if and only if* σ *is a neutral simplex.*

Two Sperner-like theorems can immediately be derived by imposing the right labelling constraints:

Theorem 1.4 (Octahedral Sperner with octahedral labels). *Let* T *be a triangulation of* $\Diamond^n$. *Let* $\lambda : V(T) \to \operatorname{ext}(\Diamond^n)$ *be a label function such that for all boundary vertices* $x = (x_1, \ldots, x_n) \in V(T) \cap \partial(\Diamond^n)$, *for all* $1 \leq i \leq n$, *if* $x_i \geq 0$ *(respectively, if* $x_i \leq 0$*), then* $\lambda(x) \neq -e_i$ *(respectively,* $\lambda(x) \neq e_i$*). Then* T *contains a complementary edge.*

Theorem 1.5 (Cubical Sperner with cubical labels). *Let* T *be a triangulation of* $\Box^n$. *Let* $\lambda : V(T) \to \operatorname{ext}(\Box^n)$ *be a label function such that for all vertices* $x = (x_1, \ldots, x_n) \in V(T)$, *for all* $1 \leq i \leq n$, *if* $x_i \in \{-1, 1\}$, *then* $\lambda(x)_i = x_i$. *Then* T *contains a neutral simplex.*

Theorem 1.4 is a special case of a theorem originally conjectured by Atanassov and proven by De Loera et al. [1,9]. Theorem 1.5 is implied by a result of Kuhn [8], in which the vertices of a triangulation of $\Box^n$ are labelled using only $(n+1)$ labels. Allowing the domain and codomain to differ yields several additional fixed-point theorems, such as the following:

Theorem 1.6 (Cubical Sperner with octahedral labels). *Let* T *be a triangulation of* $\Box^n$. *Let* $\lambda : V(T) \to \operatorname{ext}(\Diamond^n)$ *be a label function with the property that for all vertices* $x = (x_1, \ldots, x_n) \in V(T)$, *for all* $1 \leq i \leq n$, *if* $x_i \in \{-1, 1\}$, *then* $\lambda_i(x) \neq -x_i e_i$. *Then* T *contains a complementary edge.*

Theorem 1.7 (Octahedral Sperner with cubical labels). *Let T be a triangulation of $\lozenge^n$. Let $\lambda : V(T) \to \text{ext}(\square^n)$ be a label function with the property that for all vertices $x \in V(T)$, for all $v \in \text{ext}(\square^n)$, if $v^T x = 1$, then $\lambda(x) \neq -v$. Then T contains a neutral simplex.*

Theorem 1.6 is due to Freund [4]. Theorem 1.7 appears to be novel, though it is related to a general result of Freund [5] that claims a conclusion too strong to hold here (details in full paper). Although the combinatorial methods of Freund and of Kuhn appear insufficient to prove Theorem 1.7, it can be proven using standard techniques via Brouwer's fixed-point theorem (see full version).

To state Tucker-like theorems, we require a notion of antipodality:

Definition 1.8. Let T be a topological triangulation of B^n with $X = \bigcup_{\sigma \in T} \sigma$. T is said to be *antipodally symmetric on the boundary* if, for all simplices $\sigma \in \partial(T)$, the reflected simplex $-\sigma$ also lies in $\partial(T)$.

Tucker's lemma [14] can then be stated as follows:

Theorem 1.9 (Tucker's lemma). *Let T be a triangulation of B^n that is antipodally symmetric on the boundary of the domain X. Let $\lambda : V(T) \to \text{ext}(\lozenge^n)$ be a label function such that $\lambda(v) = -\lambda(-v)$ for all $v \in \partial(X)$. Then T contains a complementary edge.*

We can establish similar theorems using labels from other codomains homeomorphic to B^n. By extending Ky Fan's theorem [3] to a broader class of label sets, we establish a combinatorial proof of the following in the full version:

Theorem 1.10 (Tucker's lemma with cubical labels). *Let T be a triangulation of B^n that is antipodally symmetric on the boundary of the domain X. Let $\lambda : V(T) \to \text{ext}(\square^n)$ be a label function such that $\lambda(v) = -\lambda(-v)$ for all $v \in \partial(X)$. Then T contains a neutral simplex.*

2 Geometric proofs of Sperner-like theorems

We now describe a technique that enables us to explicitly construct geometric reductions between some of combinatorial fixed-point theorems discussed above. We illustrate our technique through an example:

Theorem 2.1. *Tucker's theorem implies Theorem* 1.4 *(octahedral Sperner with octahedral labels).*

Proof sketch. Let T be a triangulation of $\lozenge^n$ with label function $\lambda : V(T) \to \text{ext}(\lozenge^n)$ satisfying the conditions of Theorem 1.4. Let $X = 2\lozenge^n$

be a dilated copy of the n-dimensional octahedron, so that $\Diamond^n$ lies entirely within the interior of X. Our key idea is to extend the triangulation T and label function λ to a triangulation T^* of X and a label function $\lambda^* : V(T^*) \to \text{ext}(\Diamond^n)$ so that the following properties hold:

1. $T \subset T^*$, and $\lambda^*(v) = \lambda(v)$ for each vertex v in $V(T)$.
2. T^* is antipodally symmetric on the boundary, and $\lambda^*(v) = -\lambda^*(-v)$ for each $v \in V(T^*) \cap \partial(X)$.
3. There are no complementary edges in $T^* \setminus T$.

If we can construct such a T^* and λ^*, then Theorem 1.4 immediately follows from Tucker's lemma, since T^* must contain a complementary edge if property (2) is true, and this complementary edge must then lie in T by properties (1) and (3). Further details are provided in the full version. □

A similar argument can also be used to show that Tucker's theorem implies Theorem 1.6 (Cubical Sperner with octahedral labels). Indeed, we can also use this technique to show the equivalence of Theorem 1.6 and Theorem 1.4. Unfortunately, this style of geometric argument relies crucially on a labelling scheme in which negations are permitted. Accordingly, it appears that additional insight is required in order for it to be possible to use a geometric construction of this nature to directly prove Sperner's lemma from a Tucker-like theorem.

References

[1] K. T. Atanassov, *On Sperner's lemma*, Studia Sci. Math. Hungar. **32** (1996), no. 1-2, 71–74.

[2] X. Chen, X. Deng and S-H. Teng, *Settling the complexity of computing two-player nash equilibria*, Journal of the ACM (JACM) **56** (2009), no. 3, 14.

[3] K. Fan, *A generalization of Tucker's combinatorial lemma with topological applications*, Ann. of Math. (2) **56** (1952), 431–437.

[4] R. M. Freund, *Variable dimension complexes. II. A unified approach to some combinatorial lemmas in topology*, Math. Oper. Res. **9** (1984), no. 4, 498–509.

[5] R. M. Freund, *Combinatorial analogs of brouwer's fixed-point theorem on a bounded polyhedron*, JCTB **47** (1989), no. 2, 192–219.

[6] R. M. Freund and M. J. Todd, *A constructive proof of Tucker's combinatorial lemma*, J. Combin. Theory Ser. A **30** (1981), no. 3, 321–325.

[7] E. GRANT and W. MA, *A geometric approach to combinatorial fixed-point theorems*, available at http://arxiv.org/abs/1305.6158.

[8] H. W. KUHN, *Some combinatorial lemmas in topology*, IBM J. Res. Develop. **4** (1960), 508–524.

[9] J. A. DE LOERA, E. PETERSON and F. E. SU, *A polytopal generalization of Sperner's lemma*, J. Combin. Theory Ser. A **100** (2002), no. 1, 1–26.

[10] J. MATOUŠEK, "Using the Borsuk-Ulam Theorem", Springer-Verlag, Berlin, 2003.

[11] K. L. NYMAN *and* F. E. SU, *A borsuk-ulam equivalent that directly implies sperner's lemma*, unpublished manuscript (2012).

[12] M. R. RAHMAN, *Survey on topological methods in distributed computing*.

[13] F. E. SU, *Borsuk-Ulam implies Brouwer: a direct construction*, Amer. Math. Monthly **104** (1997), no. 9, 855–859.

[14] A. W. TUCKER, *Some topological properties of disk and sphere*, Proc. First Canadian Math. Congress, Montreal, 1945, University of Toronto Press, 1946, . 285–309.

[15] M. YANNAKAKIS, *Equilibria, fixed points, and complexity classes*, Computer Science Review **3** (2009), no. 2, 71–85.

[16] RADE T ŽIVALJEVIĆ, *Oriented matroids and Ky Fan's theorem*, Combinatorica **30** (2010), no. 4, 471–484.

Hamiltonicity

Proof of a conjecture of Thomassen on Hamilton cycles in highly connected tournaments

Daniela Kühn[1], John Lapinskas[1], Deryk Osthus[1] and Viresh Patel[1]

Abstract. A conjecture of Thomassen from 1982 states that for every k there is an $f(k)$ so that every strongly $f(k)$-connected tournament contains k edge-disjoint Hamilton cycles. A classical theorem of Camion, that every strongly connected tournament contains a Hamilton cycle, implies that $f(1) = 1$. So far, even the existence of $f(2)$ was open. In this paper, we prove Thomassen's conjecture by showing that $f(k) = O(k^2 \log^2 k)$. This is best possible up to the logarithmic factor. As a tool, we show that every strongly $10^4 k \log k$-connected tournament is k-linked (which improves a previous exponential bound). The proof of the latter is based on a fundamental result of Ajtai, Komlós and Szemerédi on asymptotically optimal sorting networks.

Main result. A *tournament* is an orientation of a complete graph and a Hamilton cycle in a tournament T is a (consistently oriented) cycle which contains all the vertices of T. T is *strongly connected* if, for every pair of vertices x and y of T, there are directed paths from x to y and from y to x. T is *strongly k-connected* if $|T| > k$ and T remains strongly connected after the removal of any $k-1$ vertices.

Hamilton cycles in tournaments have a long and rich history. For instance, one of the most basic results about tournaments is Camion's theorem, which states that every strongly connected tournament has a Hamilton cycle [7]. This is strengthened by Moon's theorem [16], which implies that such a tournament is also pancyclic, i.e. contains cycles of all possible lengths. Many related results have been proved; the monograph by Bang-Jensen and Gutin [4] gives an overview which also includes many recent results.

In 1982, Thomassen [18] made a very natural conjecture on how to guarantee not just one Hamilton cycle, but many edge-disjoint ones: he conjectured that for every k there is an $f(k)$ so that every strongly $f(k)$-connected tournament contains k edge-disjoint Hamilton cycles (see also the recent surveys [3, 13]). This turned out to be surprisingly difficult: not even the existence of $f(2)$ was known so far. Our main result shows that $f(k) = O(k^2 \log^2 k)$.

[1] School of Mathematics, University of Birmingham, Birmingham B15 2TT, United Kingdom. Email: d.kuhn@bham.ac.uk, jal129@bham.ac.uk, d.osthus@bham.ac.uk, v.patel.3@bham.ac.uk

Theorem 1 ([12]). *There exists $C > 0$ such that for all $k \in \mathbb{N}$ with $k \geq 2$ every strongly $Ck^2 \log^2 k$-connected tournament contains k edge-disjoint Hamilton cycles.*

We also have a construction which shows that $f(k) \geq (k-1)^2/4$, and so our bound on the connectivity is asymptotically close to best possible. Thomassen [18] observed that $f(2) > 2$ and conjectured that $f(2) = 3$. He also observed that one cannot weaken the assumption in Theorem 1 by replacing strong connectivity with strong edge-connectivity.

We have made no attempt to optimize the value of the constant C in Theorem 1: one can take $C := 10^{12}$ for $k \geq 20$. Rather than proving Theorem 1 directly, we deduce it as an immediate consequence of two further results, which are both of independent interest: we show that every sufficiently highly connected tournament is highly linked (see Theorem 3) and that every highly linked tournament contains many edge-disjoint Hamilton cycles (see Theorem 2).

Linkedness in tournaments. Given sets A, B of size k in a strongly k-connected digraph D, Menger's theorem implies that D contains k vertex-disjoint paths from A to B. A k-linked digraph is one in which we can even specify the initial and final vertex of each such path. More precisely a digraph D is *k-linked* if $|D| \geq 2k$ and whenever $x_1, \ldots, x_k, y_1, \ldots, y_k$ are $2k$ distinct vertices of D, there exist vertex-disjoint paths $P_1, \ldots, P_k$ such that P_i is a path from x_i to y_i.

Theorem 2 ([12]). *There exists $C' > 0$ such that for all $k \in \mathbb{N}$ with $k \geq 2$ every $C'k^2 \log k$-linked tournament contains k edge-disjoint Hamilton cycles.*

Similarly as for Theorem 1, the bound in Theorem 2 is asymptotically close to best possible. Moreover, one can take $C' := 10^7$ for all $k \geq 20$. (As mentioned earlier, we have made no attempt to optimise the value of this constant.)

It is not clear from the definition that every (very) highly connected tournament is also highly linked. In fact, for general digraphs this is far from true: Thomassen [20] showed that for all k there are strongly k-connected digraphs which are not even 2-linked. On the other hand, he showed that there is an (exponential) function $g(k)$ so that every strongly $g(k)$-connected tournament is k-linked [19]. Theorem 3 below shows that we can take $g(k)$ to be almost linear in k. Note that Theorem 3 together with the construction mentioned earlier (showing that $f(k) \geq (k-1)^2/4$) shows that Theorem 2 is asymptotically best possible up to logarithmic terms.

Theorem 3 ([12]). *For all $k \in \mathbb{N}$ with $k \geq 2$ every tournament that is strongly $10^4 k \log k$-connected is k-linked.*

For small k, the constant 10^4 can easily be improved. The proof of Theorem 3 is based on a fundamental result of Ajtai, Komlós and Szemerédi [1,2] on the existence of asymptotically optimal sorting networks. Though their result is asymptotically optimal, it is not clear whether this is the case for Theorem 3. In fact, for the case of (undirected) graphs, a deep result of Bollobás and Thomason [6] states that every $22k$-connected graph is k-linked (this was improved to $10k$ by Thomas and Wollan [17]). Thus one might believe that a similar relation also holds in the case of tournaments:

Conjecture 4 ([12]). *There exists $C > 0$ such that for all $k \in \mathbb{N}$ every strongly Ck-connected tournament is k-linked.*

Similarly, we believe that the logarithmic terms can also be removed in Theorems 1 and 2:

Conjecture 5 ([12]).

(i) *There exists $C' > 0$ such that for all $k \in \mathbb{N}$ every $C'k^2$-linked tournament contains k edge-disjoint Hamilton cycles.*
(ii) *There exists $C'' > 0$ such that for all $k \in \mathbb{N}$ every strongly $C''k^2$-connected tournament contains k edge-disjoint Hamilton cycles.*

Note that Conjectures 4 and 5(i) together imply Conjecture 5(ii).

Algorithmic aspects. Both Hamiltonicity and linkedness in tournaments have also been studied from an algorithmic perspective. Camion's theorem implies that the Hamilton cycle problem (though NP-complete in general) is solvable in polynomial time for tournaments. Chudnovsky, Scott and Seymour [8] solved a long-standing problem of Bang-Jensen and Thomassen [5] by showing that the linkedness problem is also solvable in polynomial time for tournaments. More precisely, for a given tournament on n vertices, one can determine in time polynomial in n whether it is k-linked and if yes, one can produce a corresponding set of k paths (also in polynomial time). Fortune, Hopcroft and Wyllie [10] showed that for general digraphs, the problem is NP-complete even for $k = 2$.

We can use the result in [8] to obtain an algorithmic version of Theorem 2. More precisely, given a $C'k^2 \log k$-linked tournament on n vertices, one can find k edge-disjoint Hamilton cycles in time polynomial in n (where k is fixed).

Note that this immediately results in an algorithmic version of Theorem 1.

Related results and spanning regular subgraphs. Our construction showing that $f(k) \geq (k-1)^2/4$ actually suggests that the 'bottleneck' to finding k edge-disjoint Hamilton cycles is the existence of a k-regular subdigraph. Indeed, the construction shows that if the connectivity of a tournament T is significantly lower than in Theorem 1, then T may not even contain a spanning k-regular subdigraph. There are other results which exhibit this phenomenon: if T is itself regular, then Kelly's conjecture from 1968 states that T itself has a Hamilton decomposition. Kelly's conjecture was proved very recently (for large tournaments) by Kühn and Osthus [14].

Erdős raised a 'probabilistic' version of Kelly's conjecture: for a tournament T, let $\delta^0(T)$ denote the minimum of the minimum out-degree and the minimum in-degree. He conjectured that for almost all tournaments T, the maximum number of edge-disjoint Hamilton cycles in T is exactly $\delta^0(T)$. In particular, this would imply that with high probability, $\delta^0(T)$ is also the degree of a densest spanning regular subdigraph in a random tournament T. This conjecture of Erdős was proved by Kühn and Osthus [15], based on the main result in [14].

It would be interesting to obtain further conditions which relate the degree of the densest spanning regular subdigraph of a tournament T to the number of edge-disjoint Hamilton cycles in T. For undirected graphs, one such conjecture was made in [11]: it states that for any graph G satisfying the conditions of Dirac's theorem, the number of edge-disjoint Hamilton cycles in G is exactly half the degree of a densest spanning even-regular subgraph of G. An approximate version of this conjecture was proved by Ferber, Krivelevich and Sudakov [9], see e.g. [11,15] for some related results.

The methods used in the current paper are quite different from those used e.g. in the above results. A crucial ingredient is the construction of highly structured dominating sets.

1 Sketch of the proof of Theorem 2

In this section, we give an outline of the proof of Theorem 2. An important idea is the notion of a 'covering edge'. Let T be a tournament, let $x \in V(T)$, and suppose C is a cycle in T covering $T - x$. If $yz \in E(C)$ and $yx, xz \in E(T)$, then we can replace yz by yxz in C to turn C into a Hamilton cycle. We call yz a *covering edge* for x. More generally, if $S \subseteq V(T)$ and C is a cycle in T spanning $V(T) - S$, then if C contains a covering edge for each $x \in S$ then we can turn C into a Hamilton cycle by using all these covering edges. Note that this idea still works if C covers some part of S.

The following consequence of the Gallai-Milgram theorem is another important tool: suppose that G is an oriented graph on n vertices with

$\delta(G) \geq n - \ell$. Then the vertices of G can be covered with ℓ vertex-disjoint paths. We use this as follows: suppose we are given a highly linked tournament T and have already found i edge-disjoint Hamilton cycles in T. Then the Gallai-Milgram theorem implies that we can cover the vertices of the remaining oriented graph by a set of $2i + 1$ vertex-disjoint paths. Very roughly, the aim is to link together these paths using the high linkedness of the original tournament T.

As above, suppose that we have already found i edge-disjoint Hamilton cycles in a highly linked tournament T. Let T' be the oriented subgraph of T obtained by removing the edges of these Hamilton cycles. Set $t = 2i + 1$ and suppose that we also have the following 'linked dominating structure', $L \subseteq T'$, which is found at the outset of the proof using the assumption that T is highly linked. L consists of the vertices and edges of:

- small disjoint transitive out-dominating sets $A_1, \dots, A_t$;
- small disjoint transitive in-dominating sets $B_1, \dots, B_t$;
- a set of short vertex-disjoint paths $P_1, \dots, P_t$, where each P_ℓ is a path from the sink of B_ℓ to the source of A_ℓ.

Here, we define A_ℓ to be a transitive out-dominating set if $T[A_\ell]$ is transitive and if every vertex of $V(T) \setminus A_\ell$ receives an edge from A_ℓ. Transitive in-dominating sets are similarly defined.

Since $\delta(T' - V(L)) \geq n - 1 - 2i = n - t$, the Gallai-Milgram theorem implies that we can cover the vertices of $T' - V(L)$ with t vertex-disjoint paths $Q_1, \dots, Q_t$. Now, using the dominating sets in our 'linked dominating structure', we can extend $Q_1, \dots, Q_t$ into L, and if this is done in the right way, then we can link up all the paths $Q_1, \dots, Q_t$ and $P_1, \dots, P_t$ into a single cycle C which covers all vertices outside $V(L)$ (and some of the vertices inside $V(L)$). In our construction, we will ensure that the paths P_ℓ contain a set of covering edges for $V(L)$. So C also contains covering edges for $V(L)$, and so we can transform C into a Hamilton cycle as discussed earlier.

References

[1] M. Ajtai, J. Komlós and E. Szemerédi, *An $O(n \log n)$ sorting network*, Proc. 15th Ann. ACM Symp. on Theory of Computing (1983), 1–9.

[2] M. Ajtai, J. Komlós and E. Szemerédi, *Sorting in $C \log N$ parallel steps*, Combinatorica **3** (1983), 1–19.

[3] J. Bang-Jensen, *Problems and conjectures concerning connectivity, paths, trees and cycles in tournament-like digraphs*, Discrete Mathematics **309** (2009), 5655–5667.

[4] J. BANG-JENSEN and G. GUTIN, "Digraphs: Theory, Algorithms and Applications", 2nd edition, Springer 2008.

[5] J. BANG-JENSEN and C. THOMASSEN, *A polynomial algorithm for the 2-path problem for semicomplete digraphs*, SIAM Journal Discrete Mathematics **5** (1992), 366–376.

[6] B. BOLLOBÁS and A. THOMASON, *Highly linked graphs*, Combinatorica **16** (1996), 313–320.

[7] P. CAMION, *Chemins et circuits hamiltoniens des graphes complets*, C. R. Acad. Sci. Paris **249** (1959), 2151–2152.

[8] M. CHUDNOVSKY, A. SCOTT and P. SEYMOUR, *Disjoint paths in tournaments*, preprint.

[9] A. FERBER, M. KRIVELEVICH and B. SUDAKOV, *Counting and packing Hamilton cycles in dense graphs and oriented graphs*, preprint.

[10] S. FORTUNE, J. E. HOPCROFT and J. WYLLIE, The directed subgraph homeomorphism problem, *J. Theoret. Comput. Sci.* **10** (1980), 111–121.

[11] D. KÜHN, J. LAPINSKAS and D. OSTHUS, *Optimal packings of Hamilton cycles in graphs of high minimum degree*, Combinatorics, Probability and Computing **22** (2013), 394–416.

[12] D. KÜHN, J. LAPINSKAS, D. OSTHUS and V. PATEL, *Proof of a conjecture of Thomassen on Hamilton cycles in highly connected tournaments*, preprint.

[13] D. KÜHN and D. OSTHUS, *A survey on Hamilton cycles in directed graphs*, European J. Combinatorics **33** (2012), 750–766.

[14] D. KÜHN and D. OSTHUS, *Hamilton decompositions of regular expanders: a proof of Kelly's conjecture for large tournaments*, Advances in Mathematics **237** (2013), 62–146.

[15] D. KÜHN and D. OSTHUS, *Hamilton decompositions of regular expanders: applications*, preprint.

[16] J. W. MOON, *On subtournaments of a tournament*, Canadian Mathematical Bulletin **9** (1966), 297–301.

[17] R. THOMAS and P. WOLLAN, *An improved extremal function for graph linkages*, European J. of Combinatorics **26** (2005), 309–324.

[18] C. THOMASSEN, *Edge-disjoint Hamiltonian paths and cycles in tournaments*, Proc. London Math. Soc. **45** (1982), 151–168.

[19] C. THOMASSEN, *Connectivity in tournaments*, In: "Graph Theory and Combinatorics, A volume in honour of Paul Erdős", B. Bollobás (ed.), Academic Press, London (1984), 305–313.

[20] C. THOMASSEN, *Note on highly connected non-2-linked digraphs*, Combinatorica **11** (1991), 393–395.

Proof of the 1-factorization and Hamilton decomposition conjectures

Béla Csaba[1], Daniela Kühn[2], Allan Lo[2], Deryk Osthus[2] and Andrew Treglown[3]

Abstract. We prove the following results (via a unified approach) for all sufficiently large n:

(i) [*1-factorization conjecture*] Suppose that n is even and $D \geq 2\lceil n/4 \rceil - 1$. Then every D-regular graph G on n vertices has a decomposition into perfect matchings. Equivalently, $\chi'(G) = D$.

(ii) [*Hamilton decomposition conjecture*] Suppose that $D \geq \lfloor n/2 \rfloor$. Then every D-regular graph G on n vertices has a decomposition into Hamilton cycles and at most one perfect matching.

(iii) [*Optimal packings of Hamilton cycles*] Suppose that G is a graph on n vertices with minimum degree $\delta \geq n/2$. Then G contains at least $(n-2)/8$ edge-disjoint Hamilton cycles.

According to Dirac, (i) was first raised in the 1950's. (ii) and (iii) answer questions of Nash-Williams from 1970. All of the above bounds are best possible.

1 Introduction

In a sequence of four papers [5, 6, 11, 12], we provide a unified approach towards proving three long-standing conjectures for all sufficiently large graphs. Firstly, the 1-factorization conjecture, which can be formulated as an edge-colouring problem; secondly, the Hamilton decomposition conjecture, which provides a far-reaching generalization of Walecki's result [15] that every complete graph of odd order has a Hamilton decomposition and thirdly, a best possible result on packing edge-disjoint Hamilton cycles in Dirac graphs. The latter two were raised by Nash-Williams [17–19] in 1970. A key tool is the recent result of Kühn and Osthus [13] that every dense even-regular robustly expanding graph has a Hamilton decomposition.

1.1 The 1-factorization conjecture

Vizing's theorem states that for any graph G of maximum degree Δ, its edge-chromatic number $\chi'(G)$ is either Δ or $\Delta + 1$. In general, it is a very difficult problem to determine which graphs G attain the (trivial) lower bound Δ – much of the recent book [22] is devoted to the subject. For regular graphs G, $\chi'(G) = \Delta(G)$ is equivalent to the existence of a 1-factorization: a *1-factorization* of a graph G consists of a set of edge-

[1] Bolyai Institute, University of Szeged, H-6720 Szeged, Hungary. Email: bcsaba@math.u-szeged.hu

[2] School of Mathematics, University of Birmingham, Birmingham B15 2TT, United Kingdom. Email: d.kuhn@bham.ac.uk, s.a.lo@bham.ac.uk, d.osthus@bham.ac.uk

[3] School of Mathematical Sciences, Queen Mary, University of London, London E1 4NS, United Kingdom. Email: a.treglown@qmul.ac.uk

disjoint perfect matchings covering all edges of G. The long-standing 1-factorization conjecture states that every regular graph of sufficiently high degree has a 1-factorization. It was first stated explicitly by Chetwynd and Hilton [1,3] (who also proved partial results). However, they state that according to Dirac, it was already discussed in the 1950's.

Theorem 1.1. *There exists an $n_0 \in \mathbb{N}$ such that the following holds. Let $n, D \in \mathbb{N}$ be such that $n \geq n_0$ is even and $D \geq 2\lceil n/4 \rceil - 1$. Then every D-regular graph G on n vertices has a 1-factorization. Equivalently, $\chi'(G) = D$.*

The bound on the degree in Theorem 1.1 is best possible. To see this, suppose first that $n = 2 \pmod 4$. Consider the graph which is the disjoint union of two cliques of order $n/2$ (which is odd). If $n = 0 \pmod 4$, consider the graph obtained from the disjoint union of cliques of orders $n/2-1$ and $n/2+1$ (both odd) by deleting a Hamilton cycle in the larger clique.

Note that Theorem 1.1 implies that for every regular graph G on an even number of vertices, either G or its complement has a 1-factorization. Also, Theorem 1.1 has an interpretation in terms of scheduling round-robin tournaments (where n players play all of each other in $n-1$ rounds): one can schedule the first half of the rounds arbitrarily before one needs to plan the remainder of the tournament.

The best previous result towards Theorem 1.1 is due to Perkovic and Reed [20], who proved an approximate version, *i.e.* they assumed that $D \geq n/2 + \varepsilon n$. This was generalized by Vaughan [23] to multigraphs of bounded multiplicity. Indeed, he proved an approximate version of the following multigraph version of the 1-factorization conjecture which was raised by Plantholt and Tipnis [21]: Let G be a regular multigraph of even order n with multiplicity at most r. If the degree of G is at least $rn/2$ then G is 1-factorizable.

In 1986, Chetwynd and Hilton [2] made the following 'overfull subgraph' conjecture, which also generalizes the 1-factorization conjecture. Roughly speaking, this says that a dense graph satisfies $\chi'(G) = \Delta(G)$ unless there is a trivial obstruction in the form of a dense subgraph H on an odd number of vertices. Formally, we say that a subgraph H of G is *overfull* if $e(H) > \Delta(G)\lfloor |H|/2 \rfloor$ (note this requires $|H|$ to be odd).

Conjecture 1.2. A graph G on n vertices with $\Delta(G) \geq n/3$ satisfies $\chi'(G) = \Delta(G)$ if and only if G contains no overfull subgraph.

This conjecture is still wide open – partial results are discussed in [22], which also discusses further results and questions related to the 1-factorization conjecture.

1.2 The Hamilton decomposition conjecture

Rather than asking for a 1-factorization, Nash-Williams [17,19] raised the more difficult problem of finding a Hamilton decomposition in an even-regular graph. Here, a *Hamilton decomposition* of a graph G consists of a set of edge-disjoint Hamilton cycles covering all edges of G. A natural extension of this to regular graphs G of odd degree is to ask for a decomposition into Hamilton cycles and one perfect matching (*i.e.* one perfect matching M in G together with a Hamilton decomposition of $G - M$). The following result solves the problem of Nash-Williams for all large graphs.

Theorem 1.3. *There exists an $n_0 \in \mathbb{N}$ such that the following holds. Let $n, D \in \mathbb{N}$ be such that $n \geq n_0$ and $D \geq \lfloor n/2 \rfloor$. Then every D-regular graph G on n vertices has a decomposition into Hamilton cycles and at most one perfect matching.*

Again, the bound on the degree in Theorem 1.3 is best possible. Previous results include the following: Nash-Williams [16] showed that the degree bound in Theorem 1.3 ensures a single Hamilton cycle. Jackson [8] showed that one can ensure close to $D/2 - n/6$ edge-disjoint Hamilton cycles. Christofides, Kühn and Osthus [4] obtained an approximate decomposition under the assumption that $D \geq n/2 + \varepsilon n$. Under the same assumption, Kühn and Osthus [14] obtained an exact decomposition (as a consequence of their main result in [13] on Hamilton decompositions of robustly expanding graphs).

Note that Theorem 1.3 does not quite imply Theorem 1.1, as the degree threshold in the former result is slightly higher.

A natural question is whether one can extend Theorem 1.3 to sparser (quasi)-random graphs. Indeed, for random regular graphs of bounded degree this was proved by Kim and Wormald [9] and for (quasi-)random regular graphs of linear degree this was proved in [14] as a consequence of the main result in [13]. However, the intermediate range remains open.

1.3 Packing Hamilton cycles in graphs of large minimum degree

Although Dirac's theorem is best possible in the sense that the minimum degree condition $\delta \geq n/2$ is best possible, the conclusion can be strengthened considerably: a remarkable result of Nash-Williams [18] states that every graph G on n vertices with minimum degree $\delta(G) \geq n/2$ contains $\lfloor 5n/224 \rfloor$ edge-disjoint Hamilton cycles. He raised the question of finding the best possible bound, which we answer below for all large graphs.

Theorem 1.4. *There exists an $n_0 \in \mathbb{N}$ such that the following holds. Suppose that G is a graph on $n \geq n_0$ vertices with minimum degree $\delta \geq n/2$. Then G contains at least $(n-2)/8$ edge-disjoint Hamilton cycles.*

The following construction (which is based on a construction of Babai, see [17]) shows that the bound is best possible for $n = 8k + 2$, where $k \in \mathbb{N}$. Consider the graph G consisting of one empty vertex class A of size $4k$, one vertex class B of size $4k + 2$ containing a perfect matching and no other edges, and all possible edges between A and B. Thus G has order $n = 8k + 2$ and minimum degree $4k + 1 = n/2$. Any Hamilton cycle in G must contain at least two edges of the perfect matching in B, so G contains at most $\lfloor |B|/4 \rfloor = k = (n-2)/8$ edge-disjoint Hamilton cycles.

A more general question is to ask for the number of edge-disjoint Hamilton cycles one can guarantee in a graph G of minimum degree δ. This number has been determined exactly by Kühn, Lapinskas and Osthus [10] unless G is close to one of the extremal graphs for Dirac's theorem (*i.e.* unless G is close to the complete balanced bipartite graph or close to the union of two disjoint copies of a clique). In particular, the number of edge-disjoint Hamilton cycles one can guarantee is known exactly whenever $\delta \geq n/2 + \varepsilon n$. This improves earlier results of Christofides, Kühn and Osthus [4] as well as Hartke and Seacrest [7]. Actually, our proof of Theorem 1.4 also settles the cases when G is close to the extremal graphs for Dirac's theorem. So altogether this solves the problem for all values of δ.

2 Overview of the proofs of Theorems 1.1 and 1.3

The proofs develop methods established by Kühn and Osthus [13], who proved a generalization of Kelly's conjecture that every regular tournament has a Hamilton decomposition (for large tournaments). For all three of our main results, we split the argument according to the structure of the graph G under consideration:

(i) G is close to the complete balanced bipartite graph $K_{n/2,n/2}$;
(ii) G is close to the union of two disjoint copies of a clique $K_{n/2}$;
(iii) G is a 'robust expander'.

Informally, a graph G is a robust expander if for every set $S \subseteq V(G)$ which is not too large or too small, its neighbourhood is substantially larger than $|S|$, even if we delete a small proportion of the edges of G. In other words, G is an expander graph which is 'locally resilient'. The main result of [13] states that every dense regular robust expander has a Hamilton decomposition. This immediately implies Theorems 1.1 and 1.3 in Case (iii).

Suppose we are going to prove Theorem 1.3 in the case when D is even. So our aim is to decompose G into $D/2$ edge-disjoint Hamilton cycles. As mentioned above, we may assume that G is in either Case (i) or Case (ii). In [6], we find an approximate Hamilton decomposition of G in both cases, *i.e.* a set of edge-disjoint Hamilton cycles covering almost all edges of G. However, one does not have any control over the 'leftover' graph H, which makes a complete decomposition seem infeasible. This problem was overcome in [13] by introducing the concept of a 'robustly decomposable graph' G^{rob}. Roughly speaking, this is a sparse regular graph with the following property: given *any* very sparse regular graph H with $V(H) = V(G^{\mathrm{rob}})$ which is edge-disjoint from G^{rob}, one can guarantee that $G^{\mathrm{rob}} \cup H$ has a Hamilton decomposition. This leads to a natural (and very general) strategy to obtain a decomposition of G:

(1) find a (sparse) robustly decomposable graph G^{rob} in G and let G' denote the leftover;
(2) find an approximate Hamilton decomposition of G' and let H denote the (very sparse) leftover;
(3) find a Hamilton decomposition of $G^{\mathrm{rob}} \cup H$.

It is of course far from clear that one can always find such a graph G^{rob}, especially in Case (ii) where G is close to being disconnected. In [5], we find G^{rob} for Case (i). In [11, 12], we find G^{rob} for Case (ii).

References

[1] A. G. Chetwynd and A. J. W. Hilton, *Regular graphs of high degree are 1-factorizable*, Proc. London Math. Soc. **50** (1985), 193–206.

[2] A. G. Chetwynd and A. J. W. Hilton, *Star multigraphs with three vertices of maximum degree*, Math. Proc. Cambridge Philosophical Soc. **100** (1986), 303–317.

[3] A. G. Chetwynd and A. J. W. Hilton, *1-factorizing regular graphs of high degree—an improved bound*, Discrete Math. **75** (1989), 103–112.

[4] D. Christofides, D. Kühn and D. Osthus, *Edge-disjoint Hamilton cycles in graphs*, J. Combin. Theory B **102** (2012), 1035–1060.

[5] B. Csaba, D. Kühn, A. Lo, D. Osthus and A. Treglown, *Proof of the 1-factorization and Hamilton decomposition conjectures II: the bipartite case*, preprint.

[6] B. Csaba, D. Kühn, A. Lo, D. Osthus and A. Treglown, *Proof of the 1-factorization and Hamilton decomposition conjectures III: approximate decompositions*, preprint.

[7] S. G. Hartke and T. Seacrest, *Random partitions and edge-disjoint Hamiltonian cycles*, preprint.

[8] B. Jackson, *Edge-disjoint Hamilton cycles in regular graphs of large degree*, J. London Math. Soc. **19** (1979), 13–16.

[9] J. H. Kim and N. C. Wormald, *Random matchings which induce Hamilton cycles and Hamiltonian decompositions of random regular graphs*, J. Combin. Theory B **81** (2001), 20–44.

[10] D. KÜHN, J. LAPINSKAS and D. OSTHUS, *Optimal packings of Hamilton cycles in graphs of high minimum degree*, Combin. Probab. Comput. **22** (2013), 394–416.

[11] D. KÜHN, A. LO and D. OSTHUS, *Proof of the 1-factorization and Hamilton decomposition conjectures IV: exceptional systems for the two cliques case*, preprint.

[12] D. KÜHN, A. LO, D. OSTHUS and A. TREGLOWN, *Proof of the 1-factorization and Hamilton decomposition conjectures I: the two cliques case*, preprint.

[13] D. KÜHN and D. OSTHUS, *Hamilton decompositions of regular expanders: a proof of Kelly's conjecture for large tournaments*, Adv. in Math. **237** (2013), 62–146.

[14] D. KÜHN and D. OSTHUS, *Hamilton decompositions of regular expanders: applications*, preprint.

[15] E. LUCAS, "Récréations Mathématiques", Vol. 2, Gautheir-Villars, 1892.

[16] C. ST. J. A. NASH-WILLIAMS, "Valency Sequences which Force Graphs to have Hamiltonian Circuits", University of Waterloo Research Report, Waterloo, Ontario, 1969.

[17] C. ST. J. A. NASH-WILLIAMS, *Hamiltonian lines in graphs whose vertices have sufficiently large valencies*, In: "Combinatorial Theory and its Applications", III (Proc. Colloq., Balatonfüred, 1969), North-Holland, Amsterdam, 1970, 813–819.

[18] C. ST. J. A. NASH-WILLIAMS, *Edge-disjoint Hamiltonian circuits in graphs with vertices of large valency*, In: "Studies in Pure Mathematics" (Presented to Richard Rado), Academic Press, London, 1971, 157–183.

[19] C. ST. J. A. NASH-WILLIAMS, *Hamiltonian arcs and circuits*, In: "Recent Trends in Graph Theory" (Proc. Conf., New York, 1970), Springer, Berlin, 1971, 197–210.

[20] L. PERKOVIC and B. REED, *Edge coloring regular graphs of high degree*, Discrete Math. **165/166** (1997), 567–578.

[21] M. J. PLANTHOLT and S. K. TIPNIS, *All regular multigraphs of even order and high degree are 1-factorable*, Electron. J. Combin. **8** (2001), R41.

[22] M. STIEBITZ, D. SCHEIDE, B. TOFT and L.M. FAVRHOLDT, "Graph Edge Coloring: Vizing's Theorem and Goldberg's Conjecture", Wiley 2012.

[23] E. VAUGHAN, *An asymptotic version of the multigraph 1-factorization conjecture*, J. Graph Theory **72** (2013), 19–29.

Regular hypergraphs: asymptotic counting and loose Hamilton cycles

Andrzej Dudek[1], Alan Frieze[2], Andrzej Ruciński[3] and Matas Šileikis[4]

Abstract. We present results from two papers by the authors on analysis of d-regular k-uniform hypergraphs, when k is fixed and the number n of vertices tends to infinity. The first result is approximate enumeration of such hypergraphs, provided $d = d(n) = o(n^\kappa)$, where $\kappa = \kappa(k) = 1$ for all $k \geq 4$, while $\kappa(3) = 1/2$. The second result is that a random d-regular hypergraph contains as a dense subgraph the uniform random hypergraph (a generalization of the Erdős-Rényi uniform graph), and, in view of known results, contains a loose Hamilton cycle with probability tending to one.

1. Regular k-graphs and k-multigraphs. We consider *k-uniform hypergraphs* (or *k-graphs*, for short) on the vertex set $V = [n] := \{1, \ldots, n\}$, that is, families of k-element subsets of V. A k-graph H is *d-regular*, if the degree of every vertex $v \in V$, $\deg_H(v) := \deg(v) := |\{e \in H : v \in e\}|$ equals d.

Let $\mathcal{H}^{(k)}(n, d)$ be the class of all d-regular k-graphs on $[n]$. Note that each $H \in \mathcal{H}^{(k)}(n, d)$ has $M := nd/k$ edges (throughout, we implicitly assume that k divides nd). Let $\mathbb{H}^{(k)}(n, d)$ be a k-graph chosen from $\mathcal{H}^{(k)}(n, d)$ uniformly at random. We treat d as a function of n (possibly constant) and study $\mathcal{H}^{(k)}(n, d)$ as well as $\mathbb{H}^{(k)}(n, d)$ as n tends to infinity.

By a *k-multigraph* on the vertex set $[n]$ we mean a multiset of k-element multisubsets of $[n]$. An edge is called a *loop* if it contains more than one copy of some vertex and otherwise it is called *a proper edge*. A k-multigraph is *simple*, if it is a k-graph.

A standard tool to study regular (hyper)graphs is the so called *configuration model* of a random k-multigraph (see [10] for $k = 2$; its generalization to every k is straightforward). We use a slightly different model yielding the same distribution of k-multigraphs. Let $\mathcal{S} \subset [n]^{nd}$ be the

[1] Department of Mathematics, Western Michigan University, Kalamazoo, MI.
Email: andrzej.dudek@wmich.edu

[2] Department of Mathematical Sciences, Carnegie Mellon University, Pittsburgh, PA.
Email: alan@random.math.cmu.edu

[3] Department of Discrete Mathematics, Adam Mickiewicz University, Poznań, Poland.
Email: andrzej@mathcs.emory.edu

[4] Department of Mathematics, Uppsala University, Sweden. Email: matas.sileikis@gmail.com

family of all sequences in which every value $i \in [n]$ occurs precisely d times. Let $\mathbf{Y} = (Y_1, \ldots, Y_{nd})$ be a sequence chosen from $\mathcal{S}$ uniformly at random, and define $\mathbb{H}_*^{(k)}(n,d)$ as a k-multigraph with the edge set

$$\{Y_{ki+1} \ldots Y_{ki+k} : i = 0, \ldots, nd-1\}.$$

2. The switching. The essential tool in both papers presented here is the so called *switching* technique, introduced by McKay [8] for asymptotic enumeration of regular graphs. McKay and Wormald [9] improved McKay's result by applying a more advanced version of switching, which we extend to hypergraphs as follows.

Let us view sequences $\mathcal{S}$ as ordered k-multigraphs (that is, k-multigraphs with an ordering of edges). Let $\mathcal{E}_l \subset \mathcal{S}$ be the family of sequences with no multiple edges and exactly l loops, but no loops with less than $k-1$ distinct vertices. Thus, every loop in such a k-multigraph has only one multiple vertex, which has multiplicity two.

The *switching* is an operation which maps a sequence $\mathbf{x} \in \mathcal{E}_l$ to $\mathbf{y} \in \mathcal{E}_{l-1}$ as follows. Choose a loop f and two proper edges e_1, e_2. Select a vertex $v \in e_1 \setminus e_2$ as well as a vertex $w \in e_2 \setminus e_1$. Suppose that u is the multiple vertex of f. Swap v with one copy of u and w with the other. The effect of this is that edges f, e_1, e_2 in $\mathbf{x}$ are replaced with the following three edges in $\mathbf{y}$

$$f \setminus \{u,u\} \cup \{v,w\}, \quad e_1 \setminus \{v\} \cup \{u\}, \quad e_2 \setminus \{w\} \cup \{u\}.$$

3. Counting Regular Hypergraphs. In [5] we approximately count d-regular k-graphs. Since $|\mathcal{S}| = (nd)!/(d!)^n$ and every every simple k-graph is given by exactly $M!(k!)^M$ sequences in $\mathcal{S}$, the number of d-regular k-graphs is precisely

$$|\mathcal{H}^{(k)}(n,d)| = \frac{(nd)!}{M!(k!)^M(d!)^n}\mathbb{P}\left(\mathbb{H}_*^{(k)}(n,d) \text{ is simple}\right).$$

Therefore the problem of asymptotic enumeration reduces to a the analysis of the probability. For graphs, that is, $k=2$, this has been well studied (see [10]). For general k but fixed d, Cooper, Frieze, Molloy and Reed [1] showed that the probability converges to $\exp\{-(k-1)(d-1)/2\}$. In [5] we extend this to the following formula. Let $\kappa(k) = 1$, if $k \geq 4$ and $\kappa(3) = 1/2$.

Theorem 1. *For $k \geq 3$ and $1 \leq d = o(n^{\kappa(k)})$ we have*

$$\mathbb{P}\left(\mathbb{H}_*^{(k)}(n,d) \textit{ is simple}\right) = \exp\left\{-\frac{(k-1)(d-1)}{2} + O\left(\frac{d^2}{n} + \sqrt{\frac{d}{n}}\right)\right\}$$

The proof of Theorem 1 is simplified by the fact (shown in [5]) that a randomly chosen sequence $\mathbf{Y} \in \mathcal{S}$ with probability tending to one belongs to $\mathcal{E}_l$ with a reasonably small l. This allows us to reduce the analysis of the probability to estimating the ratios $|\mathcal{E}_l|/|\mathcal{E}_{l-1}|$. This is done by bounding the number of ways one can apply switching to a sequence. The proof works because, as it turns out, the number of possible switchings depends essentially on the number of loops, but not on the structure of the sequence.

4. Hamilton Cycles in Regular Hypergraphs. Let us recall that for integer $m \in [0, \binom{n}{k}]$, $\mathbb{H}^{(k)}(n, m)$ is the random graph chosen uniformly at random among k-graph on $[n]$ with precisely m edges.

Our main result in [6] is that we can couple $\mathbb{H}^{(k)}(n, d)$ and $\mathbb{H}^{(k)}(n, m)$ so that the latter is a subgraph of the former with probability tending to one.

Theorem 2. *For every $k \geq 3$, there are positive constants c and C such that if $d \geq C \log n$, $d = o(n^{1/2})$ and $m = \lfloor cM \rfloor = \lfloor cnd/k \rfloor$, then one can define a joint distribution of random graphs $\mathbb{H}^{(k)}(n, d)$ and $\mathbb{H}^{(k)}(n, m)$ in such a way that*

$$\mathbb{P}\left(\mathbb{H}^{(k)}(n, m) \subset \mathbb{H}^{(k)}(n, d)\right) \to 1, \qquad n \to \infty.$$

The idea of the proof is as follows. Since m is a fraction of M, we are able to couple $\mathbb{H}^{(k)}(n, m)$ with $\mathbb{H}_*^{(k)}(n, d)$ (treated as an ordered k-multigraph) in such a way that with probability tending to one $\mathbb{H}^{(k)}(n, m)$ is contained in an initial segment of $\mathbb{H}_*^{(k)}(n, d)$, which we colour *red*. Then we swap all red loops of $\mathbb{H}_*^{(k)}(n, d)$ with randomly selected non-red (*green*) proper edges. Finally, we destroy the green loops of $\mathbb{H}_*^{(k)}(n, d)$ one by one applying a randomly chosen switching which involves green edges only. This does not destroy the previously embedded copy of $\mathbb{H}^{(k)}(n, m)$. Moreover, it transforms $\mathbb{H}_*^{(k)}(n, d)$ into a k-graph $\tilde{\mathbb{H}}^{(k)}(n, d)$, which is distributed approximately as $\mathbb{H}^{(k)}(n, d)$, that is, almost uniformly. Theorem 2 then follows by a (maximal) coupling of $\tilde{\mathbb{H}}^{(k)}(n, d)$ and $\mathbb{H}^{(k)}(n, d)$.

A *loose Hamilton cycle* on a vertex set V is a set of edges $e_1, \ldots, e_s$ such that for some cyclic order of V each e_i consists of k consecutive vertices and $|e_i \cap e_{i+1}| = 1$ for $i = 1, \ldots, s$, with $e_{s+1} = e_1$. For $k = 2$ this coincides with the standard notion of a Hamilton cycle.

Asymptotic hamiltonicity for graphs has been intensely investigated since 1978 and rather recently was established in full generality for every $d \geq 3$, both fixed and growing with n (see [2] and references in it).

As for general k, this has been known for some simpler models of random k-graphs. From results of Frieze [7], Dudek and Frieze [3] as well as Dudek, Frieze, Loh and Speiss [4], it follows that $\mathbb{H}^{(k)}(n,m)$ contains a loose Hamilton cycle when the expected degree of a vertex grows faster than $\log n$. This, combined with Theorem 2, implies the following fact.

Corollary 3. *Suppose that $d = o(n^{1/2})$. If $k = 3$ and $d \geq C \log n$ for large constant C or $k \geq 4$ and $\log n = o(d)$, then*

$$\mathbb{P}\left(\mathbb{H}^{(k)}(n,d) \text{ contains a loose Hamilton cycle}\right) \to 1, \qquad n \to \infty.$$

References

[1] C. Cooper, A. Frieze, M. Molloy and B. Reed, *Perfect matchings in random r-regular, s-uniform hypergraphs*, Combin. Probab. Comput. **5** (1) (1996), 1–14.

[2] C. Cooper, A. Frieze and B. Reed, *Random regular graphs of non-constant degree: connectivity and Hamiltonicity*, Combin. Probab. Comput. **11** (3) (2002), 249–261.

[3] A. Dudek and A. Frieze, *Loose Hamilton cycles in random uniform hypergraphs*, Electron. J. Combin. **18** (1) (2011), Paper 48, pp. 14.

[4] A. Dudek, A. Frieze, P.-S. Loh and S. Speiss, *Optimal divisibility conditions for loose Hamilton cycles in random hypergraphs*, Electron. J. Combin. **19** (4) (2012), Paper 44, pp. 17.

[5] A. Dudek, A. Frieze, A. Ruciński and M. Šileikis, *Approximate counting of regular hypergraphs*, preprint, 2013. http://arxiv.org/abs/1303.0400.

[6] A. Dudek, A. Frieze, A. Ruciński and M. Šileikis, *Loose hamilton cycles in regular hypergraphs*, preprint, 2013. http://sites.google.com/site/matassileikis/DFRS20132.pdf.

[7] A. Frieze, *Loose Hamilton cycles in random 3-uniform hypergraphs*, Electron. J. Combin. **17** (1) (2010), Note 28, pp. 4.

[8] B. D. McKay, *Asymptotics for symmetric* 0-1 *matrices with prescribed row sums*, Ars Combin. **19** (A) (1985), 15–25.

[9] B. D. McKay and N. C. Wormald, *Uniform generation of random regular graphs of moderate degree*, J. Algorithms **11** (1) (1990), 52–67.

[10] N. C. Wormald, *Models of random regular graphs*, In: "Surveys in combinatorics, 1999 (Canterbury)", volume 267 of London Math. Soc. Lecture Note Ser., Cambridge Univ. Press, Cambridge, 1999, 239–298.

Triangles

Dynamic concentration of the triangle-free process

Tom Bohman[1] and Peter Keevash[2]

Abstract. The triangle-free process begins with an empty graph on n vertices and iteratively adds edges chosen uniformly at random subject to the constraint that no triangle is formed. We determine the asymptotic number of edges in the maximal triangle-free graph at which the triangle-free process terminates. We also bound the independence number of this graph, which gives an improved lower bound on Ramsey numbers: we show $R(3, t) > (1/4 - o(1))t^2/\log t$, which is within a $4 + o(1)$ factor of the best known upper bound. Furthermore, we determine which bounded size subgraphs are likely to appear in the maximal triangle-free graph produced by the triangle-free process: they are precisely those triangle-free graphs with maximal average density at most 2.

1 Introduction

Constrained random graph processes provide an interesting class of random graph models and a natural source for constructions in graph theory. Although the dependencies introduced by the constraints make such processes difficult to analyse, the evidence to date suggests that they are particularly useful for producing graphs of interest for certain extremal problems. Here we consider the triangle-free random graph process, which is defined by sequentially adding edges, starting with the empty graph, chosen uniformly at random subject to the constraint that no triangle is formed.

This process was introduced by Bollobás and Erdős (see [7]), and first analysed by Erdős, Suen and Winkler [10], using a differential equations method introduced by Ruciński and Wormald [17] for the analysis of the constrained graph process known as the 'd-process'. One motivation for their work was that their analysis of the triangle-free process led to the best lower bound on the Ramsey number $R(3, t)$ known at that time. The Ramsey number $R(s, t)$ is the least number n such that any graph on

[1] Department of Mathematical Sciences, Carnegie Mellon University, Pittsburgh, PA 15213, USA. Email: tbohman@math.cmu.edu. Research supported in part by NSF grants DMS-1001638 and DMS-1100215.

[2] School of Mathematical Sciences, Queen Mary, University of London, Mile End Road, London E1 4NS, UK. Email: p.keevash@qmul.ac.uk. Research supported in part by ERC grant 239696 and EPSRC grant EP/G056730/1.

n vertices contains a complete graph with s vertices or an independent set with t vertices. In general, very little is known about these numbers, even approximately. The upper bound $R(3,t) = O(t^2/\log t)$ was obtained by Ajtai, Komlós and Szemerédi [1], but for many years the best known lower bound, due to Erdős [9], was $\Omega(t^2/\log^2 t)$. The order of magnitude was finally determined by Kim [13], who showed that $R(3,t) = \Omega(t^2/\log t)$. He employed a semi-random construction that is loosely related to the triangle-free process, thus leaving open the question of whether the triangle-free process itself achieves this bound; this was conjectured by Spencer [19] and proved by Bohman [3]. There is now a large literature on the general H-free process, obtained by replacing 'triangle' by any fixed graph H in the definition; see [6,8,15,16,22–25]. However, the theory is still in its early stages: we conjectured that our lower bound for H strictly 2-balanced, given in [6], gives the correct order of magnitude for the length of the process, but so far this has only been proved for some special graphs.

In this paper we specialise to the triangle-free process, where we can now give an asymptotically optimal analysis. Our improvement on previous analyses of this process exploits the self-correcting nature of key statistics of the process.

Let G be the maximal triangle-free graph at which the triangle-free process terminates.

Theorem 1.1. *Whp every vertex of G has degree* $(1+o(1))\sqrt{\frac{1}{2}n\log n}$.

We also obtain the following bound on the size of any independent set in G.

Theorem 1.2. *Whp G has independence number at most* $(1+o(1))\sqrt{2n\log n}$.

An immediate consequence is the following new lower bound on Ramsey numbers. The best known upper bound is $R(3,t) < (1+o(1))\frac{t^2}{\log t}$, due to Shearer [18].

Theorem 1.3. $R(3,t) > \left(\frac{1}{4} - o(1)\right)\frac{t^2}{\log t}$.

These results are predicted by a simple heuristic: the graph $G(i)$ after i steps of the triangle-free process should resemble the Erdős-Rényi random graph $G_{n,p}$ with $i = n^2p/2$, with the exception that $G_{n,p}$ has many triangles while $G(i)$ has none. We also show that this heuristic extends to all small subgraph counts; in particular, we answer the question of which subgraphs appear in G. Suppose H is a graph with at least 3 vertices.

The *average density* of H is $d(H) = \frac{|E_H|}{|V_H|}$. The *maximum average density* $m(H)$ of H is the maximum of $d(H')$ over nonempty subgraphs H' of H.

Theorem 1.4. *Let H be a triangle-free graph with at least* 3 *vertices.*

(i) *If $m(H) \leq 2$ then $\mathbb{P}(H \subseteq G) = 1 - o(1)$.*
(ii) *If $m(H) > 2$ then $\mathbb{P}(H \subseteq G) = o(1)$.*

Thus, the small subgraphs that are likely to appear in G are exactly the same as the triangle-free subgraphs that appear in the corresponding $G_{n,p}$.

2 Overview of the proof

We are guided by the heuristic that $G(i)$ resembles $G_{n,p}$ with $i = n^2p/2$. We introduce a continuous time that scales as $t = in^{-3/2}$. Note that $p = 2tn^{-1/2}$. We define $Q(i)$ to be the number of open *ordered* pairs in $G(i)$. This variable is crucial to our understanding of the process: we have $Q(0) = n^2 - n$, and the process ends when $Q(i) = 0$. How do we expect $Q(i)$ to evolve? If $G(i)$ resembles $G_{n,p}$ then for any pair uv we should have $\mathbb{P}(uv \in O(i)) \approx \left(1 - p^2\right)^{n-2} \approx e^{-np^2} = e^{-4t^2}$. We set $q(t) = e^{-4t^2}n^2$ and expect to have $Q(i) \approx q(t)$ for most of the evolution of the process. This is exactly what we prove.

2.1 Strategy

We use dynamic concentration inequalities for a carefully chosen ensemble of random variables associated with the process. We show $V(i) \approx v(t)$ for all variables V in the ensemble, for some smooth function $v(t)$, which we refer to as the *scaling* of V. Here $V(i)$ denotes the value of V after i steps of the process. For each V we define a *tracking variable* $\mathcal{T}V(i)$ and show that $\mathcal{D}V(i) = V(i) - \mathcal{T}V(i)$ satisfies $|\mathcal{D}V(i)| < e_V(t)v(t)$, for some error functions $e_V(t)$. We use $\mathcal{T}V(i)$ rather than $v(t)$ so that we can isolate variations in V from variations in other variables that have an impact on V.

The improvement to earlier analysis of the process comes from 'self-correction', *i.e.* the mean-reverting properties of the system of variables. We take $e_V(t) = f_V(t) + 2g_V(t)$, where we think of $f_V(t)$ as the 'main error term' and $g_V(t)$ as the 'martingale deviation term'. We usually have $g_V \ll f_V$, but there are some exceptions when t is small and hence $f_V(t)$ is too small. We require $g_V(t)v(t)$ to be 'approximately non-increasing' in t, in that $g_V(t')v(t') = O(g_V(t)v(t))$ for all $t' \geq t$. We define the *critical window* $W_V(i) = [(f_V(t) + g_V(t))v(t), (f_V(t) + 2g_V(t))v(t)]$.

We prove the *trend hypothesis*: $\mathcal{Z}V(i) := |\mathcal{D}V(i)| - e_V(t)v(t)$ is a supermartingale when $|\mathcal{D}V(i)| \in W_V(i)$. The trend hypothesis will follow from the *variation equation* for $e_V(t)$, which balances the changes in $\mathcal{D}V(i)$ and $e_V(t)v(t)$. Since errors can transfer from one variable to another, each variation equation is a differential inequality that can involve many of the error functions.

We track the process up to the time $t_{max} = \frac{1}{2}\sqrt{(1/2-\varepsilon)\log n}$. If the tracking fails, then there is some $i^* \le i_{max}$ and a variable V such that $\mathcal{D}V(i)$ enters $W_V(i')$ from below at some step $i' < i^*$, stays in $W_V(i)$ for $i' \le i \le i^*$ then goes above $W_V(i^*)$ at step i^*. During this time $\mathcal{Z}V(i)$ is a supermartingale, with $\mathcal{Z}V(i') \le -g_V(t')v(t')$ and $\mathcal{Z}V(i^*) \ge 0$, so we have an increase of at least $g_V(t')v(t')$ against the drift of the supermartingale. We can estimate the probability of this event using Freedman's martingale inequality [11], provided that we have good estimates on $Var_V(t) = Var(\mathcal{Z}V(i) \mid \mathcal{F}_{i-1})$ and $N_V(t) = |\mathcal{Z}V(i+1) - \mathcal{Z}V(i)|$; we refer to this as the *boundedness hypothesis*. Thus it suffices to verify the trend and boundedness hypotheses for all variables.

2.2 Variables

All definitions are with respect to the graph $G(i)$. Sometimes we use a variable name to also denote the set that it counts, *e.g.* $Q(i)$ is the number of ordered open pairs, and also denotes the set of ordered open pairs. We usually omit (i) and (t) from our notation, *e.g.* Q means $Q(i)$ and q means $q(t)$. We use capital letters for variable names and the corresponding lower case letter for the scaling. We express scalings using the (approximate) edge density and open pair density, namely $p = 2in^{-2} = 2tn^{-1/2}$ and $\hat{q} = e^{-4t^2}$.

The next most important variable in our analysis, after the variable Q defined above, is the variable Y_{uv} which, for a fixed pair of vertices uv, is the number of vertices w such that uw is an open pair and vw is an edge. It is natural that Y_{uv} should play an important role in this analysis, as when the pair uv is added as an edge, the number of open edges that become closed is exactly $Y_{uv} + Y_{vu}$. The motivation for introducing the ensembles of variables defined below is as follows: control of the global variables is needed to get good control of Q, control of the stacking variables is needed to get good control of Y_{uv}, and controllable variables play a crucial role in our analysis of the stacking variables.

The *global variables* consist of Q, R and S, where $Q = 2|O(i)|$ is the number of ordered open pairs, R is the number of ordered triples with 3 open pairs, and S is the number of ordered triples abc where ab is an edge and ac, bc are open pairs.

The *stacking variables* are built from four basic building blocks: X_u is the number of vertices ω such that $u\omega$ is open, Y_u is the number of vertices ω such that $u\omega$ is an edge, X_{uv} is the number of vertices w such that uw and vw are open pairs, Y_{uv} is the number of vertices w such that uw is an open pair and vw is an edge. We defer the exact definition to the full version of the paper, but roughly speaking, the idea is that the relative errors in these variables decrease as the number of steps increases, so that after a large constant number of steps they are essentially global.

Finally, we formulate a very general condition under which we have some control on a variable. Suppose Γ is a graph, J is a spanning subgraph of Γ and $A \subseteq V_\Gamma$. We refer to (A, J, Γ) as an *extension*. Suppose that $\phi : A \to [n]$ is an injective mapping. We define the *extension variables* $X_{\phi,J,\Gamma}(i)$ to be the number of injective maps $f : V_\Gamma \to [n]$ such that f restricts to ϕ on A, $f(e) \in E(i)$ for every $e \in E_J$ not contained in A, and $f(e) \in O(i)$ for every $e \in E_\Gamma \setminus E_J$ not contained in A. We introduce the abbreviations $V = X_{\phi,J,\Gamma}$, $n(V) = |V_\Gamma| - |A|$, $e(V) = e_J - e_{J[A]}$, and $o(V) = (e_\Gamma - e_J) - (e_{\Gamma[A]} - e_{J[A]})$. The *scaling* is $v = x_{A,J,\Gamma} = n^{n(V)} p^{e(V)} \hat{q}^{o(V)}$. We expect $V \approx v$, provided there is no subextension that is 'sparse', in that it has scaling much smaller than 1. Given $A \subseteq B \subseteq B' \subseteq V_\Gamma$ we define $S_B^{B'} = S_B^{B'}(J, \Gamma)$ to equal

$$n^{|B'|-|B|} p^{e_{J[B']}-e_{J[B]}} \hat{q}^{(e_{\Gamma[B']}-e_{J[B']})-(e_{\Gamma[B]}-e_{J[B]})}.$$

Let $t' \geq 1$. We say that V is *controllable at time* t' if $J \neq \Gamma$ (*i.e.* at least one pair is open) and for $1 \leq t \leq t'$ and $A \subsetneq B \subseteq V_\Gamma$ we have $S_A^B(J, \Gamma) \geq n^\delta$, where $\delta > 0$ is a fixed global parameter that is sufficiently small given ε. (This condition is essentially identical to the condition needed to prove concentration of subgraphs counts in $G_{n,p}$ using Kim-Vu polynomial concentration [14].)

3 Concluding remarks

We have determined $R(3, t)$ to within a factor of $4 + o(1)$, so we should perhaps hazard a guess for its asymptotics: we are tempted to believe the construction rather than the bound, *i.e.* that $R(3, t) \sim t^2/4 \log t$. We should note that we only have an upper bound on the independence number of the graph G produced by the triangle-free process. So, formally speaking, the triangle-free process could produce a graph that gives a better lower bound on $R(3, t)$. But we believe that this is not the case; that is, we conjecture that the bound on the independence number in Theorem 1.2 is asymptotically best possible.

Our method for establishing self-correction builds on ideas used recently by Bohman, Frieze and Lubetzky [5] for an analysis of the triangle-removal process (see also [4] for a simpler context). Furthermore, the results of this paper have also been obtained independently and simultaneously by Fiz Pontiveros, Griffiths and Morris; their proof also exploits self-correction, but is different to ours in some important ways.

Another natural direction for future research is to provide an asymptotically optimal analysis in greater generality for the H-free process. No doubt the technical challenges will be formidable, given the difficulties that arise in the case of triangles. But on an optimistic note, it is encouraging that one can build on two different proofs of this case.

References

[1] M. AJTAI, J. KOMLÓS and E. SZEMERÉDI, *A note on Ramsey numbers*, J. Combin. Theory Ser. A **29** (1980), 354–360.

[2] N. ALON and J. SPENCER, "The Probabilistic Method", second edition, Wiley, New York, 2000.

[3] T. BOHMAN, *The triangle-free process*, Adv. Math. **221** (2009), 1653–1677.

[4] T. BOHMAN, A. FRIEZE and E. LUBETZKY, *A note on the random greedy triangle-packing algorithm*, J. Combinatorics **1** (2010), 477–488.

[5] T. BOHMAN, A. FRIEZE and E. LUBETZKY, *Random triangle removal*, arXiv:1203.4223.

[6] T. BOHMAN and P. KEEVASH, *The early evolution of the H-free process*, Invent. Math. **181** (2010), 291–336.

[7] B. BOLLOBÁS and O. RIORDAN, *Random graphs and branching processes*, In: "Handbook of Large-scale Random Networks", Bolyai Soc. Math. Stud. **18**, Springer, Berlin, 2009, 15–115.

[8] B. BOLLOBÁS and O. RIORDAN, *Constrained graph processes*, Electronic J. Combin. **7** (2000), R18.

[9] P. ERDŐS, *Graph theory and probability, II*, Canad. J. Math. **13** (1961), 346–352.

[10] P. ERDŐS, S. SUEN and P. WINKLER, *On the size of a random maximal graph*, Random Structures Algorithms **6** (1995), 309–318.

[11] D. A. FREEDMAN, *On tail probabilities for martingales*, Ann. Probability **3** (1975), 100–118.

[12] S. GERKE and T. MAKAI, *No dense subgraphs appear in the triangle-free graph process*, Electron. J. Combin. **18** (2011), R168.

[13] J. H. KIM, *The Ramsey number $R(3,t)$ has order of magnitude $t^2/\log t$*, Random Structures Algorithms **7** (1995), 173–207.

[14] J. H. KIM and V. H. VU, *Concentration of multivariate polynomials and its applications*, Combinatorica **20** (2000) 417–434.

[15] D. OSTHUS and A. TARAZ, *Random maximal H-free graphs*, Random Structures Algorithms **18** (2001), 61–82.

[16] M. PICOLLELLI, *The final size of the C_4-free process*, Combin. Probab. Comput. **20** (2011), 939–955.

[17] A. RUCIŃSKI and N. WORMALD, *Random graph processes with degree restrictions*, Combin. Probab. Comput. **1** (1992), 169–180.

[18] J. SHEARER, *A note on the independence number of triangle-free graphs*, Disc. Math. **46** (1983), 83–87.

[19] J. SPENCER, *Maximal trianglefree graphs and Ramsey $R(3, k)$*, unpublished manuscript.

[20] J. SPENCER, *Asymptotic lower bounds for Ramsey functions*, Disc. Math. **20** (1997), 69–76.

[21] J. SPENCER, *Counting extensions*, J. Combin. Theory Ser. A **55** (1990), 247–255.

[22] L. WARNKE, *Dense subgraphs in the H-free process*, Disc. Math. **333** (2011), 2703–2707.

[23] L. WARNKE, *When does the K_4-free process stop?* Random Structures Algorithms, to appear.

[24] G. WOLFOVITZ, *Lower bounds for the size of random maximal H-free graphs*, Electronic J. Combin. **16**, 2009, R4.

[25] G. WOLFOVITZ, *Triangle-free subgraphs in the triangle-free process*, Random Structures Algorithms **39** (2011), 539–543.

Subcubic triangle-free graphs have fractional chromatic number at most 14/5

Zdeněk Dvořák[1], Jean-Sébastien Sereni[2] and Jan Volec[3]

Abstract. We show that every subcubic triangle-free graph has fractional chromatic number at most 14/5, thus confirming a conjecture of Heckman and Thomas [A new proof of the independence ratio of triangle-free cubic graphs. Discrete Math. 233 (2001), 233–237].

1 Introduction

One of the most celebrated results in Graph Theory is the Four-Color Theorem (4CT), which states that every planar graph is 4-colorable. It was proved by Appel and Hacken [3,4] in 1977 and, about twenty years later, Robertson, Sanders, Seymour and Thomas [17] found a new (and much simpler) proof. However, both of the proofs require a computer assistance, and finding a fully human-checkable proof is still one of the main open problems in Graph Theory. An immediate corollary of the 4CT implies that every n-vertex planar graph contains an independent set of size $n/4$ (this statement is sometimes called the Erdős-Vizing conjecture). Although this seems to be an easier problem than the 4CT itself, no proof without the 4CT is known. The best known result that does not use the 4CT is due to Albertson [1], who showed the existence of an independent set of size $2n/9$.

An intermediate step between the 4CT and the Erdős-Vizing conjecture is the fractional version of the 4CT — every planar graph is fractionally 4-

[1] Computer Science Institute of Charles University, Prague, Czech Republic. E-mail: rakdver@iuuk.mff.cuni.cz Supported by the Center of Excellence – Inst. for Theor. Comp. Sci., Prague, project P202/12/G061 of Czech Science Foundation.

[2] Centre National de la Recherche Scientifique (LORIA), Nancy, France.
E-mail: sereni@kam.mff.cuni.cz. This author's work was partially supported by the French *Agence Nationale de la Recherche* under reference ANR 10 JCJC 0204 01.

[3] Mathematics Institute and DIMAP, University of Warwick, Coventry CV4 7AL, UK. E-mail: honza@ucw.cz. This author's work was supported by a grant of the French Government.

This research was supported by LEA STRUCO and the Czech-French bilateral project MEB 021115 (French reference PHC Barrande 24444XD).

colorable. In fact, fractional colorings were introduced in 1973 [12] as an approach for either disproving, or giving more evidence to the 4CT. For a real number k, a graph G is fractionally k-colorable, if for every assignment of weights to its vertices there is an independent set that contains at least $(1/k)$-fraction of the total weight. In particular, every fractionally k-colorable graph on n vertices contains an independent set of size at least n/k. The existence of independent sets of certain ratios in *subcubic* graphs, *i.e.*, graphs with maximum degree at most 3, led Heckman and Thomas to pose the following two conjectures.

Conjecture 1.1 (Heckman and Thomas [10]). Every subcubic triangle-free graph is fractionally 14/5-colorable.

Conjecture 1.2 (Heckman and Thomas [11]). Every subcubic triangle-free planar graph is fractionally 8/3-colorable.

Note that a graph is called *triangle-free* if it does not contain a triangle as a subgraph. Here we want to announce the confirmation of Conjecture 1.1. The manuscript of our result is available on arXiv [5].

Unlike for general planar graphs, colorings of triangle-free planar graphs are well understood. Already in 1959, Grötzsch [8] proved that every triangle-free planar graph is 3-colorable. Therefore, such a graph on n vertices has to contain an independent set of size $n/3$. In 1976, Albertson, Bollobás and Tucker [2] conjectured that a triangle-free planar graph also has to contain an independent set of size strictly larger than $n/3$.

Their conjecture was confirmed in 1993 by Steinberg and Tovey [19], even in a stronger sense: such a graph admits a 3-coloring where at least $\lfloor n/3 \rfloor + 1$ vertices have the same color. On the other hand, Jones [13] found an infinite family of triangle-free planar graphs with maximum degree four and no independent set of size $\lfloor n/3 \rfloor + 2$. However, if the maximum degree is at most three, then Albertson *et al.* [2] conjectured that an independent set of size much larger than $n/3$ exists. Specifically, they asked whether there is a constant $s \in \left(\frac{1}{3}, \frac{3}{8}\right]$, such that every subcubic triangle-free planar graph contains an independent set of size sn. We note that for $s > 3/8$ the statement would not be true.

The strongest possible variant of this conjecture, *i.e.*, for $s = 3/8$, was finally confirmed by Heckman and Thomas [11]. However, for $s = 5/14$, it was implied by a much earlier result of Staton [18], who actually showed that every subcubic triangle-free (but not necessarily planar) graph contains an independent set of size $5n/14$. Jones [14] then found a simpler proof of this result; an even simpler one is due to Heckman and Thomas [10]. On the other hand, Fajtlowicz [6] observed that one cannot prove anything larger than $5n/14$. As we already mentioned, our main result is the strengthening of Staton's theorem to the fractional (weighted) version.

This conjecture attracted a considerable amount of attention and it spawned a number of interesting works in the last few years. In 2009, Hatami and Zhu [9] showed that for every graph that satisfies the assumptions of Conjecture 1.1, the fractional chromatic number is at most $3 - 3/64 \approx 2.953$. (The fractional chromatic number of a graph is the smallest number k such that the graph is fractionally k-colorable.) The result of Hatami and Zhu is the first to establish that the fractional chromatic number of every subcubic triangle-free graph is smaller than 3. In 2012, Lu and Peng [16] improved the bound to $3 - 3/43 \approx 2.930$. There are also two very recent improvements on the upper bound. The first one is due to Ferguson, Kaiser and Král' [7], who showed that the fractional chromatic number is at most $32/11 \approx 2.909$. The other one is due to Liu [15], who improved the upper bound to $43/15 \approx 2.867$.

2 Main result

We start with another definition of a fractional coloring that will be used to state our main result. It is equivalent to the one mentioned in the previous section by Linear Programming Duality.

Let G be a graph. A *fractional k-coloring* is an assignment of measurable subsets of the interval $[0, 1]$ to the vertices of G such that each vertex is assigned a subset of measure $1/k$ and the subsets assigned to adjacent vertices are disjoint. The *fractional chromatic number of* G is the infimum over all positive real numbers k such that G admits a fractional k-coloring. Note that for finite graphs, such a real k always exists, the infimum is in fact a minimum, and its value is always rational. We let $\chi_f(G)$ be this minimum.

A *demand function* f is a function from $V(G)$ to $[0, 1]$ with rational values. Let μ be the Lebesgue measure on real numbers. An *f-coloring* of G is an assignment φ of measurable subsets of $[0, 1]$ to the vertices of G such that $\mu(\varphi(v)) \geq f(v)$ for every $v \in V(G)$ and such that $\varphi(u) \cap \varphi(v) = \varnothing$ whenever u and v are two adjacent vertices of G. Note that for a rational number r, the graph G has fractional chromatic number at most r if and only if it has an f_r-coloring for the function f_r that assigns $1/r$ to every vertex of G.

A graph H is *dangerous* if H is either a 5-cycle or the graph K_4' obtained from K_4 by subdividing both edges of its perfect matching twice, see Figure 2.1. The vertices of degree two of a dangerous graph are called *special*. Let G be a subcubic graph and let B be a subset of its vertices. Let H be a dangerous induced subgraph of G. A special vertex v of H

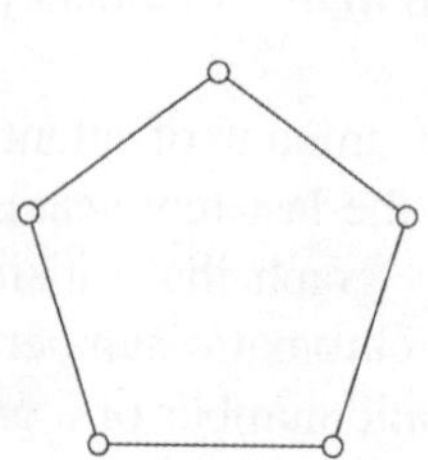
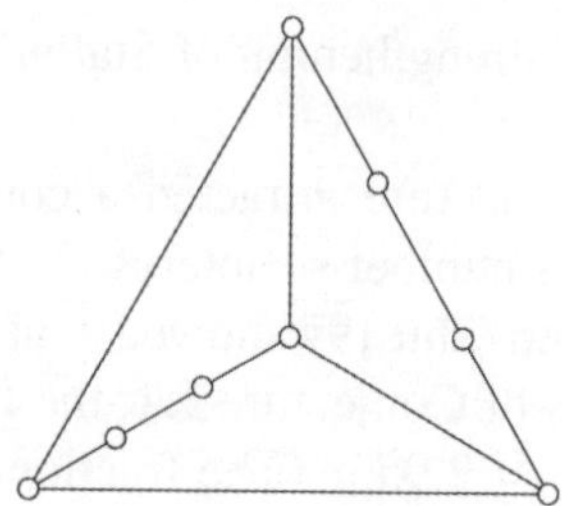

Figure 2.1. Dangerous graphs.

is B-*safe* if either $v \in B$ or v has degree three in G. If G is a subcubic graph, a set $B \subseteq V(G)$ is called a *nail* if every vertex in B has degree at most two and every dangerous induced subgraph of G contains at least two B-safe special vertices. For a subcubic graph G and its nail B, let f_B^G be the demand function defined as follows: if $v \in B$, then $f_B^G(v) = (7 - \deg_G(v))/14$; otherwise $f_B^G(v) = (8 - \deg_G(v))/14$.

In order to show that every subcubic triangle-free graph has fractional chromatic number at most $14/5$, we show the following stronger statement.

Theorem 2.1. *If G is a subcubic triangle-free graph and $B \subseteq V(G)$ is a nail, then G has an f_B^G-coloring.*

We point out that the motivation for the formulation of Theorem 2.1 as well as for some parts of its proof comes from the work of Heckman and Thomas [10], in which an analogous strengthening is used to prove the existence of an independent set of size $5n/14$.

3 Conclusion

In order to prove Theorem 2.1, we used several equivalent definitions of fractional colorings. As a consequence, unlike the result of Heckman and Thomas [10], our proof is not constructive and the following question remains open.

Problem 3.1. Does there exist a polynomial-time algorithm to find a fractional $14/5$-coloring of a given input subcubic triangle-free graph?

References

[1] M. ALBERTSON, *A lower bound for the independence number of a planar graph*, J. Combin. Theory Ser. B **20** (1976), 84–93.

[2] M. ALBERTSON, B. BOLLOBÁS and S. TUCKER, *The independence ratio and maximum degree of a graph*, In: "Proceedings of the

Seventh Southeastern Conference on Combinatorics", Graph Theory, and Computing (1976), Utilitas Math., Congressus Numerantium, No. XVII, 43–50.

[3] K. APPEL and W. HAKEN, *Every planar map is four colorable. I. Discharging*, Illinois J. Math. **21** (1977), 429–490.

[4] K. APPEL, W. HAKEN and J. KOCH, *Every planar map is four colorable. II. Reducibility*, Illinois J. Math. **21** (1977), 491–567.

[5] Z. DVOŘÁK, J.-S. SERENI and J. VOLEC, *Subcubic triangle-free graphs have fractional chromatic number at most* 14/5, submitted for publication, arXiv:1301.5296.

[6] S. FAJTLOWICZ, *On the size of independent sets in graphs*, Congr. Numer. **21** (1978), 269–274.

[7] D. FERGUSON, T. KAISER and D. KRÁL', *The fractional chromatic number of triangle-free subcubic graphs*, to appear in European Journal of Combinatorics.

[8] H. GRÖTZSCH, *Zur Theorie der diskreten Gebilde. VII. Ein Dreifarbensatz für dreikreisfreie Netze auf der Kugel*, Wiss. Z. Martin-Luther-Univ. Halle-Wittenberg. Math.-Nat. Reihe **8** (1958/1959), 109–120.

[9] H. HATAMI and X. ZHU, *The fractional chromatic number of graphs of maximum degree at most three*, SIAM J. Discrete Math. **23** (2009), 1162–1175.

[10] C. C. HECKMAN and R. THOMAS, *A new proof of the independence ratio of triangle-free cubic graphs*, Discrete Math. **233** (2001), 233–237.

[11] C. C. HECKMAN and R. THOMAS, *Independent sets in triangle-free cubic planar graphs*, J. Combin. Theory Ser. B **96** (2006), 253–275.

[12] A. J. W. HILTON, R. RADO and S. H. SCOTT, *A ($<$ 5)-colour theorem for planar graphs*, Bull. London Math. Soc. **5** (1973), 302–306.

[13] K. F. JONES, *Independence in graphs with maximum degree four*, J. Combin. Theory Ser. B **37** (1984), 254–269.

[14] K. F. JONES, *Size and independence in triangle-free graphs with maximum degree three*, J. Graph Theory **14** (1990), 525–535.

[15] C.-H. LIU, *An upper bound on the fractional chromatic number of triangle-free subcubic graphs*, submitted for publication.

[16] L. LU and X. PENG, *The fractional chromatic number of triangle-free graphs with* $\Delta \leq 3$, Discrete Math. **312** (2012), 3502–3516.

[17] N. ROBERTSON, D. P. SANDERS, P. SEYMOUR and R. THOMAS, *The four-colour theorem*, J. Combin. Theory Ser. B **70** (1997), 2–44.

[18] W. STATON, *Some Ramsey-type numbers and the independence ratio*, Trans. Amer. Math. Soc. **256** (1979), 353–370.

[19] R. STEINBERG and C. TOVEY, *Planar Ramsey numbers*, J. Combin. Theory Ser. B **59** (1993), 288–296.

Henneberg steps for triangle representations

Nieke Aerts[1] and Stefan Felsner[2]

Abstract. Which plane graphs admit a straight line representation such that all faces have the shape of a triangle? In previous work we have studied necessary and sufficient conditions based on flat angle assignments, *i.e.*, selections of angles of the graph that have size π in the representation. A flat angle assignment that fullfills these conditions is called good. The complexity for checking whether a graph has a good flat angle assignment remains unknown.

In this paper we deal with extensions of good flat angle assignments. We show that if G has a good flat angle assignment and G^+ is obtained via a planar Henneberg step of type 2, then G^+ also admits a good flat angle assignment. A similar result holds for certain combinations of Henneberg type 1 steps followed by a type 2 step. As a consequence we obtain a large class of pseudo-triangulations that admit drawings such that all faces have the shape of a triangle. In particular, every 3-connected, plane generic circuit admits a good flat angle assignment.

1 Introduction

In this paper we study a representation of planar graphs in the classical setting, *i.e.*, vertices are presented as points in the Euclidean plane and edges as straight line segments. We are interested in the class of planar graphs that admit a representation in which all faces are triangles. Note that in such a representation each face f has exactly $\deg(f) - 3$ incident vertices that have an angle of size π in f. Conversely each vertex has at most one angle of size π. In [2] we have studied necessary and sufficient conditions based on flat angle assignments, *i.e.*, selections of angles of the graph that have size π in the representation. Flat angle assignments that fullfill these conditions are called *good*. The complexity for checking whether a graph has a good flat angle assignment remains unknown.

A *pseudo-triangle* is a simple polygon with precisely three convex angles, all other vertices of the polygon admit a concave angle at the inte-

[1] Institut für Mathematik, Technische Universität Berlin, Germany. Email: aerts@math.tu-berlin.de

[2] Institut für Mathematik, Technische Universität Berlin, Germany.
Email: felsner@math.tu-berlin.de

The full version of this paper can be found online [1].
Partially supported by DFG grant FE-340/7-2 and ESF EuroGIGA project GraDR.

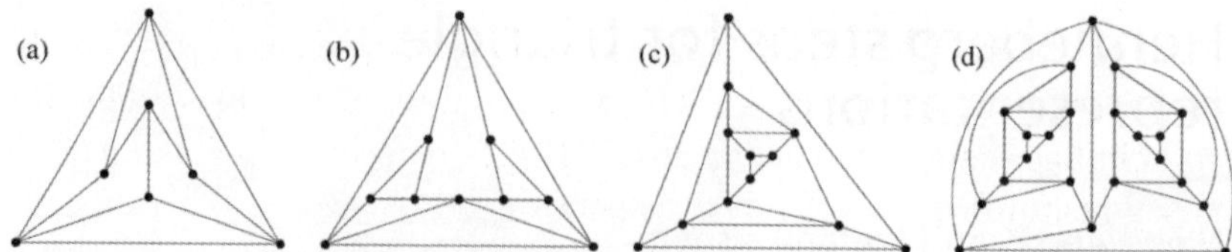

Figure 1.1. (a) A pseudo-triangulation that does not induce an SLTR, (b) a Laman graph that has an SLTR but can not be constructed using only the two steps we give in Section 2, (c) a planar Laman graph that has no SLTR for this embedding, (d) a planar Laman graph that has no SLTR.

rior of the polygon. A *pseudo-triangulation* (PT) is a planar graph with a drawing such that all faces are pseudo-triangles. An example of a PT is given in Figure 1.1 (a).

A pseudo-triangulation is *pointed* if each vertex has an angle of size $> \pi$. A pointed pseudotriangulation with n vertices must have exactly $2n - 3$ edges. Indeed pointed pseudotriangulations have the Laman property: they have $2n - 3$ edges, and subgraphs induced by k vertices have at most $2k - 3$ edges. Laman graphs, and hence also pointed pseudotriangulations, are minimally rigid graphs. A detailed survey on pseudotriangulations has been given by Rote et al. [8].

Pseudotriangulations induce an assignment of big angles to vertices. This assignment is closely related to a flat angle assignment.

A **Straight Line Triangle Representation** (SLTR) of a graph G is a plane drawing of G such that all edges are straight line segments and all faces are triangles (*e.g.* Figure 1.1 (b)). Throughout this paper $G = (V, E)$ will be a plane, internally 3-connected graph. Three vertices which are the corners of the outer face in an SLT Representation of the graph are given and we call these vertices suspensions. A plane graph G with suspensions s_1, s_2, s_3 is said to be *internally 3-connected* when the addition of a new vertex v_∞ in the outer face, that is made adjacent to the three suspension vertices, yields a 3-connected plane graph.

A *flat angle assignment* (FAA) of a graph is a mapping from a subset U of the non-suspension vertices to faces such that, the vertex is incident to the face and,

[C_v] Every vertex of U is assigned to at most one face,

[C_f] For every face f, precisely $|f| - 3$ vertices are assigned to f.

An FAA is called *good* (GFAA) when it induces an SLTR. In [2] we have shown that an FAA is good if and only if it induces a contact family of pseudosegments Σ which has the following property:

[C_P] Every subset S of Σ with $|S| \geq 2$ has at least three free points.

Informally, pseudosegments arise from merging the edges that are incident to an assigned angle of a vertex, the vertex will be an interior point of the pseudosegment. Since a vertex is assigned at most once, the pseudosegments do not cross. The pseudosegments will be stretched to straight line segments to obtain an SLTR. Let p be an endpoint of a pseudosegment in $S \subseteq \Sigma$. If p is a suspension vertex, then p is a free point for S. When p is not a suspension then it is a free point for S if: p is incident to the unbounded region of S, it has at least one neighbor not in S and it is not an interior point for a pseudosegment in S.

The drawback of this characterization is that we are not aware of an efficient way to test whether a given graph has an FAA that is good.

A *combinatorial pseudo-triangulation* (CPT) is an assignment of the labels *big* and *small* to the angles around each vertex. Each vertex has at most one angle labeled *big* and each inner face has precisely three incident angles labeled *small*, the outer face has all angles labeled *big*. For an interior angle labeled *big*, let the incident vertex be assigned to the incident face, and a vertex is not assigned if it has no angle labeled *big*. Three vertices of the outer face are chosen to be the suspensions, the other vertices are assigned to the outer face. Hence a CPT induces an FAA and the similarly an FAA induces a CPT.

A CPT does not always induce a PT, Orden et al. have shown that the *generalized Laman condition* is necessary and sufficient for a CPT to induce a PT [7].

Lemma 1.1 (Generalized Laman Condition). *Let G be the graph of a pseudo-triangulation of a planar point set in general position. Every subset of x not assigned vertices plus y assigned vertices of G, with $x + y \geq 2$ spans a subgraph with at most $3x + 2y - 3$ edges.*

Proposition 1.2. *A GFAA of an internally 3-connected plane graph satisfies the generalized Laman Condition.*

An FAA that is *not* good may also satisfy the generalized Laman Condition (*e.g.* Figure 1.1 (a)), therefore this condition is necessary but not sufficient. Every Laman graph can be constructed from an edge by Henneberg steps [6,9]. A graph $G = (V, E)$, with $|E| = 2|V| - 3$ that has an SLTR, has a CPT that induces a PT by Prop. 1.2 and by the result of Haas et al. it must be a Laman graph. Therefore it must have a Henneberg construction. In the next section we will investigate how to use this construction such that a GFAA can be extended along the steps.

2 Construction steps

It has been shown that planar Laman graphs admit a planar Henneberg construction [5]. Since we consider plane graphs (with a given set of suspension vertices), we consider the Henneberg steps in a plane setting. Recall that the graph must be internally 3-connected.

Henneberg type 2 step
Given a graph G and a GFAA ψ of G. A Henneberg Type 2 step (HEN$_2$) subdivides an edge uv and connects the new vertex x to a third vertex w (see Figure 2.1). The face f, incident to uv and w is splitted into f_u (the face incident to u) and f_v. The other face incident to uv is denoted with f_x. The resulting graph is denoted G^+. We will construct an assignment ψ^+ for G^+ and proof that ψ^+ is a GFAA.

There are three vertices not assigned to f under ψ, we will call them *corners* of f. We consider two cases, firstly f_u is incident to all corners of f, secondly, f_u is incident to precisely two corners of f. Note that if w is a corner of f it will be a corner for both f_u and f_v. The vertices different from u, v, w, x, that are assigned to f under ψ, will be assigned in the trivial way under ψ^+, *i.e.*, such a vertex is assigned to f_u resp. f_v, if in G^+ it is incident to f_u resp. f_v.

Case 1: *f_u is incident to all corners of f.* If u or w is assigned to f under ψ, it is assigned to f_u under ψ^+. The vertex v is assigned to f_x and x to f_u under ψ^+.

Case 2: *f_u is incident to precisely two corners of f.* If u or w is assigned to f under ψ, it is assigned to f_u under ψ^+, if v was assigned to f it is assigned to f_v under ψ^+ and x is assigned to f_x.

This yields an assignment ψ^+ for G^+.

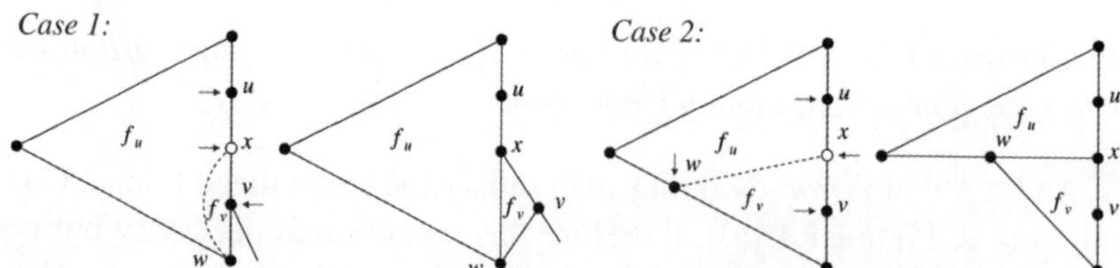

Figure 2.1. Updating the assignment after a HEN$_2$ step. The white vertex and dashed edge represent the step.

Theorem 2.1. *Given a 3-connected, plane graph G with a GFAA ψ. Let G^+ be the result of a* HEN$_2$ *step applied to G and let ψ^+ be the updated assignment. Then ψ^+ is a GFAA and G^+ admits an SLTR.*

Proof. It is easy to see that ψ^+ satisfies C_v and C_f and hence is an FAA.

We consider the induced families of pseudosegments, Σ and Σ^+ of ψ and ψ^+ respectively. Since ψ is a Good FAA, we know that every subset

S of Σ has at least three free points or S has cardinality at most one. Let $S \subseteq \Sigma^+$ of cardinality at least two, and for every pseudosegment p of Σ^+ which is not a pseudosegment of Σ consider $p \in S$ and show that S has at least three free points by using that there is an equivalent set under Σ which has at least three free points.

In both cases there are three pseudosegments that have changed, we will only discuss Case 1 here. Let s_x resp. s_v be the pseudosegment that has x resp. v as interior point and let s_w the pseudosegment containing the edge vw.

- If $s_x \in S$ then replace s_x by the pseudosegment s'_x of Σ that has u and v as interior points.
- If $s_v \in S$ then replace s_v by the pseudosegment s'_v of Σ that ends in v and contains all the edges, except vx, of s_v.
- If $s_w \in S$ then delete s_w.

Now we have a set $S' \in \Sigma$, thus S' has three free points unless $|S'| = 1$.

- If $s_x \in S$ then s_x contributes the same free points to S as s'_x to S'.
- If $s_v \in S$ then if v was a free point for S', x is for S. Hence s_v contributes the same number of free points to S as s'_v to S'.
- If $s_w \in S$ then if $|S'| = 1$, s_w contributes at least one free point to S and it covers no other points, thus S has three free points, or, if $|S'| > 1$ then S' has at least three free points, adding s_w does not cover any of them and therefore S has at least three free points.

We conclude that in Case 1 every set S of cardinality at least two has at least three free points. The argumentation for Case 2 is similar and it follows that ψ^+ is a GFAA. □

A graph $G = (V, E)$ is a *generic circuit* if $|E| = 2|V| - 2$ and subgraphs induced by k vertices have at most $2k - 3$ edges. The generic circuit with the least number of vertices is K_4.

Theorem 2.2. *Every 3-connected, plane, generic circuit admits an SLTR.*

Proof. A 3-connected, generic circuit can be constructed with HEN_2 steps from K_4 (Berg and Jordán [3]) and K_4 admits an SLTR. Every plane 3-connected generic circuit can be constructed with HEN_2 steps from K_4 such that all intermediate graphs are plane. By Theorem 2.1 we have that every 3-connected, plane generic circuit admits an SLTR. □

Henneberg combination step

For connectivity reasons single HEN_1 steps are not compatible with SLTRs. However certain sequences of HEN_1 steps followed by a HEN_2 step allow for the extension of a GFAA. Next we will describe such a combination step denoted HEN_{1^n2}.

Let f be a face with $n+1$ vertices. The first HEN_1 step stacks a new vertex v_0 over an edge of f. The vertices $v_0, \dots, v_{n-1}$ are introduced by the n HEN_1 steps in such a way that $v_0, \dots, v_{n-1}$ is a path and each vertex of f is a neighbor of some v_i. The final HEN_2 step subdivides an edge which is incident to some v_i (not v_{n-1}) and connects the new vertex v_n to v_{n-1}. Of course the construction has to maintain planarity.

Theorem 2.3. *Given a 3-connected, plane graph G with a GFAA ψ. Let G_n be the result of a HEN_{1^n2} step applied to G and ψ_n be the updated assignment. Then ψ_n is a GFAA and G_n admits an SLTR.*

Due to the lack of space we have left out the algorithm that decides how to update the assignment, this and the proof of Theorem 2.3 can be found in the full version.

3 Conclusion and open problems

We have given two construction steps such that a GFAA can be extended along these steps and the extended assignment is also a GFAA. However, this does not define the class of Laman graphs that have an SLTR. Therefore the problem: Is the recognition of graphs that have an SLTR (GFAA) in P? is still open, even for graphs in which all non-suspension vertices have to be assigned.

The class of 3-connected quadrangulations is well-defined, *e.g.* Brinkmann et al. give a characterization using two expansion steps [4]. Adding a diagonal edge in the outer face of a plane, 3-connected quadrangulation yields a Laman graph. One of the expansion steps (denoted P_3 in [4]) is a Henneberg Combination step, hence a GFAA can be extended along this step. It would be interesting to know if a GFAA could also be extended along the other expansion step (denoted P_1 in [4]).

Adding an edge in a plane graph that has a GFAA requires only minor changes to the GFAA of the original graph to obtain a GFAA for the resulting graph. An interesting question arises: Does every graph that admits an SLTR in which not every non-suspension vertex admits a straight angle, have a spanning Laman subgraph that admits an SLTR?

References

[1] N. AERTS and S. FELSNER, *Henneberg Steps for Triangle Representations*. http://page.math.tu-berlin.de/ aerts/pubs/ptsltr.pdf.

[2] N. AERTS and S. FELSNER, *Straight Line Triangle Representations*. http://page.math.tu-berlin.de/ aerts/pubs/sltr.pdf.

[3] A. R. BERG and T. JORDÁN, *A proof of Connelly's conjecture on 3-connected circuits of the rigidity matroid*, J. Comb. Theory, Ser. B **88** (2003), 77–97.

[4] G. BRINKMANN, S. GREENBERG, C. S. GREENHILL, B. D. MCKAY, R. THOMAS, and P. WOLLAN, *Generation of simple quadrangulations of the sphere*, Discrete Math. **305** (2005), 33–54.

[5] R. HAAS, D. ORDEN, G. ROTE, F. SANTOS, B. SERVATIUS, H. SERVATIUS, D. L. SOUVAINE, I. STREINU and W. WHITELEY, *Planar minimally rigid graphs and pseudo-triangulations*, Comput. Geom. **31** (1-2) (2005), 31–61.

[6] L. HENNEBERG, "Die Graphische Statik der Starren Systeme", Teubner, 1911, Johnson Reprint 1968.

[7] D. ORDEN, F. SANTOS, B. SERVATIUS and H. SERVATIUS, *Combinatorial pseudo-triangulations*, Discrete Math. **307** (1-3) (2007), 554–566.

[8] G. ROTE, F. SANTOS and I. STREINU, *Pseudo-triangulations – a survey*, In: "Surveys on Discrete and Computational Geometry—Twenty Years Later", E. Goodman, J. Pach and R. Pollack (eds.), Contemporary Mathematics, American Mathematical Society, December 2008.

[9] W. WHITELEY, *Some Matroids from Discrete Applied Geometry*, AMS Contemporary Mathematics (1997), 171–313.

Cycles and Girth of Graphs

Cycle-continuous mappings – order structure

Robert Šámal[1]

Abstract. Given two graphs, a mapping between their edge-sets is *cycle-continuous*, if the preimage of every cycle is a cycle. Answering a question of DeVos, Nešetřil, and Raspaud, we prove that there exists an infinite set of graphs with no cycle-continuous mapping between them. Further extending this result, we show that every countable poset can be represented by graphs and existence of cycle-continuous mappings between them.

1 Introduction

Many questions at the core of graph theory can be formulated as questions about cycles or more generally about flows on graphs. Examples are the cycle double cover conjecture, the Berge-Fulkerson conjecture, and Tutte's 3-flow, 4-flow, and 5-flow conjectures. For a detailed treatment of this area the reader may refer to [7] or [8].

As an approach to these problems Jaeger [4] and DeVos, Nešetřil, and Raspaud [2] defined a notion of graph morphism continuous with respect to group-valued flows. In this paper we restrict ourselves to the case of $\mathbb{Z}_2$-flows, that is to cycles. Thus, the following is the principal notion we study in this paper:

Given graphs (parallel edges or loops allowed) G and H, a mapping $f : E(G) \to E(H)$ is called *cycle-continuous*, if for every cycle $C \subseteq E(H)$, the preimage $f^{-1}(C)$ is a cycle in G. We emphasize, that by a *cycle* we understand (as is common in this area) a set of edges such that every vertex is adjacent with an even number of them. For shortness we sometimes call cycle-continuous mappings just *cc* mappings.

The fact that f is a *cc* mapping from G to H is denoted by $f : G \xrightarrow{cc} H$. If we just need to say that there exists a *cc* mapping from G to H, we write $G \xrightarrow{cc} H$. With the definition covered, we mention the main conjecture describing the properties of *cc* mappings.

[1] Computer Science Institute, Charles University. Email: samal@iuuk.mff.cuni.cz. Supported by grant GA ČR P201/10/P337 and by grant LL1201 ERC CZ of the Czech Ministry of Education, Youth and Sports.

Conjecture 1.1 (Jaeger). For every bridgeless graph G we have $G \xrightarrow{cc}$ Pt, where Pt denotes the Petersen graph.

If true, this would imply many conjectures in the area. To illustrate this, suppose we want to find a 5-tuple of cycles in a graph G covering each of its edges twice (this is conjectured to exist by 5-Cycle double cover conjecture). Further, suppose $f : G \xrightarrow{cc} \mathrm{Pt}$. We can use $C_1, \ldots, C_5$ — a 5-tuple of cycles in the Petersen graph double-covering its edges— and then it is easy to check that $f^{-1}(C_1), \ldots, f^{-1}(C_5)$ have the same property in G.

DeVos *et al.* [2] study this notion further and ask the following question about the structure of cycle-continuous mappings. We say that graphs G, G' are cc-incomparable if there is no cc mapping between them, that is $G \overset{cc}{\not\to} G'$ and $G' \overset{cc}{\not\to} G$.

Question 1.2 ([2]). Is there an infinite set $\mathcal{G}$ of bridgeless graphs such that every two of them are cc-incomparable?

DeVos *et al.* [2] also show that arbitrary large sets of cc-incomparable graphs exist. Their proof is based on the notion of critical snarks and on Lemma 3.1; these will be crucial also for our proof. We will show, that the answer to Conjecture 1.2 is positive, even in a stronger form. Thus, the following are our main results.

Theorem 1.3. *There is an infinite set $\mathcal{G}$ of cubic bridgeless graphs such that every two of them are cc-incomparable.*

Theorem 1.4. *Every countable (finite or infinite) poset can be represented by a set of graphs and existence of cycle-continuous mappings between them.*

2 Properties of cycle-continuous mappings

2.1 Properties of a 2-join

In this and the next section we will describe two common constructions of *snarks* (*i.e.*, 3-regular bridgeless graphs, that are not 3-edge-colorable). While the constructions are known (see, *e.g.*, [8]), the relation to cycle-continuous mappings has not been investigated elsewhere, and is crucial to our result. The first construction can be informally described as adding a "gadget" on an edge of a graph. Formally, let G_1, G_2 be graphs, and let $e_i = x_i y_i$ be an edge of G_i. We delete edge e_i from G_i (for $i = 1, 2$), and connect the two graphs by adding two new edges $x_1 x_2$ and $y_1 y_2$. The resulting graph will be called the *2-join* of the graphs G_1, G_2 (some authors call this a 2-sum); it will be denoted by $G_1 \oplus_2 G_2$. We note that the resulting graph depends on our choice of the edges $x_i y_i$, but for our purposes this coarse description will suffice.

Lemma 2.1. *For every graphs G_1, G_2 we have $G_i \overset{cc}{\to} G_1 \oplus_2 G_2$ for $i \in \{1, 2\}$.*

Lemma 2.2. *Let G_1, G_2 be any graphs. Let K be an edge-transitive graph. Then $G_1 \oplus_2 G_2 \overset{cc}{\to} K$ if and only if $G_1 \overset{cc}{\to} K$ and $G_2 \overset{cc}{\to} K$.*

As an immediate corollary we get the following classical result about snarks and 2-joins:

Corollary 2.3. *If G_1, G_2 are bridgeless cubic. Then $G_1 \oplus_2 G_2$ is a snark whenever at least one of G_1, G_2 is a snark.*

Another easy corollary of Lemma 2.2 is that minimal counterexample (if it exists) to Conjecture 1.1 does not contain a nontrivial 2-edge-cut.

Corollary 2.4. *Let G_1, G_2 be cubic bridgeless graphs. If $G_1 \oplus_2 G_2 \overset{cc}{\not\to} \mathrm{Pt}$ then $G_i \overset{cc}{\not\to} \mathrm{Pt}$ for some $i \in \{1, 2\}$.*

2.2 Properties of a 3-join

A *3-join* (also called 3-sum) is a method to create new snarks – ones that contain nontrivial 3-edge cuts. One way to view this is that we replace a vertex in a graph by a "gadget" created from another graph. To be more precise, we consider graphs G_1 and G_2, delete a vertex u_i of each G_i, and add a matching between neighbors of former vertices u_1 and u_2. The resulting (cubic) graph in general depends on our choice of u_i's, and of the matching. We use $G_1 \oplus_3 G_2$ to denote (any of) the resulting graph(s); we call in the *3-join of G_1 and G_2*. Connecting edges of the 3-join are the three edges we added to connect G_1 and G_2.

Lemma 2.5. *For any graphs G_1, G_2 we have $G_i \overset{cc}{\to} G_1 \oplus_3 G_2$ for $i = 1, 2$. We shall call the cycle-continuous mapping from G_i to $G_1 \oplus_3 G_2$ a* natural inclusion.

Lemma 2.6. *Let G_1, G_2 be any graphs. Let K be a cyclically 4-edge-connected cubic graph that is 2-transitive. Then $G_1 \oplus_3 G_2 \overset{cc}{\to} K$ if and only if $G_1 \overset{cc}{\to} K$ and $G_2 \overset{cc}{\to} K$.*

As an immediate corollary we get the following classical result about snarks and 3-joins:

Corollary 2.7. *Let G_1, G_2 be cubic bridgeless graphs. Then $G_1 \oplus_3 G_2$ is a snark, iff at least one of G_1, G_2 is a snark.*

As another easy application, we observe that minimal counterexample (if it exists) to Conjecture 1.1 does not contain a nontrivial 3-edge-cut.

Corollary 2.8. *Let G_1, G_2 be cubic bridgeless graphs. If $G_1 \oplus_3 G_2 \overset{cc}{\not\to} \mathrm{Pt}$ then $G_i \overset{cc}{\not\to} \mathrm{Pt}$ for some $i \in \{1, 2\}$.*

The above notwithstanding, we proceed to study the structure of cycle-continuous mapping in graphs with 3-edge-cuts, for two reasons: first we believe, it provides insights that might be useful in further progress towards solving Conjecture 1.1, second, we find it has an independent interest.

Lemma 2.9. *Let G_1, G_2 be cc-incomparable snarks. Then $G_1 \oplus_3 G_2 \overset{cc}{\not\to} G_i$ for each $i \in \{1, 2\}$.*

3 The proof

3.1 Critical snarks

For our construction we will need the following notion of criticality of snarks. It appears in Nedĕla *et al.* [5] and in DeVos *et al.* [2]; see also [1], where these graphs are called flow-critical snarks.

Recall a graph G is a snark if $G \overset{cc}{\not\to} K_2^3$, where K_2^3 is a graph formed by two vertices and three parallel edges. We say G is a *critical snark* if for every edge e of G we have $G - e \overset{cc}{\to} K_2^3$. (Equivalently [1], $G/e \overset{cc}{\to} K_2^3$.)

The following lemma is the basis of our control over cycle-continuous mappings between graphs in our construction.

Lemma 3.1 ([2]). *Let G, H be cyclically 4-edge-connected cubic graphs, both of which are critical snarks, suppose that $|E(G)| = |E(H)|$. Then $G \overset{cc}{\to} H$ iff $G \cong H$. Moreover, every cycle-continuous mapping is a bijection that is induced by an isomorphism of G and H.*

Lemma 3.2. *There are two snarks B_1, B_2 with 18 vertices, that are critical and nonisomorphic. Moreover, none of B_1, B_2 is vertex transitive.*

Proof. It is well-known that the two Blanuša snarks on 18 vertices satisfy these requirements. □

3.2 Tree of snarks

Let $\mathcal{G} = \{G_1, \ldots, G_n\}$ be a family of critical snarks of the same size, so that for $i \neq j$ graphs G_i and G_j are not isomorphic (equivalently: $G_i \overset{cc}{\not\to} G_j$ and $G_j \overset{cc}{\not\to} G_i$).

Let T be a tree with a vertex coloring (not necessarily proper) $c : V(T) \to [n]$. We denote by $T(\mathcal{G})$ a family of graphs that can be obtained by replacing each $v \in V(T)$ by a copy of $G_{c(v)}$ and performing a 3-join on each edge; see Fig. 3.1 for an illustration. There are in general many graphs that can be constructed in this way, depending on which vertices one chooses for the 3-join operations.

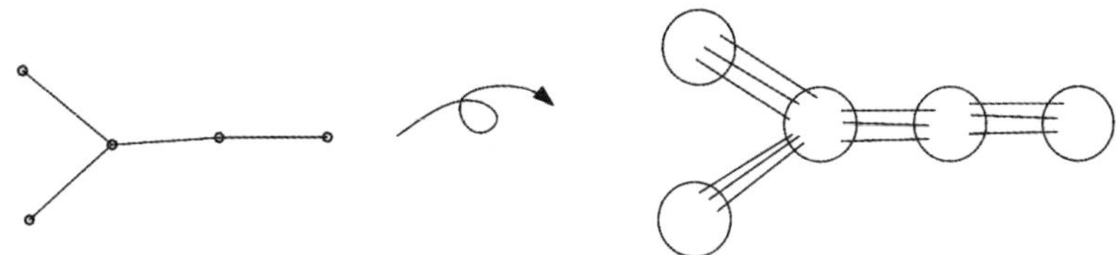

Figure 3.1. Illustration of the "tree-snark" construction.

More precisely, for each $v \in V(T)$ we fix a bijection r_v from $N_T(v)$ to an *independent set* A_v in $G_{c(v)}$, we also specify an ordering of edges going out of vertices of A_v. Next, we split each vertex w in A_v into three degree 1 vertices; these will be denoted by w_1, w_2, w_3. For each edge uv of T we identify vertices $r_u(v)_i$ with $r_v(u)_i$ for $i = 1, 2, 3$. Finally, we suppress all vertices of degree 2.

If H is a graph in $T(\mathcal{G})$ and v a vertex of T, we let H_v denote a "copy" of $G_{c(v)}$: subgraph of H consisting of a copy of $G_{c(v)} - A_v$ together with the incident edges and neighboring vertices in H. Further, we let ι_v denote the natural inclusion of $G_{c(v)}$ into H, which maps bijectively on H_v.

We define $\bar{H}_v$ to be the graph H with all edges outside of H_v contracted. In other words, $\bar{H}_v$ is truly an isomorphic copy of $G_{c(v)}$. Further, $H_{u,v}$ will denote the three edges in the intersection $H_u \cap H_v$.

The following theorem is the key to our construction.

Theorem 3.3. *Let T_1, T_2 be two trees and let $c_i : V(T_i) \to [n]$ be arbitrary colorings. Let $\mathcal{G}$ be as above.*

Suppose $H_i \in T_i(\mathcal{G})$ for $i = 1, 2$. Every cc mapping $g : H_1 \xrightarrow{cc} H_2$ is guided by a homomorphism $f : T_1 \to T_2$ of reflexive colored graphs: There is a mapping $f : V(T_1) \to V(T_2)$ such that

- *$c_2(f(v)) = c_1(v)$ (f respects colors), and*
- *if uv is an edge of T_1, then $f(u)f(v)$ is an edge of T_2 or $f(u) = f(v)$. In the first case, g maps $H_{u,v}$ to $H_{f(u),f(v)}$. In the second one, $H_{u,v}$ is mapped to some $H_{f(u),v'}$.*

Moreover, g induces a mapping $(\bar{H}_1)_v$ to $(\bar{H}_2)_{f(v)}$ that is cycle-continuous.

As a corollary we obtain our first result, that already answers Question 1.2.

Corollary 3.4. *There is an infinite set of cc-incomparable graphs.*

Proof. Let T_n be a path with vertices $\{0, 1, \dots, n\}$ colored as $1(2)^{n-1}1$. We let $\mathcal{G} = \{B_1, B_2\}$, where as in Lemma 3.2, B_i's denote the Blanuša snarks on 18 vertices. We fix vertices a, b of B_2 so that no automorphism of B_2 sends a to b. For all vertices $v \in V(T_n)$ of degree 2 we create r_v so, that $r_v(v-1) = a$ and $r_v(v+1) = b$. We do not specify A_0 nor A_n, neither the order of edges adjacent to a or b. We let H_n denote any of $T_n(\mathcal{G})$.

Consider H_m, H_n, suppose that $g : H_m \xrightarrow{cc} H_n$ is cc mapping. Let $f : V(T_m) \to V(T_n)$ be the mapping guaranteed by Theorem 3.3. As f respects colors, we have $\{f(0), f(m)\} \subseteq \{0, n\}$. Next, define $G_i = (H_m)_i$, and $G'_j = (H_n)_j$. By Theorem 3.3 again, g is cc mapping $G_i \xrightarrow{cc} G'_{f(i)}$. As G_i and $G'_{f(i)}$ are isomorphic to B_2, Lemma 3.2 implies that $f(i+1) = f(i)+1$. It follows that $m = n$, which finishes the proof. □

3.3 Representing posets by cycle-continuous mappings

Question 1.2 should be understood as a question about how complicated is the structure of cc mappings. Next, we provide even further indication, that the structure is complicated indeed.

Corollary 3.5. *Every countable (finite or infinite) poset can be represented by a set of graphs and existence of cc mappings between them.*

Proof. (Sketch) We use the result of Hubička and Nešetřil [3], claiming that arbitrary countable posets can be represented by finite directed paths and existence of homomorphisms between them.

Thus, we only need to find a mapping m that to directed paths assigns cubic bridgeless graphs, so that P_1 hom P_2 iff $m(P_1) \xrightarrow{cc} m(P_2)$. We choose vertices a, b in the Blanuša snark B_2 so that no automorphism of B_2 maps a to b. Then we replace each directed edge by a copy of B_2 "from a to b" and perform a 3-join operation in-between each pair of adjacent edges. Formally, let P be a path with edges (from one end to the other) $e_1, \dots, e_m$. We let $t(i)$ be the index of the edge at the tail of e_i – that is $t(i)$ is either $i-1$ (if e_i goes forward with respect to our labeling) or $i+1$. Note that $t(i)$ may be undefined for $i \in \{1, m\}$. Similarly, we define $h(i)$ to be the index of the edge adjacent to e_i at its head. We will use the construction from Section 3.2. Our tree T will be a path with vertices $1, \dots, m$ all colored by 1, our set of snarks will consist just of the second Blanuša snark, $\mathcal{G} = \{G_1 = B_2\}$. We define $r_i(t(i)) = a$, and $r_i(h(i)) = b$, whenever $t(i)$ ($h(i)$, resp.) are defined. We choose an ordering of edges going out of a, and b; we keep this fixed for all vertices of all paths. Then we let $m(P)$ be the graph in $T(\mathcal{G})$ determined by the above described choices. (The rest of the proof is omitted in the extended abstract.) □

4 Concluding remarks

While being a resolution to Question 1.2, none of the family of examples we gave does violate Conjecture 1.1:

Theorem 4.1. *If $H \in T(\mathcal{G})$ and for every $G \in \mathcal{G}$ we have $G \stackrel{cc}{\to} \mathrm{Pt}$ then $H \stackrel{cc}{\to} \mathrm{Pt}$.*

Still, the presented results illustrate the complexity of *cc* mappings. To better understand their structure, we suggest the following questions:

Question 4.2. Does the poset of cubic cyclically 4-edge-connected graphs and *cc* mappings between them have infinite antichains? Does it contain every countable poset as a subposet? How about cyclically 5-edge-connected graphs?

For the next question, recall that in a poset $(X, \leq)$ an interval (a, b) is the set $\{x \in X : a < x < b\}$ (we must have $a < b$ for this definition to make sense, otherwise we call (a, b) degenerate interval).

Question 4.3. In the poset of graphs and *cc* mappings between them, is every non-degenerate interval nonempty? Does every non-degenerate interval contain infinite antichain? Does every non-degenerate interval contain every countable poset?

Note, that if Conjecture 1.1 is true, then (Pt, K_2) is an empty but non-degenerated interval. Is there some other?

We also briefly note the more general definition of flow-continuous mappings, that extends the notion of cycle-continuous mappings: a mapping $f : E(G) \to E(H)$ is called M-flow-continuous (for an abelian group M) if for every M-flow φ on H, the composition $\varphi \circ f$ is an M-flow on G. For detailed discussion, see [2] or [6]. We only mention here, that cycle-continuous mappings are exactly $\mathbb{Z}_2$-flow-continuous ones, and that Corollaries 3.4 and 3.5 extend trivially to $\mathbb{Z}$-flow-continuous mappings.

References

[1] C. Nunes da Silva and C. L. Lucchesi, *Flow-critical graphs*, The IV Latin-American Algorithms, Graphs, and Optimization Symposium, Electron. Notes Discrete Math., Vol. 30, Elsevier Sci. B. V., Amsterdam, 2008, pp. 165–170.

[2] M. DeVos, J. Nešetřil and A. Raspaud, *On edge-maps whose inverse preserves flows and tensions*, Graph Theory in Paris: "Proceedings of a Conference in Memory of Claude Berge", J. A. Bondy, J. Fonlupt, J.-L. Fouquet, J.-C. Fournier and J. L. Ramirez Alfonsin (eds.), Trends in Mathematics, Birkhäuser, 2006.

[3] J. Hubička and J. Nešetřil, *Finite paths are universal*, Order **22** (2005), no. 1, 21–40.

[4] F. Jaeger, *On graphic-minimal spaces*, Ann. Discrete Math. **8** (1980), 123–126, Combinatorics 79 (Proc. Colloq., Univ. Montréal, Montreal, Que., 1979), Part I.

[5] R. Nedela and M. Škoviera, *Decompositions and reductions of snarks*, Journal of Graph Theory **22** (3) (1996), 253–279.

[6] R. Šámal, "On XY mappings", Ph.D. thesis, Charles University, 2006.

[7] P. D. Seymour, *Nowhere-zero flows*, Handbook of combinatorics, Vol. 1, 2, Elsevier, Amsterdam, 1995, Appendix: Colouring, stable sets and perfect graphs, pp. 289–299.

[8] C.-Q. Zhang, "Integer Flows and Cycle Covers of Graphs", Monographs and Textbooks in Pure and Applied Mathematics, vol. 205, Marcel Dekker Inc., New York, 1997.

On the structure of graphs with given odd girth and large minimum degree

Silvia Messuti[1] and Mathias Schacht[1]

Abstract. We study the structure of graphs with high minimum degree conditions and given odd girth. For example, the classical work of Andrásfai, Erdős, and Sós implies that every n-vertex graph with odd girth $2k+1$ and minimum degree bigger than $\frac{2n}{2k+1}$ must be bipartite. We consider graphs with a weaker condition on the minimum degree. Generalizing results of Häggkvist and of Häggkvist and Jin for the cases $k = 2$ and 3, we show that every n-vertex graph with odd girth $2k + 1$ and minimum degree bigger than $\frac{3n}{4k}$ is homomorphic to the cycle of length $2k+1$. This is best possible in the sense that there are graphs with minimum degree $\frac{3n}{4k}$ and odd girth $2k + 1$ which are not homomorphic to the cycle of length $2k + 1$. Similar results were obtained by Brandt and Ribe-Baumann.

1 Introduction

We consider finite and simple graphs without loops and for any notation not defined here we refer to the standard textbooks. In particular, we denote by K_r the complete graph on r vertices, by C_r a cycle of length r, where the length of a cycle or of a path denotes its number of edges. A *homomorphism* from a graph G into a graph H is a mapping $\varphi\colon V(G) \to V(H)$ with the property that $\{\varphi(u), \varphi(w)\} \in E(H)$ whenever $\{u, w\} \in E(G)$. We say that G is *homomorphic* to H if there exists a homomorphism from G into H. Furthermore, a graph G is a *blow-up* of a graph H, if there exists a surjective homomorphism φ from G into H, but for any supergraph of G on the same vertex set the mapping φ is not a homomorphism into H anymore. In particular, a graph G is homomorphic to H if and only if it is a subgraph of a suitable blow-up of H. Moreover, we say a blow-up G of H is *balanced* if the homomorphism φ signifying

[1] Fachbereich Mathematik, Universität Hamburg, Bundesstraße 55, D-20146 Hamburg, Germany. Email: silvia.messuti@math.uni-hamburg.de, schacht@math.uni-hamburg.de

Research supported through the Heisenberg-Programme of the DFG.

that G is a blow-up has the additional property that $|\varphi^{-1}(u)| = |\varphi^{-1}(u')|$ for all vertices u and u' of H.

Homomorphisms can be used to capture structural properties of graphs. For example, a graph is k-colourable if and only if it is homomorphic to K_k. Many results in extremal graph theory establish relationships between the minimum degree of a graph and the existence of a given subgraph. The following theorem of Andrásfai, Erdős and Sós [2] is a classical result of that type.

Theorem 1.1 (Andrásfai, Erdős & Sós). *For every integer $r \geq 3$ and for every n-vertex graph G the following holds. If G has minimum degree $\delta(G) > \frac{3r-7}{3r-4}n$ and G contains no copy of K_r, then G is $(r-1)$-colourable.*

In the special case $r = 3$, Theorem 1.1 states that every triangle-free n-vertex graph with minimum degree greater than $2n/5$ is bipartite, *i.e.*, it is homomorphic to K_2. Several extensions of this result and related questions were studied. One line of research (see, *e.g.*, [4,7,9,10]) concerned the question for which minimum degree condition a triangle-free graph G is homomorphic to a graph H of bounded size, which is triangle-free itself. In particular, Häggkvist [7] showed that triangle-free graphs $G = (V, E)$ with $\delta(G) > 3|V|/8$ are homomorphic to C_5. In other words, such a graph G is a subgraph of suitable blow-up of C_5. This can be viewed as an extension of Theorem 1.1 for $r = 3$, since balanced blow-ups of C_5 show that the degree condition $\delta(G) > 2|V|/5$ is sharp there. Strengthening the assumption of triangle-freeness to graphs of higher odd girth, allows us to consider graphs with a more relaxed minimum degree condition. In this direction Häggkvist and Jin [8] showed that graphs $G = (V, E)$ which contain no odd cycle of length three and five and with minimum degree $\delta(G) > |V|/4$ are homomorphic to C_7.

We generalize those results to arbitrary odd girth. We say a graph G has *odd girth* at least g, if the shortest cycle with odd length has length at least g.

Theorem 1.2. *For every integer $k \geq 2$ and for every n-vertex graph G the following holds. If G has minimum degree $\delta(G) > \frac{3n}{4k}$ and G has odd girth at least $2k+1$, then G is homomorphic to C_{2k+1}.*

Note that the degree condition given in Theorem 1.2 is best possible as the following example shows. For an even integer $r \geq 6$ we denote by M_r the so-called *Möbius ladder* (see, *e.g.*, [6]), *i.e.*, the graph obtained by adding all diagonals to a cycle of length r, where a diagonal connects vertices of distance $r/2$ in the cycle. One may check that M_{4k} has odd girth $2k+1$, but it is not homomorphic to C_{2k+1}. Moreover, M_{4k} is 3-regular and, consequently, balanced blow-ups of M_{4k} show that the degree condition in Theorem 1.2 is best possible when n is divisible by $4k$.

We also remark that Theorem 1.2 implies that every graph with odd girth at least $2k+1$ and minimum degree bigger than $\frac{3n}{4k}$ contains an independent set of size at least $\frac{kn}{2k+1}$. This answers affirmatively a question of Albertson, Chan, and Haas [1]. Similar results were obtained by Brandt and Ribe-Baumann.

2 Sketch of the proof

In the proof of Theorem 1.2 we consider an edge-maximal graph and show that it is either a bipartite graph or a blow-up of a $(2k+1)$-cycle. We say that a graph G with odd girth at least $2k+1$ is *edge-maximal* if adding any edge to G yields an odd cycle of length at most $2k-1$. We denote by $\mathcal{G}_{n,k}$ all edge-maximal n-vertex graphs satisfying the assumptions of the main theorem, *i.e.*, for integers $k \geq 2$ and n we set

$$\mathcal{G}_{n,k} = \{G = (V, E)\colon\ |V| = n\,,\ \delta(G) > \tfrac{3n}{4k}\,,$$
$$\text{and } G \text{ is edge-maximal with odd girth } 2k+1\}\,.$$

The proof of the theorem relies on two lemmas, Lemmas 2.1 and 2.3 below, which state that certain configurations cannot occur in such edge-maximal graphs.

Lemma 2.1. *Let Φ denote the graph obtained from C_6 by adding one diagonal. For all $k \geq 2$ and n, every $G \in \mathcal{G}_{n,k}$ does not contain an induced copy of Φ.*

Proof (sketch). Suppose, contrary to the assertion, that $G = (V, E)$ contains Φ in an induced way. Since G is edge-maximal, the non-existence of a diagonal must be forced by the existence of an even path which, together with the missing diagonal, would yield an odd cycle of length at most $2k-1$. One can show that such a path must have length exactly $2k-2$ and that it must be edge-disjoint from Φ. Since there are two missing diagonals and since one can show that the related paths are also disjoint, the resulting configuration Φ' has $4k$ vertices. Finally one shows that no vertex in G can be joined to four vertices of Φ', which leads to a contradiction to the minimum degree condition of G. □

We remark that the above lemma can also be deduced from [8, Lemma 2], where is shown that $G \in \mathcal{G}_{n,k}$ cannot contain a cycle of length $4k$ with two consecutive diagonals. The next lemma states that graphs $G \in \mathcal{G}_{n,k}$ contain no graph from the following family, which can be viewed as tetrahedra with three faces formed by cycles of length $2k + 1$.

Definition 2.2. For $k \geq 2$ we denote by $\mathcal{T}_k$ those subdivisions T of K_4 satisfying

(i) three triangles of K_4 are subdivided into cycles of length $2k + 1$
(ii) two of the three edges contained in two of those triangles are subdivided into paths of length at least two.

We remark that in similar context "odd subdivisions" of K_4 appeared in [5].

Lemma 2.3. *For all integers $k \geq 2$ and n and for every $G \in \mathcal{G}_{n,k}$ we have that G does not contain any $T \in \mathcal{T}_k$ as a (not necessarily induced) subgraph.*

Proof (sketch). Similarly to the previous lemma, one can show that if such a $T \in \mathcal{T}_k$ is contained in G, then we get a contradiction to the minimum degree condition. In fact, (i) in Definition 2.2 implies that all four triangles of K_4 must be subdivided into a cycle of odd length in T. Since all these cycles must have length at least $2k + 1$ it follows that T consists of at least $4k$ vertices. Then some case analysis shows that any vertex in G can be joined to at most three vertices in T, contradicting the assumption on the minimum degree of G. □

In the proof of the main theorem, we assume that G is not bipartite and show that G is a blow-up of a $(2k + 1)$-cycle. In particular, we show that if a vertex of G is not contained in a maximal blow-up, then it gives rise to one of the forbidden configurations of Lemmas 2.1 and 2.3.

Proof of Theorem 1.2 *(sketch).* Suppose G is not bipartite. The edge-maximality of G implies that it contains a cycle of length C_{2k+1}. Let B be a vertex-maximal blow-up of a $(2k + 1)$-cycle contained in G with vertex classes $A_0, \ldots, A_{2k}$. We show $B = G$. Suppose that there exists a vertex $x \in V \backslash V(B)$. Owing to the odd girth assumption on G, the vertex x can have neighbours in at most two of the vertex classes of B and if there are two such classes, then within B each vertex in one class has distance two from the vertices in the other class.

Suppose first that x has neighbours in two classes A_{i-1} and A_{i+1}. If we are able to prove that x is adjacent to all the vertices in the two classes, then x can be included in A_i. If this is not the case, then by symmetry we may assume that there exists some vertex $b_{i-1} \in A_{i-1}$ which is not a neighbour of x. Fix vertices $a_{i-2} \in A_{i-2}$ and $a_i \in A_i$ arbitrarily and let $a_{i-1} \in A_{i-1}$ and $a_{i+1} \in A_{i+1}$ be neighbours of x. This fixes a cycle of length six in G, namely $xa_{i+1}a_ib_{i-1}a_{i-2}a_{i-1}x$, with one diagonal $\{a_{i-1}, a_i\}$. Due to Lemma 2.1, there must be at least one more diagonal and one can easily show that this diagonal must be $\{b_{i-1}, x\}$, since $\{a_{i+1}, a_{i-2}\}$ is a "shortcut" in the blow-up which would create a cycle of length $2k - 1$ in G.

Suppose that all neighbours of x in B are in A_i. Let $a_i \in A_i$ be a neighbour of x and fix a cycle $a_0a_1 \ldots a_{2k}$ in B containing a_i. Due to the edge-maximality, the non-existence of the edges $\{x, a_{i-2}\}$ and $\{x, a_{i+2}\}$ is forced by two paths which, together with the missing edges, would create short odd cycles. One can check that such paths have length exactly $2k-2$ and, together with the fixed cycle, they form a graph $T \in \mathcal{T}_k$, contradicting Lemma 2.3. Since G is connected due to its edge-maximality, this concludes the proof of Theorem 1.2. □

3 Conluding remarks

Extremal case in Theorem 1.2. A more careful analysis yields that the unique n-vertex graph with odd girth at least $2k + 1$ and minimum degree exactly $\frac{3n}{4k}$, which is not homomorphic to C_{2k+1}, is the balanced blow-up of the Möbius ladder M_{4k}. In fact, the proofs of Lemmas 2.1 and 2.3 can be adjusted such that they either exclude the existence of Φ resp. T in G or they yield $M_{4k} \subseteq G$. In the former case, one can repeat the proof of Theorem 1.2 based on those lemmas and obtains that G is homomorphic to C_{2k+1}. In the latter case, one uses the degree assumption to deduce that G is isomorphic to a balanced blow-up of M_{4k}.

Open questions. It would be interesting to study the situation, when we further relax the degree condition in Theorem 1.2. It seems plausible that if G has odd girth at least $2k + 1$ and $\delta(G) \geq (\frac{3}{4k} - \varepsilon)n$ for sufficiently small $\varepsilon > 0$, then the graph G is homomorphic to M_{4k}. In fact, this could be true until $\delta(G) > \frac{4n}{6k-1}$. At this point blow-ups of the $(6k - 1)$-cycle with all chords connecting two vertices of distance $2k$ in the cycle added show that this is best possible. For $k = 2$ such a result was proved in [4] and for $k = 3$ it appeared in [3].

For $\ell \geq 2$ and $k \geq 3$ let $F_{\ell,k}$ be the graph obtained from a cycle of length $(2k-1)(\ell-1)+2$ by adding all chords which connect vertices with distance of the form $j(2k - 1) + 1$ in the cycle for some $j = 1, \ldots, \lfloor(\ell - 1)/2\rfloor$. Note that $F_{2,k} = C_{2k+1}$ and $F_{3,k} = M_{4k}$. For every $\ell \geq 2$ the graph $F_{\ell,k}$ is ℓ-regular, has odd girth $2k + 1$, and it has chromatic number three. Moreover, $F_{\ell+1,k}$ is not homomorphic to $F_{\ell,k}$, but contains it as a subgraph.

A possible generalization of the known results would be the following: if an n-vertex graph G has odd girth at least $2k+1$ and minimum degree bigger than $\frac{\ell n}{(2k-1)(\ell-1)+2}$, then it is homomorphic to $F_{\ell-1,k}$. This is known to be false for $k=2$ and $\ell>10$, since such a graph G may contain a copy of the Grötzsch graph which (due to having chromatic number four) is not homomorphically embeddable into any $F_{\ell,k}$. In some sense this is the only exception for $k=2$ and $\ell>10$, since adding the condition $\chi(G)\leq 3$ makes the statement true for $k=2$ (see, *e.g.*, [4]). It is not known what happens for $k>2$.

This discussion motivates the following extension of a result of Łuczak for triangle-free graphs from [10]. For fixed k the density of $F_{\ell,k}$ tends to $\frac{1}{2k-1}$ as $\ell\to\infty$. Is it true that every n-vertex graph with odd girth at least $2k+1$ and minimum degree at least $(\frac{1}{2k-1}+\varepsilon)n$ can be mapped homomorphically into a graph H which also has odd girth at least $2k+1$ and $V(H)$ is bounded by a constant $C=C(\varepsilon)$ independent of n? In [10] Łuczak proved this for $k=2$.

References

[1] M. O. Albertson, L. Chan and R. Haas, *Independence and graph homomorphisms*, J. Graph Theory **17** (1993), 581–588.

[2] B. Andrásfai, P. Erdős and V. T. Sós, *On the connection between chromatic number, maximal clique and minimal degree of a graph*, Discrete Math. **8** (1974), 205–218.

[3] St. Brandt and E. Ribe-Baumann, *Graphs of odd girth 7 with large degree*, Electron. Notes Discrete Math., vol. 34, Elsevier Sci. B. V., Amsterdam, 2009, pp. 89–93.

[4] C. C. Chen, G. P. Jin and K. M. Koh, *Triangle-free graphs with large degree*, Combin. Probab. Comput. **6** (1997), 381–396.

[5] A. M. H. Gerards, *Homomorphisms of graphs into odd cycles*, J. Graph Theory **12** (1988), 73–83.

[6] R. K. Guy and F. Harary, *On the Möbius ladders*, Canad. Math. Bull. **10** (1967), 493–496.

[7] R. Häggkvist, *Odd cycles of specified length in nonbipartite graphs*, Graph theory (Cambridge, 1981), North-Holland Math. Stud., vol. 62, North-Holland, Amsterdam, 1982, pp. 89–99.

[8] R. Häggkvist and G. P. Jin, *Graphs with odd girth at least seven and high minimum degree*, Graphs Combin. **14** (1998), 351–362.

[9] G. P. Jin, *Triangle-free four-chromatic graphs*, Discrete Math. **145** (1995), 151–170.

[10] T. Łuczak, *On the structure of triangle-free graphs of large minimum degree*, Combinatorica **26** (2006), 489–493.

On the order of cages with a given girth pair

Julian Salas[1,2] and Camino Balbuena[1]

1 Cages with a given girth pair

The cage problem asks for the construction of regular simple graphs with specified degree k, girth g and minimum order $n(k; g)$, (see [5] for a complete survey). This problem was first considered by Tutte [10]. In 1963, Erdös and Sachs [4] proved that $(k; g)$-cages exist for any given values of k and g.

Counting the numbers of vertices in the distance partition with respect to a vertex when g is odd, and with respect to an edge when g is even, yields the lower bound $n_0(k; g)$ on the order of a $(k; g)$-cage. For $k \geq 3$ and $g \geq 5$ the order $n(k; g)$ of a cage is bounded by

$$n(k; g) \geq n_0(k; g) = \begin{cases} 1 + k \displaystyle\sum_{i=0}^{(g-3)/2} (k-1)^i & g \text{ odd} \\ 2 \displaystyle\sum_{i=0}^{(g-2)/2} (k-1)^i & g \text{ even} \end{cases} \tag{1.1}$$

This bound is called the Moore bound, it is known that the order of a $(k; g)$-cage $n(k; g) = n_0(k; g)$ only for $g = 6, 8, 12$ and $k = q + 1$ with q a prime power; and for $g = 5$ and $k = 3, 7, 57$ (*cf.* [1,3]). Therefore

[1] Departament de Matemàtica Aplicada III, Universitat Politècnica de Catalunya, Campus Nord, Jordi Girona 1 i 3, 08034 Barcelona, Spain. Email: m.camino.balbuena@upc.edu

[2] Institut d'Investigació en Intel·ligència Artificial, IIIA, Consejo Superior de Investigaciones Científicas, CSIC, Campus Universitat Autònoma de Barcelona, 08193 Bellaterra, Spain. Email: julian.salas@iiia.csic.es

This research was supported by the Ministry of Education and Science, Spain, and the European Regional Development Fund (ERDF) under project MTM2008-06620-C03-02; and under the Catalonian Government project 1298 SGR2009. Partial support by the Spanish MEC (project ARES – CONSOLIDER INGENIO 2010 CSD2007-00004) is acknowledged.

Biggs and Ito [2] defined the excess of a cage to be the number $n(k; g) - n_0(k; g)$ and proved the following theorem.

Theorem 1.1 ([2]). *[2] Let G be a a $(k; g)$-cage of girth $g = 2m \leq 6$ and excess e. If $e \leq k - 2$ then G is bipartite and its diameter is $m + 1$.*

By allowing a *girth pair* $g < h$ (one even and the other odd), in [6], Harary and Kovács introduced the concept of a *$(k; g, h)$-cage* as the smallest k-regular graph with girth pair g, h. They obtained the bound $n(k; g, h) \leq 2n(k; h)$ that relates the order of a cage with girth pair with the order of a cage, and showed that $n(k; h - 1, h) \leq n(k; h)$, i.e. in general the bound $n(k; g, h) \leq 2n(k; h)$ is not the best and stated the following conjecture.

Conjecture 1.2 ([6]). $n(k; g, h) \leq n(k; h)$, for all $k, g \geq 3$.

Xu, Wang, Wang [11] proved the strict inequality, $n(k; h - 1, h) < n(k; h)$. Kovács proved that the the Möbius ladder of order $2(h - 1)$ is the unique minimal $(3; 4, h)$-graph [7]. Campbell [8] studied the size of smallest cubic graphs with girth pair $(6, b)$ and constructed the cages for the exact values $(3; 6, 7)$, $(3; 6, 9)$ and $(3; 6, 11)$.

We obtain that the conjecture $n(k; g, h) < n(k; h)$ holds for all $(k; g, h)$-cages when g is odd. For g even, we settle the conjecture for cages of small excess, i.e. such that $n(k; g) - n_0(k; g) \leq k - 1$, also we prove that $n(k; g, h) < n(k; h)$ for h sufficiently large, in both cases, for g even and g odd, under the assumption that $(k; g)$-cages are bipartite for g even. Notice that the cages of even girth and small excess are known to be bipartite [2], furthermore it is conjectured that all cages with even girth are bipartite [9, 12].

2 Notation

For any graph of girth $g \geq 4$ even, $uv \in E(G)$ and $0 \leq l \leq \frac{g}{2} - 1$, let us denote the sets

$$B^l_{uv} = \{x \in V(G) : d(x, u) = l \text{ and } d(x, v) = l + 1\} \text{ and } \overline{B}^l_{uv} = \bigcup_{i=0}^{l} B^i_{uv}.$$

Observe that $B^0_{uv} = \{u\} = \overline{B}^0_{uv}$ and $B^1_{uv} = N(u) - v$ while $\overline{B}^1_{uv} = (N(u) - v) \cup \{u\}$. Moreover, note that $B^l_{uv} \neq B^l_{vu}$ and $\overline{B}^l_{uv} \neq \overline{B}^l_{vu}$. Denote $T^l_{uv} = G[\overline{B}^l_{uv} \cup \overline{B}^l_{vu}]$ and observe that if $l \leq \frac{g}{2} - 2$, where g is the girth of G, then T^l_{uv} is the tree rooted in the edge uv of depth l. When $l = \frac{g}{2} - 1$ the subgraph T^l_{uv} may not be a tree, it can contain edges between vertices in B^l_{uv} and vertices in B^l_{vu}.

We will denote the set of cycles in G by $\mathcal{C}(G) = \{\alpha : \alpha \text{ is a cycle in } G\}$.

Let G be a $(k; g)$-cage of even girth $g = 2m$. The *excess* e of G with respect to an edge $uv \in E(G)$ is the cardinality of the set $X = V(G) \setminus T_{uv}^{m-1}$. Note that the order of T_{uv}^{m-1} is the same for every edge $uv \in E(G)$. Thus, $e = |X| = n(k; g) - n_0(k; g)$.

3 Constructions for g odd

Deleting from G the sets $\overline{B}_{uv}^{(\frac{h-g-1}{2}-1)} \cup \overline{B}_{vu}^{(\frac{h-g+1}{2}-1)}$ and completing the degrees of the remaining vertices we obtain Theorem 3.2. In order to keep cycles of length h after the deletion we proved the following lemma.

Lemma 3.1. *Let G be a $(k; h)$-cage with $k \geq 3$ and even girth $h \geq 6$. Then G contains a cycle β of length h such that $V(\beta) \cap B_{uv}^{\frac{h}{4}-1} = \emptyset$ or $V(\beta) \cap B_{vu}^{\frac{h}{4}-1} = \emptyset$.*

Theorem 3.2. *Let $h \geq 6$ even and $k \geq 3$. Suppose that there is a bipartite $(k; h)$-cage. If $g \geq 5$ is an odd number such that $\frac{h}{2} + 1 \leq g < h$, then*

$$n(k; g, h) \leq n(k; h) - 2 \sum_{i=0}^{\frac{h-g-3}{2}} (k-1)^i - (k-1)^{\frac{h-g-1}{2}}.$$

Theorem 3.3. *Let $h \geq 6$ even and $k \geq 3$. Suppose that there is a bipartite $(k; h)$-cage. If g is an odd number such that $g < h$, then $n(k; g, h) < n(k; h)$.*

4 Constructions for g even and h odd

First of all we introduce a construction that we will use later for breaking short odd cycles while preserving the regularity and the even girth.

Definition 4.1. Let G, H be two vertex-disjoint graphs, $uv \in E(G)$ and $st \in E(H)$. We will define a new graph $G^{uv}\Gamma_{st}H$, that we will call the *insertion* of (G, uv) into (H, st) by letting:

- $V(G^{uv}\Gamma_{st}H) = V(G) \cup V(H)$
- $E(G^{uv}\Gamma_{st}H) = (E(G) \setminus \{uv\}) \cup (E(H) \setminus \{st\}) \cup \{us, vt\}$.

See Figure 4.1, for an example illustrating this definition.

The first basic result with respect to g even and h odd is the following theorem, the bound is proved inserting a graph (G, uv) into a copy $(G', u'v')$, and performing some edge operations in order to obtain cycles of length $2g - 1$.

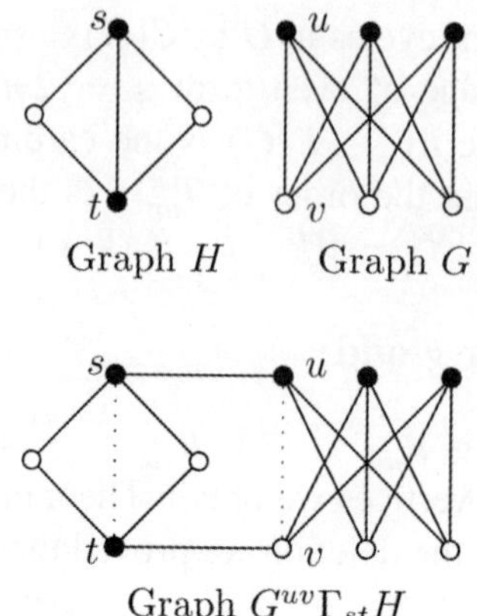

Figure 4.1. The insertion $G^{uv}\Gamma_{st}H$.

Theorem 4.2. *Let $k \geq 3$ and $g \geq 6$ even. Then $n(k; g, 2g-1) \leq 2n(k; g)$ provided that there is a bipartite $(k; g)$-cage.*

To prove the corresponding result for $n(k; g, g+r)$ we will use the following remark.

Remark 4.3. Let G, H be graphs with girths g, h, respectively, such that $g \leq h$, and let $G^{uv}\Gamma_{st}H$ be the insertion of (G, uv) into (H, st). Then the set of cycles in $G^{uv}\Gamma_{st}H$ is:

$$\begin{aligned} \mathcal{C}(G^{uv}\Gamma_{st}H) = {} & (\mathcal{C}(G) \setminus \{\alpha \in \mathcal{C}(G) : uv \in E(\alpha)\}) \\ & \cup (\mathcal{C}(H) \setminus \{\beta \in \mathcal{C}(H) : st \in E(\beta)\}) \\ & \cup \{\gamma = P_1 vt P_2 su : \ P_1 \text{ is a } uv\text{-path in } G - uv \\ & \qquad \text{and } P_2 \text{ is a } ts\text{-path in } H - st\}. \end{aligned}$$

This means that if there were cycles of lengths c_1 and c_2 in graphs G and H that used the edges uv and st, respectively, they are removed in the new graph $G^{uv}\Gamma_{st}H$ and new cycles of length $c_1 + c_2$ are created.

Theorem 4.4. *Let $k \geq 3$, g even such that $6 \leq g$ and r an odd number such that $1 \leq r \leq g-3$. Then $n(k; g, g+r) \leq 4n(k; g)$, provided that there is a bipartite $(k; g)$-cage.*

By applying the insertion to the graph obtained in Theorem 4.4 on specific edges, we obtain the following lemma.

Lemma 4.5. *Let $k \geq 3$, $g \geq 6$ even and suppose that there is a bipartite $(k; g)$-cage. Then $n(k; g, mg+r) \leq 4n(k; g) + k(m-1)n(k; g)$, for $m \geq 1$ and r any odd number such that $1 \leq r \leq g-1$. In particular when $r = g-1$, from Theorem 4.2, we have $n(k; g, (m+1)g-1) \leq 2mn(k; g)$.*

Theorem 4.6. *Suppose there is a bipartite $(k; g)$-cage with degree $k \geq 3$ and even girth $g \geq 6$. Then $n(k; g, h) < n(k, h)$, for h sufficiently large.*

So, we have proved Conjecture 1.2 for even girth g in general but asymptotically. For specific values of k and g, the Conjecture 1.2 can be completely settled, as we will show next.

Corollary 4.7. *For every $(k; g)$-cage of even girth g, degree $k \geq 3$ and excess $e \leq k - 2$, we have $n(k + 1; g, h) < n(k + 1; h)$.*

References

[1] E. BANNAI and T. ITO, *On finite Moore graphs*, J. Fac. Sci. Tokyo, Sect. 1A **20** (1973), 191–208.
[2] N. BIGGS and T. ITO, *Graphs with even girth and small excess*, Math. Proc. Cambridge Philos. Soc. **88** (1980) 1–10.
[3] R. M. DAMERELL, *On Moore graphs*, Proc. Cambridge Phil. Soc. **74** (1973), 227–236.
[4] P. ERDŐS and H. SACHS, *Regulare Graphen gegebener Taillenweite mit minimaler Knotenzahl*, Wiss. Z. Uni. Halle (Math. Nat.) **12** (1963), 251–257.
[5] G. EXOO and R. JAJCAY, *Dynamic Cage Survey*, Electron. J. Combin. **15** (2008), #DS16.
[6] F. HARARY and P. KOVÁCS, *Regular graphs with given girth pair*, J. Graph Theory **7** (1983), 209–218.
[7] P. KOVÁCS, *The minimal trivalent graphs with given smallest odd cycle*, Discrete Math. **54** (1985), 295–299.
[8] C. M. CAMPBELL, *On cages for girth pair* $(6, b)$, Discrete Math. **177** (1997), 259–266.
[9] T. PISANSKI, M. BOBEN, D. MARUSIC, A. ORBANIC and A. GRAOVAC, *The* 10*-cages and derived configurations*, Discrete Math. **275** (2004), 265–276.
[10] W. T. TUTTE, *A family of cubical graphs*, Proc. Cambridge Philos. Soc. (1947), 459–474.
[11] B.-G. XU, P. WANG and J.-F. WANG, *On the Monotonicity of $(k; g, h)$-graphs*, Acta Mathematicae Applicatae Sinica, English Series **18** (3) (2002), 477–480.
[12] P. K. WONG, *Cages-a survey*, J. Graph Theory **6** (1982), 1–22.

Enumerations, Lattices and Tableaux

Directed and multi-directed animals on the King's lattice

Axel Bacher[1]

Abstract. We define the *directed King's lattice* to be the square lattice with diagonal (next nearest neighbor) bonds and with the preferred directions $\{\leftarrow, \nwarrow, \uparrow, \nearrow, \rightarrow\}$. We enumerate directed animals on this lattice using a bijection with Viennot's heaps of pieces. We also define and enumerate a superclass of directed animals, the elements of which are called multi-directed animals. This follows Bousquet-Mélou and Rechnitzer's work on the directed triangular and square lattices. Our final results show that directed and multi-directed animals asymptotically behave similarly to the ones on the triangular and square lattices.

1 Introduction

An animal on a lattice is a finite and connected set of vertices. The enumeration of animals (up to a translation) is a longstanding problem in statistical physics and combinatorics. The problem, however, is extremely difficult, and little progress has been made [12,16]. A more realistic goal, therefore, is to enumerate natural subclasses of animals.

The class of *directed animals* is one of the most classical of these subclasses. Directed animals have been enumerated in a variety of lattices; let us cite, non-exhaustively, the square and triangular lattices [2,9,11,14, 18], Bousquet-Mélou and Conway's lattices $\mathcal{L}_n$ [4,8], and the "strange" or n-decorated lattices [3,7] (Figure 1.1). Unsolved lattices include, notably, the honeycomb lattice [13].

The class of *multi-directed animals* is a superclass of directed animals, first introduced by Klarner [15] on the square and triangular lattices. Bousquet-Mélou and Rechnitzer [5] clarified Klarner's definition and introduced a variant class on the square lattice. Moreover, they gave closed expressions for the generating function of multi-directed animals and showed that it is not D-finite.

The goal of this paper is to enumerate directed and multi-directed animals on a new lattice. We call *King's lattice* the square lattice with added

[1] LIPN, Université Paris Nord. Email: bacher@lipn.univ-paris13.fr

diagonal bonds, or next nearest neighbor bonds. We also consider the preferred orientations $\{\leftarrow, \nwarrow, \uparrow, \nearrow, \rightarrow\}$ (Figure 1.1, right). Directed animals on the King's lattice are a superclass of directed animals on Bousquet-Mélou and Conway's lattice $\mathcal{L}_3$, which has arcs $\{\nwarrow, \uparrow, \nearrow\}$ [4].

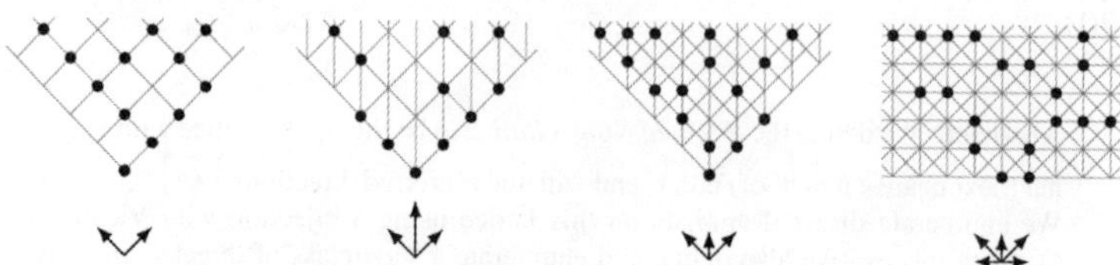

Figure 1.1. Directed animals on a selection of lattices. From left to right: the square lattice, the triangular lattice, the lattice $\mathcal{L}_3$, and the King's lattice.

Several techniques have been used to enumerate directed animals on the various lattices. Among them are direct bijections with other combinatorial objects [11], comparison with gas models [1,3,9,17] and the use of Viennot's theory of heaps of pieces [2,5,8,20,21]. In this paper, we use the last method; we show that directed animals on the King's lattice are in bijection with heaps of segments, already defined in [6].

2 Animals on the King's lattice and heaps of segments

2.1 Definitions

Definition 2.1. We call *segment* a closed real interval of the form $[i, j]$, where i and j are integers such that $j > i$. Two segments are called *concurrent* if they intersect, even by a point. A *heap of segments* is a finite sequence of segments, considered up to commutation of non-concurrent segments.

The heaps of segments described here are the same as in [6], except that the segment reduced to a point is not allowed. More information on heaps of pieces in general can be found in [20]. Graphically, a heap is built by dropping segments in succession; a segment either falls on the ground or on another segment concurrent to it. Examples are shown in Figures 2.1 and 2.2.

2.2 Directed animals and pyramids of segments

Let A be an animal; we say that a site t of A is *connected* to another site s if there exists a directed path (*i.e.* respecting the preferred directions of the lattice) from s to t visiting only sites of A. We say that the animal A is *directed* of source s if every site t of A is connected to s. The source s

is not unique; it may be any of the bottommost sites of A (see Figure 2.1, left). By convention, we call *source* of A the leftmost bottom site.

In Figure 2.1 is illustrated a bijection between directed animals and *pyramids of segments* (or heaps with only one segment lying on the ground). This bijection works identically to the classical bijection between directed animals on the square lattice and strict pyramids of dimers [2, 20].

Figure 2.1. Left: a directed animal on the King's lattice (represented, for clarity, as a polyomino on the dual lattice) with its source circled. Right: the pyramid of segments obtained by replacing each maximal sequence of ℓ consecutive sites by a segment of length ℓ.

2.3 Multi-directed animals and connected heaps of segments

Let A be an animal. For any abscissa i, we denote by $b(i)$ the ordinate of the bottommost site of A at abscissa i (or $b(i) = +\infty$ if there is no site of A at abscissa i). We call *source* of A a site that realizes a local minimum of b and *keystone* of A a site that realizes a local maximum. In case several consecutive sites realize a minimum or maximum, the source or keystone is the leftmost one (Figure 2.2, left). This is a purely arbitrary choice that does not alter the definition.

Definition 2.2. Let A be an animal. The animal A is said *multi-directed* if it satisfies the two conditions:

- for every site t of A, there exists a source s such that t is connected to s;
- for every keystone t of A, there exist two sources s_ℓ and s_r, to the left and to the right of t respectively, such that t is connected to both s_ℓ and s_r. Moreover, the directed paths connecting t to s_ℓ and s_r do not go through a keystone at the same height as t.

As a directed animal has only one source and no keystone, every directed animal is multi-directed. Multi-directed animals are in bijection with *connected heaps of segments* (or heaps without an empty column). A multi-directed animal and its corresponding heap are depicted in Figure 2.2.

Definition 2.2 can be adapted in the directed square and triangular lattices; the animals thus defined are in bijection with *connected heaps of*

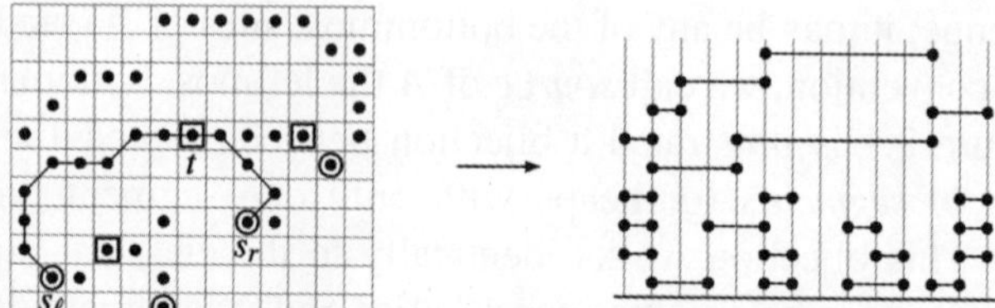

Figure 2.2. Left: a multi-directed animal with four sources (circled) and three keystones (boxed). The directed paths connecting one keystone, denoted by t, to the sources s_ℓ and s_r are shown. Right: the corresponding connected heap of segments, with has four minimal pieces (one for each source of the animal).

dimers. Bousquet-Mélou and Rechnitzer also defined multi-directed animals in bijection with connected heaps of dimers in [5], in a slightly different way. Our definition of multi-directed animals has the advantages of being more intrinsic and of having a vertical symmetry.

3 Enumeration of directed animals

In this section, we use the bijection with pyramids of segments to enumerate directed animals on the King's lattice. We call *half-animal* a directed animal with no site on the left side of its source. The associated pyramids are called *half-pyramids*. We adapt Bétéma and Penaud's methods [2] to decompose the pyramids of segments, which yields the following result.

Theorem 3.1. *The generating functions* $S(t)$ *and* $D(t)$ *of half-animals and animals are:*

$$S(t) = \frac{1 - 3t - \sqrt{1 - 6t + t^2}}{4t}; \quad D(t) = \frac{1}{4}\left(\frac{1+t}{\sqrt{1 - 6t + t^2}} - 1\right).$$

The decomposition of the half-pyramids is sketched in Figure 3.1. Interestingly, the generating function $S(t)$ is already known in combinatorics. Its coefficients are the little Schröder numbers, **A001003** in the OEIS [19]. The coefficients of $D(t)$ also appear as **A047781**. This is remindful of the triangular lattice, where the half-animals are enumerated by the Catalan numbers [2].

4 Enumeration of multi-directed animals

In this section, we enumerate multi-directed animals, or, equivalently, connected heaps of segments. To do this, we adapt the *Nordic decomposition*, invented by Viennot to enumerate connected heaps of dimers [21].

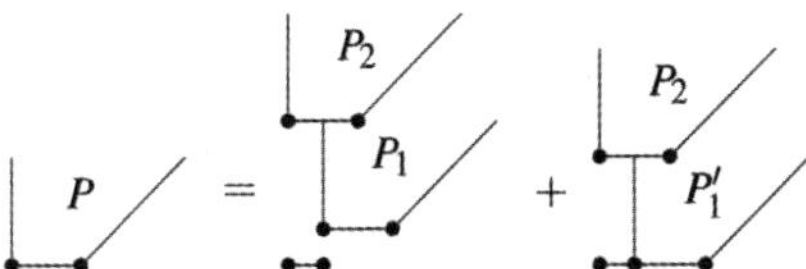

Figure 3.1. Sketch of the two cases in the decomposition of half-pyramids. The generating function of the possible heaps P_1 and P_2 is $1 + S(t)$, while the generating function of the possible heaps P_1' is $S(t)$. This shows the identity $S = t(1 + S)^2 + tS(1 + S)$, from which we derive the value of $S(t)$.

Theorem 4.1. *Let $M = M(t)$ be the generating function of multi-directed animals. Let $S = S(t)$, $D = D(t)$ be the power series defined in Theorem 3.1, $R = S + t(1 + S)$ and $Q = 2(1 - t)S - t$. The generating function M is given by:*

$$M = \frac{D}{1 - \sum_{k \geq 0} S(1+S)^k \frac{QR^k}{1 - QR^k}}.$$

5 Asymptotic results

Finally, we derive asymptotic results from Theorems 3.1 and 4.1.

Theorem 5.1. *Let D_n and M_n be the number of directed and multi-directed animals of area n, respectively. As n tends to infinity, we have:*

$$D_n \sim \kappa \left(3 + \sqrt{8}\right)^n n^{-1/2}; \qquad M_n \sim \lambda \mu^n,$$

with $\mu = 6.475\ldots$. The average width of directed animals grows like $\sqrt{n}$, while the average width of multi-directed animals grows like n. Finally, the series $M(t)$ is not D-finite.

The results on directed animals are a straightforward application of singularity analysis [10, Theorem VI.4]. The results on multi-directed animals are more involved. Similar results already exist on the square and triangular lattices, including the non-D-finiteness of the series $M(t)$ [5].

References

[1] M. Albenque, *A note on the enumeration of directed animals via gas considerations*, Ann. Appl. Probab. **19** (5) (2009), 1860–1879.

[2] J. Bétréma and J.-G. Penaud, *Modèles avec particules dures, animaux dirigés et séries en variables partiellement commutatives*, ArXiv Mathematics e-prints, 2001. arXiv:math/0106210.

[3] M. BOUSQUET-MÉLOU, *New enumerative results on two-dimensional directed animals*, In: "Proceedings of the 7th Conference on Formal Power Series and Algebraic Combinatorics (Noisy-le-Grand, 1995)", volume 180, 1998, 73–106.

[4] M. BOUSQUET-MÉLOU and A. R. CONWAY, *Enumeration of directed animals on an infinite family of lattices*, J. Phys. A **29** (13) (1996), 3357–3365.

[5] M. BOUSQUET-MÉLOU and A. RECHNITZER, *Lattice animals and heaps of dimers*, Discrete Math. **258** (1-3) (2002), 235–274.

[6] M. BOUSQUET-MÉLOU and X. G. VIENNOT, *Empilements de segments et q-énumération de polyominos convexes dirigés*, J. Combin. Theory Ser. A **60** (2) (1992), 196–224.

[7] A. R. CONWAY, R. BRAK and A. J. GUTTMANN, *Directed animals on two-dimensional lattices*, J. Phys. A: Math. Gen. **26** (1993), 3085–3091.

[8] S. CORTEEL, A. DENISE and D. GOUYOU-BEAUCHAMPS, *Bijections for directed animals on infinite families of lattices*, Ann. Comb. **4** (3-4) (2000), 269–284.

[9] D. DHAR, *Equivalence of the two-dimensional directed-site animal problem to Baxter's hard-square lattice-gas model*, Phys. Rev. Lett. **49** (14) (1982), 959–962.

[10] P. FLAJOLET and R. SEDGEWICK, "Analytic Combinatorics", Cambridge University Press, Cambridge, 2009.

[11] D. GOUYOU-BEAUCHAMPS and G. VIENNOT, *Equivalence of the two-dimensional directed animal problem to a one-dimensional path problem*, Adv. in Appl. Math. **9** (3) (1988), 334–357.

[12] A. J. GUTTMANN, *On the number of lattice animals embeddable in the square lattice*, Journal of Physics A: Mathematical and General **15** (6) (1987), 1987, 1982.

[13] A. J. GUTTMANN and A. R. CONWAY, *Hexagonal lattice directed site animals*, In: "Statistical Physics on the Eve of the 21st Century", volume 14 of Ser. Adv. Statist. Mech., World Sci. Publ., River Edge, NJ, 1999, 491–504.

[14] V. HAKIM and J. P. NADAL, *Exact results for 2D directed animals on a strip of finite width*, J. Phys. A **16** (7) (1983), L213–L218.

[15] D. A. KLARNER, *Cell growth problems*, Canad. J. Math. **19** (1967), 851–863.

[16] D. A. KLARNER and R. L. RIVEST, *A procedure for improving the upper bound for the number of n-ominoes*, Canad. J. Math. **25** (1973), 585–602.

[17] Y. Le Borgne and J.-F. Marckert, *Directed animals and gas models revisited*, Electron. J. Combin. **14** (1) (2007), Research Paper 71, 36 pp. (electronic).

[18] J. P. Nadal, B. Derrida and J. Vannimenus, *Directed lattice animals in 2 dimensions: numerical and exact results*, J. Physique **43** (11) (1982), 1561–1574.

[19] N. J. A. Sloane, *The on-line encyclopedia of integer sequences*, Notices Amer. Math. Soc. **50** (8) (2003), 912–915. http://oeis.org.

[20] G. X. Viennot, *Heaps of pieces, I: basic definitions and combinatorial lemmas*, Combinatoire Énumérative, G. Labelle and P. Leroux (eds.), Springer-Verlag, Berlin, 1986, 321–350.

[21] X. Viennot, *Multi-directed animals, connected heaps of dimers and Lorentzian triangulations*, In: "Journal of Physics", volume 42 of Conferences Series, 2006, 268–280.

Results and conjectures on the number of standard strong marked tableaux

Susanna Fishel[1] and Matjaž Konvalinka[2]

Abstract. Standard strong marked tableaux play a role for k-Schur functions similar to the role standard Young tableaux play for Schur functions. We discuss results and conjectures toward an analogue of the hook length formula.

1 Introduction

In 1988, Macdonald [5] introduced a new class of polynomials and conjectured that they expand positively in terms of Schur functions. This conjecture, proved by Haiman [2], has led to an enormous amount of work, including the development of the k-Schur functions. These were defined in [3] and led to a refinement of the Macdonald conjecture. The k-Schur functions have since been found to arise in other contexts; for example, as the Schubert cells of the cohomology of affine Grassmannian permutations, and they are related to the quantum cohomology of the affine permutations.

One of the intriguing features of standard Young tableaux is the hook-length formula, which enumerates them. It has many different proofs and generalizations, see *e.g.* [6, Chapter 7] and [1].

In this extended abstract, we partially succeed in finding an analogue of the hook-length formula for standard strong marked tableaux (or starred tableaux for short), which are a natural generalization of standard Young tableaux in the context of k-Schur functions. For a fixed n, the shape of a starred tableau is necessarily an n-core, a partition for which all hook-lengths are different from n. In [4, Proposition 9.17], a formula is given for the number of starred tableaux for $n = 3$, which can be rewritten as

[1] School of Mathematical and Statistical Sciences, Arizona State University. Email: fishel@math.asu.edu

[2] Department of Mathematics, University of Ljubljana. Email: matjaz.konvalinka@fmf.uni-lj.si

S. Fishel, was partially supported by NSF grant # 1200280 and Simons Foundation grant # 209806. M. Konvalinka, was partially supported by Research Program L1–069 of the Slovenian Research Agency.

$m!/\prod_{\substack{i,j\in\lambda \\ h_{ij}<3}} h_{ij}$, where m is the number of boxes of λ with hook-length $<$ 3. This is reminiscent of the classical hook-length formula. The authors left the enumeration for $n > 3$ as an open problem.

The main result (Theorem 3.1) of this extended abstract implies that for each n, there exist $(n-1)!$ rational numbers (which we call *correction factors*) so that the number of starred tableaux of shape λ for any n-core λ can be easily computed. In fact, Theorem 3.1 gives a t-analogue of this result. The theorem is "incomplete" in the sense that we were not able to find explicit formulas for the (weighted) correction factors. We have, however, been able to state some of their properties (some conjecturally), the most interesting of them being the unimodality conjecture (Conjecture 3.4). Another result of interest is a new, alternative description of strong marked covers via simple triangular arrays of integers which we call *residue tables* and *quotient tables* (Theorem 4.2). We think these tables are of great importance in the theory of k-Schur functions.

2 Preliminaries

Here we introduce notation and review some constructions. Please see [6] for the definitions of integer partitions, ribbons, hook lengths, the hook-length formula, standard and semistandard Young tableaux, etc., which we omit in this extended abstract.

2.1 Cores and bounded partitions

Let n be a positive integer. An *n-core* is a partition λ such that $h_{ij}^{\lambda} \neq n$ for all $(i,j) \in \lambda$. Core partitions were introduced to describe when two ordinary irreducible representations of the symmetric group belong to the same block. There is a simple bijection between $(k+1)$-cores and k-bounded partitions. Given a $(k+1)$-core λ, let π_i be the number of boxes in row i of λ with hook-length $\leq k$. The resulting $\pi = (\pi_1, \ldots, \pi_\ell)$ is a k-bounded partition, we denote it $\mathfrak{b}(\lambda)$. Conversely, given a k-bounded partition π, move from the last row of π upwards, and in row i, shift the π_i boxes of the diagram of π to the right until their hook-lengths are $\leq k$. The resulting $(k+1)$-core is denoted $\mathfrak{c}(\pi)$. In this extended abstract, $n = k+1$.

Example 2.1. The reader can check that for $k = 5$ and $\lambda = 953211$, $\mathfrak{b}(\lambda) = 432211$, and $\mathfrak{c}(\pi) = 75221$ for $k = 6$ and $\pi = 54221$.

Of particular importance are k-bounded partitions π for which $m_i(\pi) \leq k-i$ for all $i = 1, \ldots, k$. We call such partitions *k-irreducible partitions*, see [3]. The number of k-irreducible partitions is $k!$.

2.2 Strong marked and starred tableaux

The Frame-Thrall-Robinson hook-length formula shows how to compute f_λ, the number of standard Young tableaux of shape λ: we have $f_\lambda = \frac{|\lambda|!}{\prod_{(i,j)\in\lambda} h^\lambda_{ij}}$. There exists a well-known weighted version of this formula, see *e.g.* [6, Corollary 7.21.5]. Our goal is to find an analogue of this form in the setting of k-bounded partitions and k-Schur functions.

The *strong n-core poset* $\mathcal{C}_n$ is the subposet of $\mathcal{Y}$ induced by the set of all n-core partitions. That is, its vertices are n-core partitions and $\lambda \leq \mu$ in $\mathcal{C}_n$ if $\lambda \subseteq \mu$. The cover relations are trickier to describe in $\mathcal{C}_n$ than in $\mathcal{Y}$, see *e.g.* [4, Proposition 9.5]

The rank of an n-core is the number of boxes of its diagram with hook-length $< n$. If $\lambda \lessdot \mu$ and μ/λ consists of m ribbons, we say that μ *covers* λ *in the strong order with multiplicity* m. A *strong marked cover* is a triple (λ, μ, c) such that $\lambda \lessdot \mu$ and that c is the content of the head of one of the ribbons. We call c the *marking* of the strong marked cover. A *strong marked horizontal strip of size r and shape* μ/λ is a sequence $(\nu^{(i)}, \nu^{(i+1)}, c_i)_{i=0}^{r-1}$ of strong marked covers such that $c_i < c_{i+1}$, $\nu^{(0)} = \lambda$, $\nu^{(r)} = \mu$. If λ is an n-core, a *strong marked tableau* T of shape λ is a sequence of strong marked horizontal strips of shapes $\mu^{(i+1)}/\mu^{(i)}$, $i = 0, \ldots, m-1$, such that $\mu^{(0)} = \emptyset$ and $\mu^{(m)} = \lambda$. The *weight* of T is the composition $(r_1, \ldots, r_m)$, where r_i is the size of the strong marked horizontal strip $\mu^{(i)}/\mu^{(i-1)}$. If all strong marked horizontal strips are of size 1, we call T a *standard strong marked tableau* or a *starred tableau* for short. For a k-bounded partition π, denote the number of starred tableaux of shape $\mathfrak{c}(\pi)$ by $F^{(k)}_\pi$. The next figure illustrates $F^{(3)}_{211} = 6$.

1*	2*	3*
4*		
4		

1*	2*	4*
3*		
4		

1*	2*	4
3*		
4*		

1*	3*	4*
2*		
4		

1*	3*	4
2*		
4*		

1*	4	4*
2*		
3*		

If λ is a k-bounded partition that is also a $k+1$-core ($\lambda_1 + \ell(\lambda) \leq k+1$), then strong marked covers on the interval $[\emptyset, \lambda]$ are equivalent to the covers in the Young lattice, strong marked tableaux of shape λ are equivalent to semistandard Young tableaux of shape λ, and starred tableaux of shape λ are equivalent to standard Young tableaux of shape λ.

3 Main results and conjectures

For a starred tableau T, define the *descent set of* T, $D(T)$, as the set of all i for which the marked box at i is strictly above the marked box at $i+1$. Define the *major index of* T, $\mathrm{maj}(T)$, by $\sum_{i\in D(T)} i$. For a k-bounded partition π, define the polynomial $F_\pi^{(k)}(t) = \sum_T t^{\mathrm{maj}(T)}$, where the sum is over all starred tableaux of shape $\mathfrak{c}(\pi)$. Clearly $F_\pi^{(k)} = F_\pi^{(k)}(1)$.

Theorem 3.1. *Let π be k-bounded, $\pi = \langle k^{a_1+1\cdot w_1}, \ldots, 1^{a_k+k\cdot w_k}\rangle$ for $0 \le a_i < i$. If $\sigma = \langle k^{a_1}, \ldots, 1^{a_k}\rangle$ and $(j) = 1 + \cdots + t^{j-1}$, then*

$$F_\pi^{(k)}(t) = \frac{t^{\sum_{i=1}^k w_i \binom{i}{2}(k-i+1)}(|\pi|)!\, F_\sigma^{(k)}(t)}{(|\sigma|)!\, \prod_{j=1}^k (j)^{\sum_{i=1}^k w_i \min\{i,j,k+1-i,k+1-j\}}}.$$

The theorem implies that in order to compute $F_\pi^{(k)}(t)$ (and, by plugging in $t = 1$, $F_\pi^{(k)}$) for all k-bounded partitions π, it suffices to compute $F_\sigma^{(k)}(t)$ (resp., $F_\sigma^{(k)}$) only for k-irreducible partitions σ.

Example 3.2. For $k = 3$, we have, among other formulas,

$$F^{(3)}_{3^{w_1}2^{1+2w_2}1^{1+3w_3}}(t) = \frac{t^{2w_2+3w_3+1}\cdot(\mathbf{3w_3+4w_2+3w_1+3})!}{(\mathbf{2})^{w_1+2w_2+w_3}\cdot(\mathbf{3})^{w_1+w_2+w_3+1}}.$$

For a k-bounded partition π, let $H_\pi^{(k)}(t) = \prod (\boldsymbol{h_{ij}})$, where the product is over all boxes (i, j) of the $(k+1)$-core $\mathfrak{c}(\pi)$ with hook-lengths at most k, and let $H_\pi^{(k)} = H_\pi^{(k)}(1)$ be the product of all hook-lengths $\le k$ of $\mathfrak{c}(\pi)$. Furthermore, if b_j is the number of boxes in the j-column of $\mathfrak{c}(\pi)$ with hook-length at most k, write $b_\pi^{(k)} = \sum_j \binom{b_j}{2}$.

Let $C_\sigma^{(k)}(t) = F_\sigma^{(k)}(t)H_\sigma^{(k)}(t)/(t^{b_\sigma^{(k)}}(|\sigma|)!C_\sigma^{(k)}(t))$ be the *weighted correction factors* of a k-irreducible partition σ. By (and in the notation of) Theorem 3.1, we can express $F_\pi^{(k)}(t)$ (for all k-bounded partitions π) in another way which is reminiscent of the classical hook-length formula:

$$F_\pi^{(k)}(t) = \frac{t^{b_\sigma^{(k)}+\sum_{i=1}^k w_i \binom{i}{2}(k+1-i)}(|\pi|)!\, C_\sigma^{(k)}(t)}{H_\sigma^{(k)}(t)\cdot \prod_{j=1}^k (j)^{\sum_{i=1}^k w_i \min\{i,j,k+1-i,k+1-j\}}}.$$

For $k \le 3$, all weighted correction factors are 1. For $k = 4$, all but four of the 24 weighted correction factors are 1, and the ones different from 1 are $\frac{1+2t+t^2+t^3}{(\mathbf{2})(\mathbf{3})}$, $\frac{1+t+2t^2+t^3}{(\mathbf{2})(\mathbf{3})}$, $\frac{1+2t+2t^2+2t^3+t^4}{(\mathbf{3})^2}$, $\frac{1+t+3t^2+t^3+t^4}{(\mathbf{3})^2}$ for partitions 2211, 321, 3211 and 32211.

For a k-bounded partition π, denote by $\partial_k(\pi)$ the boxes of $\mathfrak{c}(\pi)$ with hook-length $\le k$. If $\partial_k(\pi)$ is not connected, we say that π *splits* into k-bounded partitions $(\pi^i)_{i=1}^m$, where the connected components of $\partial_k(\pi)$ are horizontal translates of $\partial_k(\pi^i)$. It turns out that $C_\pi^{(k)}(t) = \prod_{i=1}^m C_{\pi^i}^{(k)}(t)$.

Conjecture 3.3. For a k-irreducible partition π, the weighted correction factor is 1 if and only if π splits into $\pi^1, \ldots, \pi^l$, where each π^i is a k-bounded partition that is also a $(k+1)$-core.

Conjecture 3.4. For a k-irreducible partition σ, we can write $1 - C_\sigma^{(k)}(t)$ as $\frac{P_1(t)}{P_2(t)}$, where $P_1(t)$ is a unimodal polynomial with non-negative integer coefficients and $P_2(t) = \prod_{i=1}^{k-1}(j)^{w_j}$ for some integers $w_j \geq 0$.

4 Strong covers and k-bounded partitions

Our proof of Theorem 3.1, omitted in the extended abstract, closely follows one of the possible proofs (via quasisymmetric functions) of the classical (non-weighted and weighted) hook-length formula, see *e.g.* [6, Section 7.21]. Note, however, that the truly elegant proofs (*e.g.* [1]) are via induction. In this section, we show the first steps toward such a proof. In the process, we present a new description of strong marked covers in terms of bounded partitions (previous descriptions included cores—at least implicitly, via k-conjugation—affine permutations and abacuses). See the definition of residue and quotient tables below, and Theorem 4.2.

Identify $\pi = \langle k^{p_1}, \ldots, 1^{p_k} \rangle$ with $p = (p_1, \ldots, p_k)$. Given i, j, m, $0 \leq m < i \leq j \leq k$, define $p^{i,j,m}$ as follows: $p_{i-1}^{i,j,m} = p_{i-1} + m$, $p_i^{i,j,m} = p_i - m$ for $i \neq j$, $p_j^{i,j,m} = p_j - m - 1$ for $i \neq j$, $p_i^{i,i,m} = p_i - 2m - 1$, $p_{j+1}^{i,j,m} = p_{j+1} + m + 1$, $p_h^{i,j,m} = p_h$ for $h \neq i-1, i, j, j+1$. See Example 4.3.

Define upper-triangular arrays $\mathcal{R} = (r_{ij})_{1 \leq i \leq j \leq k}$ (*residue table*), $\mathcal{Q} = (q_{ij})_{1 \leq i \leq j \leq k}$ (*quotient table*) by

- $r_{jj} = p_j \bmod j$, $r_{ij} = (p_i + r_{i+1,j}) \bmod i$ for $i < j$,
- $q_{jj} = p_j \operatorname{div} j$, $q_{ij} = (p_i + r_{i+1,j}) \operatorname{div} i$ for $j < i$.

Example 4.1. Take $k = 4$, $p = (1, 3, 2, 5)$. Then $\mathcal{R} = \begin{matrix} 0000 \\ 111 \\ 20 \\ 1 \end{matrix}$, $\mathcal{Q} = \begin{matrix} 1222 \\ 121 \\ 01 \\ 1 \end{matrix}$.

Theorem 4.2. *Take $p = (p_1, \ldots, p_k)$ and $1 \leq i \leq j \leq k$. If $r_{ij} < r_{i+1,j}, \ldots, r_{jj}$, then p covers $p^{i,j,r_{ij}}$ in the strong order with multiplicity $q_{ij} + \cdots + q_{jj}$. Furthermore, these are precisely all strong covers.*

Example 4.3. For $k = 4$, $p = (1, 3, 2, 5)$, we have $r_{ij} < r_{i+1,j}, \ldots, r_{jj}$ for all $(i, j) \neq (1, 4), (2, 4)$. Therefore p covers, for example, $p^{1,2,0} = (1, 2, 3, 5)$ with multiplicity 3, and $p^{2,3,1} = (2,2,0,7)$ with multiplicity 2.

For a k-bounded partition π, we clearly have $F_\pi^{(k)} = \sum_\tau m_{\tau\pi} F_\tau^{(k)}$, where the sum is over all k-bounded τ that are covered by π, and $m_{\tau\pi}$ is the multiplicity of the cover. Therefore the theorem can be used to prove Theorem 3.1 for $t = 1$ for small values of k by induction: all we have to do is check $k!$ equalities. The authors did that for $k \leq 8$.

References

[1] C. GREENE, A. NIJENHUIS and H. S. WILF, *A probabilistic proof of a formula for the number of Young tableaux of a given shape*, Adv. in Math. **31** (1) (1979), 104–109. MR 521470 (80b:05016)

[2] M. HAIMAN, *Hilbert schemes, polygraphs and the Macdonald positivity conjecture*, J. Amer. Math. Soc. **14** (4) (2001), 941–1006. (electronic). MR 1839919 (2002c:14008)

[3] L. LAPOINTE, A. LASCOUX and J. MORSE, *Tableau atoms and a new Mac- donald positivity conjecture*, Duke Math. J. **116** (1) (2003), 103–146. MR 1950481 (2004c:05208)

[4] T. LAM, L. LAPOINTE, J. MORSE and M. SHIMOZONO, *Affine insertion and Pieri rules for the a ne Grassmannian*, Mem. Amer. Math. Soc. **208** (977) (2010), xii+82. MR 2741963

[5] *I. G. Macdonald*, "Symmetric Functions and Hall Polynomials", second ed., Oxford Mathematical Monographs, The Clarendon Press Oxford University Press, New York, 1995, With contributions by A. Zelevinsky, Oxford Science Publications. MR 1354144 (96h:05207)

[6] R. P. STANLEY, "Enumerative Combinatorics", Vol. 2, Cambridge Studies in Advanced Mathematics, Vol. 62, Cambridge University Press, Cambridge, 1999, With a foreword by Gian-Carlo Rota and appendix 1 by Sergey Fomin. MR 1676282 (2000k:05026)

On independent transversals in matroidal Latin rectangles

Ron Aharoni[1], Daniel Kotlar[2] and Ran Ziv[2]

A *Latin rectangle* is an $m \times n$ matrix in which the entries of each row and each column are all distinct numbers. In a row-Latin rectangle only the rows are required to consist of distinct entries. A *partial transversal* in a Latin rectangle is a set of distinct numbers such that no two of them are in the same row or column. A well-known conjecture of Ryser, Brualdi and Stein [2,8] asserts that every $n \times n$ Latin rectangle has a partial transversal of size $n - 1$. Woolbright [9], and independently and Brouwer, de Vries and Wieringa [1], proved the existence of a partial transversal of size $n - \sqrt{n}$ and Shor and Hatami [5,7] improved the bound to $n - O(\log^2 n)$.

Chappell [3], and later on the current authors [6], generalized the notion of a Latin rectangle, or Latin square, to matroids, by replacing "distinct" with "independent". Thus, a matroidal Latin rectangle is an $m \times n$ array over the ground set of a matroid $\mathcal{M}$, whose rows and columns are independent sets. A matroidal row-Latin rectangle is defined analogously. An *independent partial transversal* is an independent set whose elements belong to different rows and columns.

Chappell [3] proved that every $(2n - 1) \times n$ matroidal row-Latin rectangle has an independent partial transversal of size n. This generalized an earlier result of Drisko [4]. Drisko gave an example where $(2n - 2)$ rows are not enough to ensure an independent transversal of size n.

We suggest dealing with the more general problem of finding a rainbow set in the intersection of two matroids: given two matroids $\mathcal{M}$ and $\mathcal{N}$ on the same ground set S, an *independent matching* is a subset of S in $\mathcal{M} \cap \mathcal{N}$. A partial rainbow set for a family of sets $\mathcal{F}$ is a set of representatives for some sub-family of $\mathcal{F}$. We conjecture that any n independent matchings, each of size n, have a partial rainbow independent matching of size $n - 1$.

[1] Department of Mathematics, Technion, Haifa 32000, Israel. Email: ra@tx.technion.ac.il

[2] Computer Science Department, Tel-Hai College, Upper Galilee 12210, Israel.
Email: dannykot@telhai.ac.il, vaksler@cs.bgu.ac.il

In the current paper we present the results for two matroids $\mathcal{M}$ and $\mathcal{N}$, defined on the same ground set S. The first theorem generalizes the result mentioned above of Woolbright [9] and Brouwer, de Vries and Wieringa [1]:

Theorem 1. *Any n independent matchings, each of size n, have a partial rainbow independent matching of size $n - \sqrt{n}$.*

The next theorem generalizes the results mentioned above of Drisko [4] and Chappell [3]:

Theorem 2. *Any $2n - 1$ independent matchings, each of size n, have a partial rainbow independent matching of size n.*

In both proofs we assume that a maximal rainbow matching has size smaller than the one we claim to exist, and we generate a larger rainbow matching using the notion of *colorful alternating path*. In the process of constructing the path we use some new basis exchange properties of matroids.

References

[1] A. E. BROUWER, A. J. DE VRIES and R. M. A. WIERINGA, *A lower bound for the length of partial transversals in a Latin square*, Nieuw Arch. Wiskd. **24** (1978), 330–332.

[2] R. A. BRUALDI and H. J. RYSER, "Combinatorial Matrix Theory", Cambridge University Press, 1991.

[3] G. G. CHAPPELL, *A matroid generalization of a result on row-latin rectangles*, Journal of Combinatorial Theory, Series A **88** (1999), 235–245.

[4] A. A. DRISKO, *Proof of the Alon-Tarsi conjecture for $n = 2^r p$*, The Electronic Journal of Combinatorics (1998), no. R28.

[5] P. HATAMI and P. W. SHOR, *A lower bound for the length of a partial transversal in a Latin square*, J. Combin. Theory A **115** (2008), 1103–1113.

[6] D. KOTLAR and R. ZIV, *On the length of a partial independent transversal in a matroidal Latin square*, Electron. J. Combin. **19** (2012), no. 3.

[7] P. W. SHOR, *A lower bound for the length of a partial transversal in a Latin square*, Journal of Combinatorial Theory, Series A **33** (1982), 1–8.

[8] S. K. STEIN, *Transversals of Latin squares and their generalizations*, Pacific Journal of Mathematics **59** (1975), 567–575.

[9] D. E. WOOLBRIGHT, *An $n \times n$ Latin square has a transversal with at least $n - \sqrt{n}$ distinct elements*, J. Combin. Theory A **24** (1978), 235–237.

Multivariate Lagrange inversion formula and the cycle lemma

Axel Bacher[1] and Gilles Schaeffer[2]

Abstract. We give a multitype extension of the cycle lemma of (Dvoretzky and Motzkin 1947). This allows us to obtain a combinatorial proof of the multivariate Lagrange inversion formula that generalizes the celebrated proof of (Raney 1963) in the univariate case, and its extension in (Chottin 1981) to the two variable case.

Until now, only the alternative approach of (Joyal 1981) and (Labelle 1981) via labelled arborescences and endofunctions had been successfully extended to the multivariate case in (Gessel 1983), (Goulden and Kulkarni 1996), (Bousquet et al. 2003), and the extension of the cycle lemma to more than 2 variables was elusive.

The cycle lemma has found a lot of applications in combinatorics, so we expect our multivariate extension to be quite fruitful: as a first application we mention economical linear time exact random sampling for multispecies trees.

1 Introduction

For any power series $g(x)$ with $g(0) \neq 0$, there exists a unique power series $f(t)$ solution of the equation $f = tg(f)$. The Lagrange inversion formula says that the nth coefficient of $f(t)$ is $\frac{1}{n}[x^{n-1}]g(x)^n$. This formula is now known as a fundamental tool to derive tree enumeration results. The two simpler and most classical examples are:

- if $g(x) = \exp(x)$ then $f(t)$ is the exponential generating function of Cayley trees, so that the number of rooted Cayley trees with n nodes is equal to $\frac{1}{n}\left[\frac{x^{n-1}}{(n-1)!}\right]\exp(nx) = n^{n-2}$;
- if $g(x) = (1+x)^2$ then $f(t)$ is the ordinary generating function of binary trees, so that the number of rooted binary trees with n nodes is equal to $\frac{1}{n}[x^{n-1}](1+x)^{2n} = \frac{1}{n}\binom{2n}{n-1} = \frac{1}{n+1}\binom{2n}{n}$.

These examples are in a sense generic: the bijection between doubly rooted Cayley trees and endofunctions underlies Labelle's proof of the Lagrange inversion formula [11], while the cyclic lemma used in [7] to count ballot numbers underlies Raney's proof [16].

[1] Université Paris Nord. Email: bacher@lipn.univ-paris13.fr

[2] CNRS / École Polytechnique. Email: gilles.schaeffer@lix.polytechnique.fr

Our interest is in the multivariate extension of the Lagrange inversion formula. Fix an integer $k \geq 1$. Let bold letters denote k-dimensional vectors; write $\mathbf{x}^{\mathbf{n}} = x_1^{n_1} \cdots x_k^{n_k}$ and $\mathbf{x}^{\mathbf{n}-\mathbf{1}} = x_1^{n_1-1} \cdots x_k^{n_k-1}$. Let $h(\mathbf{x})$ and $g_1(\mathbf{x}), \ldots, g_k(\mathbf{x})$ be power series in $\mathbf{x}$ such that for $i = 1, \ldots, k$, $g_i(\mathbf{0}) \neq 0$. Again there is a unique familly of power series $\mathbf{f}(\mathbf{t})$ solution of the system of equations $f_i = t_i g_i(f_i)$ for $i = 1, \ldots, k$ and the multivariate Lagrange inversion formula admits the two equivalent following formulations (among several others):

$$[\mathbf{t}^{\mathbf{n}}]h(\mathbf{f}(\mathbf{t})) = [\mathbf{x}^{\mathbf{n}-\mathbf{1}}]h(\mathbf{x})\mathbf{g}(\mathbf{x})^{\mathbf{n}} \det\left(\delta_{i,j} - \frac{x_i}{g_j(\mathbf{x})}\frac{\partial g_j(\mathbf{x})}{\partial x_i}\right) \tag{1.1}$$

$$= \frac{1}{n_1 \cdots n_k}[\mathbf{x}^{\mathbf{n}-\mathbf{1}}]\sum_T \partial_T(h, g_1^{n_1}, \ldots, g_k^{n_k}) \tag{1.2}$$

where the sum is over oriented 0-rooted Cayley trees (non-plane trees with arcs going toward 0) with vertices $\{0, \ldots, k\}$ and the derivative ∂_G with respect to a directed graph G with vertex set $V = \{0, \ldots, k\}$ and edge set E is defined as

$$\partial_G(f_0(\mathbf{x}), \ldots, f_k(\mathbf{x})) = \prod_{j=0}^{k}\left(\left(\prod_{(i,j)\in E}\frac{\partial}{\partial x_i}\right)f_j(\mathbf{x})\right).$$

Several variants of (1.1) are given in [8] but (1.2) appeared more recently, implicitely in [9] and explicitely in [1]. As far as we know all combinatorial proofs of the multivariate Lagrange inversion extend Joyal and Labelle's approach [10, 11]: [8] proves another variant of (1.1), [3,9] prove (1.2). A completely different approach was recently proposed in [2].

Chottin [4,5] instead proposed a remarkable extension of Raney's strategy to prove the two variable Lagrange inversion formula. Yet he failed to move to three variables and the problem of proving the multivariate Lagrange inversion formula with the cycle lemma was considered as difficult. Apart from the theoretical interest of such a proof, an extension of the cycle lemma is desirable in view of its numerous applications, to tree and map enumeration [12, Chapter 11], [13, Chapter 9], probability [14], and random sampling [6]. We present such an extension in this paper.

2 Generalized cycle lemma

Following [5] and the modern accounts of Raney's proof, our combinatorial construction is in terms of encodings of rooted plane trees by sequences of nodes. To deal with the multivariate case, we introduce colored trees.

A *colored tree* is a plane, rooted tree in which all edges have a color in the set $\{1, \dots, k\}$. A *colored bush* is a colored tree that can have "pending" edges with no node attached. We call such edges *free edges*; an edge is *occupied* if a node is attached to it.

Let ℓ be an integer with $0 \leq \ell \leq k$. An *ℓ-bush* is a colored bush such that occupied edges have color more than ℓ and free edges have color at most ℓ. In particular, a k-bush only has a single node, while a 0-bush is a colored tree. Therefore, ℓ-bushes can be seen as intermediate objects between colored nodes and colored trees.

Definition 2.1. Let $0 \leq \ell \leq k$. We denote by $\mathcal{S}_\ell$ the set of tuples of the form $S = (S_0, \dots, S_\ell, e_1, \dots, e_k)$, where $S_0, \dots, S_\ell$ are sequences of ℓ-bushes and e_i is an edge of color i in S (refered to as a *marked edge*), satisfying the following conditions.

1. The sequence S_0 has only one element; for $i = 1, \dots, \ell$, the number of elements of the sequence S_i is equal to the number of edges of color i in S.
2. Let T be the graph with vertices $0, \dots, \ell$ and an arc $i \to j$ if the edge e_j is in the sequence S_i. The graph T is a 0-rooted Cayley tree.

According to the previous remark, the objects of $\mathcal{S}_\ell$ may also be seen as intermediates between two objects. If $k = \ell$, then $S_0, \dots, S_k$ are simply sequences of nodes. If $\ell = 0$, the unique element of the sequence S_0 is a colored tree.

Theorem 2.2 (generalized cycle lemma). *There is a bijection between the sets $\mathcal{S}_\ell$ and $\mathcal{S}_{\ell-1}$ that works by attaching the elements of the sequence S_ℓ to the edges of color ℓ.*

The actual description of the bijection is given in Section 3; an example is given in Figure 2.1. Observe that for $\ell = k = 1$, the statement is equivalent to the standard Cycle Lemma: there is a bijection between p-uples of rooted plane trees with one pointed node (represented here as a unique tree with an extra node of degree p at the root) and pairs formed of a node of degree p having a marked free edge and a sequence of nodes such that the number of nodes in the sequence equals the total number of free edges.

Iterating the bijection yields a bijection between $\mathcal{S}_0$ and $\mathcal{S}_k$. Given a Cayley tree T with vertices $\{0, \dots, k\}$, the generating function of the associated subset of $\mathcal{S}_k$ is $hg_1^{n_1} \cdots g_k^{n_k}$ with the proper derivatives to mark edges (taking a derivative of $g_j^{n_j}$ with respect to x_i amounts to marking an edge of color i in the sequence S_j). It can thus be seen that the formula (1.2) is a corollary of the generalized Cycle Lemma.

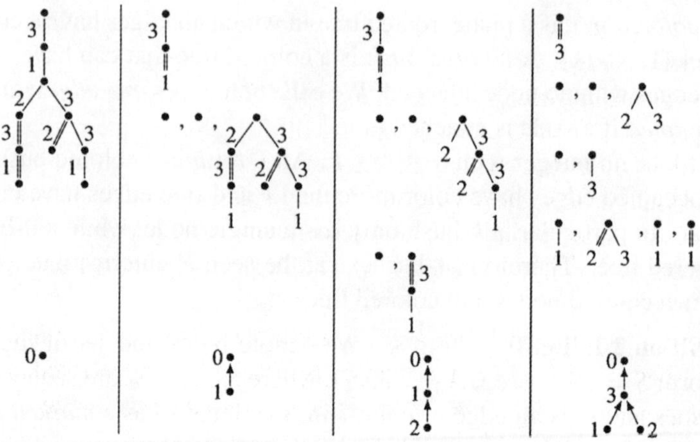

Figure 2.1. An example with $k = 3$, a tree of $\mathcal{S}_0$ and the corresponding elements of $\mathcal{S}_1$, $\mathcal{S}_2$ and $\mathcal{S}_3$ (from left to right), with their associated Cayley trees (bottom). Marked edges are represented as double lines.

3 The bijection

We now describe without proof our bijection between the sets $\mathcal{S}_\ell$ and $\mathcal{S}_{\ell-1}$, for $1 \leq \ell \leq k$, illustrated in Figure 2.1. This bijection uses *Prüfer codes* of Cayley trees [15]. The Prüfer code of a 0-rooted Cayley tree T with vertices $0, \dots, s$ is a sequence $p_1, \dots, p_s$ with $p_s = 0$; there exists a permutation σ such that the parent vertex of j in T is $p_{\sigma(j)}$ for $j = 1, \dots, s$. Moreover, every sequence corresponds to a unique tree.

Let $S = (S_0, \dots, S_\ell, e_1, \dots, e_k)$ be an element of $\mathcal{S}_\ell$; let T be the associated Cayley tree. In the following, we call *ℓ-edge* an edge of color ℓ. Since we are dealing with plane trees, there is a natural order on the set of ℓ-edges of S given by depth-first traversals of the ℓ-bushes.

Let r be the number of ℓ-edges in the sequences $S_0, \dots, S_{\ell-1}$. Since T is a Cayley tree, it does not contain the arc $\ell \to \ell$, which means that the marked edge e_ℓ is not in the sequence S_ℓ. Thus, we have $r \geq 1$. Let u be the root of the first element of S_ℓ. By Definition 2.1, the sequence S_ℓ has exactly r more elements than it has ℓ-edges. We combine the elements of S_ℓ using the Cycle Lemma; we denote by $b_1, \dots, b_r$ the resulting $\ell - 1$-bushes, so that the node u is in b_1.

First case: the vertex ℓ is a leaf of T. In this case, we attach the bushes $b_1, \dots, b_r$ to the remaining free ℓ-edges in order, with a cyclic permutation chosen so that the bush b_1 is attached to the marked edge e_ℓ.

Second case: the vertex ℓ is not a leaf of T. In the Cayley tree T, all arcs going toward ℓ correspond to marked edges in the sequence S_ℓ. Let

$m_1, \dots, m_s$ be the $\ell - 1$-bushes that we just formed containing at least one marked edge.

We break up the Cayley tree T in the following manner. For $i = 1, \dots, s$, let T_i be the forest composed of the colors of the marked edges in m_i and their descendants in T. Let T_0 be the tree composed of the remaining vertices of T with the vertex ℓ deleted; in other words, the vertices of T_0 are exactly the non-descendants of ℓ. We also assume that the order of the m_i's was chosen so that the T_i's are ordered according to their lowest label.

Now, attach the bushes $b_1, \dots, b_r$ to the free ℓ-edges in order, with a cyclic permutation chosen so that b_s is attached to the marked edge e_ℓ. For $j = 1, \dots, s$, let $p_j = i$ if m_j is attached to an edge in the sequence with an index in T_i. As the edge e_ℓ is in the sequence corresponding to the parent of ℓ in T, which is in T_0, we have $p_s = 0$.

We can therefore regard the sequence $p_1, \dots, p_s$ as the Prüfer code of a 0-rooted Cayley tree $\widetilde{T}$ with labels $\{0, \dots, s\}$. Let σ be the permutation associated to this tree and swap the bushes $m_1, \dots, m_s$ according to the permutation σ.

Let $S'_0, \dots, S'_{\ell-1}$ be the sequences resulting from this procedure. As no free ℓ-edges remain, every element of the sequences $S'_0, \dots, S'_{\ell-1}$ is an $\ell - 1$-bush. Finally, let e'_ℓ be the parent edge of the node u defined at the beginning. Let T' be the graph associated with the marked edges $e_1, \dots, e_{\ell-1}$. In the first case above, this is the tree T with the leaf ℓ deleted; in the second, it is a compound of the tree T_0 and the forests $T_1, \dots, T_s$, arranged according to the tree $\widetilde{T}$; it is therefore a Cayley tree. This shows that the tuple $(S'_0, \dots, S'_{\ell-1}, e_1, \dots, e'_\ell, \dots, e_k)$ is an element of $\mathcal{S}_{\ell-1}$.

We conclude with some comments. We use Cayley trees and Prüfer codes in a manner that may seem needlessly complicated; however, in the first stage, we see how important the condition that the graph T contains no edge $\ell \to \ell$ is, which is implied by the fact that T is a Cayley tree. The construction aims at ensuring that the graph T' describing the marked edges $e_1, \dots, e_{\ell-1}$ remains a Cayley tree.

References

[1] E. A. BENDER and L. B. RICHMOND, *A multivariate Lagrange inversion formula for asymptotic calculations*, Electron. J. Combin., 5:Research Paper 33, 4 pp. (electronic), 1998.

[2] O. BERNARDI and A. H. MORALES *Counting trees using symmetries*, arXiv:1206.0598, 2013.

[3] M. BOUSQUET, C. CHAUVE, G. LABELLE and P. LEROUX, *Two bijective proofs for the arborescent form of the Good-Lagrange formula and some applications to colored rooted trees and cacti*, Theoret. Comput. Sci. **307** (2) (2003), 277–302.

[4] L. CHOTTIN, *Une démonstration combinatoire de la formule de Lagrange à deux variables*, Discrete Math. **13** (3) (1975), 215–224.

[5] L. CHOTTIN, *Énumération d'arbres et formules d'inversion de séries formelles*, J. Combin. Theory Ser. B **31** (1) (1981), 23–45.

[6] N. DERSHOWITZ and S. ZAKS, *The cycle lemma and some applications*, European J. Combin. **11** (1) (1990), 35–40.

[7] A. DVORETZKY and TH. MOTZKIN, *A problem of arrangements*, Duke Math. J. **14** (1947), 305–313.

[8] IRA M. GESSEL, *A combinatorial proof of the multivariable Lagrange inversion formula*, J. Combin. Theory Ser. A **45** (2) (1987), 178–195.

[9] I. P. GOULDEN and D. M. KULKARNI, *Multivariable Lagrange inversion, Gessel-Viennot cancellation, and the matrix tree theorem*, J. Combin. Theory Ser. A **80** (2) (1997), 295–308.

[10] A. JOYAL, *Une théorie combinatoire des séries formelles*, Advances in Mathematics **42** (1981), 1–82.

[11] G. LABELLE, *Une nouvelle démonstration combinatoire des formules d'inversion de Lagrange*, Adv. in Math. **42** (3) (1981), 217–247.

[12] M. LOTHAIRE, "Combinatorics on Words", Cambridge University Press, 2nd edition, 1999.

[13] M. LOTHAIRE, "Applied Combinatorics on Words", Cambridge University Press, 2005.

[14] J. PITMAN, *Enumerations of trees and forests related to branching processes and random walks*, In: "Microsurveys in Discrete Probability", D. Aldous and J. Propp (eds.), volume 41 of DIMACS Ser. Discrete Math. Theoret. Comput. Sci., Amer. Math. Soc., Providence, RI, 1998.

[15] H. PRÜFER, *Neuer beweis eines satzes über permutationen*, Arch. Math. Phys. **27** (1918), 742–744.

[16] GEORGE N. RANEY, *Functional composition patterns and power series reversion*, Trans. Amer. Math. Soc. **94** (1960), 441–451.

Combinatorics and algorithms

Simplifying inclusion - exclusion formulas

Xavier Goaoc[1], Jiří Matoušek[2], Pavel Paták[3], Zuzana Safernová[4] and Martin Tancer[5]

Abstract. Let $\mathcal{F} = \{F_1, F_2, \ldots, F_n\}$ be a family of n sets on a ground set S, such as a family of balls in $\mathbb{R}^d$. For every finite measure μ on S, such that the sets of $\mathcal{F}$ are measurable, the classical *inclusion-exclusion formula* asserts that $\mu(F_1 \cup F_2 \cup \cdots \cup F_n) = \sum_{I:\emptyset \neq I \subseteq [n]} (-1)^{|I|+1} \mu\bigl(\bigcap_{i \in I} F_i\bigr)$; that is, the measure of the union is expressed using measures of various intersections. The number of terms in this formula is exponential in n, and a significant amount of research, originating in applied areas, has been devoted to constructing simpler formulas for particular families $\mathcal{F}$. We provide an upper bound valid for an arbitrary $\mathcal{F}$: we show that every system $\mathcal{F}$ of n sets with m nonempty fields in the Venn diagram admits an inclusion-exclusion formula with $m^{O(\log^2 n)}$ terms and with ± 1 coefficients, and that such a formula can be computed in $m^{O(\log^2 n)}$ expected time.

1 Introduction

One of the basic topics in introductory courses of discrete mathematics is the *inclusion-exclusion principle* (also called the *sieve formula*), which allows one to compute the number of elements of a union $F_1 \cup F_2 \cup \cdots \cup F_n$ of n sets from the knowledge of the sizes of all intersections of the F_i's.

We will consider a slightly more general setting, where we have a ground set S and a (finite) *measure* μ on S; then the inclusion-exclusion

[1] Université de Lorraine, Villers-lès-Nancy, F-54600, France. CNRS, Villers-lès-Nancy, F-54600, France, Inria, Villers-lès-Nancy, F-54600, France. A visit of this author in Prague was partially supported from Grant GRADR Eurogiga GIG/11/E023. E-Mail: goaoc@loria.fr

[2] Department of Applied Mathematics, Charles University, Malostranské nám. 25, 118 00 Praha 1, Czech Republic, and Institute of Theoretical Computer Science, ETH Zurich, 8092 Zurich, Switzerland. Supported by the ERC Advanced Grant No. 267165. Partially supported by the Charles University Grant GAUK 421511 and by Grant GRADR Eurogiga GIG/11/E023. E-mail: matousek@kam.mff.cuni.cz

[3] Department of Algebra, Charles University, Sokolovská 83, 186 75 Praha 8, Czech Republic. Partially supported by the Charles University Grant GAUK 421511 and SVV-2012-265317. E-Mail: patak@kam.mff.cuni.cz

[4] Department of Applied Mathematics, Charles University, Malostranské nám. 25, 118 00 Praha 1, Czech Republic. Supported by the ERC Advanced Grant No. 267165. Partially supported by the Charles University Grant GAUK 421511. E-Mail: zuzka@kam.mff.cuni.cz

[5] Department of Applied Mathematics, Charles University, Malostranské nám. 25, 118 00 Praha 1, Czech Republic. Supported by the ERC Advanced Grant No. 267165. Partially supported by the Charles University Grant GAUK 421511. E-Mail: tancer@kam.mff.cuni.cz

principle asserts that for every collection $F_1, F_2, \ldots, F_n$ of μ-measurable sets, we have

$$\mu\left(\bigcup_{i=1}^{n} F_i\right) = \sum_{I:\emptyset\neq I\subseteq[n]} (-1)^{|I|+1}\mu\left(\bigcap_{i\in I} F_i\right). \tag{1.1}$$

(Here, as usual, $[n] = \{1, 2, \ldots, n\}$ and $|I|$ denotes the cardinality of the set I.) This principle not only plays a fundamental role in various areas of mathematics such as probability theory or combinatorics, but it also has important algorithmic applications, e.g., the best known algorithms for several NP-hard problems including graph k-coloring [2], travelling salesman problem on bounded-degree graphs [3], dominating set [11], or partial dominating set and set splitting [8].

The inclusion-exclusion principle involves a number of summands that is exponential in n, the number of sets. In general this cannot be avoided if one wants an *exact* formula valid for *every* family $\mathcal{F}=\{F_1, F_2, \ldots, F_n\}$. Yet, since this is a serious obstacle to efficient uses of inclusion-exclusion, much effort has been devoted to finding "smaller" formulas. These efforts essentially organize along two lines of research.

The first approach gives up on exactness and tries to *approximate* efficiently the measure of the union using the measure of only *some* of the intersections. See, e.g., classical *Bonferroni inequalities* [4]. We give a short overview of this line in the full version of this paper [7].

The second line of research looks for "small" inclusion-exclusion formulas valid for *specific* families of sets. To illustrate the type of simplifications afforded by fixing the sets, consider the family $\mathcal{F} = \{F_1, F_2, F_3\}$ of Figure 1.1. As $F_1 \cap F_3 = F_1 \cap F_2 \cap F_3$, Formula (1.1) can be simplified to

$$\mu(F_1 \cup F_2 \cup F_3) = \mu(F_1) + \mu(F_2) + \mu(F_3) - \mu(F_1 \cap F_2) - \mu(F_2 \cap F_3).$$

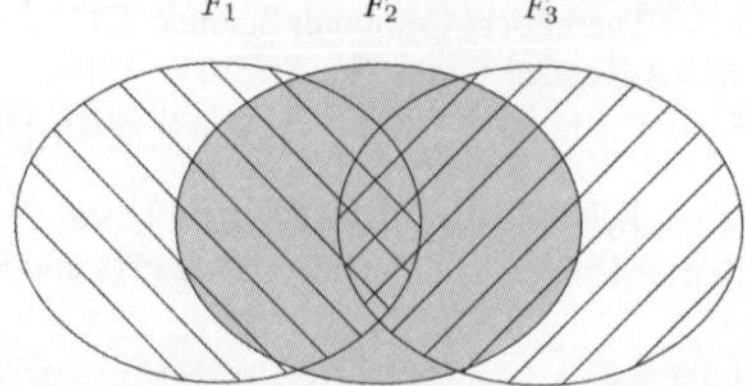

Figure 1.1. Three subsets of $\mathbb{R}^2$ admitting a simpler inclusion-exclusion formula. The ground set $F_1 \cup F_2 \cup F_3$ splits into six nonempty regions recognizable by the filling pattern.

More generally, let us consider a family $\mathcal{F} = \{F_1, F_2, \ldots, F_n\}$, and let us say that a coefficient vector

$$\boldsymbol{\alpha} = (\alpha_I)_{\emptyset \neq I \subseteq [n]} \in \mathbb{R}^{2^n - 1}$$

is an *IE-vector for* $\mathcal{F}$ if we have

$$\mu\left(\bigcup_{i=1}^{n} F_i\right) = \sum_{I:\, \emptyset \neq I \subseteq [n]} \alpha_I \mu\left(\bigcap_{i \in I} F_i\right) \tag{1.2}$$

for every finite measure μ on the ground set of $\mathcal{F}$ (with all the F_i's measurable). Given $\mathcal{F}$, we would like to find an IE-vector for $\mathcal{F}$, such that both the number of nonzero coefficients is small, and the coefficients themselves are not too large. We learned this idea from from [1] and refer to the monograph of Dohmen [5] for an overview of this line of research.

Given a specific family $\mathcal{F} = \{F_1, F_2, \ldots, F_n\}$ of sets, how small can we expect an inclusion-exclusion formula to be? To formalize the problem, we should specify how $\mathcal{F}$ is given. Let us consider the *Venn diagram* of $\mathcal{F}$. For each nonempty index set $\tau \subseteq [n]$, we define the *region* of τ, denoted by $\mathrm{reg}(\tau)$, as the set of all points that belong to the sets F_i with $i \in \tau$ and no others (see Figure 1.1);

$$\mathrm{reg}(\tau) = \left(\bigcap_{i \in \tau} F_i\right) \setminus \left(\bigcup_{i \notin \tau} F_i\right).$$

The *Venn diagram* of $\mathcal{F}$ is then the collection of all subsets of $[n]$ with non-empty regions; that is,

$$\mathcal{V} = \mathcal{V}(\mathcal{F}) := \{\tau \subseteq [n] \colon \mathrm{reg}(\tau) \neq \emptyset\}.$$

We regard the Venn diagram as a set system on the ground set $[n]$; it is a "dual" of the set system $\mathcal{F}$.

As far as inclusion-exclusion formulas are concerned, all points in a single region are equivalent; it only matters which of the regions are nonempty. Thus, in order to simplify our formulations, we can assume that $\mathcal{F}$ is *standardized*, meaning that the ground set equals the union of the F_i's and each nonempty region has exactly one point. From an algorithmic point of view, this amounts to a preprocessing step for $\mathcal{F}$, in which the part of the ground set S in each nonempty region is contracted to a single point.

Let $\mathcal{F} = \{F_1, F_2, \ldots, F_n\}$ be a family of sets and let m denote the size of $\mathcal{V}$ (which equals the size of the ground set for $\mathcal{F}$ standardized).

A linear-algebraic argument shows that *every* (finite) family $\mathcal{F}$ has an inclusion-exclusion formula with at most m terms [7, Corollary 2.4]. However, the coefficients might be exponentially large (see [7, Example 2.5]). This requires computing the measure of the intersections with enormous precision to obtain meaningful results. Thus, we prefer inclusion-exclusion formulas where not only the number of terms is small, but the coefficients are also small.

Our main result is the following general upper bound; to our knowledge, it is the first upper bound applicable for an arbitrary family.

Theorem 1.1. *Let n and m be integers and let $k = \lceil 2e \ln m \rceil \lceil 1 + \ln \frac{n}{\ln m} \rceil$. Then for every family $\mathcal{F}$ of n sets with Venn diagram of size m, there is an IE-vector $\boldsymbol{\alpha}$ for $\mathcal{F}$ that has at most $\sum_{i=1}^{k} \binom{n}{i} \leq m^{O(\ln^2 n)}$ nonzero coefficients, and in which all nonzero coefficients are ± 1's. Such an $\boldsymbol{\alpha}$ can be computed in $m^{O(\ln^2 n)}$ expected time if $\mathcal{F}$ is standardized.*

The bound in this theorem is quasi-polynomial, but not polynomial, in m and n. We do not know if a polynomial bound can be achieved with ± 1 coefficients but in the full version we show that inclusion-exclusion formulas of *linear* size are impossible in general. Specifically, we show [7, Theorem 1.2] that for infinitely many values of n, there are families of n sets on n points, for which every IE-vector has ℓ_1-norm at least $(n/2)^{3/2}$.

We only sketch proofs of the theorems above. In particular, we skip proofs of auxiliary lemmas. They are proved in the full version [7].

2 Sketch of a proof of Theorem 1.1

We consider a family $\mathcal{F} = \{F_1, F_2, \ldots, F_n\}$ of sets on a ground set S, and assume that the F_i are all distinct. Besides the Venn diagram $\mathcal{V}$, we associate yet another set system with $\mathcal{F}$, namely, the *nerve* $\mathcal{N}$ of $\mathcal{F}$:

$$\mathcal{N} = \mathcal{N}(\mathcal{F}) := \left\{ \sigma \subseteq [n] : \sigma \neq \emptyset, \bigcap_{i \in \sigma} F_i \neq \emptyset \right\}.$$

So both of $\mathcal{N}$ and $\mathcal{V}$ have ground set $[n]$, and we have $\mathcal{V} \subseteq \mathcal{N}$.

Let us enumerate the elements of $\mathcal{V}$ as $\mathcal{V} = \{\tau_1, \tau_2, \ldots, \tau_m\}$ in such a way that $|\tau_i| \leq |\tau_j|$ for $i < j$, and let us enumerate $\mathcal{N} = \{\sigma_1, \sigma_2, \ldots, \sigma_{|\mathcal{N}|}\}$ so that the sets of $\mathcal{V}$ come first, i.e., $\sigma_i = \tau_i$ for $i = 1, 2, \ldots, m$.

In the introduction, we were indexing IE-vectors for $\mathcal{F}$ by all possible subsets $I \subseteq [n]$. But if I is not in the nerve, the corresponding intersection is empty, and thus w.l.o.g. we may assume that its coefficient is zero. Thus, from now on, we will index IE-vectors $\boldsymbol{x}$ as $(x_1, \ldots, x_{|\mathcal{N}|})$, where x_j is the coefficient of $\mu(\bigcap_{i \in \sigma_j} F_i)$.

Abstract tubes. An *(abstract) simplicial complex* with vertex set $[n]$ is a hereditary system of nonempty subsets of $[n]$.[6] An *abstract tube* is a pair $(\mathcal{F}, \mathcal{K})$, where $\mathcal{F} = \{F_1, F_2, \ldots, F_n\}$ is a family of sets and $\mathcal{K}$ is a simplicial complex with vertex set $[n]$, such that for every nonempty region τ of the Venn diagram of $\mathcal{F}$, the subcomplex *induced* on $\mathcal{K}$ by τ, $\mathcal{K}[\tau] := \{\vartheta \in \mathcal{K} \colon \vartheta \subseteq \tau\}$, is contractible.[7]

As first noted by Naiman and Wynn [9, 10], if $(\mathcal{F}, \mathcal{K})$ is an abstract tube, then

$$\mu\left(\bigcup_{i=1}^{n} F_i\right) = \sum_{I \in \mathcal{K}} (-1)^{|I|+1} \mu\left(\bigcap_{i \in I} F_i\right). \tag{2.1}$$

Moreover, truncating the sum yields upper and lower bounds in the spirit of the Bonferroni inequalities ([10]; also see [5, Theorem 3.1.9]).

Small abstract tubes have been identified for families of balls [1,9,10] or halfspaces [10] in $\mathbb{R}^d$, and similar structures were found for families of pseudodisks [6]. We establish Theorem 1.1 by proving that for every family of sets there exists an abstract tube with "small" size that, in addition, can be computed efficiently. We will use the following sufficient condition guaranteeing that $(\mathcal{F}, \mathcal{K})$ is an abstract tube; it is a reformulation of [5, Theorem 4.2.5]. Let MNF($\mathcal{K}$) denote the system of all inclusion-minimal non-faces of $\mathcal{K}$, i.e., of all nonempty sets $I \subseteq [n]$ with $I \notin \mathcal{K}$ but with $I' \in \mathcal{K}$ for every proper subset $I' \subset I$.

Proposition 2.1. *Let $\mathcal{F} = \{F_1, F_2, \ldots, F_n\}$ be a family of sets with Venn diagram $\mathcal{V}$ and let $\mathcal{K}$ be a simplicial complex with vertex set $[n]$. If no set of $\mathcal{V}$ can be expressed as a union of sets in* MNF($\mathcal{K}$), *then $(\mathcal{F}, \mathcal{K})$ is an abstract tube.*

Abstract tubes from selectors. Let $\mathcal{F} = \{F_1, F_2, \ldots, F_n\}$ be a family of sets, and let $\mathcal{V}$ be the Venn diagram of $\mathcal{F}$. A *selector* for $\mathcal{V}$ is a map $w \colon \mathcal{V} \to [n]$ such that $w(\tau) \in \tau$ for every $\tau \in \mathcal{V}$. We observe that each selector for $\mathcal{V}$ provides an abstract tube for $\mathcal{F}$ (which satisfies the sufficient condition of Proposition 2.1).

Lemma 2.2. *Let $\mathcal{F} = \{F_1, F_2, \ldots, F_n\}$, $\mathcal{V} = \mathcal{V}(\mathcal{F})$, and let w be a selector for $\mathcal{V}$. We define the simplicial complex $\mathcal{K}_w = \{\sigma \in \mathcal{N}(\mathcal{F}) \colon$ for all nonempty $\vartheta \subseteq \sigma$ there is $\tau \in \mathcal{V}$ such that $w(\tau) \in \vartheta \subseteq \tau\}$. Then $(\mathcal{F}, \mathcal{K}_w)$ is an abstract tube.*

---

[6] We emphasize that we exclude an empty set from the definition of a simplicial complex. This is non-standard definition; however, it is convenient for our purposes.

[7] By *contractible* we mean contractibility in the sense of topology; roughly speaking, the topological space defined by $\mathcal{K}[\tau]$ can be continuously shrunk to a point.

Large simplices in random $\mathcal{K}_w$. Let ρ be a permutation of $[n]$. We define a selector w_ρ for $\mathcal{V}$ by taking $w(\tau)$ as the smallest element of τ in the linear ordering $\prec$ on $[n]$ given by $\rho(1) \prec \rho(2) \prec \cdots \prec \rho(n)$.

For better readability we write $\mathcal{K}_\rho$ instead of $\mathcal{K}_{w_\rho}$. We want to show that for random ρ, $\mathcal{K}_\rho$ is unlikely to contain too large simplices, and thus leads to a small inclusion-exclusion formula.

Let Γ denote the incidence matrix of $\mathcal{V}$, that is, the 0-1 matrix with m rows and n columns where $\Gamma_{ij} = 1$ if and only if $j \in \tau_i$ (if the original system $\mathcal{F}$ was standardized, then Γ is the transposition of the usual incidence matrix of $\mathcal{F}$). We also denote by Γ_ρ the matrix obtained by applying the permutation ρ to the columns of Γ: the $\rho(i)$th column of Γ_ρ is the ith column of Γ and represents the incidences between permuted $[n]$ and $\mathcal{V}$. The lemma below says that if $\mathcal{K}_\rho$ contains a large simplex, then Γ_ρ contains a particular substructure.

We say that a row R of Γ_ρ is *compatible* with a subset $I \subseteq [n]$ if R contains 1's in all columns with index in I and 0's in all columns with index smaller than $\min(I)$.

Lemma 2.3. *If $\rho(\tau) = \{i_1, i_2, \ldots, i_k\}$ for a simplex τ in $\mathcal{K}_\rho$, with $i_1 < i_2 < \ldots < i_k$, then for every $s \in \{1, 2, \ldots, k\}$ the matrix Γ_ρ contains a row compatible with $\{i_s, i_{s+1}, \ldots, i_k\}$.*

Sketch of a proof of Theorem 1.1. Let n and $m \geq 3$ be integers. Let $\mathcal{F} = \{F_1, F_2, \ldots, F_n\}$ be a family of n sets whose Venn diagram $\mathcal{V}$ has size m. We argue that if ρ is a permutation of $[n]$ chosen uniformly at random, the probability that all compatibility conditions of Lemma 2.3 are satisfied for some $\{i_1, \ldots, i_k\}$ is at most $\frac{1}{2}$ for $k := \lceil 2e \ln m \rceil \lceil 1 + \ln \frac{n}{\ln m} \rceil$. In particular, there exists a permutation ρ^* such that $\mathcal{K}_{\rho^*}$ contains no simplex of size k (or larger). Lemma 2.2 concludes the proof of Equation (2.1).

In order to actually compute a suitable coefficient vector, we choose a random permutation ρ and compute $\mathcal{K}_\rho$.

The choice of a random permutation ρ takes $O(n \ln n)$ time and n random bits. Accepting or rejecting a new simplex by brute-force testing takes $O(mn)$ time. The expected number of times we have to start over with a new permutation ρ is $O(1)$. Altogether, the expected running time of this algorithm is $O\left(\binom{n}{k} mn\right) = m^{O(\ln^2 n)}$. □

References

[1] D. ATTALI and H. EDELSBRUNNER, *Inclusion-exclusion formulas from independent complexes*, Discrete Comput. Geom. **37** (1) (2007), .

[2] A. BJÖRKLUND, T. HUSFELDT and M. KOIVISTO, *Set partitioning via inclusion-exclusion*, SIAM J. Comput. **39** (2009), 546–563.

[3] A. BJÖRKLUND, T. HUSFELDT, P. KASKI and M. KOIVISTO, *The travelling salesman problem in bounded degree graphs*, In: "Automata, Languages and Programming", Part I, volume 5125 of Lecture Notes in Comput. Sci., Springer, Berlin, 2008, 198–209.

[4] C. E. BONFERRONI, *Teoria statistica delle classi e calcolo delle probabilità*, Pubbl. d. R. Ist. Super. di Sci. Econom. e Commerciali di Firenze **8** (1936), 1–62.

[5] K. DOHMEN, "Improved Bonferroni Inequalities via Abstract Tubes", volume 1826 of Lecture Notes in Mathematics, Springer-Verlag, Berlin, 2003.

[6] H. EDELSBRUNNER and E. A. RAMOS, *Inclusion-exclusion complexes for pseudodisk collections*, Discrete Comput. Geom. **17** (1997), 287–306.

[7] X. GOAOC, J. MATOUŠEK, P. PATÁK, Z. SAFERNOVÁ and M. TANCER, *Simplifying inclusion-exclusion formulas*, preprint; http://arxiv.org/abs/1207.2591, 2012.

[8] J. NEDERLOF and J. M. M. VAN ROOIJ, *Inclusion/exclusion branching for partial dominating set and set splitting*, In: "Parameterized and Exact Computation", volume 6478 of Lecture Notes in Comput. Sci., Springer, Berlin, 2010, 204–215.

[9] D. Q. NAIMAN and H. P. WYNN, *Inclusion-exclusion-Bonferroni identities and inequalities for discrete tube-like problems via Euler characteristics*, Ann. Statist. **20** (1) (1992), 43–76.

[10] D. Q. NAIMAN and H. P. WYNN, *Abstract tubes, improved inclusion-exclusion identities and inequalities and importance sampling*, Ann. Statist. **25** (5) (1997), 1954–1983.

[11] J. M. M. VAN ROOIJ, J. NEDERLOF and T. C. VAN DIJK, *Inclusion/exclusion meets measure and conquer: exact algorithms for counting dominating sets*, In: "Algorithms—ESA 2009", volume 5757 of Lecture Notes in Comput. Sci., Springer, Berlin, 2009, 554–565.

Majority and plurality problems

Dániel Gerbner[1], Gyula O. H. Katona[1], Dömötör Pålvölgyi[2] and Balázs Patkós[1]

Abstract. Given a set of n balls each colored with a color, a ball is said to be majority, k-majority, plurality if its color class has size larger than half of the number of balls, has size at least k, has size larger than any other color class; respectively. We address the problem of finding the minimum number of queries (a comparison of a pair of balls if they have the same color or not) that is needed to decide whether a majority, k-majority or plurality ball exists and if so then show one such ball. We consider both adaptive and non-adaptive strategies and in certain cases, we also address weighted versions of the problems.

1 Introduction

Two very much investigated problems in combinatorial search theory are the so-called majority and plurality problems. In this context, we are given n balls in an urn, each colored with one color. A majority ball is one such that its color class has size strictly larger than $n/2$. A plurality ball is one such that its color class is strictly larger than any other color class. The aim is either to decide whether there exists a majority/plurality ball or even to show one (if there exists one). Note that if the number of colors is two, then the majority and the plurality problems coincide. Although there are other models (*e.g.* [6]), in the original settings a query is a pair of balls and the answer to the query tells us whether the two balls have the same color or not. Throughout the paper we consider queries of this sort.

We distinguish two types of algorithms for each problem we consider. An algorithm is *adaptive* if the ith query might depend on the answers received for the first $i-1$ queries. A *non-adaptive* algorithm is simply a set of queries that should be answered at the same time. Clearly, any non-adaptive algorithm can be viewed as an adaptive one and therefore for any kind of combinatorial search problem, the minimum number of queries required in an adaptive algorithm is not more than the minimum number of queries required in a non-adaptive algorithm.

[1] Hungarian Academy of Sciences, Alfréd Rényi Institute, of Mathematics, P.O.B. 127, Budapest H-1364, Hungary.
Email: gerbner.daniel@renyi.mta.hu, ohkatona@renyi.hu, patkos.balazs@nenyi.mta.hu

[2] Eötvös Loránd University, Department of Computer Science, Pázmány Péter sétány 1/C, Budapest H-1117, Hungary. Email: dom@cs.elte.hu

The first results concerning plurality and majority problems are due to Fisher and Salzberg [7] and Saks and Werman [8]. In [7] it is proved that if the number of possible colors is unknown, then the minimum number of queries in an adaptive search for a majority ball is $\lfloor 3n/2 \rfloor - 2$, while [8] contains the result that if the number of colors is two, then the minimum number of queries needed to find a majority ball is $n - b(n)$, where $b(n)$ is the number of 1's in the binary representation of n. The latter result was later reproved in a simpler way by Alonso, Reingold, and Schott [3] and Wiener [10].

The adaptive version of the plurality problem was first considered by Aigner, De Marco, and Montangero in [2], where they showed that for any fixed positive integer c, if the number of possible colors is at most c, then the minimum number of queries needed in an adaptive search for a plurality ball is of linear order, and the constants depend on c. Non-adaptive and other versions of the plurality problem were considered in [1].

Non-adaptive strategies were also studied by Chung, Graham, Mao and Yao [4,5]. They showed a linear upper bound for the majority problem in case the existence of a majority color is assumed. They mention a quadratic lower bound without this extra assumption. We precisely determine the minimum number of queries needed. They also obtain lower and upper bounds on the plurality problem in the non-adaptive case. We improve those bounds and find the correct asymptotics of the minimum number of queries.

1.1 Preliminaries and notation and main results

To state our results we introduce some notations. $M_c(n)$ denotes the minimum number of queries that is needed to determine if there exists a majority color and if so, then to show one ball of that color and $P_c(n)$ denotes the minimum number of queries that is needed to determine if there exists a plurality color and if so, then to show one ball of that color. In both cases the subscript c stands for the number of possible colors. The corresponding non-adaptive parameters are denoted by $M_c^*(n)$ and $P_c^*(n)$. A ball is said to be k-majority if its color class contains at least k balls. $M_c(n, k)$ denotes the minimum number of queries that is needed to determine if there exists a k-majority color and if so, then to show one ball of that color and $M_c^*(n, k)$ denotes the parameter of the non-adaptive variant.

We also consider weighted problems. Let $S = \{w(1), \dots, w(n)\}$ be a multiset of positive numbers, where $w(i)$ is considered to be the weight of the ith ball. For all weighted problems considered in the paper, we assume that the weights are known to all participants. The total weight $w = w(S)$ of the balls is $\sum_{i=1}^{n} w(i)$. The weight $w(T)$ of a subset $T \subseteq [n]$ is $\sum_{i \in T} w(i)$. A color is majority if its color class C satisfies $w(C) > w/2$ and k-majority if $w(C) \geq k$ holds. A color is said to be plurality if the weight $w(C)$ of its color class C is strictly greater than the weights of all the other color classes. The appropriate parameters are denoted by $M_c(S)$, $M_c(S,k)$, $M_c^*(S)$, $M_c^*(S,k)$ and $P_c(S)$, $P_c^*(S)$.

For a set Q of queries we define the *query graph* G_Q to be the graph where the vertices correspond to balls and two vertices are joined by an edge if and only if there exists a query in Q that asks for the comparison of the two corresponding balls. Our main results are the following two theorems.

Theorem 1.1. *Suppose $n \geq c > 2$. Then $M_c^*(n) = \lceil \lceil n/2 \rceil n/2 \rceil$.*

Theorem 1.2. *For any pair of integers n and c, the following holds:*

$$\left\lceil \frac{1}{2}\left(n - 1 - \frac{n-1}{c-1}\right) n \right\rceil \leq P_c^*(n) \leq \frac{c-2}{2(c-1)} n^2 + n.$$

2 Sketches of proofs and additional results

Instead of proving Theorem 1.1 directly, let us address the more general, weighted k-majority model. The next theorem characterizes the query graphs that solve the weighted k-majority problem provided some extra assumptions are satisfied. For simplicity, we will assume that the vertex set of the query graph is $[n]$. Given a multiset $S = \{w_1, w_2, \ldots, w_n\}$ of weights let $\mathcal{F} = \{F \subset [n] : w(F) \geq k\}$ be the family of the k-majority sets. Let $\mathcal{F}_0$ denote the subfamily of minimal sets in $\mathcal{F}$.

Theorem 2.1. *Suppose there are no 1-element sets in $\mathcal{F}$. Then*

(i) *If each member of $\mathcal{F}_0$ induces a connected subgraph of the query graph G_Q, then G_Q solves the weighted k-majority problem.*
(ii) *If $2w([n]) < (k+1)(c+1) - 2$, $c > 2$ and G_Q solves the weighted k-majority problem, then each member of $\mathcal{F}_0$ induces a connected subgraph of G_Q.*
(iii) *Considering the non-weighted version, suppose k is an integer. If $n \leq ck - k - c + 2$, $c > 2$ and G_Q solves the k-majority problem, then each member of $\mathcal{F}$,* i.e. *any set with at least k elements, induces a connected subgraph of G_Q.*

Corollary 2.2. *Suppose $c > 2$, $n > k > n/2$ and $n > 1$. Then a query graph G_Q solves the k-majority problem if and only if G_Q is $(n-k+1)$-connected.*

Theorem 1.1 can be easily deduced from Corollary 2.2. Let us note here that the upper bound of Theorem 1.1 holds also in the weighted case, but such general lower bound cannot be found without extra assumptions on the multiset S of weights. Indeed, if $w_1 > \sum_{i=2}^{n} w_i$, holds, then without any query one knows that the ball with weight w_1 is a majority ball.

Let us now turn our attention to adaptive majority problems. We will only address problems where the number of colors is two. Let $\mu(n)$ denote the largest integer l such that 2^l divides n.

Proposition 2.3. *Let $k > n/2$. Then*

$$M_2(n,k) \geq n - 1 - \mu\left(\sum_{i=k}^{n} \binom{n}{i}\right).$$

Proposition 2.4. *Let $k > n/2$. Then*

(i) *Let us fix an arbitrary ball. The number $Fix_2(n,k)$ of questions needed to determine if the fixed ball is a k-majority ball is at least $n - 1 - \mu(\sum_{i=k}^{n} \binom{n-1}{i-1})$.*

(ii) $M_2(n,k) \geq n - 2 - \mu(\sum_{i=k}^{n} \binom{n-1}{i-1})$.

Let us now consider the weighted (adaptive) majority problem with two colors. Suppose there are $p \neq 0$ ways to partition the multiset S into two parts of equal weight. Then

Proposition 2.5.

(i) *At least $n - 1 - \mu(p)$ questions are needed.*

(ii) *In case p is even, $n - 2$ questions are enough.*

Note that this means $M_2(S) = n - 1$ iff $\mu(p) = 0$. If $\mu(p) = 1$, then $M_2(S) = n - 2$, but the opposite direction is not true.

Another possible assumption about the multiset S of weights is that "every element matters", *i.e.* for every $s \in S$ there exists a coloring of $S \setminus \{s\}$ with two colors (red and blue) such that the majority color is different if we extend this coloring by giving s color red or blue. We say that a multiset S of weights is *non-slavery* if the above condition is satisfied.

Proposition 2.6. *For every non-slavery multiset S of weights the inequality $M_2(S) \geq \lfloor n/2 \rfloor$ holds.*

We now adress plurality problems. The lower bound of Theorem 1.2 immediately follows from the following lemma.

Lemma 2.7. *If Q is a set of queries that solve the problem, then the minimum degree in G_Q is larger than $n-1-\lceil\frac{n-1}{c-1}\rceil$. Furthermore, if $n-1 \equiv 1 \bmod c-1$, then the minimum degree in G_Q is larger than $n-1-\lfloor\frac{n-1}{c-1}\rfloor$.*

To obtain the upper bound of Theorem 1.2 it is enough to show a graph $G_{c,n}$ with $\frac{c-2}{2(c-1)}n^2+n$ edges such that no matter what colors the balls have, we are able to solve the problem after receiving the answers to queries corresponding to edges of $G_{c,n}$. Let $G_{c,n}$ be the $(c-1)$-partite Turán graph on n vertices with a spanning cycle added to each partite set $V_1,\ldots,V_{c-1}$.

One can improve the upper bound of Theorem 1.2 for $c=3$.

Theorem 2.8.

(i) $P_3^*(2k)=k(k+1)$,
(ii) $\frac{1}{2}(k+1)(2k+1) \le P_3^*(2k+1) \le \frac{1}{2}(k+1)(2k+1)+k-1$.

Finally, we turn our attention to the non-adaptive weighted plurality problem, *i.e.* determining $P_c^*(S)$ for a multiset S of weights. Theorem 1.2 shows that in general we cannot hope for anything better than the number of edges of the balanced complete $(c-1)$-partite graph on n vertices. Our last theorem states that for any multiset of weights the number of edges of the balanced complete c-partite graph on n vertices and a linear number of additional queries can solve the problem.

Theorem 2.9. *For any multiset S of n weights the inequality $P_c^*(S) \le \frac{c-1}{2c}n^2+n-c$ holds.*

References

[1] M. AIGNER, *Variants of the majority problem*, Discrete Applied Mathematics **137** (2004), 3–25.

[2] M. AIGNER, G. DE MARCO and M. MONTANGEROB, *The plurality problem with three colors and more*, Theoretical Computer Science **337** (2005), 319–330.

[3] L. ALONSO, E. REINGOLD and R. SCHOTT, *Determining the majority*, Inform. Process. Lett. **47** (1993), 253–255.

[4] F. CHUNG, R. GRAHAM, J. MAO and A. YAO, "Finding Favorites, Electronic Colloquium on Computational Complexity", Report No. 78, 2003.

[5] F. CHUNG, R. GRAHAM, J. MAO and A. YAO, *Oblivious and Adaptive Strategies for the Majority and Plurality Problems*, Computing and combinatorics, Lecture Notes in Computer Science 3595, Springer, Berlin (2005), 329–338.

[6] G. DE MARCO, E. KRANAKIS and G. WIENER, *Computing majority with triple queries*, Proceedings of COCOON 2011, Lecture Notes in Computer Science 6842, 604–611.

[7] M. J. FISHER and S. L. SALZBERG, *Finding a majority among n votes*, J. Algorithms **3** (1982), 375–379.

[8] M. E. SAKS and M. WERMAN, *On computing majority by comparisons*, Combinatorica **11** (1991), 383–387.

[9] R. RIVEST and J. VUILLEMIN, *On recognizing graph properties from adjacency matrices*, Theoret. Comput. Sci. **3** (1976), 371–384.

[10] G. WIENER, *Search for a majority element*, J. Stat. Plann. Inf. **100** (2002), 313–318.

Combinatorial bounds on relational complexity

David Hartman[1], Jan Hubička[1] and Jaroslav Nešetřil[1]

1 Introduction

An *ultrahomogeneous* structure is a (finite or countable) relational structure for which every partial isomorphism between finite substructures can be extended to a global isomorphism. This very strong symmetry condition implies that there are just a few ultrahomogeneous structures. For example, by [14], there are just countably many ultrahomogeneous undirected graphs. The *classification program* is one of the celebrated lines of research in the model theory, see [4, 15]. Various measures were introduced in order to modify a structure to an ultrahomogeneous one. A particularly interesting measure is the minimal arity of added relations (*i.e.* the minimal arity of an extension or lift) which suffice to produce an ultrahomogeneous structure. If these added relations are not changing the automorphism group then the problem is called the *relational complexity* and this is the subject of this paper. In the context of permutation groups, the relational complexity was defined in [5] and was recently popularized by Cherlin [2,3]. We determine the relational complexity of one of the most natural class of structures (the class of structures defined by forbidden homomorphisms). This class has a (countably) universal structure [6]. As a consequence of our main result (Theorem 3.1) we strengthen this by determining its relational complexity. Although formulated in the context of model theory this result has a combinatorial character. Full details will appear in [9].

2 Preliminaries

A *relational structure* (or simply *structure*) $\mathbf{A}$ is a pair $(A, (R^i_{\mathbf{A}} : i \in I))$, where $R^i_{\mathbf{A}} \subseteq A^{\delta_i}$ (*i.e.*, $R^i_{\mathbf{A}}$ is a δ_i-ary relation on A). The family $(\delta_i : i \in I)$ is called the *type* Δ. The type is assumed to be fixed and understood from the context thorough this paper. The class of all (countable) relational structures of type Δ will be denoted by $\mathrm{Rel}(\Delta)$. If the set A is finite we call $\mathbf{A}$ a *finite structure*. We consider only countable

[1] Czech Republic Computer Science Institute of Charles University, Prague 4.
Email: hartman@kam.mff.cuni.cz, hubicka@kam.mff.cuni.cz, nesetril@kam.mff.cuni.cz
The Computer Science Institute of Charles University (IUUK) is supported by grant ERC-CZ LL-1201 of the Czech Ministry of Education and CE-ITI P202/16/6061 of GAČR

or finite structures. We see relational structures as a generalization of digraphs and adopt standard graph theoretic terms (such as isomorphism, homomorphism or connected structures).

Let $\Delta' = (\delta'_i; i \in I')$ be a type containing type Δ. (That is $I \subseteq I'$ and $\delta'_i = \delta_i$ for $i \in I$.) Then every structure $\mathbf{X} \in \mathrm{Rel}(\Delta')$ may be viewed as a structure $\mathbf{A} = (A, (R^i_{\mathbf{A}}; i \in I)) \in \mathrm{Rel}(\Delta)$ together with some additional relations for $i \in I' \setminus I$. We will thus also write $\mathbf{X} = (A, (R^i_{\mathbf{A}}; i \in I), (R^i_{\mathbf{X}}; i \in I' \setminus I))$.

We call $\mathbf{X}$ a *lift* of $\mathbf{A}$. Note that a lift is also in the model-theoretic setting called an *expansion* (as we are expanding our relational language).

For a class $\mathcal{K}$ of relational structures, we denote by $\mathrm{Age}(\mathcal{K})$ the class of all finite structures isomorphic to an (induced) substructure of some $\mathbf{A} \in \mathcal{K}$ and call it the *age of* $\mathcal{K}$. For a structure $\mathbf{A}$, the age of $\mathbf{A}$, $\mathrm{Age}(\mathbf{A})$, is $\mathrm{Age}(\{\mathbf{A}\})$.

The classical result of Fraïssé characterize ultrahomogeneous structures in terms of their age and it can be seen as "zero instance" of problems considered in this paper. Ages of ultrahomogeneous have amalgamation property and there is 1-1 correspondence in between ultrahomogeneous structures and their ages. See *e.g.* [10] for details.

3 Relational complexity

Let $\mathbf{A}$ be a relational structure and let $\mathrm{Aut}(\mathbf{A})$ be the automorphism group of $\mathbf{A}$. A k-ary relation $\rho \subseteq A^k$ is an *invariant* of $\mathrm{Aut}(\mathbf{A})$ if $(\alpha(x_1), \ldots, \alpha(x_k)) \in \rho$ for all $\alpha \in \mathrm{Aut}(\mathbf{A})$ and all $(x_1, \ldots, x_k) \in \rho$. Let $\mathrm{Inv}_k(\mathbf{A})$ denote the set of all k-ary invariants of $\mathrm{Aut}(\mathbf{A})$ and let $\mathrm{Inv}(\mathbf{A}) = \bigcup_{k \geq 1} \mathrm{Inv}_k(\mathbf{A})$, $\mathrm{Inv}_{\leq k} = \bigcup_{1 \leq k' \leq k} \mathrm{Inv}_{k'}(\mathbf{A})$.

It easily follows that lift $(A, (R^i_{\mathbf{A}} : i \in I), \mathrm{Inv}(\mathbf{A}))$ (possibly of infinite type) is an ultrahomogeneous structure for every structure $\mathbf{A} = (A, (R^i_{\mathbf{A}} : i \in I))$. For a structure $\mathbf{A}$ the *relational complexity*, $\mathrm{rc}(\mathbf{A})$, of a $\mathbf{A}$ is the least k such that $(A, (R^i_{\mathbf{A}} : i \in I), \mathrm{Inv}_{\leq_k}(\mathbf{A}))$ is ultrahomogeneous, if such a k exist. If no such k exists, we say that the relational complexity of $\mathbf{A}$ is not finite and write $\mathrm{rc}(\mathbf{A}) = \infty$. Note that if $\mathrm{rc}(\mathbf{A})$ is less than the arity of some relation in Δ, then $\mathrm{rc}(\mathbf{A})$ may be lower than the relational complexity of $\mathrm{Aut}(\mathbf{A})$ as defined in [2].

By Fraïssé Theorem the amalgamation property can be seen as the critical property of age $\mathcal{K}$ such that there exists structure $\mathbf{U}$, $\mathrm{Age}(\mathbf{U}) = \mathcal{K}$, satisfying $\mathrm{rc}(\mathbf{A}) = 0$. We seek, for given n, the structural properties of age $\mathcal{K}$ such that there exists structure $\mathbf{U}$, $\mathrm{Age}(\mathbf{U}) = \mathcal{K}$, $\mathrm{rc}(\mathbf{A}) = n$.

Relational complexity is not interesting for rigid structures (with trivial automorphism group), where it is always 1. Such a structure exists for almost every age. We thus restrict our attention to ω-categorical structures. Recall that structure is ω-categorical if and only if it has only finitely many orbits on n-tuples, for every n, and thus also there are only finitely many invariant relations of arity n, see [10]. Moreover countable ω-categorical structure $\mathbf{U}$ contains as an induced substructure every countable structure $\mathbf{A}$, $\mathrm{Age}(\mathbf{A}) \subseteq \mathrm{Age}(\mathbf{U})$, see [1]. We say that $\mathbf{U}$ is *universal* for the class of all structures of age at most $\mathrm{Age}(\mathbf{A})$ (also called structures *younger* than $\mathbf{U}$).

There is no 1-1 correspondence in between ω-categorical structures and their ages. Consider Rado graph $\mathbf{R}$, graph $\mathbf{R}'$ created as a disjoint union of two Rado graphs, and graph $\mathbf{R}''$ created from $\mathbf{R}'$ by adding a vertex of degree 1 connected to one of vertices of $\mathbf{R}'$. It is not difficult to see that all three graphs are ω-categorical and their age is the class of finite graphs. The relational complexities are different: $\mathrm{rc}(\mathbf{R}) = 0$, $\mathrm{rc}(\mathbf{R}') = 2$, and $\mathrm{rc}(\mathbf{R}'') = 1$.

Relational complexity of a structure differs from the minimal arity of lifted relation needed to turn the structure into an ultrahomogeneous one (studied *i.e.* in [11]). $\mathbf{R}'$ can be homogenized by adding unary relation distinguishing vertices of the first copy of Rado graph; this lift is however not invariant.

Among all ω-categorical structures with a given age we can turn our attention to the "most ultrahomogeneous like" in the following sense. Structure $\mathbf{A}$ with $\mathrm{Age}(\mathbf{A}) = \mathcal{K}$ is *existentially complete* if for every structure $\mathbf{B}$, such that $\mathrm{Age}(\mathbf{B}) = \mathcal{K}$ and the identity mapping (of A) is an embedding $\mathbf{A} \to \mathbf{B}$, every existential statement ψ which is defined in $\mathbf{A}$ and true in $\mathbf{B}$ is also true in $\mathbf{A}$. By [6] for every age $\mathcal{K}$ defined by forbidden monomorphisms with ω-categorical universal structure there is also up to isomorphism unique ω-categorical, existentially complete, and ω-saturated universal structure. This in fact holds more generally. In such cases, for a given age $\mathcal{K}$, the *canonical universal structure* of age $\mathcal{K}$ is the unique ω-categorical, existentially complete, and ω-saturated structure $\mathbf{U}$ such that $\mathrm{Age}(\mathbf{U}) = \mathcal{K}$. Given an age $\mathcal{K}$ we can thus ask:

I. What is the minimal relational complexity of an ω-categorical structure $\mathbf{U}$ such that $\mathrm{Age}(\mathbf{U}) = \mathcal{K}$?
II. What is the relational complexity of the canonical universal structure of age $\mathcal{K}$?

We consider universal structures for class $\mathrm{Forb_h}(\mathcal{F})$ where $\mathcal{F}$ is a family of connected structures. $\mathrm{Forb_h}(\mathcal{F})$ denotes the class of all structures $\mathbf{A}$ for which there is no homomorphism $\mathbf{F} \to \mathbf{A}$, $\mathbf{F} \in \mathcal{F}$. Classes $\mathrm{Forb_h}(\mathcal{F})$ are among the most natural ones where the existence of a universal structure is guaranteed for every finite $\mathcal{F}$, see [6]. For such $\mathcal{F}$ we can fully answer the questions above.

For a structure $\mathbf{A} = (A, (R^i_{\mathbf{A}}, i \in I))$, the *Gaifman graph* is the graph $G_{\mathbf{A}}$ with vertices A and all those edges which are a subset of a tuple of a relation of $\mathbf{A}$, *i.e.*, $G = (A, E)$, where the neighborhood of $\{x, y\} \in E$ if and only if $x \neq y$ and there exists a tuple $\vec{v} \in R^i_{\mathbf{A}}$, $i \in I$, such that $x, y \in \vec{v}$. For a structure $\mathbf{A}$ and a subset of its vertices $B \subseteq A$, the *neighborhood* of set B is the set of all vertices of $A \setminus B$ connected in $G_{\mathbf{A}}$ by an edge to a vertex of B. We denote by $G_{\mathbf{A}} \setminus B$ the graph created from $G_{\mathbf{A}}$ by removing the vertices in B.

A *g-cut* in $\mathbf{A}$ is a subset C of A that is a vertex cut of $G_{\mathbf{A}}$. A g-cut C is *minimal g-separating* in $\mathbf{A}$ if there exists structures $\mathbf{A}_1 \neq \mathbf{A}_2$ induced by $\mathbf{A}$ on two connected components of $G_{\mathbf{A}} \setminus C$ such that C is the intersection of the neighborhood of A_1 and the neighborhood of A_2 in $\mathbf{A}$.

A family of structures is called *minimal* if and only if all structures in $\mathcal{F}$ are cores and there is no homomorphism between two structures in $\mathcal{F}$.

Theorem 3.1. *Let $\mathcal{F}$ be a finite minimal family of finite connected relational structures and $\mathbf{U}$ an ω-categorical universal structure for* $\mathrm{Forb_h}(\mathcal{F})$. *Denote by n the size of the largest minimal g-separating g-cut in $\mathcal{F}$. Then (a)* $\mathrm{rc}(\mathbf{U}) \geq n$*; (b) if $\mathbf{U}$ is the canonical universal structure for* $\mathrm{Forb_h}(\mathcal{F})$*, then* $\mathrm{rc}(\mathbf{U}) = n$.

In the rest of the paper we outline the bounds given by this theorem.

3.1 Upper bounds on relational complexity

It appears that relational complexity is closely related to the homogenization method of constructing universal structures as used in [7]. The main result of [7] is in fact a variant of Fraïssé Theorem with the amalgamation reduced to so-called local failure of amalgamation.

Amalgamation failure of a given age $\mathcal{K}$ is a triple $(\mathbf{A}, \mathbf{B}, \mathbf{C})$ such that $\mathbf{A}, \mathbf{B}, \mathbf{C} \in \mathcal{K}$, the identity mapping (of C) is an embedding $\mathbf{C} \to \mathbf{A}$ and $\mathbf{C} \to \mathbf{B}$, and there is no amalgamation of $\mathbf{A}$ and $\mathbf{B}$ over $\mathbf{C}$ in $\mathcal{K}$. (*i.e.* $(\mathbf{A}, \mathbf{B}, \mathbf{C})$ shows that $\mathcal{K}$ has no amalgamation property). Amalgamation failure is *minimal* if there is no another amalgamation failure $(\mathbf{A}', \mathbf{B}', \mathbf{C}')$ such that identity mappings are embeddings $\mathbf{A}' \to \mathbf{A}$, $\mathbf{B}' \to \mathbf{B}$ and $\mathbf{C}' \to \mathbf{C}$. By techniques of [7] we can show:

Theorem 3.2. *Let $\mathbf{U}$ be the canonical universal structure for age $\mathcal{K}$ and S the set of isomorphism types of minimal amalgamation failures of $\mathbf{U}$. If S is finite then* $\mathrm{rc}(\mathbf{U})$ *is bounded from above by the largest size of $\mathbf{C}$ such that* $(\mathbf{A}, \mathbf{B}, \mathbf{C}) \in S$.

In the special case of $\mathcal{K} = \mathrm{Age}(\mathrm{Forb_h}(\mathcal{F}))$ one can prove a stronger result. This is a consequence of [12].

Theorem 3.3. *Let $\mathcal{F}$ be a (finite or infinite) family of connected structures such that there exists $\mathbf{U}$, the canonical universal structure for* $\mathrm{Age}(\mathrm{Forb_h}(\mathcal{F}))$. *Then* $\mathrm{rc}(\mathbf{U})$ *is bounded from above by the size of the largest minimal g-separating g-cut in $\mathcal{F}$.*

Examples. Theorem 3.3 can be easily applied to many families $\mathcal{F}$. For example:
1. Let $\mathcal{F}$ be family of relational trees and **U** the canonical universal structure for $\mathrm{Forb_h}(\mathcal{F})$ (if it exists). By Theorem 3.3 $\mathrm{rc}(\mathbf{U}) \leq 1$. In fact **U** can be seen as a "blown up" core of a homomorphism dual **D** (given by [8] even for some infinite families $\mathcal{F}$) where each vertex is replaced by infinitely many vertices and each edge by a random bipartite graph. In this case the bound given by Theorem 3.2 is not tight even for $\mathcal{F}$ consisting of an oriented path on 4 vertices.
2. Let $\mathcal{F}_{C_n}$ contain a single odd graph cycle on n vertices. The relational complexity the canonical universal structure for $\mathrm{Forb_h}(\mathcal{F}_{C_n})$ is at most 2.
3. Let $\mathcal{F}_{odd}$ be class of all odd graph cycles. The canonical universal structure for $\mathrm{Forb_h}(\mathcal{F}_{odd})$ is the random bipartite graph **B**. By Theorem 3.3 $\mathrm{rc}(\mathbf{B}) \leq 2$.

3.2 Lower bounds on relational complexity

We obtain the following bound:
Theorem 3.4. *Let $\mathcal{F}$ be a finite minimal family of finite connected structures and* **U** *the ω-categorical universal structure for* $\mathrm{Forb_h}(\mathcal{F})$. *Then* $\mathrm{rc}(\mathbf{U})$ *is bounded from bellow by the size of largest minimal g-separating g-cut in $\mathcal{F}$.*

We use of the following result proved by a special Ramsey-type construction. This is not a technical finesse but this is in a way necessary. It has been shown by [16, 17] that Ramsey classes are related to ultrahomogeneous structures. This connection has been elaborated in the context of topological dynamics in [13].
Theorem 3.5 ([11]). *Let $\mathcal{F}$ be a finite minimal family of finite connected relational structures and $\mathcal{K}$ a lift of class* $\mathrm{Forb_h}(\mathcal{F})$ *adding finitely many new relations of arity at most r. If $\mathcal{K}$ contains ultrahomogeneous lift* **U** *that is universal for $\mathcal{K}$ then the size of minimal g-separating g-cuts of* $\mathbf{F} \in \mathcal{F}$ *is bounded by r.*
Examples. The complexity of a ω-categorical graph universal for $\mathrm{Forb_h}(\mathcal{F}_{C_n})$, $n \geq 5$, (of graphs without odd cycles of length at most n) is at least 2. Combining with Theorem 3.3 we know that relational complexity of the canonical universal structure for the class $\mathrm{Forb_h}(\mathcal{F}_{C_n})$ is 2. On the other hand, however, this does not hold for the class $\mathcal{F}_{odd}$. There exists universal bipartite graphs of relational complexity 1. Finiteness and minimality assumptions are thus needed in Theorem 3.4.

References

[1] P. J. CAMERON, *The age of a relational structure*, In: "Directions in Infinite Graph Theory and Combatorics", R. Diestel (ed.), Topic in Discrete Math. 3, North-Holland, Amsterdam (1992), 49–67.

[2] G. L. CHERLIN, *Finite Groups and Model Theory*, In: "Proceedings of 2nd workshop on homogeneous structures", D. Hartman (ed.), Matfyz press (2012), 6–8.

[3] G. L. CHERLIN, *On the relational complexity of a finite permutation group*, in preparation, available at http://www.math.rutgers.edu/~cherlin/Paper/inprep.html.

[4] G. L. CHERLIN, "The classification of countable homogeneous directed graphs and countable homogeneous n-tournaments", Memoirs Amer. Math. Soc. 621, American Mathematical Society, Providence, RI (1998).

[5] G. L. CHERLIN, G. MARTIN and D. SARACINO, *Arities of permutation groups: Wreath products and k-sets*, J. Combinatorial Theory, Ser. A **74** (1996), 249–286.

[6] G. L. CHERLIN, S. SHELAH and N. SHI, *Universal Graphs with Forbidden Subgraphs and Algebraic Closure*, Advances in Applied Mathematics **22** (1999), 454–491.

[7] J. COVINGTON, *Homogenizable Relational Structures*, Illinois J. Mathematics **34** (4) (1990), 731–743.

[8] P. L. ERDŐS, D. PÁLVÖLGYI, C. TARDIF and G. TARDOS, *On infinite-finite tree-duality pairs of relational structures*, arXiv:1207.4402v1 (submitted) (2012).

[9] D. HARTMAN, J. HUBIČKA, J. NEŠETŘIL, *Complexities of relational structures*, Math. Slovaca, accepted for publication.

[10] W. HODGES, "Model Theory", Cambridge University Press, 1993.

[11] J. HUBIČKA and J. NEŠETŘIL, *Homomorphism and embedding universal structures for restricted classes*, arXiv:0909.4939.

[12] J. HUBIČKA and J. NEŠETŘIL, *Universal structures with forbidden homomorphisms*, arXiv:0907.4079, to appear in J. Väänänen Festschrift, Ontos.

[13] A. S. KECHRIS, V. G. PESTOV and S. TODORČEVIČ, *Fraïssé Limits, Ramsey Theory, and Topological Dynamics of Automorphism Groups*, Geom. Funct. Anal. **15** (2005), 106–189.

[14] A. H. LACHLAN and R. E. WOODROOW, *Countable ultrahomogeneous graphs*, Trans. Amer. Math. Soc. **284** (2) (1984), 431–461.

[15] A. H. LACHLAN, *Homogeneous Structures*, In: "Proc. of the ICM 1986", AMS, Providence, 1987, 314–321.

[16] J. NEŠETŘIL, *For graphs there are only four types of hereditary Ramsey Classes*, J. Combin. Theory B **46** (2) (1989), 127–132.

[17] J. NEŠETŘIL, *Ramsey Classes and Homogeneous Structures*, Combinatorics, Probablity and Computing (2005) 14, 171–189.

A combinatorial approach to colourful simplicial depth

Antoine Deza[1], Frédéric Meunier[2] and Pauline Sarrabezolles[2]

Abstract. The colourful simplicial depth conjecture states that any point in the convex hull of each of $d+1$ sets, or colours, of $d+1$ points in general position in $\mathbb{R}^d$ is contained in at least d^2+1 simplices with one vertex from each set. We verify the conjecture in dimension 4 and strengthen the known lower bounds in higher dimensions. These results are obtained using a combinatorial generalization of colourful point configurations called octahedral systems, which was suggested by Imre Bárány. We present properties of octahedral systems generalizing earlier results on colourful point configurations and exhibit an octahedral system which cannot arise from a colourful point configuration. The number of octahedral systems is also given.

1 Colourful simplicial depth

Given three blue points, three red points, and three green points in the plane such that the convex hull of each of those three monochromatic sets contains the origin $\mathbf{0}$, there exists a blue point, a red point, and a green point forming a triangle containing $\mathbf{0}$.

Generally, a *colourful point configuration* in $\mathbb{R}^d$ is a collection of $d+1$ sets of points, or colours, $\mathbf{S}_1, \ldots, \mathbf{S}_{d+1}$. A *colourful simplex* is defined as the convex hull of a subset S of $\bigcup_{i=1}^{d+1} \mathbf{S}_i$ with $|S \cap \mathbf{S}_i| \leq 1$ for $i = 1, \ldots, d+1$. The Colourful Carathéodory Theorem proven by Bárány in 1982 states that, if the origin $\mathbf{0}$ is in the convex hull of each set of a colourful point configuration, there is a colourful simplex containing $\mathbf{0}$.

Theorem 1.1 (Colourful Carathéodory's theorem [1]). *Let* $\mathbf{S}_1,\ldots,\mathbf{S}_{d+1}$ *be a colourful point configuration. If* $\mathbf{0} \in \bigcap_{i=1}^{d+1} \operatorname{conv}(\mathbf{S}_i)$*, then there is a subset* S *of* $\bigcup_{i=1}^{d+1} \mathbf{S}_i$ *with* $|S \cap \mathbf{S}_i| \leq 1$ *for* $i = 1, \ldots, d+1$*, containing* $\mathbf{0}$ *in its convex hull.*

Assuming that all points are in general position, we define $\mu(d)$ to be the minimum number of colourful simplices containing $\mathbf{0}$ over all colourful point configurations with $\mathbf{0} \in \bigcap_{i=1}^{d+1} \operatorname{conv}(\mathbf{S}_i)$. It has been recently

[1] McMaster University, Advanced Optimization Laboratory, Hamilton, Ontario, Canada. Email: deza@mcmaster.ca

[2] Université Paris Est, CERMICS, Cité Descartes, 77455 Marne-la-Vallée, Cedex 2, France. Email:frederic.meunier@enpc.fr, pauline.sarrabezolles@enpc.fr

investigated by Bárány and Matoušek [3], by Stephen and Thomas [4] and by Deza, Stephen and Xie [7]. In particular, it has been proven that $\mu(d) \leq d^2 + 1$ [2] and, in the same paper, this inequality is conjectured to be an equality. This conjecture has been proven for $d = 1, 2, 3$. It is known that $\mu(d) \geq \left\lceil \frac{(d+1)^2}{2} \right\rceil$ for $d \geq 1$, see [7].

This quantity $\mu(d)$ has been used to obtain a lower bound for the minimum number of simplices containing $\mathbf{0}$ drawn from a set of points in $\mathbb{R}^d$. We refer to [5] for a recent breakthrough on this number by Gromov.

Our main result is Theorem 1.2 which improves the known lower bounds of $\mu(d)$. The proof uses a combinatorial generalization of the colourful point configurations, called octahedral systems, defined in Section 2.

Theorem 1.2.

$$\mu(d) \geq \frac{1}{2}d^2 + \frac{7}{2}d - 8 \textit{ for } d \geq 4$$

Furthermore, we show that $\mu(4) = 17$ proving the conjecture for $d = 4$.

2 Combinatorial approach: octahedral systems

An n-uniform hypergraph is said to be *n-partite* if its vertex set is the disjoint union of n sets $V_1, \ldots, V_n$ and each edge intersects each V_i at exactly one vertex. Such a hypergraph is an $(n+1)$-tuple $(V_1, \ldots, V_n, E)$ where E is the set of edges. We consider the following combinatorial generalization suggested by Bárány to study $\mu(d)$. An *octahedral system* Ω is an n-uniform n-partite hypergraph $(V_1, \ldots, V_n, E)$ with $|V_i| \geq 2$ for $i = 1, \ldots, n$ and satisfying the following *parity condition*: the number of edges of Ω induced by $X \subseteq \bigcup_{i=1}^n V_i$ is even if $|X \cap V_i| = 2$ for $i = 1, \ldots, n$.

The Octahedral Lemma [6] states that, given a colourful point configuration $\mathbf{S}_1, \ldots, \mathbf{S}_{d+1}$ and a subset $X \subseteq \bigcup_{i=1}^{d+1} \mathbf{S}_i$ of points such that $|X \cap \mathbf{S}_i| = 2$ for $i = 1, \ldots, d+1$, there is an even number of colourful simplices generated by points of X and containing the origin $\mathbf{0}$. It shows that the hypergraph $\Omega = (V_1, \ldots, V_{d+1}, E)$, with $V_i = \mathbf{S}_i$ for $i = 1, \ldots, d+1$, and where the edges in E correspond to the colourful simplices containing $\mathbf{0}$, is an octahedral system.

An octahedral system arising from a colourful point configuration $\mathbf{S}_1, \ldots, \mathbf{S}_{d+1}$ such that $\mathbf{0} \in \bigcap_{i=1}^{d+1} \operatorname{conv}(\mathbf{S}_i)$ is without isolated vertex, *i.e.* each vertex belongs to at least one edge. Indeed, an improved version of the Colourful Carathéodory Theorem, given by Bárány [1], states that, any point of such a colourful point configuration is the vertex of at least one colourful simplex containing $\mathbf{0}$. Such a colourful point configuration and its corresponding octahedral system are shown in Figure 2.1.

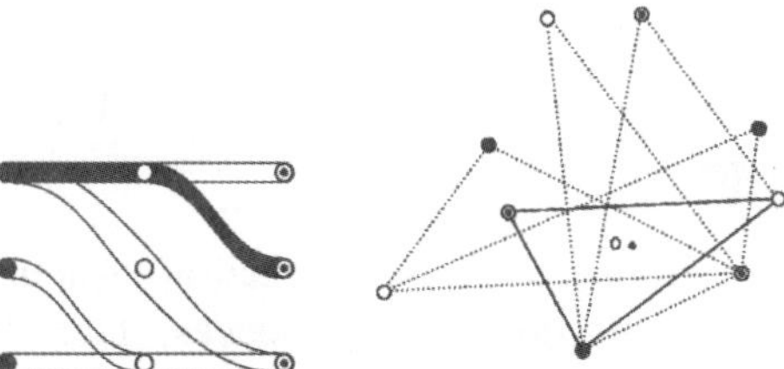

Figure 2.1. An octahedral system arising from a colourful configuration in $\mathbb{R}^2$.

Theorem 2.1 provides a bound for the number of edges of an octahedral system without isolated vertex.

Theorem 2.1. *Any octahedral system without isolated vertex and with* $|V_1| = \ldots = |V_n| = m$ *has at least* $\frac{1}{2}m^2 + \frac{5}{2}m - 11$ *edges, for* $4 \leq m \leq n$.

This theorem is proven by induction on the size $\sum_{i=1}^{n} |V_i|$ of octahedral systems. Troughout the induction, either a vertex and its incident edges can be deleted, resulting in a smaller octahedral system without isolated vertex, or the number of edges can be estimated via combinatorial arguments. Theorem 1.2 is the special case of Theorem 2.1 where $m = n = d + 1$.

We exhibit octahedral systems without isolated vertex with exactly $\sum_{i=1}^{n}(|V_i| - 2) + 2$ edges, proving that the lower bound of Theorem 2.1 cannot be improved beyond $n(m - 2) + 2$ for $m \leq n$.

3 Additional properties of octahedral systems

Proposition 3.1 generalizes a parity property given in [2], Proposition 3.2 generalizes the bound $\mu(d) \geq d + 1$ due to the improved version of Colourful Carathéodory's Theorem by Bárány [1], and Proposition 3.3 improves the result of Theorem 2.1 for $m = n = 5$ refining the same arguments.

Proposition 3.1. *An octahedral system* $\Omega = (V_1, \ldots, V_n, E)$ *with even* $|V_i|$ *for* $i = 1, \ldots, n$ *has an even number of edges.*

Proposition 3.2. *A non-trivial octahedral system* $\Omega = (V_1, \ldots, V_n, E)$ *has at least* $\min_i |V_i|$ *edges, and this bound is tight.*

Proposition 3.3. *An octahedral system without isolated vertex* $\Omega = (V_1, \ldots, V_5, E)$ *with* $|V_1| = \ldots = |V_5| = 5$ *has at least* 17 *edges.*

Proposition 3.3 gives $\mu(4) = 17$, since $\mu(d) \leq d^2 + 1$. It shows that the conjecture $\mu(d) = d^2 + 1$ holds for $d = 4$.

Theorem 3.4 determines the number of distinct octahedral systems, solving an open question raised in [6].

Theorem 3.4. *Given n disjoint finite vertex sets $V_1, \ldots, V_n$, the number of octahedral systems on $V_1, \ldots, V_n$ is $2^{\Pi_{i=1}^n |V_i| - \Pi_{i=1}^n (|V_i|-1)}$.*

The proof uses the fact that the octahedral systems form the kernel of some linear map between $\mathbb{F}_2$-vector spaces. Thus the octahedral systems form a $\mathbb{F}_2$-vector space, whose dimension is computed via the rank-nullity theorem.

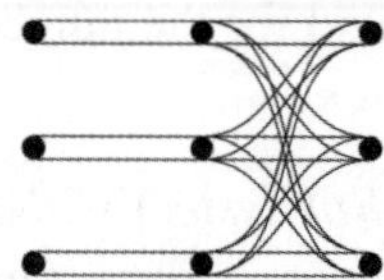

Figure 3.1. A non realisable (3, 3, 3)-octahedral system with 9 edges

Finally, we provide an octahedral system without isolated vertex that cannot arise from a colourful point configuration $\mathbf{S}_1, \ldots, \mathbf{S}_{d+1}$ in $\mathbb{R}^d$, answering Question 6 of [6], see Figure 3.1.

References

[1] I. BÁRÁNY, *A generalization of Carathéodory's theorem*, Discrete Math. **40** (2-3) (1982), 141–152.

[2] A. DEZA, S. HUANG, T. STEPHEN and T. TERLAKY, *Colourful simplicial depth*, Discrete Comput. Geom. **35** (4) (2006), 597–615.

[3] I. BÁRÁNY and J. MATOUŠEK, *Quadratically many colorful simplices*, SIAM J. Discrete Math. **21** (1) (2007), 191–198.

[4] T. STEPHEN and H. THOMAS, *A quadratic lower bound for colourful simplicial depth*, J. Comb. Optim. **16** (4) (2008), 324–327.

[5] M. GROMOV, *Singularities, Expanders and Topology of Maps. Part 2: from Combinatorics to Topology Via Algebraic Isoperimetry*, Geom. Funct. Anal. **20** (2) (2010), 416–526.

[6] G. CUSTARD, A. DEZA, T. STEPHEN and F. XIE, *Small octahedral systems*, Proceedings of the 23rd Canadian Conference on Computational Geometry (CCCG'11), 2011.

[7] A. DEZA, T. STEPHEN and F. XIE, *More colourful simplices*, Discrete Comput. Geom. **45** (2) (2011), 272–278.

Complexity and approximation of the smallest k-enclosing ball problem

Vladimir Shenmaier[1]

Abstract. Given an n-point set in Euclidean space $\mathbb{R}^d$ and an integer k, consider the problem of finding the smallest ball enclosing at least k of the points. In the case of a fixed dimension the problem is polynomial-time solvable but in the general case, when d is not fixed, the complexity status of the problem was not yet known. We prove that the problem is strongly NP-hard and describe an idea of PTAS.

The Smallest k-Enclosing Ball problem is considered:

Problem Sk-EB Given a set X of n points in Euclidean space $\mathbb{R}^d$ and an integer k. Find the smallest ball enclosing at least k of the points.

The problem has a lot of interesting interpretations of life, due to simplicity of the formulation. One of them is in the area of military affairs: given coordinates of n purposes, hit k of them in one gulp with a minimum charge.

Related work

The earliest reference of a special case of this problem occurs in the middle of the 19th century [9]. In the case of a fixed dimension, particularly Euclidean plane (the most studied case), the problem is polynomially solvable [2]. However, the running time of the best known algorithms depends exponentially on dimension [4, 6]. In the general case, when d is not fixed (belongs to a problem instance), the complexity status of the problem was not yet known. Agarwal *et al.* [1] study a very related problem "smallest enclosing ball with outliers" and present an approximation scheme (PTAS) for high dimensions based on coresets.

[1] Sobolev Institute of Mathematics, Novosibirsk, Russia. Email: shenmaier@mail.ru. This research is supported by RFBR (projects 12-01-00093, 12-01-33028 and 13-01-91370ST), Presidium RAS (project 227) and IM SBRAS (project 7B).

Our results

We prove that the problem is strongly NP-hard and unless P = NP there is no fully polynomial-time approximation scheme (FPTAS) (that does not follow from the strong NP-hardness when solution values are not integer). Also we describe a straightforward idea of PTAS that computes a $(1+\varepsilon)$-approximation in $O(n^{1/\varepsilon^2+1}d)$ time for any $\varepsilon > 0$.

1 Hardness results

We formulate a special case of Sk-EB in the form of the problem of verification of properties on a set of Boolean vectors:

Special Case Given a set X of n Boolean points, $X \subseteq \{0, 1\}^d$, an integer $k \in [1, n]$, and a real value $R > 0$. Determine wether there is a Euclidean ball of radius R enclosing at least k of the points.

Using ideas from the work of [7], we give a reduction to this special case from the following strongly NP-complete problem [8]:

Clique in Regular Graph Given a regular graph G and an integer k. Determine wether there is a complete k-vertex subgraph of G.

Theorem 1.1. *Clique in Regular Graph can be reduced to Special Case in polynomial time.*

Reduction. Let G be a regular graph on n vertices and m edges, and Δ be the degree of the vertices of G. Define a set X as the set of m-dimensional rows of the incidence matrix of G. Observe that any two points x and y of X have a distance $\sqrt{2\Delta - 2}$ if the corresponding vertices are adjacent, and $\sqrt{2\Delta}$ otherwise.

For any k-point set K in $\mathbb{R}^d$, define an average of K as $\overline{c}(K) = \sum_{x\in K} x \,/ k$, and for any $y \in \mathbb{R}^d$, define a value $f(y, K) = \sum_{x\in K} \|x - y\|^2$.

Lemma 1.2. $f(y, K) = f(\overline{c}(K), K) + k\,\|y - \overline{c}(K)\|^2$.

Proof. Indeed, $f(y, K) = \sum_{x\in K} \|x - \overline{c}(K)\|^2 - \sum_{x\in K} 2\,\langle x - \overline{c}(K), y - \overline{c}(K)\rangle + \sum_{x\in K} \|y - \overline{c}(K)\|^2$, where $\langle .\,, .\rangle$ is scalar product of vectors. The first term in this expression is equal to $f(\overline{c}(K), K)$, the latest is equal to $k\,\|y - \overline{c}(K)\|^2$, and the second is equal to zero, since $\sum\limits_{x\in K} x = k\,\overline{c}(K)$. □

For any k-point set K in $\mathbb{R}^d$, define a value $g(K) = \sum_{x\in K} \sum_{y\in K} \|x - y\|^2$.

Lemma 1.3. $g(K) = 2k\, f(\overline{c}(K), K)$.

Proof. This follows from Lemma 1.2. □

Lemma 1.4. *Let R be the radius of the smallest ball enclosing at least k of the points of X, and $A = (1 - 1/k)(\Delta - 1)$. Then $R^2 \leq A$, if there is a k-clique in the graph G, and $R^2 \geq A + 2/k^2$ otherwise.*

Proof. Suppose that a k-clique exists. Consider the point $\overline{c}(K)$, where K is the set of points corresponding to the vertices of the clique. Using Lemma 1.3, we have $f(\overline{c}(K), K) = g(K)/2k = (k^2 - k)(2\Delta - 2)/2k = k\,A$. By symmetry, distances from $\overline{c}(K)$ to all the points of K are the same. It follows that the squares of these distances are equal to the value $f(\overline{c}(K), K)/k = A$. Thus, the ball of radius $\sqrt{A}$ centered at the point $\overline{c}(K)$ covers all the points of K.

Suppose that there is no k-clique. Let c be the center of the smallest ball enclosing k points of X, and K be the k-point subset it covers. Then R is equal to the maximal distance between c and the points of K. Since the maximum of squares of distances is at least its average, we have $R^2 \geq f(c, K)/k$. On the other hand, from Lemma 1.2 it follows that $f(c, K) \geq f(\overline{c}(K), K)$. Therefore, $R^2 \geq f(\overline{c}(K), K)/k = g(K)/2k^2$. But any of $k^2 - k$ summands in the definition of $g(K)$ corresponding to the pairs of distinct points is either 2Δ or $2\Delta - 2$. And, by the assumption of nonexistence of k-clique, at least two of them equals 2Δ. Therefore, $g(K) \geq (k^2 - k)(2\Delta - 2) + 4$, and then $R^2 \geq (1 - 1/k)(\Delta - 1) + 2/k^2 = A + 2/k^2$. □

Proof of Theorem 1.1. Lemma 1.4 implies that existence of a k-clique in graph G is equivalent to existence of a ball of radius $\sqrt{A}$ enclosing k points of X. This completes the proof of Theorem 1.1. □

Corollary 1.5. *The Smallest k-Enclosing Ball problem is strongly* NP-*hard.*

In the case of optimization problems with integer-valued solutions the strong NP-hardness implies that there is no fully polynomial-time approximation scheme (FPTAS) [5, 10]. Unfortunately, the radius of the smallest k-enclosing ball is not integer in general, and so that fact needs proof.

Theorem 1.6. *For the Smallest k-Enclosing Ball problem, there is no fully polynomial-time approximation scheme (FPTAS) unless* P = NP.

Proof. By Lemma 1.4, if there is a k-clique in a graph G then the radius of the smallest ball enclosing k of the points of X is bounded by $\sqrt{A}$, and otherwise it is at least $\sqrt{A + 2/k^2} > \sqrt{A}\,(1 + 1/2Ak^2) > \sqrt{A}\,(1 + 1/2n^3)$. It follows that there is no polynomial-time algorithm that computes a $(1+1/2n^3)$-approximation unless $\mathsf{P} = \mathsf{NP}$. On the other hand, for an arbitrary polynomial $p(n)$, any FPTAS allows to compute a $(1 + 1/p(n))$-approximation in polynomial time. Therefore, existence of FPTAS is impossible. □

2 Approximation scheme

We describe an idea of a polynomial-time approximation scheme (PTAS) for Sk-EB based on the simple gradient descent type algorithm [3] for the Small Enclosing Ball problem:

Problem SEB Given a set X of n points in Euclidean space $\mathbb{R}^d$. Find the smallest ball enclosing all the points.

Observe that this is a special case of Sk-EB: $k = n$. The following approximation algorithm is considered in [3]:

Algorithm for SEB Let i be an arbitrary positive integer.
`Step 1:` Choose any point $c_1 \in K$.
`Step j, ` $j = 2, 3, \ldots, i$`:` Take a point $p_j \in K$, which is furthest away from c_{j-1}, and define $c_j = c_{j-1} + (p_j - c_{j-1})/j$.
`Output:` The ball of radius $R(c_i, K) = \max_{x \in K} \|x - c_i\|$ centered in c_i.

Observe that the points $p_1, p_2, \ldots, p_i$, where $p_1 = c_1$, are not necessary distinct. And the point c_i is equal to the average of the points $p_1, \ldots, p_i$: $c_i = \sum_{j=1}^{i} p_j / i$.

Proposition 2.1 ([3]). *Let c^* and R^* be the center and the radius of the smallest ball enclosing all the points of K. Then $\|c_i - c^*\| \le R^*/\sqrt{i}$.*

Proposition 2.1 and the triangle inequality imply that the ball of radius $R(c_i, K)$ centered in c_i is a $(1 + 1/\sqrt{i}\,)$-approximation for the problem SEB. In fact, the above algorithm is a fully polynomial-time approximation scheme (FPTAS) that computes a $(1 + \varepsilon)$-approximate solution of SEB in $O(nd/\varepsilon^2)$ time for any $\varepsilon > 0$.

Describe an algorithm for the original problem Sk-EB. Let c^* and R^* be the center and the radius of the smallest ball enclosing at least k of the points. Clearly, this ball is an optimal solution of the problem SEB on the k-point set K^* the ball covers. Then by Proposition 2.1, the average of some points $p_1, \ldots, p_i \in K^*$ is at distance $R^*/\sqrt{i}$ from the point c^*.

Initially, we have no any information about the points $p_1, \ldots, p_i$ (and about the whole set K^*), but we know that these points are in the set X. The idea of the algorithm is brute-force searching for all the sequences of length i in the set X to find that sequence $p_1, \ldots, p_i$, whose average is close to c^*.

Algorithm for Sk-EB Let i be an arbitrary positive integer and $t^i : \{1, \ldots, n^i\} \to X^i$ be a enumeration of all the sequences of length i in the set X.
`Step` s, $s = 1, \ldots, n^i$: Consider the sequence $t^i(s)$, say $t^i(s) = p_1, \ldots, p_i$. Define a point $c^i(s) = \sum_{j=1}^{i} p_j / i$, find a set $K^i(s)$ of the k points of X nearest to $c^i(s)$, and obtain the radius $R(c^i(s), K^i(s)) = \max_{x \in K^i(s)} \|x - c^i(s)\|$.
`Output:` The ball of radius R^i centered in c^i corresponding to the minimal radius $R(c^i(s), K^i(s))$ obtained at steps $s = 1, \ldots, n^i$.

Theorem 2.2. *Let R^* be the radius of the smallest ball enclosing k of the points of X. Then $R^i / R^* \le 1 + 1/\sqrt{i}$.*

Proof. As mentioned above, an optimal solution of the problem Sk-EB is also optimal for the problem SEB on the k-point set K^* this solution covers. Suppose that points $p_1, \ldots, p_i$ are chosen at steps $1, \ldots, i$ of the algorithm for SEB on the set K^*, $c_i = \sum_{j=1}^{i} p_j / i$ and $R(c_i, K^*) = \max_{x \in K^*} \|x - c_i\|$. By Proposition 2.1 and the triangle inequality, it follows that $R(c_i, K^*)/R^* \le 1 + 1/\sqrt{i}$.

On the other hand, the sequence $p_1, \ldots, p_i$ is equal to some sequence $t^i(s)$ in the algorithm for Sk-EB. Then $c^i(s) = c_i$ and $R(c^i(s), K^i(s)) = R(c_i, K^i(s)) \le R(c_i, K^*)$ since the set $K^i(s)$ consists of the nearest points to c_i. Therefore, $R^i \le R(c_i, K^*)$ and we have $R^i / R^* \le 1 + 1/\sqrt{i}$. □

Estimate the running time of the algorithm. Since a choice of the k points of X nearest to $c^i(s)$ takes at most $O(n)$ operations (*e.g.* using the algorithm for the kth smallest number from n [11]), and all the operations over d-dimensional points take a time $O(d)$, the running time of the algorithm for Sk-EB is bounded by $O(n^{i+1} d)$.

Observe that for any $\varepsilon > 0$, we can take the parameter $i = 1/\varepsilon^2$ to compute a $(1 + \varepsilon)$-approximation for Sk-EB. In this case the running time is bounded by $O(n^{1/\varepsilon^2 + 1} d)$. Thus, this algorithm is actually a polynomial-time approximation scheme (PTAS) for the Smallest k-Enclosing Ball problem.

References

[1] P. K. AGARWAL, S. HAR-PELED and K. R. VARADARAJAN, *Geometric approximation via coresets*, Combinatorial and Computational Geometry, MSRI **52** (2005), 1–30.

[2] A. AGGARWAL, H. IMAI, N. KATOH and S. SURI, *Finding k points with minimum diameter and related problems*, J. Algorithms **12** (1991), 38–56.

[3] M. BADOIU and K. L. CLARKSON, *Smaller core-sets for balls* In: "Proc. 14th ACM-SIAM Symposium on Discrete Alg.", 203, 801–802.

[4] D. EPPSTEIN and J. ERICKSON, *Iterated nearest neighbors and finding minimal polytopes*, Disc. Comp. Geom. **11** (1994), 321–350.

[5] M. R. GAREY and D. S. JOHNSON, *"Strong" NP-completeness results: motivation, examples, and implications*, J. ACM. **25** (3) (1978), 499–508.

[6] S. HAR-PELED and S. MAZUMDAR, *Fast algorithms for computing the smallest k-enclosing disc* Algorithmica **41** (3) (2005), 147–157.

[7] A. V. KEL'MANOV and A. V. PYATKIN, *NP-completeness of some problems of choosing a vector subset*, J. Applied and Industrial Math. **5** (3) (2011), 352–357.

[8] C. H. PAPADIMITRIOU, "Computational Complexity", New-York: Addison-Wesley, 523 pages, 1994.

[9] J. J. Sylvester: *A Question in the Geometry of Situation*, Quart. J. Math. 1:79, 1857.

[10] V. VAZIRANI Approximation Algorithms. *Berlin: Springer-Verlag*, page 71, 2001.

[11] H. WIRTH, "Algorithms + Data Structures = Programs", New Jersey: Prentice Hall, 366 pages, 1976.

Testing uniformity of stationary distribution

Sourav Chakraborty[1], Akshay Kamath[1] and Rameshwar Pratap[1]

Abstract. In this paper, we prove that for a regular directed graph whether the uniform distribution on the vertices of the graph is a stationary distribution, depends on a local property of the graph, namely if (u, v) is an directed edge then outdegree (u) is equal to indegree (v). This result also has an application to the problem of testing, whether the stationary distribution obtained by random walk on a directed graph is uniform or "far" from being uniform. We reduce this problem to testing Eulerianity in the orientation model.

1 Introduction

Markov chains are one of the most important and most studied structures in Theoretical Computer Science. The most important characteristics of a Markov chain are its stationary distribution and its mixing time. In particular, one often wants to know if a given distribution is a stationary distribution of a given Markov chain. In this paper, we focus on the Markov chain obtained by a random walk on a directed graph. Stationary distribution of a Markov chain is a global property of the graph, hence whether a particular distribution is a stationary distribution of a Markov chain depends on the global structure of that Markov chain. We prove that contrary to normal perception, if the graph is regular then whether the uniform distribution on the vertices of the graph is a stationary distribution depends on a local property of the graph. The following theorem, which is the main result of this paper, is a statement about that local property. (See [3] for full version of this paper.)

Theorem 1.1. *If $\overrightarrow{G} = (V, \overrightarrow{E})$ is a digraph such that the total degree (that is Indegree(v) + Outdegree(v)) for every vertex $v \in V$ is the same, then the uniform distribution on the vertices of $\overrightarrow{G}$ is a stationary distribution (for the Markov chain generated by a random walk on $\overrightarrow{G}$) if and only if the graph have the following properties:*
1. *For all $v \in V$, Indegree$(v) \neq 0$ and Outdegree$(v) \neq 0$,*
2. *For every edge $(u, v) \in \overrightarrow{E}$, Outdegree$(u)$ = Indegree(v).*

[1] Chennai Mathematical Institute, Chennai, India. Email: sourav@cmi.ac.in, adkamath@cmi.ac.in, rameshwar@cmi.ac.in

2 Preliminaries

2.1 Graph notations

Throughout this paper, we will be dealing with directed graphs (possibly with multiple edges between any two vertices) in which each edge is directed only in one direction. We will call them **oriented graphs**. We will denote the oriented graph by $\vec{G} = (V, \vec{E})$ and the underlying undirected graph (that is when the direction on the edges are removed) by $G = (V, E)$. For a vertex $v \in V$, the in-degree and the out-degree of v in $\vec{G}$ are denoted by $d^-(v)$ and $d^+(v)$ respectively. An oriented graph $\vec{G} = (V, \vec{E})$ is called a degree-Δ oriented graph if for all $v \in V$, $d^-(v) + d^+(v) = \Delta$. In this paper, we will be focusing on degree-Δ oriented graphs.

2.2 Markov chains

A Markov chain is a stochastic process on a set of states given by a transition matrix. Let S be the set of states with $|S| = n$. Then, the transition matrix T is a $n \times n$ matrix with entries from positive real; the rows and columns are indexed by the states; the u, v-th entry $T_{u,v}$ of the matrix denotes the probability of transition from state u to state v. Since T is stochastic, $\sum_v T_{u,v}$ must be 1. A distribution $\mu : S \to \mathbb{R}^+$ on the vertices is said to be stationary if for all vertices v,

$$\sum_v \mu(u) T_{u,v} = \mu(v).$$

If $\vec{G}$ is an oriented graph then a random walk on $\vec{G}$ defines a Markov chain, where, the states are the vertices of the graph; the probability to traverse an edge is given by the quantity $p_{u,v} = \frac{1}{d^+(u)}$; and hence, the transition probability $T_{u,v}$ from vertex u to vertex v is $p_{u,v}$ times the number of edges between u and v. The uniform distribution on the vertices of $\vec{G}$ is a stationary distribution for this Markov chain if and only if for all $v \in V$,

$$\sum_{u:(u,v)\in \vec{E}} p_{u,v} = 1 = \sum_{w:(v,w)\in \vec{E}} p_{v,w}.$$

3 Structure of graphs with uniform stationary distribution

The following Theorem is a rephrasing of Theorem 1.1.

Theorem 3.1. *Let $\vec{G} = (V, \vec{E})$ be a degree-Δ oriented graph, then the uniform distribution on the vertices of $\vec{G}$ is a stationary distribution (for* *the Markov chain generated by a random walk on* $\overrightarrow{G}$*) if and only if for all* $v \in V$*, both* $d^-(v), d^+(v) \neq 0$ *and for all* $(u, v) \in \overrightarrow{E}$*,*

$$d^+(u) = d^-(v)$$

Proof. First of all, note that the uniform distribution is a stationary distribution for $\overrightarrow{G}$, iff for all $v \in V$

$$\sum_{u:(u,v)\in \overrightarrow{E}} p_{u,v} = 1 = \sum_{w:(v,w)\in \overrightarrow{E}} p_{v,w},$$

where $p_{u,v}$ is the transition probability (defined in subsection 2.2) from vertex u to vertex v. Thus, if the graph $\overrightarrow{G}$ has the property that for all $(u, v) \in \overrightarrow{E}$, $d^+(u) = d^-(v)$, then note that

$$\sum_{u:(u,v)\in \overrightarrow{E}} p_{u,v} = \sum_{u:(u,v)\in \overrightarrow{E}} \frac{1}{d^+(u)} = \sum_{u:(u,v)\in \overrightarrow{E}} \frac{1}{d^-(v)} = 1,$$

the last equality holds because the summation is over all the edges entering v (which is non-empty) and thus have $d^-(v)$ number of items in the summation.

Similarly, we can also prove that $\sum_{w:(v,w)\in \overrightarrow{E}} p_{v,w} = 1$. Thus, we have proved this direction.

Now let us prove the other direction, that is, let us assume that the uniform distribution is a stationary distribution for the Markov chain. Note that, if the uniform distribution is a stationary distribution then there is a path from u to v if and only if u and v are in the same strongly connected component of $\overrightarrow{G}$. This is because the uniform distribution is a stationary distribution if and only if for every cut $C = V_1 \cup V_2$ where $V_2 = (V \setminus V_1)$, we have

$$\sum_{(u,v)\in \overrightarrow{E}, \text{ and } u\in V_1, v\in V_2} p_{u,v} = \sum_{(u,v)\in \overrightarrow{E}, \text{ and } u\in V_2, v\in V_1} p_{u,v}.$$

In other words, if a stationary distribution is uniform then it implies that every connected component in the undirected graph is strongly connected in the directed graph.

Let $v_0, v_1, v_2, \ldots, v_t$ be a sequence of vertices such that the following conditions are satisfied. We call such a sequence as "degree-alternating" sequence of vertices.

– For all $i \geq 0$, $(v_{i+1}, v_i) \in \overrightarrow{E}$
– For all $i \geq 0$, $d^+(v_{2i+1}) = \min\left\{d^+(w) \ : \ (w, v_{2i}) \in \overrightarrow{E}\right\}$ and
– For all $i > 0$, $d^+(v_{2i}) = \max\left\{d^+(w) \ : \ (w, v_{2i-1}) \in \overrightarrow{E}\right\}$.

Claim 3.2. Let $\{v_i\}$ be a "degree-alternating" sequence of vertices. If we define a new sequence $\{S\}$ of positive integers as: for all $k \geq 0$, $s_{2k} = d^-(v_{2k})$ and $s_{2k+1} = d^+(v_{2k+1})$, then this sequence of positive integers is a non-increasing sequence. Moreover, if v_i and v_{i+1} are two consecutive vertices in the sequence such that $d^-(v_{i+1}) \neq d^+(v_i)$ then $s_{i+1} < s_i$.

Using this claim, we would finish the proof of Theorem 3.1. (Due to limitation of space we are unable to give the proof of this claim in this abstract.) Let there be one vertex $w \in V$ such that $d^+(u) \neq d^-(w)$ for some edge $(u, w) \in \overrightarrow{E}$. Let w' be the vertex such that $(w', w) \in \overrightarrow{E}$ and $d^+(w') = \min\{d^+(u) : (u, w) \in \overrightarrow{E}\}$.

Since we have already argued that in the graph every connected component has to be strongly connected, we can create an infinite sequence of vertices such that w and w' appears consecutively and infinitely often. Now by Claim 3.2, it means that the sequence $\{S\}$ is a non-increasing sequence that decreases infinitely many times. But this cannot happen as all the numbers in the sequence $\{S\}$ represent in-degree or out-degree of vertices and hence, are always finite integers and can never be negative. Thus, if one vertex $w \in V$ such that $d^+(u) \neq d^-(w)$ for some edge $(u, w) \in \overrightarrow{E}$, then we hit a contradiction.

Thus, for all edges $(u, v) \in \overrightarrow{E}$, $d^+(u) = d^-(v)$. □

From Theorem 3.1 we can also obtain the following corollary. Both Theorem and Corollary has an application to property testing. We briefly present this application in the next section.

Corollary 3.3. *Let $\overrightarrow{G} = (V, \overrightarrow{E})$ be a connected degree-Δ oriented graph. Then the uniform distribution of vertices is a stationary distribution for the random walk markov chain on $\overrightarrow{G}$, if and only if the following conditions apply:*

1. *If $G = (V, E)$ is non-bipartite, then the graph $\overrightarrow{G}$ is Eulerian.*
2. *If G is bipartite with bipartition $V_1 \cup V_2 = V$ then $|V_1| = |V_2|$ and in-degree of all vertices in one partition will be same and it will be equal to out-degree of all vertices in other partition.*

4 Application to property testing

In property testing, the goal is to look at a very small fraction of the input and tell whether the input has a certain property or it is "far" from satisfying the property. Here "far" means that one has to change at least ϵ fraction of the input to make the input satisfy the property. Theorem 1.1 also has an application to the problem of testing whether a given distribution is uniform or "far" from being uniform. More precisely, if the distribution is the stationary distribution of the *lazy random walk*[2] on a directed graph and the graph is given as an input, then how many bits of the input graph do one need to query in order to decide whether the distribution is uniform or "far" from it? We consider this problem in the **orientation model** (see [2]). In the orientation model, the underlying graph $G = (V, E)$ is known in advance. Each edge in E is oriented (that is directed in exactly one direction). The orientation of the edges has to be queried. The graph is said to be "ϵ-far" from satisfying the property P if one has to reorient at least ϵ fraction of the edges to make the graph have the property.

We reduced this problem to testing Eulerianity in the orientation model. And using result from [1] on query complexity of testing Eulerianity, we obtain bounds on the query complexity for testing whether the stationary distribution is uniform. We briefly discuss this as follows:

Given a degree-Δ oriented graph $\vec{G} = (V, \vec{E})$, we say that the graph has the property $\mathcal{P}$ if for all $(u, v) \in \vec{E}$, we have $d^+(u) = d^-(v)$.

Since the underlying undirected graph is known in advance, we have the connected components. If the graph $\vec{G}$ is "ϵ-far" from satisfying the property $\mathcal{P}$, then there is at least one connected component of $\vec{G}$ that is also "ϵ-far" from satisfying the property $\mathcal{P}$. Thus, we can do testing connected-component wise and *w.l.o.g.*, we can assume that the graph $\vec{G}$ is connected.

From Corollary 3.3, if $\vec{G}$ is non-bipartite then we have to test whether $\vec{G}$ is Eulerian. Since we can determine whether graph is bipartite or not just by looking at the underlying undirected graph, if $\vec{G}$ is non-bipartite then we use the Eulerianity testing algorithm from [1].

Now let $\vec{G}$ be bipartite. Let the bipartition be V_L and V_R. If $|V_L| \neq |V_R|$ then the graph surely does not satisfies property $\mathcal{P}$. From Corollary 3.3, if $|V_L| = |V_R|$ then the graph must have the property that the out-degree of all vertices in V_L must be equal to the in-degree of all vertices in V_R and vice versa. Let v be a vertex in V_L and $d^-(v) = k_1$ and $d^+(v) = k_2$. Now consider any bipartite directed graph $\vec{G^*} = (V, \vec{E^*})$ with bipartition V_L and V_R that satisfies the following conditions:

- The underlying undirected graphs of $\vec{G}$ and $\vec{G^*}$ are exactly same
- $\forall v \in V_L$, $d^-_{\vec{G^*}}(v) = k_2$, $d^+_{\vec{G^*}}(v) = k_1$ and $\forall v \in V_R$, $d^-_{\vec{G^*}}(v) = k_1$, $d^+_{\vec{G^*}}(v) = k_2$.

[2] A lazy random walk always converges to a unique stationary distribution.

Now consider the graph $\overrightarrow{G^{\oplus}} = (V, \overrightarrow{E} + \overrightarrow{E^{*}})$ obtained by superimposing $\overrightarrow{G}$ and $\overrightarrow{G^{*}}$. Clearly, if $\overrightarrow{G}$ has the property $\mathcal{P}$ then $\overrightarrow{G^{\oplus}}$ is Eulerian, and farness from having property $\mathcal{P}$ is also true by following lemma:

Lemma 4.1. *If $\overrightarrow{G}$ is "ϵ-far" from having property $\mathcal{P}$ then $\overrightarrow{G^{\oplus}}$ is "$\frac{\epsilon}{2}$-far" from being Eulerian.*

Now, all we have to test is whether the new graph $\overrightarrow{G^{\oplus}}$ is Eulerian or "$\frac{\epsilon}{2}$-far" from being Eulerian. Note that every query to $\overrightarrow{G^{\oplus}}$ can be simulated by a single query to G. Thus, we can now use the Eulerian testing algorithm from [1].

5 Conclusion

The result holds only for graphs where the in-degree plus out-degree of all the vertices are the same. It would be interesting to see if one can make a similar statement for general graphs.

References

[1] E. FISCHER, O. LACHISH, I. NEWMAN, A. MATSLIAH and O. YAHALOM, *On the query complexity of testing orientations for being Eulerian*, In: "APPROX-RANDOM", 2008, 402-415.

[2] S. HALEVY, O. LACHISH, I. NEWMAN and D. TSUR, *Testing properties of constraint-graphs*, In: "IEEE Conference on Computational Complexity", 2007, 264-277.

[3] S. CHAKRABORTY, A. KAMATH and R. PRATAP, "Testing Uniformity of Stationary Distribution", CoRR, abs/1302.5366, 2013.

On a covering problem in the hypercube

Lale Özkahya[1] and Brendon Stanton[2]

Abstract. In this paper, we address a particular variation of the Turán problem for the hypercube. Alon, Krech and Szabó (2007) asked "In an n-dimensional hypercube, Q_n, and for $\ell < d < n$, what is the size of a smallest set, S, of Q_ℓ's so that every Q_d contains at least one member of S?" Likewise, they asked a similar Ramsey type question: "What is the largest number of colors that we can use to color the copies of Q_ℓ in Q_n such that each Q_d has all the colors represented on the copies of Q_ℓ's." We find upper and lower bounds for each of these questions and provide constructions of the set S above for some specific cases.

1 Introduction

For graphs Q and P, let $\text{ex}(Q, P)$ denote the *generalized Turán number*, *i.e.*, the maximum number of edges in a P-free subgraph of Q. The *n-dimensional hypercube*, Q_n, is the graph whose vertex set is $\{0, 1\}^n$ and whose edge set is the set of pairs that differ in exactly one coordinate. For a graph G, we use $n(G)$ and $e(G)$ to denote the number of vertices and the number of edges of G, respectively.

In 1984, Erdős [9] conjectured that

$$\lim_{n\to\infty} \frac{\text{ex}(Q_n, C_4)}{e(Q_n)} = \frac{1}{2}.$$

Note that this limit exists, because the function above is non-increasing for n and bounded. The best upper bound $\text{ex}(Q_n, C_4)/e(Q_n) \leq 0.6068$ was recently obtained by Balogh, Hu, Lidický and Liu [2] by improving the bound 0.62256 given by Thomason and Wagner [17]. Brass, Harborth and Nienborg [4] showed that the lower bound is $\frac{1}{2}(1 + 1/\sqrt{n})$, when $n = 4^r$ for integer r, and $\frac{1}{2}(1 + 0.9/\sqrt{n})$, when $n \geq 9$.

Erdős [9] also asked whether $o(e(Q_n))$ edges in a subgraph of Q_n would be sufficient for the existence of a cycle C_{2k} for $k > 2$. The value

[1] Department of Mathematics, Hacettepe University, 06800 Beytepe Ankara Turkey. Email: ozkahya@illinoisalumni.org

[2] Department of Mathematics, Iowa State University, Ames, Iowa 50011 USA. Email: brendon.m.stanton@gmail.

of $\text{ex}(Q_n, C_6)/e(Q_n)$ is between 1/3 and 0.3755 given by Conder [7] and Balogh et al. [2], respectively. On the other hand, nothing is known for the cycle of length 10. Except C_{10}, the question of Erdős is answered positively by showing that $\text{ex}(Q_n, C_{2k}) = o(e(Q_n))$ for $k \geq 4$ in [5, 8] and [11].

A generalization of Erdős' conjecture above is the problem of determining $\text{ex}(Q_n, Q_d)$ for $d \geq 3$. As for $d = 2$, the exact value of $\text{ex}(Q_n, Q_3)$ is still not known. The best lower bound for $\text{ex}(Q_n, Q_3)/e(Q_n)$ has been $1 - (5/8)^{0.25} \approx 0.11086$ due to Graham, Harary, Livingston and Stout [12] until Offner [15] improved it to 0.1165. The best upper bound is $\text{ex}(Q_n, Q_3)/e(Q_n) \leq 0.25$ due to Alon, Krech and Szabó [1]. They also gave the best bounds for $\text{ex}(Q_n, Q_d)$, $d \geq 4$, as

$$\Omega\left(\frac{\log d}{d2^d}\right) = 1 - \frac{\text{ex}(Q_n, Q_d)}{e(Q_n)} \leq \begin{cases} \frac{4}{(d+1)^2} & \text{if } d \text{ is odd,} \\ \frac{4}{d(d+2)} & \text{if } d \text{ is even.} \end{cases} \tag{1.1}$$

These Turán problems are also asked when vertices are removed instead of edges and most of these problems are also still open. In a recent paper, Bollobás, Leader and Malvenuto [3] discuss open problems on the vertex-version and their relation to Turán problems on hypergraphs.

Here, we present results on a similar dual version of the hypercube Turán problem that is asked by Alon, Krech and Szabó in [1]. For $\ell < d$, we call a collection of Q_ℓ's a *(d, ℓ)-covering set* if removing this collection leaves Q_n Q_d-free. Let $f^{(\ell)}(n, d)$ denote the minimum size of a (d, ℓ)-covering set of Q_n. Determining this function when $\ell = 1$ is equivalent to the determination of $\text{ex}(Q_n, Q_d)$, since $\text{ex}(Q_n, Q_d) + f^{(1)}(n, d) = e(Q_n)$ and the best bounds for $f^{(1)}(n, d)$ are given in [1] as (1.1). In [1], also the Ramsey version of this problem is asked as follows. For $\ell < d$, a coloring of the copies of Q_ℓ's is called *d, ℓ-polychromatic* if each Q_d has all the colors represented on the copies of Q_ℓ's. Let $pc^{(\ell)}(n, d)$ be the largest number of colors for which there exists a d, ℓ-polychromatic coloring of Q_n. Trivially, $pc^{(\ell)}(n, d) \leq \binom{d}{\ell} 2^{d-\ell}$.

We define $c^{(\ell)}(n, d)$ as

$$c^{(\ell)}(n, d) = \frac{f^{(\ell)}(n, d)}{2^{n-\ell}\binom{n}{\ell}}. \tag{1.2}$$

One can observe that

$$c^{(\ell)}(n, d) \leq \frac{1}{pc^{(\ell)}(n, d)}, \tag{1.3}$$

since any color class used in a d, ℓ-polychromatic coloring is a (d, ℓ)-covering set of Q_n. Note that the following limits exist, since $c^{(\ell)}(n, d)$

is non-decreasing, $pc^{(\ell)}(n,d)$ is non-increasing and both are bounded.

$$c_d^{(\ell)} = \lim_{n\to\infty} c^{(\ell)}(n,d), \quad p_d^{(\ell)} = \lim_{n\to\infty} pc^{(\ell)}(n,d).$$

We obtain bounds on $p_d^{(\ell)}$ as follows.

Theorem 1.1. *For integers $n > d > \ell$, let $0 < r \le \ell+1$ such that $r = d+1 \pmod{\ell+1}$. Then*

$$\begin{aligned} e^{\ell+1}\left(\frac{d+1}{\ell+1}\right)^{\ell+1} &\ge \binom{d+1}{\ell+1} \ge p_d^{(\ell)} \\ &\ge \left\lceil\frac{d+1}{\ell+1}\right\rceil^r \left\lfloor\frac{d+1}{\ell+1}\right\rfloor^{\ell+1-r} \approx \left(\frac{d+1}{\ell+1}\right)^{\ell+1}. \end{aligned} \tag{1.4}$$

Note that a trivial lower bound on $f^{(\ell)}(n,d)$ is given by dividing the number of Q_d's in Q_n to the number of Q_d's a single Q_ℓ can cover at most. Thus, by (1.2), for all n,

$$c^{(\ell)}(n,d) \ge \left\lceil \frac{2^{n-d}\binom{n}{d}}{\binom{n-\ell}{n-d}} \right\rceil \cdot \frac{1}{2^{n-\ell}\binom{n}{\ell}} = \left(2^{d-\ell}\binom{d}{\ell}\right)^{-1}. \tag{1.5}$$

By (1.5), we have the lower bound in the following corollary. The upper bound in Corollary 1.2 is implied by (1.3) and Theorem 1.1.

Corollary 1.2. *For integers $n > d > \ell$ and $r = d - \ell \pmod{\ell+1}$,*

$$\left(2^{d-\ell}\binom{d}{\ell}\right)^{-1} \le c_d^{(\ell)} \le \left\lceil\frac{d+1}{\ell+1}\right\rceil^{-r} \left\lfloor\frac{d+1}{\ell+1}\right\rfloor^{-(\ell+1-r)}.$$

The determination of the exact values of p_d^ℓ and c_d^ℓ remains open. When d and ℓ have a bounded difference from n, we obtain the upper bound in Theorem 1.3, which is a constant factor of the lower bound in (1.5).

Theorem 1.3. *Let $n - d$ and $n - \ell$ be fixed finite integers, where $d > \ell$. Then, for sufficiently large n,*

$$c^{(\ell)}(n,d) \le \left\lceil \frac{r\log(n-\ell)}{\log\left(\frac{r^r}{r^r - r!}\right)} \right\rceil \frac{1+o(1)}{2^{d-\ell}\binom{d}{l}},$$

where $r = n - d$.

Finally, we show an exact result for $c^{(\ell)}(n,d)$ when $d = n-1$.

Theorem 1.4. *For integers $n - 1 > \ell$,*

$$c^{(\ell)}(n, n-1) = \frac{\left\lceil \frac{2n}{n-\ell} \right\rceil}{2^{n-\ell}\binom{n}{\ell}}.$$

References

[1] N. ALON, A. KRECH and T. SZABÓ, *Turán's theorem in the hypercube*, SIAM J. Discrete Math. **21** (2007), 66–72.

[2] J. BALOGH, P. HU, B. LIDICKÝ and H. LIU, *Upper bounds on the size of 4- and 6-cycle-free subgraphs of the hypercube*, arxiv.org/abs/1201.0209v2

[3] B. BOLLOBÁS, T. LEADER and C. MALVENUTO, *Daisies and other Turán Problems*, Combin., Probab. and Comp. **20** (2011), 743–747.

[4] P. BRASS, H. HARBORTH and H. NIENBORG, *On the maximum number of edges in a C_4-free subgraph of Q_n*, J. Graph Theory **19** (1995), 17–23.

[5] C. CHUNG, *Subgraphs of a hypercube containing no small even cycles*, J. Graph Theory **16** (1992), 273–286.

[6] S. M. CIOABĂ, A. KÜNDGEN, C. M. TIMMONS and V. V. VYSOTSKY, *Covering complete r-graphs with spanning complete r-partite r-graphs*, Combin., Probab. and Comput. **20** (2011), 519–527.

[7] M. CONDER, *Hexagon-free subgraphs of hypercubes*, J. Graph Theory **17** (1993), 477–479.

[8] D. CONLON, *An extremal theorem in the hypercube*, Electron. J. Combin. **17** (2010), R111.

[9] *P. Erdős, On some problems in graph theory combinatorial analysis and combinatorial number theory*, Graph Theory and Combinatorics (1984), 1–17.

[10] P. ERDŐS and H. HANANI, *On a limit theorem in combinatorial analysis*, Publ. Math. Debrecen **10** (1963), 10–13.

[11] Z. FÜREDI and L. ÖZKAHYA, *On even-cycle-free subgraphs of the hypercube* J. of Combin. Theo., Ser. A **118** (2011), 1816–1819.

[12] N. GRAHAM, F. HARARY, M. LIVINGSTON and Q. STOUT, *Subcube fault tolerance in hypercubes*, Inform. and Comput. **102** (1993), 280–314.

[13] LU, LINCOLN, *Hexagon-free subgraphs in hypercube Q_n*, private communication.

[14] D. OFFNER, *Polychromatic colorings of subcubes of the hypercube*, SIAM J. Discrete Math. **22** (2008), 450–454.

[15] D. OFFNER, *Some Turán type results on the hypercube*, Discrete Math. **309** (2009), 2905–2912.

[16] V. RÖDL, *On a packing and covering problem*, European J. Combin. **6** (1985), 69–78.

[17] A. THOMASON and P. WAGNER, *Bounding the size of square-free subgraphs of the hypercube*, Discrete Math. **309** (2009), 1730–1735.

A classification of positive posets using isotropy groups of Dynkin diagrams

Marcin Gąsiorek[1] and Daniel Simson[1]

Abstract. We continue our study of positive one-peak posets (presented in our talk in EuroComb 2011). Here we present a more general approach to the classification of arbitrary positive posets J. In particular we show that the Coxeter spectral classification of positive posets can be effectively solved using the right action $* : \mathbb{M}_n(\mathbb{Z}) \times \mathrm{Gl}(n, \mathbb{Z})_D \to \mathbb{M}_n(\mathbb{Z})$, $A \mapsto A * B := B^{tr} \cdot A \cdot B$, of isotropy groups $\mathrm{Gl}(n, \mathbb{Z})_D$ of simply-laced Dynkin diagrams D. By applying recent results of the second named author in [SIAM J. Discrete Math. 27(2013)] we are able to show that, given two connected positive posets I and J with at most 8 points: (i) the incidence matrices C_I and C_J of I and J are $\mathbb{Z}$-congruent if and only if the Coxeter spectra of I and J coincide, and (ii) the matrix C_I is $\mathbb{Z}$-congruent with its transpose C_I^{tr}.

1 Preliminaries and main results

We continue our Coxeter spectral study of positive posets we started in [3–5] in relation with a Coxeter spectral classification of loop-free edge-bipartite graphs developed in [12–14] and combinatorial properties of root systems of simply-laced Dynkin diagrams, that is, the graphs:

$\mathbb{A}_n$: 1 — 2 — · · · · — n-1 — n $(n \geq 1)$;

$\mathbb{D}_n$: 1 — 3 — · · · · — n-1 — n, with 2 attached to 3 $(n \geq 3)$; $\mathbb{E}_6$: 1 — 2 — 3 — 5 — 6, with 4 attached to 3

$\mathbb{E}_7$: 1 — 2 — 3 — 5 — 6 — 7, with 4 attached to 3 $\quad$ $\mathbb{E}_8$: 1 — 2 — 3 — 5 — 6 — 7 — 8, with 4 attached to 3

We mainly study some of the Coxeter spectral analysis problems stated in [13] for loop-free edge-bipartite graphs D and in [15] for finite posets. We use the terminology and notation introduced there. In particular, by $\mathbb{N}$ we denote the set of non-negative integers, by $\mathbb{Z}$ the ring of integers, and by $\mathbb{Q} \subseteq \mathbb{R} \subseteq \mathbb{C}$ the rational, the real and the complex number field, respectively. We view $\mathbb{Z}^n$, with $n \geq 1$, as a free abelian group, and we denote by $e_1, \ldots, e_n$ the standard $\mathbb{Z}$-basis of $\mathbb{Z}^n$. We denote by $\mathbb{M}_n(\mathbb{Z})$ the $\mathbb{Z}$-algebra of all square n by n matrices, by $E \in \mathbb{M}_n(\mathbb{Z})$ the identity matrix, and by $\mathrm{Gl}(n, \mathbb{Z}) := \{A \in \mathbb{M}_n(\mathbb{Z}), \det A \in \{-1, 1\}\}$ the general

[1] Nicolaus Copernicus University, Toruń, Poland.
Email: mgasiorek@mat.umk.pl, simson@mat.umk.pl

$\mathbb{Z}$-linear group. By a finite poset $I \equiv (I, \preceq)$ we mean a partially ordered set I, with respect to a partial order relation $\preceq$. We say that I is a **one-peak poset** if it has a unique maximal element $*$. Every poset I is uniquely determined by its **incidence matrix** $C_I \in \mathbb{M}_m(\mathbb{Z})$, $m = |I|$, that is, the integer square matrix

$$C_I = [c_{ij}]_{i,j \in I} \in \mathbb{M}_m(\mathbb{Z}) \equiv \mathbb{M}_I(\mathbb{Z})$$

with $c_{ij} = 1$, for $i \preceq j$, and $c_{ij} = 0$, for $i \npreceq j$.

The matrix $G_I := \frac{1}{2}(C_I + C_I^{tr}) \in \mathbb{M}_I(\mathbb{Q})$ is called the **symmetric Gram matrix** of a poset I. A poset I is said to be **positive** (resp. **non-negative**), if the symmetric Gram matrix is positive definite (resp. positive semi-definite), see [15].

Following the main idea of the Coxeter spectral analysis of loop-free edge-bipartite graphs (signed graphs [17]) presented in [13,14], we study finite posets I by means of the **Coxeter spectrum** $\mathbf{specc}_I \subseteq \mathbb{C}$, that is, the set $\mathbf{specc}_I$ of all $m = |I|$ eigenvalues of the **Coxeter matrix** $\mathrm{Cox}_I := -C_I \cdot C_I^{-tr} \in \mathbb{M}_m(\mathbb{Z}) \equiv \mathbb{M}_I(\mathbb{Z})$ of I, or equivalently, the set $\mathbf{specc}_I$ of all $m = |I|$ roots of the **Coxeter polynomial** $\mathrm{cox}_I(t) := \det(t \cdot E - \mathrm{Cox}_I) \in \mathbb{Z}[t]$, introduced in [11].

We study finite posets I, J, up to two $\mathbb{Z}$-congruences $\sim_\mathbb{Z}$ and $\approx_\mathbb{Z}$, where $I \sim_\mathbb{Z} J$ iff the symmetric Gram matrices G_I and G_J are $\mathbb{Z}$-congruent, and $I \approx_\mathbb{Z} J$ iff the incidence matrices C_I and C_J are $\mathbb{Z}$-congruent, that is, $C_J = C_I * B := B^{tr} \cdot C_I \cdot B$, for some $B \in \mathrm{Gl}(m, \mathbb{Z})$. We recall from [11, 15] that if I is non-negative, the Coxeter spectrum $\mathbf{specc}_I$ lies on the unit circle $S^1 := \{z \in \mathbb{C};\ |z| = 1\}$ and all points $z \in \mathbf{specc}_I$ are roots of unity. Moreover, non-negative I is positive if and only if $1 \notin \mathbf{specc}_I$. By [10, 11, 15], if $I \approx_\mathbb{Z} J$ then $I \sim_\mathbb{Z} J$ and $\mathbf{specc}_I = \mathbf{specc}_J$, but the converse implication does not hold in general.

One of the main questions of the Coxeter spectral analysis of connected positive posets is if the congruence $I \approx_\mathbb{Z} J$ holds if and only if $\mathbf{specc}_I = \mathbf{specc}_J$. We have proved in [3–5] that this is the case for positive one-peak posets. It is done by computing a complete list of positive one-peak posets and then a case by case inspection.

In the present notes we give an alternative proof of this fact for positive one-peak posets. Moreover, we show that this is also the case for a class of arbitrary (not necessarily one-peak) positive posets I and J. Here we do it by a reduction to a combinatorial problem for $\mathrm{Gl}(m, \mathbb{Z})_D$-orbits in the set $\mathbf{Mor}_D \subset \mathrm{Gl}(m, \mathbb{Z})$ of matrix morsifications of a simply-laced Dynkin diagram D, where $\mathrm{Gl}(m, \mathbb{Z})_D$ is the isotropy group of D studied in [13, 14]. Our main results are the following theorems.

Theorem 1.1. *Assume that I and J are connected finite posets.*

(a) *I is positive if and only if $I \sim_{\mathbb{Z}} DI$, where DI is a simply-laced Dynkin diagram uniquely determined by I.*
(b) *If I and J are positive, with $|I| \leq 8$ and $|J| \leq 8$, then the congruence $I \approx_{\mathbb{Z}} J$ holds if and only if $\mathbf{specc}_I = \mathbf{specc}_J$.*
(c) *If I is positive and $m := |I| \leq 8$ then there exists a $\mathbb{Z}$-invertible matrix $B \in \mathbb{M}_m(\mathbb{Z})$ such that $C_I^{tr} = B^{tr} \cdot C_I \cdot B$ and $B^2 = E$.*

Outline of proof of (a). The "only if" part follows by applying definitions. The "if" part is a consequence of the inflation algorithm $I \mapsto DI$ [13, Algorithm 3.5] (see also [8, 15]) that reduces (in a finite number of steps) a positive connected poset I to a simply-laced Dynkin diagram $D := DI$ such that $G_D = B^{tr} \cdot G_I \cdot B$, for some $B \in \mathrm{Gl}(m, \mathbb{Z})$ and G_D is the symmetric Gram matrix of the Dynkin diagram D, viewed as a poset with a fixed orientation of edges.

The proof of (b) and (c) is outlined in the following section. Its main idea is similar to that one used in the proof of [13, Theorem 2.7].

Our Coxeter spectral study of finite posets, edge-bipartite graphs and matrix morsifications is inspired by their important application in the representation theory of posets, finite groups, finite-dimensional algebras over a field K, and cluster K-algebras, see [1, 8, 13, 14]. We also use ideas of the spectral graph theory, a graph coloring technique, and algebraic methods in graph theory, see [2].

2 Outline of the proof via matrix morsifications

In the proof of Theorem 1.1, given a simply-laced Dynkin diagram D, with $n \geq 2$ vertices, we use the isotropy group of D

$$\mathrm{Gl}(n, \mathbb{Z})_D = \{B \in \mathrm{Gl}(n, \mathbb{Z});\ G_D * B = G_D\} \subseteq \mathbb{M}_n(\mathbb{Z})$$

and its action $* : \mathbb{M}_n(\mathbb{Q}) \times \mathrm{Gl}(n, \mathbb{Z})_D \to \mathbb{M}_n(\mathbb{Q}),\ (A, B) \mapsto A * B := B^{tr} \cdot A \cdot B$ on $\mathbb{M}_n(\mathbb{Q})$ and on $\mathbb{M}_n(\mathbb{Z})$. By [14], the matrix Weyl group $\mathbb{W}_D$ of D is a subgroup of $\mathrm{Gl}(n, \mathbb{Z})_D$. Our main aim in this section is to show that the proof of Theorem 1.1 reduces to a computation of some properties of $\mathrm{Gl}(n, \mathbb{Z})_D$-orbits on the set $\mathbf{Mor}_D$ of matrix morsifications for simply-laced Dynkin diagrams D. Here we mainly apply the technique and results given in [9], and [12, 13].

Following [10]- [14], an integral (resp. rational) **matrix morsification** of D with $n \geq 2$ vertices and the symmetric Gram matrix $G_D \in \mathbb{M}_n(\mathbb{Q})$, is a non-singular matrix $A \in \mathbb{M}_n(\mathbb{Z})$ $\big($resp. $A \in \mathbb{M}_n(\mathbb{Q})\big)$ such that $A + A^{tr} = 2 \cdot G_D$ and the $\mathbb{Z}$-invertible **Coxeter matrix** $\mathrm{Cox}_A := -A \cdot A^{-tr}$

has integer coefficients. By $\mathbf{Mor}_D \subseteq \widehat{\mathbf{Mor}}_D \subseteq \mathbb{M}_n(\mathbb{Q})$ we denote the sets of all integral and all rational morsifications of D. By [14], the map

$$* : \widehat{\mathbf{Mor}}_D \times \mathrm{Gl}(n,\mathbb{Z})_D \to \widehat{\mathbf{Mor}}_D, \quad (A, B) \mapsto A * B := B^{tr} \cdot A \cdot B,$$

is an action of the isotropy group $\mathrm{Gl}(n,\mathbb{Z})_D$ of D, and the subset $\mathbf{Mor}_D \subseteq \widehat{\mathbf{Mor}}_D$ is $\mathrm{Gl}(n,\mathbb{Z})_D$-invariant. Moreover, for $A \in \widehat{\mathbf{Mor}}_D$, $\det A \in \mathbb{Q}$, the **Coxeter number** $\mathbf{c}_A \geq 2$ (i.e., a minimal integer $r \geq 2$ such that $\mathrm{Cox}_A^r = E$), and the **Coxeter polynomial** $\mathrm{cox}_A(t) := \det(t \cdot E - \mathrm{Cox}_A) \in \mathbb{Z}[t]$ of A are $\mathrm{Gl}(n,\mathbb{Z})_D$-invariant.

Given a positive finite poset I, with $n = |I|$ and DI the unique simply-laced Dynkin diagram such that $I \sim_{\mathbb{Z}} DI$ as in Theorem 1.1 (a), we fix a matrix $M_I \in \mathrm{Gl}(n,\mathbb{Z})$ defining the congruence $DI \sim_{\mathbb{Z}} I$, that is, the equality $G_{DI} = M_I^{tr} \cdot G_I \cdot M_I$ holds. By [13, 14], the matrix $C_I * M_I := M_I^{tr} \cdot C_I \cdot M_I$ lies in $\mathbf{Mor}_{DI}$ and we have $\det C_I * M_I = \det C_I = 1$, $\mathrm{cox}_{C_I * M_I}(t) = \mathrm{cox}_{C_I}(t)$, and the Coxeter numbers $\mathbf{c}_I$ and $\mathbf{c}_{C_I * M_I}$ of I and of the morsification $C_I * M_I$ coincide.

Assume that I and J are connected positive posets such that $\mathbf{specc}_I = \mathbf{specc}_J$. Then $n = |I| = |J|$, $C_I * M_I \in \mathbf{Mor}_{DI}$, $C_J * M_J \in \mathbf{Mor}_{DJ}$, $\mathrm{cox}_{C_I * M_I}(t) = \mathrm{cox}_I(t) = \mathrm{cox}_J(t) = \mathrm{cox}_{C_J * M_J}(t)$ and $\det C_I * M_I = \det C_I = 1 = \det C_J = \det C_j * M_J$. Hence we conclude, as in [13], that $DI = DJ$.

Now we show that $I \approx_{\mathbb{Z}} J$. We set $D := DI = DJ$ and we use the isotropy group $\mathrm{Gl}(n,\mathbb{Z})_D$ of the Dynkin diagram D. Consider the right action of $\mathrm{Gl}(n,\mathbb{Z})_D$ on $\widehat{\mathbf{Mor}}_D$. Denote by $F(t) \in \mathbb{Z}[t]$ the Coxeter polynomial $F(t) := \mathrm{cox}_I(t) = \mathrm{cox}_J(t) = \mathrm{cox}_{C_I * M_I}(t) = \mathrm{cox}_{C_J * M_J}(t)$. By rather lengthy computer calculation, we compute the set $\widehat{\mathbf{Mor}}_D$, the Coxeter polynomials $\mathrm{cox}_A(t)$, with $A \in \widehat{\mathbf{Mor}}_D$, the isotropy group $\mathrm{Gl}(n,\mathbb{Z})_D$, and we show that there exists a matrix $A_F \in \mathbf{Mor}_D^F := \{A \in \mathbf{Mor}_D;\ \mathrm{cox}_A(t) = F(t)\}$, such that $\mathbf{Mor}_D^F = A_F * \mathrm{Gl}(n,\mathbb{Z})_D$ if D has at most 8 vertices (see Section 3).

Since the matrices $C_I * M_I$ and $C_J * M_J$ lie in $\mathbf{Mor}_D^F = A_F * \mathrm{Gl}(n,\mathbb{Z})_D$ then there exists a matrix $B \in \mathrm{Gl}(n,\mathbb{Z})_D$ such that $C_I * M_I = B^{tr} \cdot (C_J * M_J) \cdot B$, that is, $C_I * M_I = (C_J * M_J) * B$. It follows that $C_I = C_J * (M_J B M_I^{-1})$, that is, $I \approx_{\mathbb{Z}} J$. Hence we conclude that the implications in (b) of Theorem 1.1 hold.

To prove (c), we note that $C_I^{tr} = C_J$, where $J = I^{op}$ is the poset opposite to I. Since $\mathrm{cox}_{I^{op}}(t) = \mathrm{cox}_I(t)$, then (b) applies and there is a $\mathbb{Z}$-invertible matrix B such that $C_I^{tr} = B^{tr} \cdot C_I \cdot B$. It remains to show that the matrix B can be chosen such that $B^2 = E$. Fortunately, the technique presented earlier allows us to reduce the problem to an analogous problem for matrix morsifications of $D = DI$. However, a complete proof is rather lengthy and combinatorial (it is similar to that one in [14], see also [3]- [6]). Details will be presented in a subsequent paper.

3 Our calculation technique and concluding remarks

In this section we give few concluding remarks on the proof of Theorem 1.1 and algorithmic techniques used to calculate the isotropy groups of Dynkin diagrams, see also [7].

Remark 3.1. Our technique discussed in Section 2 gives an alternative proof of the classification of one-peak positive posets presented in [4], where a complete list of such a posets I with $|I| \leq 8$ is given. We recall a complete list of one-peak positive posets of type $\mathbb{E}_6$, see [3,4].

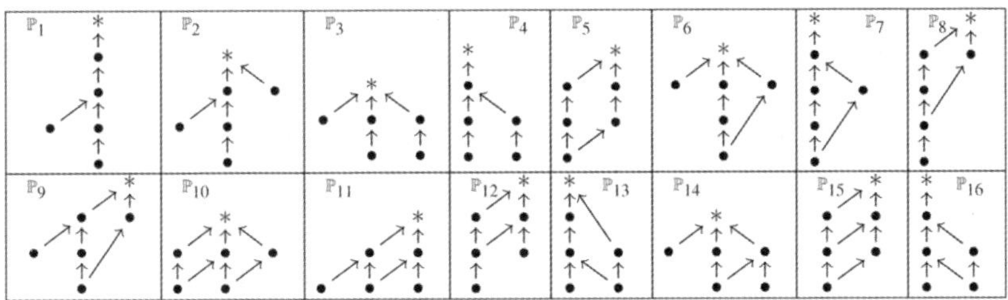

Remark 3.2. By applying our technique, we have computed a complete list of all (not necessarily one-peak) posets of type $\mathbb{E}_6$, consisting of 43 posets: 16 posets with one maximal element, 18 posets with two maximal elements and 9 posets with three maximal elements. The last part of this list looks as follows:

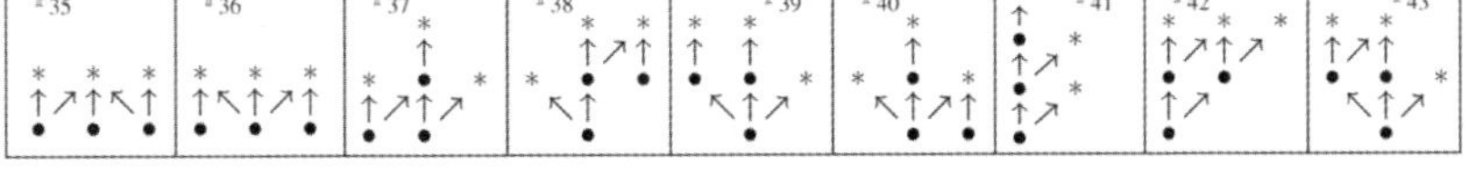

Remark 3.3. Our approach requires a numeric computation of the isotropy group $\mathrm{Gl}(n,\mathbb{Z})_D$ of any Dynkin diagram D with n vertices. This is a challenging computational task as the number of matrices in the $\mathrm{Gl}(n,\mathbb{Z})_D$ grows quickly.

D	$\lvert D\rvert$	$\lvert\mathrm{Gl}(n,\mathbb{Z})_D\rvert$
$\mathbb{A}_6$	6	10080
$\mathbb{A}_7$	7	80640
$\mathbb{A}_8$	8	725760

D	$\lvert D\rvert$	$\lvert\mathrm{Gl}(n,\mathbb{Z})_D\rvert$
$\mathbb{D}_6$	6	46080
$\mathbb{D}_7$	7	645120
$\mathbb{D}_8$	8	10321920

D	$\lvert D\rvert$	$\lvert\mathrm{Gl}(n,\mathbb{Z})_D\rvert$
$\mathbb{E}_6$	6	103680
$\mathbb{E}_7$	7	2903040
$\mathbb{E}_8$	8	696729600

It is difficult not only to compute the $\mathrm{Gl}(n,\mathbb{Z})_D$ group but also to choose a good storage schema of such a big set of data. Here we present the results of our calculations. We will discuss the technical details in the subsequent paper.

To prove Theorem 1.1 we compute the set $\widehat{\mathbf{Mor}}_D \subset \mathbb{M}_n(\mathbb{Q})$ of all rational morsifications of Dynkin diagram D with at $n \leq 8$ vertices (see [14, Note added in proof]) and the isotropy group $\mathrm{Gl}(n,\mathbb{Z})_D$. As a result of computations we prove the following key part of our technique.

Corollary 3.4. *If D is a Dynkin diagram with $n \leq 8$ vertices, then there exists a matrix $A_F \in \mathbf{Mor}_D^F := \{A \in \mathbf{Mor}_D;\ \mathrm{cox}_A(t) = F(t)\} \subseteq \mathbb{M}_n(\mathbb{Z})$ such that $\mathbf{Mor}_D^F = A_F * \mathrm{Gl}(n,\mathbb{Z})_D$.*

References

[1] I. Assem, D. Simson and A. Skowroński, "Elements of the Representation Theory of Associative Algebras. Volume 1. Techniques of Representation Theory", vol. 65. Cambridge Univ. Press, Cambridge-New York, 2006, pp. x+458.

[2] D. Cvetković, P. Rowlinson and S. Simić, "An Introduction to the Theory of Graph Spectra", vol. 75. Cambridge University Press, Cambridge, 2010, pp. xii+364.

[3] M. Gąsiorek and D. Simson, *Programming in Python and an algorithmic description of positive wandering on one-peak posets*, Electronic Notes in Discrete Mathematics **38**(2011), 419–424, doi:10.1016/j.endm.2011.09.068.

[4] M. Gąsiorek and D. Simson, *One-peak posets with positive quadratic Tits form, their mesh translation quivers of roots, and programming in Maple and Python*, Linear Algebra Appl. **436** (2012), 2240–2272, doi:10.1016/j.laa.2011.10.045.

[5] M. Gąsiorek and D. Simson, *A computation of positive one-peak posets that are Tits-sincere*, Colloq. Math. **127** (2012), 83–103, doi:10.4064/cm127-1-6.

[6] R. A. Horn and V. V. Sergeichuk, *Congruences of a square matrix and its transpose*, Linear Algebra Appl. **389** (2004), 347–353, doi:10.1016/j.laa.2004.03.010.

[7] J. E. Humphreys, "Introduction to Lie Algebras and Representation Theory", vol. 9. Springer-Verlag, New York, 1978, pp. xii+171. Second printing, revised.

[8] J. Kosakowska, *Inflation algorithms for positive and principal edge-bipartite graphs and unit quadratic forms*, Fund. Inform. **119** (2012), 149–162.

[9] M. SATO, *Periodic Coxeter matrices and their associated quadratic forms*, Linear Algebra Appl. **406** (2005), 99–108; doi: 10.1016/j.laa. 2005.03.036.

[10] D. SIMSON, *Mesh geometries of root orbits of integral quadratic forms*, J. Pure Appl. Algebra **215** (2010), 13–34, doi: 10.1016//j.jpaa. 2010.02.029.

[11] D. SIMSON, *Integral bilinear forms, Coxeter transformations and Coxeter polynomials of finite posets*, Linear Algebra Appl. **433** (2010), 699–717.

[12] D. SIMSON. *Mesh algorithms for solving principal Diophantine equations, sand-glass tubes and tori of roots*, Fund. Inform. **109** (2011), 425–462.

[13] D. SIMSON *A Coxeter-Gram classification of simply-laced edge-bipartite graphs*, SIAM J. Discrete Math. **27** (2013), 827–854, doi: 10.1137/110843721.

[14] D. SIMSON, *Algorithms determining matrix morsifications, Weyl orbits, Coxeter polynomials and mesh geometries of root orbits for Dynkin diagrams*, Fund. Inform. **123** (2013), 447–490.

[15] D. SIMSON and K. ZAJĄC, *A framework for Coxeter spectral classification of finite posets and their mesh geometries of roots*, Intern. J. Math. Mathematical Sciences, Volume 2013, Article ID 743734, 22 pages. doi:10.1155/2013/743734.

[16] D. SIMSON and K. ZAJĄC, *An inflation algorithm and a toroidal mesh algorithm for edge-bipartite graphs*, Electronic Notes in Discrete Mathematics, 2013, to appear.

[17] T. ZASLAVSKY, *Signed graphs*, Discrete Appl. Math. **4** (1982), 47–74.

Posters

Enumeration and classification of self-orthogonal partial Latin rectangles by using the polynomial method

Raúl M. Falcón[1]

The current paper deals with the enumeration and classification of the set $\mathrm{SOPLR}_{r,n}$ of self-orthogonal $r \times r$ partial Latin rectangles based on n symbols. It can be identified with the set of zeros of the zero-dimensional ideal $I_{r,n} = \langle x_{ijk} \cdot \left(1 - x_{ijk}\right),\ x_{ijk} \cdot x_{i'jk},\ x_{ijk} \cdot x_{ij'k},\ x_{ijk} \cdot x_{ijk'} \colon i \in [r],\ j \in [s],\ k \in [n],\ i' \in [r] \setminus [i],\ j' \in [s] \setminus [j],\ k' \in [n] \setminus [k] \rangle \cup \langle x_{ijp} \cdot x_{klp} \cdot x_{jiq} \cdot x_{lkq} \colon i, j, k, l \in [r],\ p, q \in [n],\ (i, j) \neq (k, l) \rangle \subseteq \mathbb{Q}[x_{111}, \ldots, x_{rrn}]$. Moreover, $|\mathrm{SOPLR}_{r,n}| = \dim_{\mathbb{Q}}(\mathbb{Q}[x_{111}, \ldots, x_{rrn}]/I_{r,n})$. In particular, we obtain the following data:

	$\lvert\mathrm{SOPLR}_{r,n}\rvert$				
	r				
n	1	2	3	4	5
1	2	5	24	147	1,050
2	3	21	407	13,701	660,447
3	4	73	5,086	850,567	256,344,232
4	5	209	47,373	35,805,129	*
5	6	501	333,236	1,035,763,371	*
6	7	1,045	1,826,659	21,134,413,357	*
7	8	1,961	8,103,642	314,221,824,351	*
8	9	3,393	30,148,121	*	*
9	10	5,509	96,972,688	*	*

Given $s \in [n] \cup \{0\}$, let $\sigma_{r,s}$ be the number of partial Latin rectangles of $\mathrm{SOPLR}_{r,n}$ which contain exactly s symbols in their cells. In particular, $|\mathrm{SOPLR}_{r,n}| = \sum_{s=0}^{n} \binom{n}{s} \cdot \sigma_{r,s}$. If $\mathfrak{P}_{s,P}$ denote the main class of $P \in \mathrm{SOPLR}_{r,s}$, then $\sigma_{r,s} = \sum_{P \in \mathfrak{P}_{r,s;s}} |\mathfrak{P}_{s,P}| = \sum_{P \in \mathfrak{P}_{r,s;s}} \frac{2 \cdot r! \cdot s!}{|\mathfrak{I}_s(P,P)| + |\mathfrak{I}_s(P,P^t)|}$, where given two partial Latin rectangles $P, Q \in \mathrm{SOPLR}_{r,s}$, the set $\mathfrak{I}_s(P, Q)$ denotes the set of isotopisms which transform P into Q, which can be identified with the set of zeros of $I_{\mathfrak{I}_s(P,Q)} = \langle 1 - \sum_{j \in [r]} x_{ij} \colon i \in [r] \rangle + \langle 1 - \sum_{j \in [n]} y_{ij} \colon i \in [s] \rangle + \langle 1 - \sum_{i \in [r]} x_{ij} \colon j \in [r] \rangle + \langle 1 -$

[1] School of Building Engineering. University of Seville.
Department of Applied Mathematics I.
Avda. Reina Mercedes 4 A, 41012 - Seville (Spain). Email: rafalgan@us.es

$\sum_{i\in[n]} y_{ij} : j \in [s] \rangle + \langle x_{ij} \cdot (1 - x_{ij}) : i, j \in [r] \rangle + \langle y_{ij} \cdot (1 - y_{ij}) : i, j \in [s] \rangle + \langle x_{ik} \cdot x_{jl} \cdot (y_{p_{ij}q_{kl}} - 1) : i, j, k, l \in [r]$, such that $p_{ij}, q_{kl} \in [s] \rangle + \langle x_{ik} \cdot x_{jl} : i, j, k, l \in [r]$, such that ($p_{ij} = \emptyset$ and $q_{kl} \in [s]$) or ($p_{ij} \in [s]$ and $q_{kl} = \emptyset$) $\rangle$. It is then verified that:

i) $|\mathrm{SOPLR}_{1,n}| = n + 1$.
ii) $|\mathrm{SOPLR}_{2,n}| = n^4 - 2n^3 + 5n^2 + 1$.
ii) $|\mathrm{SOPLR}_{3,n}| = n^9 - 15n^8 + 122n^7 - 604n^6 + 1973n^5 - 4201n^4 + 5640n^3 - 4240n^2 + 1347n + 1$.

Polynomial graph invariants from homomorphism numbers

Delia Garijo[1], Andrew J. Goodall[2] and Jaroslav Nešetřil[2]

The number of homomorphisms $\hom(G, K_k)$ from a graph G to the complete graph K_k is the value of the chromatic polynomial of G at a positive integer k. This motivates the following:

Definition. A sequence of graphs $(H_{\mathbf{k}})$, $\mathbf{k}=(k_1,\ldots,k_h)\in\mathbb{N}^h$, is *strongly polynomial* if for every graph G there is a polynomial $p(G;x_1,\ldots,x_h)$ such that $\hom(G, H_{\mathbf{k}}) = p(G;k_1,\ldots,k_h)$ for every $\mathbf{k}\in\mathbb{N}^h$.

Many important graph polynomials $p(G)$ are determined by strongly polynomial sequences of graphs $(H_{\mathbf{k}})$: *e.g.* the Tutte polynomial, the Averbouch-Godlin-Makowsky polynomial (which includes the matching polynomial) and the Tittmann-Averbouch-Godlin polynomial (which includes the independence polynomial).

We give a new construction of strongly polynomial sequences, which among other things offers a natural generalization of the above polynomials. We start with a simple graph H given as a spanning subgraph of the closure of a rooted tree T. For each $\mathbf{k}=(k_s : s\in V(T))\in\mathbb{N}^{|V(T)|}$ we use the tree T to recursively construct a graph $T^{\mathbf{k}}(H)$, in which, for each $s\in V(T)$, we create k_s isomorphic copies of the subtree T_s of T rooted at s, all pendant from the same vertex as T_s, while propagating adjacencies of H in the closure of T to these copies of T_s.

Theorem 1. *The sequence $(T^{\mathbf{k}}(H))$ is strongly polynomial.*

Define $\beta(H)$ to be the minimum value of $|V(T)|$ such that H is a subgraph of the closure of $T^{\mathbf{k}}(T)$. For example, $\beta(K_{1,\ell}) = 2$, $\beta(P_{2\ell}) =$

[1] University of Seville, Seville, Spain. Email: dgarijo@us.es. Partially supported by JA-FQM164.

[2] IUUK, Charles University, Prague, Czech Republic. Email: andrew@iuuk.mff.cuni.cz, nesetril@iuuk.mff.cuni.cz. Supported by CE-ITI P202/12/G061, and by Project ERCCZ LL1201 Cores.

2ℓ, $\beta(P_{2\ell-1}) = \ell$, and $\beta(K_\ell) = \ell$. We have tree-depth $\mathrm{td}(H) \le \beta(H)$ and $\beta(H) = |V(H)|$ if H has no involutive automorphisms.

Theorem 2. *Let $\mathcal{H}$ be a family of simple graphs such that $\{\beta(H) : H \in \mathcal{H}\}$ is bounded. Then $\mathcal{H}$ can be partitioned into a finite number of subsequences of strongly polynomial sequences of graphs.*

An Erdős–Ko–Rado theorem for matchings in the complete graph

Vikram Kamat[1] and Neeldhara Misra[1]

We consider the following higher-order analog of the Erdős–Ko–Rado theorem [1]. For positive integers r and n with $r \leq n$, let $\mathcal{M}_n^r$ be the family of all matchings of size r in the complete graph K_{2n}. For any edge $e \in E(K_{2n})$, the family $\mathcal{M}_n^r(e)$, which consists of all sets in $\mathcal{M}_n^r$ containing e is called the star centered at e. We prove the following result:

Theorem 1. *For $r < n$, if $\mathcal{A} \subseteq \mathcal{M}_n^r$ is an intersecting family of r-matchings, then $|\mathcal{A}| \leq \phi(n, r)$ with equality holding if and only if $\mathcal{A} = \mathcal{M}_n^r(e)$ for some $e \in E$.*

We note that the case $r = n$ is settled (as part of a stronger theorem for uniform set partitions) by Meagher and Moura [3]. To prove Theorem 1, we use an analog of Katona's cycle method [2]. As in Katona's original proof of the Erdős–Ko–Rado theorem, the main challenge is to come up with a class of objects over which to carry out the double counting argument. We use the notion of Baranyai partitions to construct these objects.

References

[1] P. ERDŐS, C. KO and R. RADO, *Intersection theorems for systems of finite sets*, Quart. J. Math Oxford Ser. (2) **12** (1961), 313–320.

[2] G. O. H. KATONA, *A simple proof of the Erdős–Ko–Rado theorem*, J. Combin. Theory Ser. B **12** (1972), 183–184.

[3] K. MEAGHER and L. MOURA, *Erdős–Ko–Rado theorems for uniform set-partition systems*, Electronic Journal of Combinatorics **12** (2005), Paper 40, 12 pp.

[1] Department of Computer Science & Automation, Indian Institute of Science, Bangalore – 560 012, India. Email: vkamat@csa.iisc.ernet.in, mail@neeldhara.com

A constrained path decomposition of cubic graphs and the path number of cacti

Fábio Botler[1] and Yoshiko Wakabayashi[1]

Kotzig (1957) proved that a cubic graph has a perfect matching if and only if it has a 3-path decomposition (that is, a partition of the edge set into paths of length 3). This result was generalized by Jaeger, Payan, and Kouider (1983), who proved that a $(2k+1)$-regular graph with a perfect matching can be decomposed into bistars. (A bistar is a graph obtained from two disjoint stars by joining their centers with an edge.) In another direction, Heinrich, Liu and Yu (1999) proved that a $3m$-regular graph G admits a balanced 3-path decomposition if and only if G contains an m-factor.

We generalize the result of Kotzig by proving the following result.

Theorem 1. *A cubic n-vertex graph G has a matching of size k if and only if G has a minimum path decomposition into paths of length* 2, 3 *or* 4, *with $2k - \frac{n}{2}$ paths of length* 3.

The *path number* of a graph G, denoted by $\mathrm{pn}(G)$, is the minimum number of edge-disjoint paths needed to partition the edge set of G. According to Lovász (1968), Erdős asked about this parameter, and Gallai conjectured that $\mathrm{pn}(G) \leq (n+1)/2$ for every n-vertex connected graph.

This parameter is not known for most of the graphs. Lovász (1968) proved that $\mathrm{pn}(G) = n/2$ for an n-vertex graph G without even degree vertices. We present here a formula for the path number of cacti, which generalizes the path number of trees. We recall that a cactus is a connected graph in which any two simple cycles have at most one vertex in common.

[1] Institute of Mathematics and Statistics, University of São Paulo, Brazil.
Email: fbotler@ime.usp.br, yw@ime.usp.br

Research partially supported by CNPq (Proc. 477203/2012-4, 303987/2010-3), FAPESP (Proc. 2011/08033-0) and Project USP MaCLinC/NUMEC, Brazil – {fbotler,yw}@ime.usp.br

Theorem 2. *Let G be a union of vertex-disjoint cacti. Let o be the number of odd-degree vertices of G, let c_ℓ be the number of cycles of G that cointain exactly one vertex of degree greater than 2, and let c_i be the number of cycles of G that contain no vertex of degree greater than 2. Then the path number of G is given by*

$$\mathrm{pn}(G) = o/2 + c_\ell + 2c_i\,.$$

On push chromatic number of planar graphs and planar p-cliques

Sagnik Sen[1]

An *oriented graph* $\vec{G}$ is a directed graph without cycles of length 1 or 2. *Pushing* a vertex v of an oriented graph $\vec{G}$ is to change the orientation of all its arcs (replacing the arc $\vec{xy}$ by $\vec{yx}$) incident to v. If we can obtain $\vec{G}_2$ by pushing some vertices of $\vec{G}_1$ then, the two graphs are in an equivalence relation called *push relation*. A *push graph* $[\vec{G}]$ is an equivalance class of oriented graphs ($\vec{G}$ is an element of the class) with respect to push relation. A *homomorphism* of an oriented graph $\vec{G}_1$ to another oriented graph $\vec{G}_2$ is a mapping f from $V(\vec{G}_1)$ to $V(\vec{G}_2)$ such that, if $\vec{uv}$ is an arc of $\vec{G}_1$, then $\vec{f(u)f(v)}$ is an arc of $\vec{G}_2$. The *Push chromatic number* $\chi_p(\vec{G})$ of an oriented graph $\vec{G}$ is the minimum order of an oriented graph $\vec{H}$ such that, some element of $[\vec{G}]$ admits homomorphism to some element of $[\vec{H}]$. Push graph and the notion of homomorphim and chromatic number of push graph has been introduced in [1]. We define a *push clique* or simply *p-clique* to be an oriented graph $\vec{G}$ such that $\chi_p(\vec{G}) = |V(\vec{G})|$.

Theorem 1.

(i) *For an oriented planar graph* $\vec{H}$, $9 \leq \chi_p(\vec{H}) \leq 40$.
(ii) *For a girth* 4, *oriented planar graph* $\vec{H}$, $6 \leq \chi_p(\vec{H}) \leq 20$.
(iii) *For a girth* 5, *oriented planar graph* $\vec{H}$, $4 \leq \chi_p(\vec{H}) \leq 8$.
(iv) *For a girth* 6, *oriented planar graph* $\vec{H}$, $4 \leq \chi_p(\vec{H}) \leq 7$.
(v) *For a girth* 8, *oriented planar graph* $\vec{H}$, $\chi_p(\vec{H}) = 4$.

Theorem 2. *The maximum order of a planar p-clique is* 8.

References

[1] W. F. KLOSTERMEYER and G. MACGILLIVRAY, *Homomorphisms and oriented colorings of equivalence classes of oriented graphs*, Discrete Mathematics **274** (2012), 161–172.

---

[1] Univ. Bordeaux, LaBRI, UMR5800, F-33400 Talence, France. CNRS, LaBRI, UMR5800, F-33400 Talence, France. Email: sagnik.sen@labri.fr. This work is supported by ANR GRATEL project ANR-09-blan-0373-01

Firefighting with general weights

Vitor Costa[1], Simone Dantas[1], Mitre C. Dourado[2], Lucia D. Penso[3] and Dieter Rautenbach[3]

At the 25th Manitoba Conference on Combinatorial Mathematics and Computing in Winnipeg 1995 Hartnell introduced the firefighter game modelling the containment of the spreading of an undesired property within a network. An initial configuration of the game consists of a pair (G, r) where G is a finite, simple, and undirected graph and r is a burned vertex of G. The game proceeds in rounds. In each round, first at most one vertex of G that is not burned is defended and then all vertices of G that are neither burned nor defended and have a burned neighbor are burned. Once a vertex is burned or defended, it remains so for the rest of the game. The game ends with the first round, in which no further vertex is burned. All vertices of G that are not burned at the end of the game are saved.

Here we study a generalization for weighted graphs, where the weights can be positive as well as negative. The objective of the player is to maximize the total weight of the saved vertices of positive weight minus the total weight of the burned vertices of negative weight, that is, the player should save vertices of positive weight and let vertices of negative weight burn. Allowing negative weights drastically changes the character of the game.

Our contributions are two hardness results and two greedy approximation algorithms for trees. We prove that weighted firefighter is hard al-

[1] Instituto de Matemática e Estatística, Universidade Federal Fluminense, Niterói, RJ, Brazil. Email: vitorsilcost@mat.uff.br, sdantas@im.uff.br

[2] Instituto de Matemática, Universidade Federal do Rio de Janeiro, Rio de Janeiro, RJ, Brazil. Email: mitre@dcc.ufrj.br

[3] Institute of Optimization and Operations Research, University of Ulm, Ulm, Germany. Email: lucia.penso@uni-ulm.de, dieter.rautenbach@uni-ulm.de

We acknowledge partial support by CNPq, CAPES, FAPERJ, and the CAPES DAAD Probral project "Cycles, Convexity, and Searching in Graphs".

ready for binary trees, which stands in contrast to the fact that unweighted firefighter is easy for binary trees. Furthermore, we show that weighted firefighter remains hard even if we allow arbitrarily many defended vertices per round. Our two greedy algorithms achieve approximation factors of $\frac{1}{3}$ and $\frac{1}{2}$.

Nowhere-zero flows on signed regular graphs

Eckhard Steffen[1] and Michael Schubert[1]

A signed graph (G, σ) is a graph G together with a function $\sigma : E(G) \to \{\pm 1\}$, which is called a signature of G. The set $N_\sigma = \{e : \sigma(e) = -1\}$ is the set of negative edges of (G, σ) and $E(G) - N_\sigma$ the set of positive edges. We study flows on signed graphs, and $F_c((G, \sigma))$ $(F((G, \sigma)))$ denotes the circular (integer) flow number of (G, σ).

Bouchet [1] conjectured that $F((G, \sigma)) \leq 6$ for every flow-admissible signed graph. This conjecture is equivalent to its restriction on cubic graphs. We prove this conjecture for flow-admissible cubic graphs that have three 1-factors such that any two of them induce a hamiltonian circuit of G. In particular, every flow-admissible uniquely 3-edge-colorable cubic graph has a nowhere-zero 6-flow.

For a graph G and $X \subseteq E(G)$ let $\Sigma_X(G)$ be the set of signatures σ of G, for which (G, σ) is flow-admissible and $N_\sigma \subseteq X$. Define $\mathcal{S}_X(G) = \{r :$ there is a signature $\sigma \in \Sigma_X(G)$ such that $F_c((G, \sigma)) = r\}$ to be the X-flow spectrum of G. The $E(G)$-flow spectrum is the flow spectrum of G and it is denoted by $\mathcal{S}(G)$. If we restrict our studies on integer-valued flows, then $\overline{\mathcal{S}}_X(G)$ denotes the integer X-flow spectrum of G.

We study the integer flow spectrum of signed cubic graphs G. There are cubic graphs whose integer flow spectrum does not contain 5 or 6. But we show, that $\{3, 4\} \subseteq \overline{\mathcal{S}}(G)$, for every bridgeless cubic graph $G \neq K_2^3$, where K_2^3 is the unique cubic graph on two vertices. We construct an infinite family of bridgeless cubic graphs with integer flow spectrum $\{3, 4, 6\}$.

We further study the flow spectrum of $(2t + 1)$-regular graphs $(t \geq 1)$. In [2] is proven that a $(2t + 1)$-regular graph G is bipartite if and only if $F_c((G, \emptyset)) = 2 + \frac{1}{t}$. Furthermore, if G is not bipartite, then $F_c((G, \emptyset)) \geq 2 + \frac{2}{2t-1}$. We extend this kind of result to signed $(2t + 1)$-regular graphs.

[1] Paderborn Institute for Advanced Studies in Computer Science and Engineering, 33102 Paderborn, Germany. Email: es@upb.de, mischub@upb.de

Let $r \geq 2$ be a real number and G be a graph. A set $X \subseteq E(G)$ is r-minimal if (1) there is a signature σ of G such that $F_c((G,\sigma)) = r$ and $N_\sigma = X$, and (2) $F_c((G,\sigma')) \neq r$ for every signature σ' of G with $N_{\sigma'} \subset X$.

We show for $(2t+1)$-regular graphs G, which have a t-factor: A set $X \subseteq E(G)$ is $(2+\frac{1}{t})$-minimal if and only if X is a minimal set such that $G - X$ is bipartite.

Furthermore, if $X \subseteq E(G)$ is a $(2+\frac{1}{t})$-minimal set and $r \in \mathcal{S}_X(G)$, then $r = 2 + \frac{1}{t}$ or $r \geq 2 + \frac{2}{2t-1}$.

References

[1] A. BOUCHET, *Nowhere-zero integral flows on bidirected graph*, J. Comb. Theory Ser. B **34** (1983), 279–292.

[2] E. STEFFEN, *Circular flow numbers of regular multigraphs*, J. Graph Theory **36** (2001), 24–34.

New transience bounds for long walks in weighted digraphs

Bernadette Charron-Bost[1], Matthias Függer[2] and Thomas Nowak[1]

Fix two nodes i and j in an edge-weighted diagraph and form the following sequence: Let $a(n)$ be the maximum weight of walks from i to j of length n; if no such walk exists, $a(n) = -\infty$. It is known that, if G is strongly connected, the sequence $a(n)$ is always eventually periodic with linear defect, *i.e.*, after the *transient*, $a(n + p) = a(n) + p \cdot \lambda$. In fact, the ratio λ is the largest mean weight of cycles in G. We call these cycles *critical*. Periodicity stems from the fact that the weights of critical cycles eventually dominate the maximum weight walks.

In this paper, we show two new asymptotically tight upper bounds on transients in weighted digraphs, taking into account the graph parameters cyclicity and girth. The previously best bound of Hartmann and Arguelles (Math. Oper. Res. 24, 1999) is, in general, incomparable with both our bounds. The significant benefit of our two new bounds is that each of them turns out to be linear in the number of nodes in various classes of weighted digraphs, whereas Hartmann and Arguelles' bound is intrinsically at least quadratic. In particular, our bounds are linear for (bi-directional) trees.

We hence prove the following two upper bounds on the transient in strongly connected digraphs with N nodes. They both contain the (unweighted) index of convergence $\mathrm{ind}(G)$, *i.e.*, the transient of G when choosing all weights to be equal. The term $\|G\|$ denotes the difference of the maximum and minimum weight of edges in G. The mean weight of critical cycles is denoted by λ, and λ_{nc} denotes the largest mean weight of cycles that have no node on critical cycles. Denote by G_c the subgraph induced by the critical cycles.

[1] École polytechnique, F-91128 Palaiseau, France.
Email: charron@lix.polytechnique.fr, nowak@lix.polytechnique.fr

[2] Vienna University of Technology, 1040 Wien, Österreich. Email: matthias.fuegger@tuwien.ac.at

The full paper is arXiv:1209.3342 [cs.DM]

Theorem 1 (Repetitive bound). *Denoting by $\hat{g}$ the maximum girth of strongly connected components of G_c, the transient between any two nodes is at most*

$$\max\left\{\frac{\|G\|\cdot\big(3N-2+\operatorname{ind}(G)\big)}{\lambda-\lambda_{\mathrm{nc}}}\,,\ (\hat{g}-1)+2\,\hat{g}\cdot(N-1)\right\}\ .$$

Theorem 2 (Explorative bound). *Denoting by $\hat{\gamma}$ and $\widehat{\operatorname{ind}}$ the maximum cyclicity and maximum index of strongly connected components of G_c, respectively, the transient between any two nodes is at most*

$$\max\left\{\frac{\|G\|\cdot\big(3N-2+\operatorname{ind}(G)\big)}{\lambda-\lambda_{\mathrm{nc}}}\,,\ (\hat{\gamma}-1)+2\,\hat{\gamma}\cdot(N-1)+\widehat{\operatorname{ind}}\right\}\ .$$

Complexity of determining the irregular chromatic index of a graph

Julien Bensmail[1]

An edge colouring ϕ of a graph G is *locally irregular* if each colour class of ϕ induces a graph whose every adjacent vertices have distinct degrees. The least number $\chi'_{irr}(G)$ of colours used by a locally irregular edge colouring of G (if any) is referred to as the *irregular chromatic index* of G.

Locally irregular edge colouring was introduced as another type of edge colouring permitting to distinguish the adjacent vertices of a graph. This notion is thus related to several other notions of this field, like *vertex-colouring edge-weighting* [1] and *detectable colouring* of graphs [2]. In particular, it was shown that a locally irregular edge colouring is also a vertex-colouring edge-weighting or detectable colouring in some situations [3]. As for these two types of edge colouring, it is conjectured that three colours suffice to obtain a locally irregular edge colouring of any graph whose irregular chromatic index is defined [3].

We here focus on the complexity of the following decision problem.

$$k\text{-LIEC} = \{\text{A graph } G : \text{is it true that } \chi'_{irr}(G) \leq k?\}$$

The problem 1-LIEC is in P, while the relevance of studying k-LIEC for any $k \geq 3$ depends on the correctness of the conjecture mentioned above. We show that 2-LIEC is NP-hard, even when restricted to planar graphs with maximum degree at most 6. This result is proved by reduction from 1-IN-3 SAT.

References

[1] M. KAROŃSKI, T. ŁUCZAK and A. THOMASON, *Edge weights and vertex colours*, J. Combin. Theory, Ser, B **91** (1) (2004), 151-157.

[1] LaBRI, F-33405 Talence, France. Email: jbensmai@labri.fr

[2] L. Addario-Berry, R. E. L. Aldred, K. Dalal and B. A. Reed, *Vertex colouring edge partitions*, J. Combin. Theory, Ser. B **94** (2) (2005), 237–244.

[3] O. Baudon, J. Bensmail, J. Przybyło and M. Woźniak, *On decomposing regular graphs into locally irregular subgraphs*, Preprint MD 065, http://www.ii.uj.edu.pl/preMD/index.php, 2013.

CRM Series
Publications by the Ennio De Giorgi Mathematical Research Center Pisa

The Ennio De Giorgi Mathematical Research Center in Pisa, Italy, was established in 2001 and organizes research periods focusing on specific fields of current interest, including pure mathematics as well as applications in the natural and social sciences like physics, biology, finance and economics. The CRM series publishes volumes originating from these research periods, thus advancing particular areas of mathematics and their application to problems in the industrial and technological arena.

Published volumes

1. Matematica, cultura e società 2004 (2005). ISBN 88-7642-158-0
2. Matematica, cultura e società 2005 (2006). ISBN 88-7642-188-2
3. M. Giaquinta, D. Mucci, *Maps into Manifolds and Currents: Area and* $W^{1,2}$-, $W^{1/2}$-, BV*-Energies*, 2006. ISBN 88-7642-200-5
4. U. Zannier (editor), *Diophantine Geometry*. Proceedings, 2005 (2007). ISBN 978-88-7642-206-5
5. G. Métivier, *Para-Differential Calculus and Applications to the Cauchy Problem for Nonlinear Systems*, 2008. ISBN 978-88-7642-329-1
6. F. Guerra, N. Robotti, *Ettore Majorana. Aspects of his Scientific and Academic Activity*, 2008. ISBN 978-88-7642-331-4
7. Y. Censor, M. Jiang, A. K. Louisr (editors), *Mathematical Methods in Biomedical Imaging and Intensity-Modulated Radiation Therapy (IMRT)*, 2008. ISBN 978-88-7642-314-7
8. M. Ericsson, S. Montangero (editors), *Quantum Information and Many Body Quantum systems*. Proceedings, 2007 (2008). ISBN 978-88-7642-307-9
9. M. Novaga, G. Orlandi (editors), *Singularities in Nonlinear Evolution Phenomena and Applications*. Proceedings, 2008 (2009). ISBN 978-88-7642-343-7

– Matematica, cultura e società 2006 (2009). ISBN 88-7642-315-4

10. H. HOSNI, F. MONTAGNA (editors), *Probability, Uncertainty and Rationality*, 2010. ISBN 978-88-7642-347-5
11. L. AMBROSIO (editor), *Optimal Transportation, Geometry and Functional Inequalities*, 2010. ISBN 978-88-7642-373-4
12*. O. COSTIN, F. FAUVET, F. MENOUS, D. SAUZIN (editors), *Asymptotics in Dynamics, Geometry and PDEs; Generalized Borel Summation*, vol. I, 2011. ISBN 978-88-7642-374-1, e-ISBN 978-88-7642-379-6
12**. O. COSTIN, F. FAUVET, F. MENOUS, D. SAUZIN (editors), *Asymptotics in Dynamics, Geometry and PDEs; Generalized Borel Summation*, vol. II, 2011. ISBN 978-88-7642-376-5, e-ISBN 978-88-7642-377-2
13. G. MINGIONE (editor), *Topics in Modern Regularity Theory*, 2011. ISBN 978-88-7642-426-7, e-ISBN 978-88-7642-427-4
– Matematica, cultura e società 2007-2008 (2012). ISBN 978-88-7642-382-6
14. A. BJORNER, F. COHEN, C. DE CONCINI, C. PROCESI, M. SALVETTI (editors), *Configuration Spaces*, Geometry, Combinatorics and Topology, 2012. ISBN 978-88-7642-430-4, e-ISBN 978-88-7642-431-1
15 A. CHAMBOLLE, M. NOVAGA E. VALDINOCI (editors), *Geometric Partial Differential Equations*, 2013. ISBN 978-88-7642-343-7, e-ISBN 978-88-7642-473-1
16 J. NEŠETŘIL AND M. PELLEGRINI (editors), *The Seventh European Conference on Combinatorics, Graph Theory and Applications*, EuroComb 2013. ISBN 978-88-7642-474-8, e-ISBN 978-88-7642-475-5

Volumes published earlier

Dynamical Systems. Proceedings, 2002 (2003)
Part I: *Hamiltonian Systems and Celestial Mechanics*.
ISBN 978-88-7642-259-1
Part II: *Topological, Geometrical and Ergodic Properties of Dynamics*.
ISBN 978-88-7642-260-1

Matematica, cultura e società 2003 (2004). ISBN 88-7642-129-7

Ricordando Franco Conti, 2004. ISBN 88-7642-137-8

N.V. KRYLOV, *Probabilistic Methods of Investigating Interior Smoothness of Harmonic Functions Associated with Degenerate Elliptic Operators*, 2004. ISBN 978-88-7642-261-1

Phase Space Analysis of Partial Differential Equations. Proceedings, vol. I, 2004 (2005). ISBN 978-88-7642-263-1

Phase Space Analysis of Partial Differential Equations. Proceedings, vol. II, 2004 (2005). ISBN 978-88-7642-263-1